Advances in Neurobiology

Volume 44

Advances in Neurobiology covers basic research in neurobiology and neurochemistry. It provides in-depth, book-length treatment of some of the most important topics in neuroscience including molecular and pharmacological aspects. The main audiences of the series are basic science researchers and graduate students as well as clinicians including neuroscientists (neurobiologists and neurochemists) and neurologists. *Advances in Neurobiology* is indexed in PubMed, Google Scholar, and the Thompson Reuters Book Citation Index.

Hideaki Soya
Editor

Exercise Brain Stimulation for Cognitive Function and Mental Health

Springer

Editor
Hideaki Soya
Laboratory of Exercise Biochemistry and Neuroendocrinology, Institute of Health and Sport Sciences
University of Tsukuba
Tsukuba, Ibaraki, Japan

Sport Neuroscience Division, Kokoro Division, Advanced Research Initiative for Human High Performance (ARIHHP), Institute of Health and Sport Sciences
Institute of Health and Sport Sciences, University of Tsukuba
Tsukuba, Ibaraki, Japan

Center for Cybernics Research
University of Tsukuba
Tsukuba, Ibaraki, Japan

ISSN 2190-5215 ISSN 2190-5223 (electronic)
Advances in Neurobiology
ISBN 978-981-95-0065-9 ISBN 978-981-95-0066-6 (eBook)
https://doi.org/10.1007/978-981-95-0066-6

This Springer imprint is published by the registered company Springer Nature Singapore Pte Ltd.
The registered company address is: 152 Beach Road, #21-01/04 Gateway East, Singapore 189721, Singapore

Preface

Over the past several decades, modern lifestyles have been increasingly shaped by technological advancement, automation, and convenience. These changes, while beneficial in many ways, have drastically reduced levels of physical activity across the globe. Today, one in four adults and nearly 80% of adolescents fall below recommended activity thresholds, contributing to a rise in noncommunicable diseases that affect not only physical health but also brain function and mental well-being.

This issue is particularly pressing in aging societies, where the prevalence of cognitive decline and dementia is escalating. Simultaneously, the mental health crisis—including depression and suicide—among youth continues to grow. Mounting evidence suggests that physical exercise exerts protective and restorative effects on brain health, making it a potentially powerful non-pharmacological tool for supporting cognitive and emotional functions. While there is a growing body of research focusing on the neurological and psychological effects of exercise, the precise mechanisms by which exercise influences the brain remain insufficiently understood and, thus, difficult to translate into practical and applicable knowledge.

This book was conceived to address that gap. While many existing volumes approach the cognitive effects of exercise from psychological or behavioral standpoints, few provide a systematic neuroscientific exploration of its effects on brain structure, function, and neurochemical pathways. Since the seminal discovery of exercise-induced brain-derived neurotrophic factor (BDNF) by Cotman and colleagues in 1995, the field has progressed rapidly. Yet much of this knowledge remains scattered and disjointed, particularly between studies on animal models and those on humans. The lack of cohesive translational research frameworks has limited our ability to fully understand and apply exercise as a neurobiological intervention. The present volume repositions physical exercise not merely as an essential aspect of good health but as a dynamic modulator of brain function. It brings together leading researchers across the fields of neuroscience, physiology, and psychiatry to examine how exercise shapes cognition and emotion at cellular, network, and systemic levels. Special focus is placed on brain regions such as the prefrontal cortex and hippocampus, which are central to executive control and memory.

The book is organized into three interconnected sections: Translational Approaches—integrating findings from both animal and human research to bridge basic and applied knowledge; Brain-Body Communication—exploring how peripheral signals such as myokines and hepatokines mediate the effects of exercise on the brain; and Emerging Frontiers—introducing novel methodologies and modulatory factors (e.g., music, hypoxia) that may enhance or contextualize the effects of exercise. Across these themes, contributors present not only established findings but also areas of ongoing debate and exploration, including optimal exercise parameters, age-specific responses, and potential applications for neuropsychiatric conditions. Thus, this book offers a critical and timely synthesis of a fast-evolving field.

We believe that this volume will serve as both a reference and a catalyst for researchers, clinicians, educators, and students interested in understanding how physical activity contributes to overall brain health. We hope that it will encourage further research and practical applications, supporting a shift in how we conceptualize exercise: not simply as movement but as medicine for the brain. Advances in technology have indeed made daily life more convenient, but they have also led to a significant decline in physical activity. Today, about one in four adults—and around 80% of adolescents—are physically inactive, posing serious risks to both their physical and mental health. Given the increasing rates of dementia in aging societies and depression-related suicides among youth, understanding how physical activity supports brain function and mental well-being is a pressing and extremely important issue. This book fills a significant gap in the literature by exploring how exercise activates the brain to enhance cognitive function and mental health, drawing upon and consolidating the latest scientific insights.

Lastly, we would like to take this opportunity to express our heartfelt gratitude to all the authors who have contributed here. We are confident that this book will serve as both a reference and a catalyst for researchers, clinicians, educators, and students who are interested in understanding how physical activity and exercise affect the cognitive brain and contribute to overall health. This book explores how exercise activates the brain and improves cognitive function and mental health based on the latest scientific knowledge. We strongly believe that access to such knowledge will benefit all those involved in related research and practice. We also extend our thanks to Springer Science for their kind and continuous support and invaluable insights throughout all stages of bringing this work together. Finally, we wish to acknowledge the invited researchers who could not participate in this endeavor. Many of them have played an important role in advancing the field of exercise and brain health through the development of new ideas and the implementation of key experiments. In particular, it is with great sadness that we report the passing of Professor Carl Wayne Cotman (5 April 1940–5 August 2024) from UC Irvine. His contributions to neuroscience have had a profound impact, and his invaluable work will continue to inspire the field.

In-house Editorial staff, in particular Dr. Yudai Yamazaki, Dr. Masahiro Okamoto, and Dr. Ryuta Kuwamizu, worked tirelessly on important tasks such as the detailed correspondence with authors and publishers necessary for editing, and their efforts were instrumental in successfully making this publication a reality. I am glad to have this opportunity to express my heartfelt gratitude to them.

Tsukuba, Ibaraki, Japan Hideaki Soya

Contents

Contributors

Ignacio Torres Aleman Achucarro Basque Center for Neuroscience, Leioa, Spain

Ciberned, Madrid, Spain

Ikerbasque Foundation for Science, Bilbao, Spain

Soichi Ando Graduate School of Informatics and Engineering, The University of Electro-Communications, Chofu, Tokyo, Japan

Daniel D. Callow University of Maryland, College Park, MD, USA

Hyukki Chang Department of Sport and Exercise Science, Seoul Women's University, Seoul, Republic of Korea

Yu-Kai Chang Department of Physical Education and Sports Sciences, National Taiwan Normal University, Taipei, Taiwan

Social Emotional Education and Development Center, National Taiwan Normal University, Taipei, Taiwan

Institute of Sport Science and Innovations, Lithuanian Sports University, Kaunas, Lithuania

Feng-Tzu Chen Department of Kinesiology, National Tsing Hua University, Hsinchu, Taiwan

Nai-Chi Chen Department of Physical Education and Sports Sciences, National Taiwan Normal University, Taipei, Taiwan

Joon-Yong Cho Exercise Biochemistry Laboratory, Korea National Sport University, Seoul, Republic of Korea

Estrella Fernandez de Sevilla Ciberned, Madrid, Spain

Cajal Institute, CSIC, Madrid, Spain

Toshihiko Fujimoto Institute of Excellence in Higher Education, Tohoku University, Sendai, Miyagi, Japan

Takemune Fukuie School of Nursing and Social Services, Health Sciences University of Hokkaido, Tobetsu-cho, Ishikari-gun, Hokkaido, Japan

Toshiaki Hata Laboratory of Exercise Biochemistry and Neuroendocrinology, Institute of Health and Sport Sciences, University of Tsukuba, Tsukuba, Ibaraki, Japan

Sport Neuroscience Division, Kokoro Division, Advanced Research Initiative for Human High Performance (ARIHHP), Institute of Health and Sport Sciences, University of Tsukuba, Tsukuba, Ibaraki, Japan

Taichi Hiraga Laboratory of Exercise Biochemistry and Neuroendocrinology, Institute of Health and Sport Sciences, University of Tsukuba, Tsukuba, Ibaraki, Japan

Sport Neuroscience Division, Kokoro Division, Advanced Research Initiative for Human High Performance (ARIHHP), Institute of Health and Sport Sciences, University of Tsukuba, Tsukuba, Ibaraki, Japan

Chen-Sin Hung Department of Physical Education and Sports Sciences, National Taiwan Normal University, Taipei, Taiwan

Dong-Joo Hwang Exercise Biochemistry Laboratory, Korea National Sport University, Seoul, Republic of Korea

Sport Science Institute, Korea National Sport University, Seoul, Republic of Korea

Kazuki Hyodo Physical Fitness Research Institute, Meiji Yasuda Life Foundation of Health and Welfare, Tokyo, Japan

Koshiro Inoue Center for Education in Liberal Arts and Sciences/School of Rehabilitation Sciences, Health Sciences University of Hokkaido, Hokkaido, Japan

Morimasa Kato Department of Health and Nutrition, Yamagata Prefectural Yonezawa University of Nutrition Sciences, Yamagata, Japan

Sylwester Kujach Department of Physiology, Faculty of Physical Education, Gdansk University of Physical Education and Sport, Gdańsk, Poland

Department of Physiology, Faculty of Medicine, Medical University of Gdańsk, Gdańsk, Poland

Yu-Min Kuo Department of Cell Biology and Anatomy and Institute of Basic Medical Sciences, College of Medicine, National Cheng Kung University, Tainan, Taiwan

Ryuta Kuwamizu Sport Neuroscience Division, Kokoro Division, Advanced Research Initiative for Human High Performance (ARIHHP), Institute of Health and Sport Sciences, University of Tsukuba, Tsukuba, Ibaraki, Japan

Laboratory of Exercise Biochemistry and Neuroendocrinology, Institute of Health and Sport Sciences, University of Tsukuba, Tsukuba, Ibaraki, Japan

Radosław Laskowski Department of Physiology, Faculty of Physical Education, Gdansk University of Physical Education and Sport, Gdańsk, Poland

Min-Chul Lee Department of Sports Medicine, College of Health Science, CHA University, Pocheon, South Korea

Takashi Matsui Institute of Health and Sport Sciences, University of Tsukuba, Tsukuba, Ibaraki, Japan

Sport Neuroscience Division, Kokoro Division, Advanced Research Initiative for Human High Performance (ARIHHP), Institute of Health and Sport Sciences, University of Tsukuba, Tsukuba, Ibaraki, Japan

Onanong Mee-inta Department of Cell Biology and Anatomy and Institute of Basic Medical Sciences, College of Medicine, National Cheng Kung University, Tainan, Taiwan

Chakri Naruebodindra Medical Institute, Faculty of Medicine Ramathibodi Hospital, Mahidol University, Samut Prakan, Thailand

Hyo Youl Moon Department of Education, Seoul National University, Seoul, South Korea

Learning Sciences Research Institute, Seoul National University, Seoul, South Korea

Institute on Aging, Seoul National University, Seoul, South Korea

Takeshi Nishijima Sport Neuroscience Laboratory, Department of Health Promotion Sciences, Graduate School of Human Health Sciences, Tokyo Metropolitan University, Tokyo, Japan

Genta Ochi Faculty of Health Sciences, Department of Health and Sports, Niigata University of Health and Welfare, Niigata, Niigata, Japan

Masahiro Okamoto Laboratory of Exercise Biochemistry and Neuroendocrinology, Institute of Health and Sport Sciences, University of Tsukuba, Tsukuba, Ibaraki, Japan

Sport Neuroscience Division, Kokoro Division, Advanced Research Initiative for Human High Performance (ARIHHP), Institute of Health and Sport Sciences, University of Tsukuba, Tsukuba, Ibaraki, Japan

Robert Antoni Olek Department of Athletics, Strength and Conditioning, Poznań University of Physical Education, Poznań, Poland

Gabriel S. Pena University of Maryland, College Park, MD, USA

Jaime Pignatelli Ciberned, Madrid, Spain

Cajal Institute, CSIC, Madrid, Spain

Jeremy J. Purcell University of Maryland, College Park, MD, USA

Yoshiyuki Sankai Center for Cybernics Research, University of Tsukuba, Tsukuba, Ibaraki, Japan

Daisuke Sato Advanced Research Initiative for Human High Performance (ARIHHP), Institute of Health and Sport Sciences, University of Tsukuba, Tsukuba, Ibaraki, Japan

Takeru Shima Department of Health and Physical Education, Cooperative Faculty of Education, Gunma University, Maebashi, Gunma, Japan

J. Carson Smith University of Maryland, College Park, MD, USA

Hideaki Soya Laboratory of Exercise Biochemistry and Neuroendocrinology, Institute of Health and Sport Sciences, University of Tsukuba, Tsukuba, Ibaraki, Japan

Sport Neuroscience Division, Kokoro Division, Advanced Research Initiative for Human High Performance (ARIHHP), Institute of Health and Sport Sciences, Institute of Health and Sport Sciences, University of Tsukuba, Tsukuba, Ibaraki, Japan

Center for Cybernics Research, University of Tsukuba, Tsukuba, Ibaraki, Japan

Mizuki Sudo Shinjyuku Satellite Office, Physical Fitness Research Institute, Meiji Yasuda Life Foundation of Health and Welfare, Shinjuku, Tokyo, Japan

Kazuya Suwabe Faculty of Health and Sport Sciences, Ryutsu Keizai University, Ryugasaki, Ibaraki, Japan

Laboratory of Exercise Biochemistry and Neuroendocrinology, Institute of Health and Sport Sciences, University of Tsukuba, Tsukuba, Ibaraki, Japan

Center for Cybernics Research, University of Tsukuba, Tsukuba, Ibaraki, Japan

Kenji Suzuki Center for Cybernics Research, University of Tsukuba, Tsukuba, Ibaraki, Japan

Manabu Tashiro Research Center for Accelerator and Radioisotope of Science, Sendai, Miyagi, Japan

Henriette van Praag Stiles Nicholson Brain Institute and Charles E. Schmidt College of Medicine, Florida Atlantic University, Jupiter, FL, USA

Tzu-Feng Wang Department of Cell Biology and Anatomy and Institute of Basic Medical Sciences, College of Medicine, National Cheng Kung University, Tainan, Taiwan

Yuhki Yamada Laboratory of Exercise Biochemistry and Neuroendocrinology, Institute of Health and Sport Sciences, University of Tsukuba, Tsukuba, Ibaraki, Japan

Koya Yamashiro Department of Health and Sports, Niigata University of Health and Welfare, Niigata, Japan

Institute for Human Movement and Medical Sciences, Niigata University of Health and Welfare, Niigata, Japan

Yudai Yamazaki Laboratory of Exercise Biochemistry and Neuroendocrinology, Institute of Health and Sport Sciences, University of Tsukuba, Tsukuba, Ibaraki, Japan

Michael A. Yassa Department of Neurobiology and Behavior, University of California, Irvine, CA, USA

Center for the Neurobiology of Learning and Memory, University of California, Irvine, CA, USA

Sport Neuroscience Division, Kokoro Division, Advanced Research Initiative for Human High Performance (ARIHHP), Institute of Health and Sport Sciences, University of Tsukuba, Tsukuba, Ibaraki, Japan

Center for Cybernics Research, University of Tsukuba, Tsukuba, Ibaraki, Japan

Jang Soo Yook Division of Humanities and Social Sciences, Pohang University of Science and Technology (POSTECH), Pohang, Gyeongsangbuk-do, South Korea

Jonathan Zegarra-Valdivia Achucarro Basque Center for Neuroscience, Leioa, Spain

Ciberned, Madrid, Spain

Universidad Señor de Sipán, Chiclayo, Perú

Global Brain Health Institute, University of California, San Francisco, CA, USA

Part I
Translational Approaches

Treadmill Running Model for Rodents as a Translational Approach Based on Human Physiological Responses

Masahiro Okamoto, Toshiaki Hata, Taichi Hiraga, and Hideaki Soya

Abstract The concept of "Exercise is Medicine" highlights the preventive effects of physical activity on lifestyle-related diseases, dementia, and mental disorders. However, human studies face limitations in isolating exercise-specific effects due to uncontrolled variables such as diet and living conditions, as well as the constraint of non-invasive methodologies. Animal models offer a valuable alternative, allowing for strict control of experimental conditions and detailed assessment of physiological and neural responses to exercise. While voluntary wheel running has been commonly used, it lacks standardization in exercise intensity and type. To bridge this gap, we developed a treadmill-based rodent exercise model that enables precise control over exercise parameters, including intensity, duration, and distance. By incorporating physiological markers—such as blood lactate and oxygen consumption—commonly used in humans, we succeeded in evaluating rodent fitness and establishing exercise paradigms analogous to those used in human studies. Our findings demonstrate that even light-intensity exercise can significantly enhance brain activation and memory, which may be particularly relevant for aging or low-fitness populations. This approach enables the exploration of shared neurobiological

M. Okamoto (✉) · T. Hata · T. Hiraga
Laboratory of Exercise Biochemistry and Neuroendocrinology, Institute of Health and Sport Sciences, University of Tsukuba, Tsukuba, Ibaraki, Japan

Sport Neuroscience Division, Kokoro Division, Advanced Research Initiative for Human High Performance (ARIHHP), Institute of Health and Sport Sciences, University of Tsukuba, Tsukuba, Ibaraki, Japan
e-mail: okamoto.masahiro.ga@u.tsukuba.ac.jp

H. Soya
Laboratory of Exercise Biochemistry and Neuroendocrinology, Institute of Health and Sport Sciences, University of Tsukuba, Tsukuba, Ibaraki, Japan

Sport Neuroscience Division, Kokoro Division, Advanced Research Initiative for Human High Performance (ARIHHP), Institute of Health and Sport Sciences, University of Tsukuba, Tsukuba, Ibaraki, Japan

Center for Cybernics Research, University of Tsukuba, Tsukuba, Ibaraki, Japan

H. Soya (ed.), *Exercise Brain Stimulation for Cognitive Function and Mental Health*, Advances in Neurobiology 44, https://doi.org/10.1007/978-981-95-0066-6_1

mechanisms and supports the advancement of translational research, facilitating the development of tailored exercise interventions aimed at promoting cognitive health.

Keywords Exercise intensity · Lactate threshold · Gas exchange threshold · Maximal oxygen consumption · Stress · Hypothalamus

1 Introduction

The concept of "Exercise is Medicine" is widely recognized, and regular physical activity has been shown to contribute to the prevention of lifestyle-related diseases while also reducing the risk of developing dementia and mental disorders [1–4]. However, studies in humans face challenges due to the influence of various factors such as dietary habits and living environments, making it difficult to isolate the pure effects of exercise. Additionally, human studies are limited to non-invasive methods, leaving gaps in understanding the precise neural mechanisms underlying exercise-induced cognitive enhancement.

To complement these limitations, the use of animal models is a valuable approach. Animal studies allow for strict control of nutritional conditions and the application of invasive measurements, enabling a detailed analysis of exercise-induced changes in brain function [5–7]. Traditionally, voluntary wheel running has been the most commonly used exercise intervention in animals [8–10]. However, this method only allows for the measurement of running distance, without standardization of exercise intensity or modality (e.g., endurance vs. interval running). Thus, while evidence supports the beneficial effects of exercise on brain function, identifying the most effective exercise parameters remains challenging.

In human exercise prescription, the frequency, intensity, time, and type (FITT) principle serves as a key framework [11]. In animal experiments, treadmills allow for precise control of running speed and duration, yet further refinements are required for appropriate exercise intensity regulation. A promising approach is to consider physiological indicators that change during exercise. For example, lactate concentration and oxygen consumption are commonly used markers of exercise intensity, and if these markers are also applicable in animal models, they could aid in developing translational exercise models for humans. However, a comprehensive review summarizing the experimental protocols and measurement methods for these physiological parameters in animal models has not yet been established.

This chapter first discusses the physiological markers used to define exercise intensity in humans, including metabolic and neuroendocrine responses (e.g., blood and other body fluid factors) as well as respiratory and circulatory responses (e.g., expired gas parameters). Subsequently, we will introduce our established methods for measuring these physiological markers in animal models and summarize the findings obtained from these approaches.

2 Physiological Responses as Indices of Exercise Intensity

2.1 Physiological Indices for Determining Exercise Intensity: Maximal Oxygen Consumption (VO_{2max}), Lactate Threshold (LT) and Gas Exchange Threshold (GET)

Effective exercise prescription requires a detailed understanding of the conditions that lead to positive outcomes, including the FITT principle [11], which corresponds to dosage and administration in pharmacotherapy. Among these components, intensity is one of the most critical factors and is classified based on physiological indicators that change in response to exercise, such as maximal oxygen consumption (VO_{2max}) and metabolic thresholds, including the lactate threshold (LT) and gas exchange threshold (GET).

VO_{2max} refers to the maximum amount of oxygen consumption (VO_2) that can be taken up by the body during exercise. It serves as an indicator of maximal aerobic capacity and is a useful physiological index for assessing physical fitness levels and exercise intensity in humans. VO_{2max} is typically measured using a graded exercise test performed on a treadmill or cycle ergometer. As exercise intensity gradually increases, VO_2 increases linearly. However, beyond a certain point, it reaches a level at which it no longer rises despite further increases in intensity. This phenomenon is referred to as a "plateau" or "leveling off," and the highest VO_2 value observed at this point is defined as VO_{2max} [12]. In human exercise prescription, it is common to classify exercise intensity based on VO_{2max}. For example, according to the guidelines established by the American College of Sports Medicine (ACSM), exercise intensity can be categorized into the following five levels: Very light: less than 37% VO_{2max}; light: 37–45% VO_{2max}; moderate: 46–63% VO_{2max}; vigorous: 64–90% VO_{2max}; and near maximal: 91% VO_{2max} or higher. Although the classification and terminology of exercise intensity may vary depending on the guidelines or research fields, this chapter follows the classification defined by the ACSM's guidelines to define and refer to exercise intensity.

In addition to the ACSM's classification of exercise intensity based on VO_{2max}, exercise intensity can also be classified using metabolic thresholds such as the LT and GET. During light-to-moderate intensity exercise, energy supply predominantly depends on the oxidative phosphorylation system, and as exercise intensity increases, the contribution of the glycolysis system also increases [13], leading to increased production of lactate as a metabolic byproduct. Although lactate is also utilized as an energy source, a balance between its production and utilization is maintained up to a certain intensity, keeping its concentration low. There is a point at which this balance breaks down and blood lactate levels rise sharply—this inflection point is defined as the LT [14, 15]. On the other hand, GET refers to the intensity at which carbon dioxide production (VCO_2) increases disproportionately relative to oxygen uptake (VO_2), despite both increasing with rising exercise intensity [16, 17]. This phenomenon may be explained by the buffering mechanism: as exercise intensity increases, lactate accumulates and releases proton ions (H^+), causing blood acidosis. To buffer this, bicarbonate ions (HCO_3^-) react with H^+, resulting in increased production of CO_2. Once LT is reached and blood lactate rises sharply, CO_2 production also surges, leading to GET. The elevated CO_2 stimulates the brainstem respiratory center, causing a marked increase in ventilation and the appearance of the ventilatory threshold (VT). Therefore, in a broader sense, GET is sometimes called the VT. LT and GET are known to occur at nearly the same intensity. In humans, they occur approximately at 45–74% of VO_{2max} [13, 18–20] which corresponds to moderate-intensity exercise. Therefore, LT and GET can be used as reference points to roughly define the boundaries between light- and vigorous-intensity exercise.

2.2 Stress Response

As exercise intensity increases, the demand for glycolytic metabolism becomes greater, accompanied by increases in both ventilation and blood flow. The hypothalamus plays a central role in regulating these physiological responses. The endocrine and autonomic nervous systems respond rapidly to metabolic demands and stressors via two major pathways: the hypothalamic–pituitary–adrenal (HPA) axis and the hypothalamic–sympathetic–adrenal medullary (SAM) axis. When various stressors, including exercise, are perceived by the paraventricular nucleus (PVN) of the hypothalamus, it releases corticotropin-releasing hormone (CRH) and arginine vasopressin (AVP) to the pituitary gland. In response, the pituitary is stimulated to secrete adrenocorticotropic hormone (ACTH) into the bloodstream [21, 22]. ACTH then stimulates the adrenal cortex to promote the secretion of glucocorticoids (cortisol in humans), thereby enhancing metabolic activity [23]. Simultaneously, activation of the SAM axis stimulates the release of catecholamines (adrenaline and noradrenaline) from sympathetic nerve terminals and/or the adrenal medulla. Catecholamines not only promote hepatic glucose production directly [24], but also support glucose homeostasis indirectly by inhibiting insulin secretion and enhancing glucagon release from the pancreas [23]. In addition, catecholamines facilitate physiological adaptation to increasing exercise intensity by elevating heart rate and ventilation, thereby optimizing oxygen delivery and energy metabolism to meet the rising physiological demands during exercise. These endocrine and autonomic nervous system responses are known to increase sharply around the LT/GET. An increase in circulating ACTH, as a result of HPA axis activation, is considered a marker of stress in the field of medical physiology [25, 26]. Therefore, moderate-to-vigorous intensity exercise that corresponds or exceeds LT or GET can be defined as a form of stress—referred to as exercise stress.

3 Method for Measuring LT, GET, and VO_{2max} Using Incremental Load Exercise

Physiological indices used to classify exercise intensity, such as the LT, GET, and VO_{2max}, have been shown to exist in rodents, similar to humans, as demonstrated by several previous studies using treadmill exercise model, including our own research. The treadmill exercise model is particularly useful because it allows for precise control over key exercise parameters, including running speed (intensity) and duration. The VO_{2max} of rats has been reported to range approximately from 60 to 85 ml/kg/min [27–29], and in our exercise model, it is around 80 ml/kg/min [30, 31]. Furthermore, both LT and GET have been reported to occur at running speeds of approximately 20 m/min [30, 32–34]. In our study, we simultaneously measured LT and GET in the same rat and confirmed that, as in humans, the two thresholds closely coincide and show a strong correlation [30]. The development of animal exercise models based on such physiological indices is an important step toward enhancing their translational value to human applications. In this section, we outline the methods used to measure LT, GET, and VO_{2max} in rodents and present an overview of the exercise model we have developed.

3.1 *Running Habituation*

While the treadmill is suitable for precisely adjusting exercise conditions, it should be noted that it represents a novel and potentially stressful environment for animals. The movement of the treadmill belt may elicit emotional responses, and without sufficient habituation, animals may display behavioral reactions that are unrelated to the intended exercise stimuli. Indeed, previous studies have shown that the effects of exercise on hippocampal brain-derived neurotrophic factor (BDNF) expression vary depending on the degree of pre-habituation [35], which may confound assessments of exercise effects—particularly those related to mental health. Therefore, standardizing the habituation protocol is a critical step in ensuring consistent and reproducible exercise performance in animal studies. In the present protocol, all animals undergo at least 1 week of acclimatization and routine handling under standard laboratory conditions to familiarize them with the experimenter. For habituation to the treadmill apparatus (KN-73, Natsume Ltd., Japan) and the act of running, animals are subjected to a 6–9 day habituation period (5–7 running sessions) during the dark cycle. Running sessions lasts a maximum of 30 min per day, with rest days provided every 2–3 days. The treadmill speed is gradually increased from 5 to 25 m/min over the habituation period [30, 32–34, 36, 37]. During treadmill running, animals are trained to maintain continuous running by gently encouraging them using a rod applied to the hind limbs or buttocks, and with mild (but aversive) electrical stimuli delivered through a shock grid located at the rear of the treadmill lane. In rare cases, animals that consistently refuse to run or receive excessive electrical shocks are excluded from the study. Throughout the procedure, the experimenter monitors the animals carefully to prevent injury, such as tail entrapment in the end of the treadmill lane.

3.2 Catheter Insertion Surgery

To measure the dynamics of blood metabolites during running exercise, a method for sequential blood sampling without hindering the exercise is required. Therefore, catheter insertion surgery is performed on the right jugular vein of each animal.

After completing the running habituation period, animals are given a rest day, and the surgery is conducted following previously established methods [32, 33] (Note: To minimize intraoperative bleeding and support postoperative recovery, it is recommended to perform the surgery during the light phase, when activity levels and hormonal levels are stable.). Animals are anesthetized with isoflurane, and a skin incision is made above the right clavicle. While minimizing damage to the underlying muscle tissue, the right jugular vein is exposed. A polyurethane catheter (3Fr BC-3P; Primetech, Japan), prefilled with 10% heparinized saline, is inserted toward the right atrium through the exposed vein. The insertion length is adjusted based on species and body size—for example, 13 mm for 12-week-old C57BL/6 J mice [33], 32–37 mm for 8-week-old Wistar rats [32, 36, 38, 39], and 37 mm for 11-week-old Long Evans rats [34]. The catheter is secured to the jugular vein with sterilized silk thread. A second ~1 cm skin incision is made at the nape of the neck, and the catheter is passed subcutaneously to exit at the right side of the neck. After confirming that blood can be drawn through the catheter, it is flushed again with 10% heparinized saline, and both incisions are sutured. A pinport (PNP3F22; Primetech, Japan) is attached to seal the catheter, and antibiotics (Mycillin Sol; Tamura Seiyaku, Japan) are administered subcutaneously to prevent infection (Fig. 1). Following surgery, animals are housed individually. During the period prior to the exercise test, the catheter is flushed once daily with 10% heparinized saline to prevent clot formation and ensure hygiene.

3.3 Measurement of LT

A resting period of 2–3 days is provided after the surgery to allow for weight recovery. Subsequently, an incremental load running exercise test is performed. To eliminate the influence of feeding on blood glucose and lactate levels, animals are fasted starting 2 h before the running test. To collect blood without direct contact with the animal, an approximately 50 cm extension tube is connected to the catheter. After the animal rests on the treadmill lane for more than 10 min, a baseline blood sample is collected, and it is confirmed that blood glucose and lactate levels are within the resting range (lactate: <1 mmol/L; glucose: around 5 mmol/L). Following this

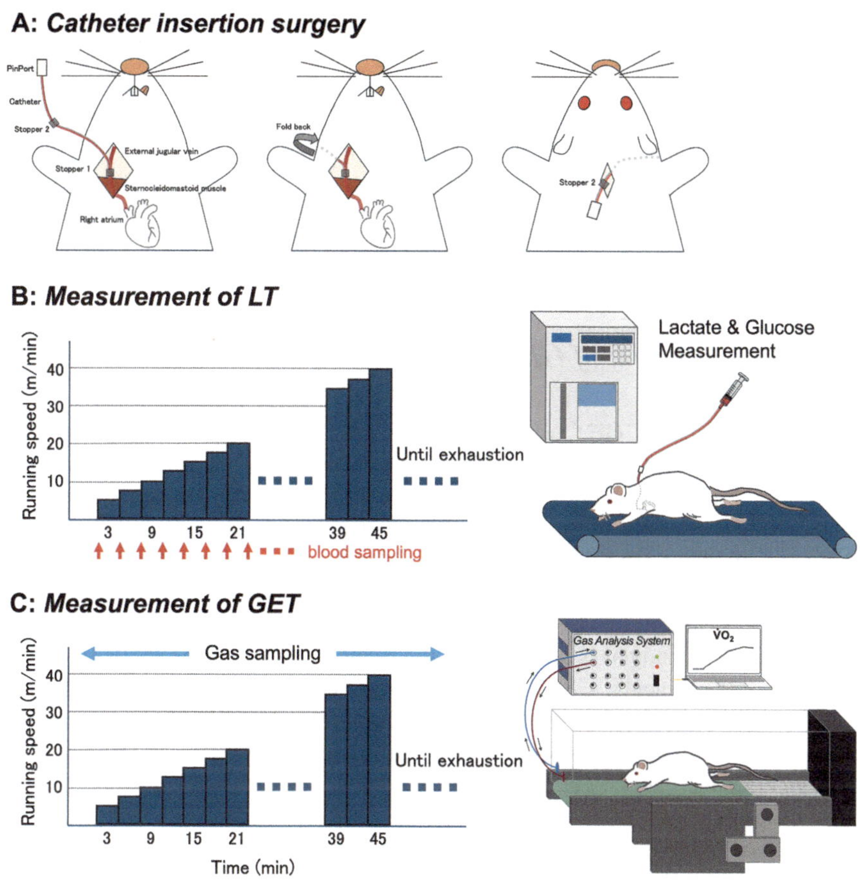

Fig. 1 Method of measuring LT and GET

confirmation, the treadmill is started at a speed of 5 m/min, and the speed is increased by 2.5 m/min every 3 min until the animal reaches voluntary exhaustion. Blood samples (~100 μL) are collected from the catheter 30 s before each speed increment to measure blood glucose and lactate levels (Fig. 1). Voluntary exhaustion is defined as the point at which the animal fails to keep pace with the treadmill and remains on the shock grid at the rear of the treadmill for at least 1 min, despite being gently encouraged by sticks or mild electrical stimulation [40, 41].

Blood glucose and lactate concentrations are measured using a glucose–lactate analyzer (2300 Stat Plus or YSI2500; Yellow Springs Instruments, USA). To determine the running speed corresponding to LT where blood lactate concentration begins to rise sharply, blood lactate dynamics are analyzed using segmented linear regression (SLR) with a single breakpoint. In our protocol, the first two workloads (at speeds of 5.0 and 7.5 m/min) are excluded from the LT analysis [30]. In the SLR model, two regression lines are fitted to the data to minimize the pooled residual sum of squares, with each line having its own slope and intercept (lower line: $Y_1 = Slope_1 \times X + Intercept_1$; upper line: $Y_2 = Slope_2 \times X + Intercept_2$). The intersection point of these two lines (X_0, Y_0) is estimated as the breakpoint, which represents the individual LT, depending on the variables analyzed. LT estimation is omitted if the change in slope between the lower and upper regression lines is less than 0.1, following the criteria of the V-slope method [15, 30, 42]. Through a series of studies, it has been shown that under similar running conditions, a speed of approximately 20 m/min corresponds to LT in rodents [30, 32–34].

3.4 *Measurement of GET*

The measurement of GET using respiratory gas analysis eliminates the need for invasive catheterization surgery, enabling non-invasive assessment of the metabolic state in animals. This approach is particularly advantageous when working with small-bodied mice, as it requires no surgical procedures and does not involve large volumes of blood. The method of running habituation is described in Sect. 3.1. Animals are habituated to running on a metabolic chamber treadmill (1010R-S; Columbus Instruments, USA) to acclimate them to the testing environment. Respiratory gas analysis is conducted using a calorimetry system (7520R; Oxymax, Columbus Instruments, USA). Air is continuously supplied to the chamber-type treadmill at a constant flow rate (mice: 0.6–0.8 L/min; rats: 2.5–3.0 L/min), and the Oxymax system calculates VO_2, VCO_2, and the respiratory exchange ratio (RER: VCO_2/VO_2) every 60 s by comparing the gas concentrations in the chamber to those of the baseline air.

After completing the running habituation, animals undergo a 2–3 day resting period before performing an incremental load running exercise test on a metabolic chamber treadmill. Prior to the exercise, a 2-h fasting period is implemented, followed by baseline measurements in the resting state for 30–60 min (rats: 20–25 mL/kg/min; mice: 50–70 mL/kg/min). As described in Sect. 3.3 for the LT measurement, in rats, running begins at a speed of 5 m/min and increases by 2.5 m/min every 3 min until the animals reach voluntary exhaustion (Fig. 1; [30]). In mice, the test starts at 3 m/min, with the speed increasing by 2.5 m/min every minute [43]. The criteria for determining voluntary exhaustion are also detailed in Sect. 3.3. During the exercise, respiratory gas dynamics are measured and recorded every minute. To determine the running speed corresponding to the GET, changes in VCO_2 over VO_2 before VO_2 drop off were examined using SLR with a single breakpoint., as referenced in Sect. 3.3. As a result, similar to LT, a running speed of approximately 20 m/min corresponds to GET in rodents under these conditions [30, 43].

Both the LT and GET reflect metabolic inflection points and are often considered physiologically synonymous. In our study, LT and GET were found to coincide closely and exhibit a strong correlation when measured in the same rat, as seen in human studies [30]; however, the running speed at GET is slightly higher than that at LT [30]. These findings suggest that although GET determined using the V-slope method is closely related to LT, it may not precisely estimate the actual onset of LT. Several human studies have investigated the relationship between blood lactate levels and gas-exchange parameters [44–47]. Notably, some researchers have reported that the workload corresponding to GET, as defined by the V-slope method, is slightly higher than that at LT [46, 47]. Therefore, our recent findings on GET and LT in rats are consistent with those reported in human studies.

3.5 Measurement of VO_{2max}

The VO_{2max} test is also conducted using a metabolic chamber treadmill connected to the Oxymax system, following stabilization of basal respiratory gas levels. While the previously described LT and GET measurement methods are effective for estimating metabolic breakpoints, they involve relatively long exercise durations and may not always bring animals to complete voluntary exhaustion. To induce fatigue more efficiently, we adopted a modified protocol based on a previously established method [29], with minor adjustments as follows. After a 5-min warm-up at 13.4 m/min with no incline, rats perform incremental-load running at a 10% incline. Running speed is increased by 6.7 m/min every 3 min until the animals reach exhaustion, determined using the same criteria described in Sect. 3.3 for the GET–LT test. Respiratory gas exchange parameters (VO_2, VCO_2, and RER) are recorded every minute using the metabolic chamber system. During the final stages of the test, VO_2 typically reaches a peak value, plateaus, and then begins to decline. VO_{2max} and RER_{max} are defined as the highest 1-min values of VO_2 and RER, respectively, regardless of the timing of load increments [28, 30]. The VO_{2max} obtained using this modified protocol (78.29 ± 5.44 ml/kg/min) is significantly higher than the peak VO_2 observed during the GET measurement (55.88 ± 4.17 ml/kg/min) [30]. This value is consistent with previously reported VO_{2max} values in rodents [27–29], supporting the validity of the modified protocol.

We successfully determined both LT and GET, which correspond to 51.78% ± 7.72% and 55.67% ± 6.90% of VO_{2max}, respectively [30]. These values closely match the estimated VO_{2max} percentages (55–65%) reported in previous studies [48], and fall within the range defined as moderate-intensity exercise by the ACSM [11]. This study provides direct evidence linking submaximal and maximal aerobic parameters in individual rats, including LT [30]. Moreover, our findings indicate that the relative intensities of these thresholds (%VO_{2max}) are comparable to those reported in humans, where LT and GET typically occur during moderate-intensity exercise (45–74% of VO_{2max}) [13, 18–20]. Collectively, these results suggest a consistent relationship between maximal and submaximal aerobic parameters across both rodent and human models.

4 Rodent Treadmill Exercise Model Based on the Physiological Indices

4.1 Classification of Exercise Intensity Based on Physiological Indices

By using the measured LT, GET, and VO_{2max} as references, it becomes possible to clearly classify exercise intensity in rodents, as has been done in humans. In our pioneering study, we focused on the fact that LT corresponds to moderate intensity and used it as a threshold to define light and vigorous intensity. Specifically, in Wistar and Long-Evans rats, since LT appeared around 20 m/min, we defined light intensity as 13.5–15 m/min, moderate intensity as 20–21.5 m/min, and vigorous-intensity as above 25 m/min [32, 34, 39, 49–53]. When rats were subjected to 30 min of treadmill exercise based on this classification, a significant increase in blood lactate levels was observed during moderate and vigorous-intensity exercise above LT, confirming the establishment of an LT-based exercise intensity model. Similarly, in C57BL/6 J mice, LT also emerged around 20 m/min [33]. During a 30-min running session, blood lactate levels remained unchanged at light-intensity (10 m/min), whereas they increased at vigorous-intensity (30 m/min).

Although such threshold-based classifications provide a bit rough categorization, the use of a treadmill metabolic chamber enables the application of a more detailed VO_{2max}-based intensity classification—such as very light, light, moderate, and vigorous—according to the ACSM's guidelines [11] even in rodents. In our recent investigation, we estimated the corresponding running speeds for Wistar rats as follows: very light: 5 m/min, light: 13.5 m/min, moderate: 20 m/min, and vigorous: 25 m/min [31]. This classification was validated by the observation that VO_2 increased in an intensity-dependent manner, while blood lactate levels increased only under moderate and vigorous intensity conditions. In the incremental exercise test used in this study, we modified the conventional VO_{2max} measurement protocol (which includes a warm-up and uses large increments in speed, as described in Sect. 3.5; [30]) by omitting the warm-up and adopting smaller speed increments. This modification was made because the conventional method resulted in VO_2 levels that already exceeded the very light to light intensity range at the beginning of the test, making it difficult to accurately estimate oxygen consumption and speed corresponding to very light to light intensity zones from VO_{2max}. Even with this modified protocol, we were able to obtain VO_{2max} values (about 80 mL/kg/min) and related parameters comparable to those obtained with the conventional method. Additionally, we successfully measured GET simultaneously, which appeared around 20 m/min.

In the following sections, we will present our research findings using this exercise model, focusing on how exercise intensity affects the stress response via the autonomic and sympathetic nervous systems, as well as the central nervous system—particularly the hypothalamus.

4.2 *Stress Hormone Dynamics During Incremental Load Exercise*

Since it has been revealed that animals also have a LT like humans, we aimed to investigate whether there are threshold responses in blood hormones, similar to LT. As sufficient blood is required for measuring blood hormones, blood sampling during exercise should be minimized. Therefore, we devised a protocol for incremental exercise tests at light intensity (sub-LT: 10 m/min), moderate intensity (approximately-LT: 20 m/min), and vigorous intensity (supra-LT: 30 m/min), conducting continuous sessions of each for 10 min [32]. Blood samples (approximately 350 μl) were collected from the catheter 30 seconds before each speed increase to examine the hormonal dynamics at different intensities. As we investigate association of physiological and neuroendocrinal response during incremental load exercise, we then determined levels of ACTH and adrenaline, biochemical markers for the activation of HPA and SAM axis, respectively. Basal ACTH concentration was same levels as sub-LT and after 10 min recovery. Interestingly, ACTH concentration increased from approximately-LT, while plasma adrenaline concentration only significantly increased with running at supra-LT [32]. This indicates that the HPA axis is highly responsive to exercise stress.

4.3 *Physiological Response to Constant-Load Exercise Based on LT/GET*

We successfully estimated the physical fitness levels and metabolic breakpoints of rodents by evaluating LT, GET, and VO_{2max}. Furthermore, we observed that these indicators are analogous to those in humans. However, in incremental-load exercise, it is important to note that the responses in high-intensity exercise may not be solely attributed to that exercise, as light-intensity and moderate-intensity exercises were performed immediately beforehand. Therefore, devising a constant-load exercise model based on these indicators may provide a detailed understanding of the effects of different intensities exercise. In our study, we defined exercise intensity as follows: below LT as light-intensity exercise (sub-LT: 15 m/min), and above LT as vigorous-intensity exercise (supra-LT: 25 m/min) [32, 36, 37, 54]. And then, we confirmed the physiological and neuroendocrinal responses to 30 min of constant-load exercise model based on this classification.

Here, we conducted blood sampling before and after constant-load exercise to assess changes in blood glucose, lactate, and ACTH concentrations. Additionally, we extracted the brain after exercise and revealed the impact on the hypothalamus neuronal activity, which regulates metabolism and stress response, using immunohistochemical staining. Following the procedures in Sects. 3.1 and 3.2, rats underwent running habituation and jugular vein catheterization. Rats were divided into three groups: sub-LT exercise group, supra-LT exercise group, and sedentary control group (waiting on the treadmill). Blood sampling (approximately 500 μl) was performed at three points: 10 min before exercise, just before the end of exercise, and 10 min after exercise. As expected, the supra-LT exercise significantly increases in blood glucose, lactate, and ACTH concentration compared to the sedentary control and sub-LT exercise. In contrast, the sub-LT exercise has no significant differences from the sedentary control in any of these parameters (Fig. 2) [32].

Neural activity in the brain was estimated by the c-Fos expression pattern. As c-Fos protein, one of the immediate early gene, is known to increase in expression approximately 90–120 min after neural activity, we performed perfusion of 4% buffered paraformaldehyde 90 min after the end of exercise. We prepared 40 μm coronal sections, including the hypothalamus, and visualized c-Fos through immunostaining. Subsequently, the sections were investigated with standard light microscopy and the c-Fos-immunoreactive (c-Fos$^+$) cells were quantified for 10 different hypothalamic sub-region: the medial preoptic nucleus (MPO), the periventricular nucleus (Pe), the suprachiasmatic nucleus (SCN), the supraoptic nucleus (SON), the parvocellular division of the paraventricular nucleus (pPVN), the anterior hypothalamic area (AH), the dorsomedial hypothalamic nucleus (DMH), the ventromedial hypothalamic nucleus (VMH), the lateral hypothalamic area (LH), the arcuate nucleus (ARC), the posterior hypothalamic area (PH), that most closely matched the Paxinos and Watson rat brain stereotaxic atlas [55]. As a result, neuronal activity in some subregions, including PVN and AH were increased in supra-LT exercise (Fig. 2) [32]. This result indicates the acute stress responses to meet the energy demands necessary for supra-LT exercise. In contrast, sub-LT exercise has no significant differences from the sedentary control group in all hypothalamic subregions [32]. Considering the results of minimal ACTH and adrenaline secretion

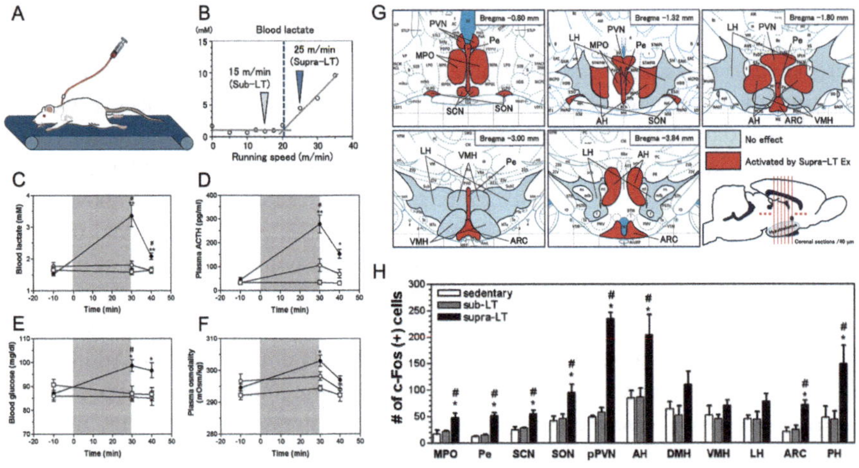

Fig. 2 Central and peripheral responses to exercise at different intensity. (Referenced by Soya et al. [32])

during sub-LT exercise, 30 min sub-LT exercise did not induce stress responses, and the impact on brain regions controlling the circulatory system and thermoregulation assumed to be minimal.

4.4 Physiological Adaptation Through Exercise Training

When verifying the effects of exercise, most studies focus on chronic effects spanning from several weeks to months. Generally, engaging in chronic endurance training leads to an improvement in physical fitness levels regardless of the effects on mental health. In fact, it is known that engaging in endurance exercise training for 20–60 min, three or more times a week, at intensities exceeding 40% VO_{2max}, enhances cardiorespiratory fitness level [11] (High-intensity interval training has also been recognized as a training method that increases cardiorespiratory fitness level [56]). Therefore, if measurements of VO_{2max}, LT, and GET are appropriate, it can be expected that moderate-to-high intensity exercise interventions based on these measurements will lead to enhancement of fitness levels. Hence, we evaluated the VO_{2max}, LT, and GET of rats after a 6-week endurance exercise training and elucidated their response.

Based on the results of constant-load exercise tests, rats were performed to either sedentary control on the treadmill, light-intensity exercise (15 m/min), or vigorous-intensity exercise (40 m/min) for 6 weeks. Training protocol consisted of running for 60 min, 5 days per week, and running speed was gradually increased. Subsequently, after a resting period, LT and GET measurements were conducted. As a result, the vigorous-intensity exercise training, but not light-intensity exercise training resulted in increased VO_{2max} as expected. At the same time, both of LT and GET were rightward shifted to higher running speeds, but in terms of percentages relative to VO_{2max}, they appeared around 50–60% of VO_{2max} [30]. This suggests that rats received adaptations similar to humans through endurance training, and indicating the appropriateness of exercise intensity based on GET and LT.

Since exercise above LT is accompanied by stress response with ACTH secretion, its long-term training may have adverse effects on both mental and physical health. Actually, we have demonstrated that 4 weeks of moderate-to-high intensity exercise (25 m/min) increases resting levels of plasma ACTH and corticosterone and increases CRH mRNA concentration in the PVN of the hypothalamus [57]. Although there are contradictory reports indicating chronic stress load induce an up-regulation of both the AVP and CRH in the resting [58–60], the change of AVP level was not observed in our protocol. While further verification is needed to understand this difference, these results suggest that at least 4 weeks of chronic exercise training may develop signs of general adaptation syndrome. Such stress loads can lead to the onset of psychiatric disorders like depression, warranting caution when implementing long-term interventions with moderate-to-high intensity exercise.

On the other hand, after 4 weeks of training, acute moderate-intensity exercise for 60 min suppresses the increase of blood lactate and glucose levels and downregulates the secretion of ACTH and corticosterone compared to the non-intervention rats [57]. This indicates a decrease in relative intensity to the same-load exercise due to training. In other words, training induces adaptation not only in metabolic responses but also in stress responses.

5 Conclusion

We have developed an animal exercise model using a treadmill to break away from the conventional approach such as wheel running and playground activities. As a result, we successfully measured the fitness levels of rodents using catheterization and respiratory gas analysis, and we achieved the development of exercise paradigms similar to those performed in humans. This allowed us to impose the arbitrary time, distance, and intensity on the previously spontaneous activity. This is a critical step in extrapolating insights from animal to humans.

While high-intensity exercise has traditionally been emphasized for improving physical fitness, our findings using this exercise model suggest that even light-intensity exercise is sufficient to enhance brain activation and memory performance. This insight may be particularly encouraging for elderly individuals or those with low-fitness level. These results underscore the importance of developing animal exercise models that closely align with human physiological responses. Such models will facilitate the identification of shared neural mechanisms responsible for exercise-induced benefits, ultimately advancing translational research aimed at promoting brain health through tailored exercise interventions.

References

1. Hillman CH, Erickson KI, Kramer AF (2008) Be smart, exercise your heart: exercise effects on brain and cognition. Nat Rev Neurosci 9:58–65
2. Erickson KI, Voss MW, Prakash RS et al (2011) Exercise training increases size of hippocampus and improves memory. Proc Natl Acad Sci USA 108:3017–3022
3. Chaddock-Heyman L, Erickson KI, Chappell MA et al (2016) Aerobic fitness is associated with greater hippocampal cerebral blood flow in children. Dev Cogn Neurosci 20:52–58
4. Soshi T, Andersson M, Kawagoe T et al (2021) Prefrontal plasticity after a 3-month exercise intervention in older adults relates to enhanced cognitive performance. Cereb Cortex 31:4501–4517
5. Yook JS, Rakwal R, Shibato J, Takahashi K, Koizumi H, Shima T, Ikemoto MJ, Oharomari LK, McEwen BS, Soya H (2019) Leptin in hippocampus mediates benefits of mild exercise by an antioxidant on neurogenesis and memory. Proc Natl Acad Sci 116:10988–10993
6. Shima T, Yoshikawa T, Onishi H (2022) Low-carbohydrate and high-protein diet suppresses working memory function in healthy mice. J Nutr Sci Vitaminol (Tokyo) 68:527–532
7. Aikawa Y, Noma Y, Agata U, Kakutani Y, Hattori S, Ogata H, Kiyono K, Omi N (2023) Running exercise and food restriction affect bone chemical properties in young female rats. Phys Act Nutr 27:62–69

8. Van Praag H, Kempermann G, Gage FH (1999) Running increases cell proliferation and neurogenesis in the adult mouse dentate gyrus. Nat Neurosci 2:266–270
9. Van Praag H, Christie BR, Sejnowski TJ, Gage FH (1999) Running enhances neurogenesis, learning, and long-term potentiation in mice. Proc Natl Acad Sci 96:13427–13431
10. Duman CH, Schlesinger L, Russell DS, Duman RS (2008) Voluntary exercise produces antidepressant and anxiolytic behavioral effects in mice. Brain Res 1199:148–158
11. Riebe D, Ehrman JK, Liguori G, Magal M (2018) ACSM's guidelines for exercise testing and prescription, 10th edn. The Canadian Chiropractic Association
12. Longstreet Taylor H, Buskirk E, Henschel A (1995) Maximal oxygen intake as an objective measure of cardio-respiratory performance. J Appl Physiol 8:73–80
13. Skinner JS, McLellan TH (1980) The transition from aerobic to anaerobic metabolism. Res Q Exerc Sport 51:234–248
14. Davis JA (1985) Anaerobic threshold: review of the concept and directions for future research. Med Sci Sports Exerc 17:6–21
15. Davis JA, Rozenek R, Decicco DM, Carizzi MT, Pham PH (2007) Comparison of three methods for detection of the lactate threshold. Clin Physiol Funct Imaging 27:381–384
16. Wasserman K, McIlroy MB (1964) Detecting the threshold of anaerobic metabolism in cardiac patients during exercise. Am J Cardiol 14:844–852
17. Wasserman K, Whipp BJ, Koyal SN, Beaver WL (1973) Anaerobic threshold and respiratory gas exchange during exercise. J Appl Physiol 35:236–243
18. Gollnick PD, Bayly WM, Hodgson DR (1986) Exercise intensity, training, diet, and lactate concentration in muscle and blood. Med Sci Sports Exerc 18:334–340
19. Gaskill SE, Ruby BC, Walker AJ, Sanchez OA, Serfass RC, Leon AS (2001) Validity and reliability of combining three methods to determine ventilatory threshold. Med Sci Sports Exerc 33:1841–1848
20. Iannetta D, Inglis EC, Mattu AT, Fontana FY, Pogliaghi S, Keir DA, Murias JM (2020) A critical evaluation of current methods for exercise prescription in women and men. Med Sci Sports Exerc 52:466–473
21. Gibbs DM (1986) Vasopressin and oxytocin: hypothalamic modulators of the stress response: a review. Psychoneuroendocrinology 11:131–139
22. Whitnall MH (1993) Regulation of the hypothalamic corticotropin-releasing hormone neurosecretory system. Prog Neurobiol 40:573–629
23. Kuo T, McQueen A, Chen TC, Wang JC (2015) Regulation of glucose homeostasis by glucocorticoids. Adv Exp Med Biol 872:99–126
24. Sigal RJ, Fisher S, Halter JB, Vranic M, Marliss EB (1996) The roles of Catecholamines in Glucoregulation in intense exercise as defined by the islet cell clamp technique. Diabetes 45:148–156
25. Pacák K (2000) Stressor-specific activation of the hypothalamic-pituitary-adrenocortical axis. Physiol Res 49:S11–S17
26. Kim EB, Susan MB, Scott B, Heddwen LB (2009) The Adrenal Medulla & Adrenal Cortex. In: Ganong's review of medical physiology, 23th edn. McGraw-Hill Medical, New York, pp 337–361
27. Bedford TG, Tipton CM, Wilson NC, Oppliger RA, Gisolfi CV (1979) Maximum oxygen consumption of rats and its changes with various experimental procedures. J Appl Physiol 47:1278–1283
28. Patch LD, Brooks GA (1980) Effects of training on VO_{2max} and VO_2 during two running intensities in rats. Pflugers Arch 386:215–219
29. Mazzeo RS, Brooks GA, Horvath SM (1984) Effects of age on metabolic responses to endurance training in rats. J Appl Physiol 57:1369–1374
30. Inoue K, Soya H, Murakumo K, Araki Y, Hiraga T, Soya S, Okamoto M (2024) Setting treadmill intensity for rat aerobic training using lactate and gas exchange thresholds. Med Sci Sports Exerc 57:434. https://doi.org/10.1249/MSS.0000000000003562

31. Hiraga T, Miyoshi K, Shimizu R, Yook JS, Okamoto M, Soya H (2025) Very-light-intensity exercise as minimal intensity threshold for activating dorsal hippocampal neurons: evidence from rat physiological exercise model. Biochem Biophys Res Commun 746:151243
32. Soya H, Mukai A, Deocaris CC, Ohiwa N, Chang H, Nishijima T, Fujikawa T, Togashi K, Saito T (2007) Threshold-like pattern of neuronal activation in the hypothalamus during treadmill running: establishment of a minimum running stress (MRS) rat model. Neurosci Res 58:341–348
33. Okamoto M, Yamamura Y, Liu Y-F et al (2015) Hormetic effects by exercise on hippocampal neurogenesis with glucocorticoid signaling. Brain Plast 1:149–158
34. Hiraga T et al (2024) Light-exercise-induced dopaminergic and noradrenergic stimulation in the dorsal hippocampus: using a rat physiological exercise model. FASEB J 38:e70215
35. Huang AM, Jen CJ, Chen HF, Yu L, Kuo YM, Chen HI (2006) Compulsive exercise acutely upregulates rat hippocampal brain-derived neurotrophic factor. J Neural Transm 113:803–811
36. Saito T, Soya H (2004) Delineation of responsive AVP-containing neurons to running stress in the hypothalamus. Am J Phys Regul Integr Comp Phys 286:R484–R490
37. Ohiwa N, Chang H, Saito T, Onaka T, Fujikawa T, Soya H (2007) Possible inhibitory role of prolactin-releasing peptide for ACTH release associated with running stress. Am J Physiol Regul Integr Comp Physiol 292:497–504
38. Takahashi K, Shima T, Soya M, Oharomari LK, Okamoto M, Soya H (2021) Differences in exercise capacity and physiological responses in Wistar rats among breeders. Exp Anim 70:508–513
39. Takahashi K, Shima T, Soya M, Yook JS, Koizumi H, Jesmin S, Saito T, Okamoto M, Soya H (2022) Exercise-induced adrenocorticotropic hormone response is cooperatively regulated by hypothalamic arginine vasopressin and corticotrophin-releasing hormone. Neuroendocrinology 112:894–903
40. Hasegawa H, Piacentini MF, Sarre S, Michotte Y, Ishiwata T, Meeusen R (2008) Influence of brain catecholamines on the development of fatigue in exercising rats in the heat. J Physiol 586:141–149
41. Matsui T, Ishikawa T, Ito H, Okamoto M, Inoue K, Lee M et al (2012) Brain glycogen supercompensation following exhaustive exercise. J Physiol 590:607–616
42. Beaver WL, Wasserman K, Whipp BJ (1985) Improved detection of lactate threshold during exercise using a log-log transformation. J Appl Physiol 59:1936–1940
43. Koizumi H, Hiraga T, Oharomari LK, Hata T, Shima T, Yook JS, Okamoto M, Mouri A, Nabeshima T, Soya H (2021) Preventive role of regular low-intensity exercise during adolescence in schizophrenia model mice with abnormal behaviors. Biochem Biophys Res Commun 534:610–616
44. Caiozzo VJ, Davis JA, Ellis JF, Azus JL, Vandagriff R, Prietto CA, McMaster WC (1982) A comparison of gas exchange indices used to detect the anaerobic threshold. J Appl Physiol 53:1184–1189
45. Gaskill SE, Ruby BC, Walker AJ et al (2001) Validity and reliability of combining three methods to determine ventilatory threshold. Med Sci Sports Exerc 33:1841–1848
46. Pallarés JG, Morán-Navarro R, Ortega JF, Fernández-Elías VE, Mora-Rodriguez R (2016) Validity and reliability of ventilatory and blood lactate thresholds in well-trained cyclists. PLoS One 11:e0163389
47. Cerezuela-Espejo V, Courel-Ibáñez J, Morán-Navarro R, Martínez-Cava A, Pallarés JG (2018) The relationship between lactate and ventilatory thresholds in runners: validity and reliability of exercise test performance parameters. Front Physiol 9:375785
48. Pilis W, Zarzeczny R, Langfort J, Kaciuba-Uściłko H, Nazar K, Wojtyna J (1993) Anaerobic threshold in rats. Comp Biochem Physiol Comp Physiol 106:285–289
49. Soya H, Nakamura T, Deocaris CC, Kimpara A, Iimura M, Fujikawa T, Chang H, McEwen BS, Nishijima T (2007) BDNF induction with mild exercise in the rat hippocampus. Biochem Biophys Res Commun 358:961–967

50. Okamoto M, Hojo Y, Inoue K, Matsui T, Kawato S, McEwen BS, Soya H (2012) Mild exercise increases dihydrotestosterone in hippocampus providing evidence for androgenic mediation of neurogenesis. Proc Natl Acad Sci USA 109:13100–13105
51. Inoue K, Okamoto M, Shibato J, Lee MC, Matsui T, Rakwal R, Soya H (2015) Long-term mild, rather than intense, exercise enhances adult hippocampal neurogenesis and greatly changes the transcriptomic profile of the hippocampus. PLoS One 10:e0128720
52. Inoue K, Hanaoka Y, Nishijima T, Okamoto M, Chang H, Saito T, Soya H (2014) Long-term mild exercise training enhances hippocampus-dependent memory in rats. Int J Sports Med 36:280–285
53. Okamoto M, Shimoda R, Amaya Y, Soya S, Soya M, Koizumi H, Nakamura K, Hiraga T, Torma F, Soya H (2024) Accelerated fear extinction by regular light-intensity exercise: a possible role of hippocampal BDNF-TrkB signaling. Med Sci Sports Exerc 56:221–229
54. Ohiwa N, Saito T, Chang H, Nakamura T, Soya H (2006) Differential responsiveness of c-Fos expression in the rat medulla oblongata to different treadmill running speeds. Neurosci Res 54:124–132
55. Paxinos G, Watson C (2009) The rat brain in stereotaxic coordinates compact, 6th edn. Academic, New York
56. Gist NH, Fedewa MV, Dishman RK, Cureton KJ (2014) Sprint interval training effects on aerobic capacity: a systematic review and meta-analysis. Sports Med 44:269–279
57. Kawashima H, Saito T, Yoshizato H, Fujikawa T, Sato Y, McEwen BS, Soya H (2004) Endurance treadmill training in rats alters CRH activity in the hypothalamic paraventricular nucleus at rest and during acute running according to its period. Life Sci 76:763–774
58. Bartanusz V, Jezova D, Bertini LT, Tilders FJH, Aubry JM, Kiss JZ (1993) Stress-induced increase in vasopressin and corticotropin-releasing factor expression in hypophysiotrophic paraventricular neurons. Endocrinology 132:895–902
59. Herman JP, Adams D, Prewitt C (1995) Regulatory changes in neuroendocrine stress-integrative circuitry produced by a variable stress paradigm. Neuroendocrinology 61:180–190
60. Sawchenko PE, Arias CA, Mortrud MT (1993) Local Tetrodotoxin blocks chronic stress effects on Corticotropin-releasing factor and vasopressin messenger ribonucleic acids in Hypophysiotropic neurons. J Neuroendocrinol 5:341–348

Neuronal Activity in Rat Hippocampus During Light-Intensity Treadmill Running

Takeshi Nishijima and Hideaki Soya

Abstract In the early 2000s, it became evident that exercise enhances brain function, particularly in the hippocampus, attracting considerable attention. However, at that time, most studies relied on the voluntary wheel running model for experiments, it was unclear whether exercise conditions affected the impact of exercise on the hippocampus. Therefore, aiming to obtain translational insights applicable to humans, we focused on exercise intensity and started with the research investigating whether light-intensity exercise activates hippocampal neural activity in rats. We established an original running model in rats comprising laser Doppler flowmetry for monitoring hippocampal cerebral blood flow (Hip-CBF) and microdialysis for drug treatment. We found that Hip-CBF increased with light-intensity treadmill running, which was elicited by hippocampal neuronal activation and subsequent N-methyl-D-aspartate/nitric oxide (NMDA/NO) signaling. In this chapter, we first retrospectively summarize what we knew and what we did not know during that time, and the impact of our findings that light-intensity exercise can evoke neuronal activity in the rat hippocampus on our subsequent research.

Keywords Hippocampus · Regional cerebral blood flow · Laser Doppler flowmetry · Exercise intensity, exercise type

---

T. Nishijima (✉)
Sport Neuroscience Laboratory, Department of Health Promotion Sciences, Graduate School of Human Health Sciences, Tokyo Metropolitan University, Tokyo, Japan
e-mail: t-nishijima@tmu.ac.jp

H. Soya
Laboratory of Exercise Biochemistry and Neuroendocrinology, Institute of Health and Sport Sciences, University of Tsukuba, Tsukuba, Ibaraki, Japan

Sport Neuroscience Division, Kokoro Division, Advanced Research Initiative for Human High Performance (ARIHHP), Institute of Health and Sport Sciences, University of Tsukuba, Tsukuba, Ibaraki, Japan

Center for Cybernics Research, University of Tsukuba, Tsukuba, Ibaraki, Japan

H. Soya (ed.), *Exercise Brain Stimulation for Cognitive Function and Mental Health*, Advances in Neurobiology 44, https://doi.org/10.1007/978-981-95-0066-6_2

1 Introduction

In 1995, a study reported that 1 week of voluntary exercise resulted in an evident increase in brain-derived neurotrophic factor (BDNF) mRNA expression within the hippocampus in mice [1]. BDNF is a pleiotropic factor that plays an indispensable role in the development and optimal functioning of neurons and contributes to preventing psychiatric disorders [2]. Furthermore, in 1999, it was disclosed that voluntary exercise for 4 weeks facilitated adult hippocampal neurogenesis and concurrently enhanced hippocampus-dependent spatial learning abilities [3, 4]. These positive effects of exercise have been replicated by numerous studies and now become well established and widely acknowledged. The remarkable aspect of these milestones is that they instigated a significant paradigm shift that the conventional understanding of the brain as an organ primarily dedicated to controlling movement changes to a recognition that it, like skeletal muscle, undergoes adaptive changes in response to physical activity. Indeed, before these studies, it was already reported that exercise leads to adaptive angiogenesis in the cerebellum [5, 6]. However, because confronting social challenges that the acceleration of aging worldwide has resulted in age-related cognitive decline and an increased prevalence of neurodegenerative diseases including dementia [7–9], the discoveries that exercise enhances the hippocampal neuroplasticity has had a profound impact in the field of exercise and neuroscience.

At present, the augmentation of brain function, particularly in the hippocampus, through exercise has been elucidated across various levels, ranging from behavioral to molecular levels. However, in the early 2000s, scientific knowledge was still insufficient to translate findings obtained from animal experiments to humans. In particular, it was entirely unclear how the effects of exercise were influenced by the conditions of the exercise, including the intensity of exercise. If the benefits can only be expected through prolonged periods of high-intensity exercise, it becomes impractical as an exercise prescription for brain health promotion. In 1999, it was also demonstrated that aerobic walking for 6 months improved cognitive function in aged adults [10], suggesting that exercise at low intensity is effective enough in promoting brain fitness. However, in basic animals, including the above-mentioned, studies predominantly performed with spontaneous wheel running as an exercise model, exercise intensity was largely overlooked. To understand the mechanisms through which long-term exercise training induces adaptive changes in the hippocampus, we deemed it crucial to elucidate the physiological responses of the hippocampus to a single bout of exercise upon consideration of exercise intensity. Therefore, in this chapter, we retrospectively examine what we knew, what we did not in the early 2000s, and why we contemplated focusing on hippocampal cerebral blood flow (Hip-CBF) during exercise [11, 12].

2 What We Knew About, but What We Didn't in the Early 2000s

2.1 Exercise and Hippocampal Neuronal Activity

In addition to promoting BDNF and neurogenesis in the hippocampus, other effects of exercise on the hippocampus had already been reported by the early 2000s, e.g., inducing trophic factors such as nerve growth factor [13] and fibroblast growth factor 2 [14], increasing resilience against stress-induced decrease in BDNF [15], and improving recovery in BDNF downregulation caused by brain injury [16]. Human prospective studies also revealed neuroprotective effects of exercise that a higher level of physical activity was associated with a low risk of cognitive impairment and dementia [17], and that exercise has antidepressant effects comparable to those of antidepressant medication in older patients with major depression [18], or even better therapeutic effects in the rate of relapse [19]. As the effectiveness of exercise in the prevention and improvement of mental disorders associated with the hippocampus became evident, the exercise-induced BDNF and neurogenesis in the hippocampus gained increasing attention as the potential mechanism underlying the neuroprotective effects of exercise on the hippocampus [20].

"Use it or lose it" is a biological basis underlying how the body adapts to physiological stimuli including exercise. Given that repeated muscle contraction serves as the trigger for adaptations in skeletal muscles, we hypothesized that enhancing hippocampal neuroplasticity through long-term exercise was manifested as cumulative effects of repeated neural activation caused by a single bout of exercise. However, while the contraction of skeletal muscles during exercise is visible, the activation of hippocampal neural activity during exercise is not visually discernible. Indeed, when we started the research in 2000, it was uncertain whether exercise evoked hippocampal neuronal activation.

BDNF expression was regulated under intracellular calcium ions influx through activated NMDA receptor, and subsequent activation of cAMP response element-binding protein (CREB) [21]. The hippocampal neurogenesis was also demonstrated to be induced by the activation of NMDA receptors [22, 23]. Furthermore, in 2003, it was reported that voluntary wheel running failed to enhance BDNF and neurogenesis in the hippocampus in NMDA receptor epsilon 1 subunit knockout mice [24]. These findings strongly supported our hypothesis that exercise evoked hippocampal neuronal activation. Unfortunately, other research groups demonstrated ahead of us that treadmill exercise activated hippocampal neuronal activation in rats by immunohistochemical examination of c-Fos, as a marker of neuronal activation [25, 26]. However, these immunohistochemical studies analyzed postmortem brain samples, it was still uncertain whether the hippocampus is activated during exercise due to the lack of established methods to measure neural activity during exercise.

2.2 rCBF Regulation: Neurovascular Coupling and Autoregulation

Brain neural activity can be indirectly measured through changes in regional cerebral blood flow (rCBF), based on the theory that rCBF increases in the activated brain area to meet the elevated energy demand since neurons cannot store energy substrates or oxygen sufficiently. This theory termed neuro-vascular coupling [27, 28] was first hypothesized by Roy and Sherrington in 1890 [29], and now is widely recognized not only as a basic physiological function to maintain brain homeostasis but also as the theoretical foundation for brain imaging techniques such as positron emission tomography (PET) and functional magnetic resonance imaging (fMRI) [30].

Neurovascular coupling is primarily regulated by the glutamatergic neurotransmission system, the most prominent excitatory neurotransmission in the brain. In brief, described in detail elsewhere [31], when glutamatergic neurons are activated, calcium ions enter the postsynaptic neurons through NMDA receptors. The influx of calcium ions activates neuronal nitric oxide synthase (nNOS), leading to the synthesis of nitric oxide (NO). NO diffuses as a signaling molecule, causing relaxation of vascular smooth muscles surrounding activated neurons and inducing vasodilation via activation of guanylate cyclase and cGMP production. In addition, substances such as epoxyeicosatrienoic acid and arachidonic acid, secreted from astrocytes upon neural activation, also play a role in regulating neurovascular coupling [31].

While the neurovascular coupling serves as a dynamic regulatory function for adjusting rCBF in response to increased energy demand, there exists a static regulatory function to maintain rCBF as constant as possible despite changes in blood pressure, called autoregulation [32, 33]. In rats, rCBF is maintained relatively constant despite changes in blood pressure within ranging from 80 to 160 mmHg of mean arterial pressure (MAP). However, autoregulation has its limits, and when the mean arterial pressure exceeds 160 mmHg, regional cerebral blood flow (rCBF) increases proportionally with the rise in arterial pressure (breakthrough of autoregulation). Conversely, transient hemorrhagic hypotension (40–90 mmHg) results in a decrease in rCBF; however, this decrease in rCBF is temporary and recovers to resting levels within approximately 3–13 s, indicating that autoregulation is considered a relatively rapid regulatory response [32].

2.3 rCBF During Exercise

Because total energy expenditure drastically increases with exercise, circulatory responses to exercise are crucial to sustaining exercise performance, which is a very important topic in the field of exercise physiology. Due to insufficient total blood volume, a significant amount of the blood pumped out from the heart is predominantly redistributed to activated skeletal muscles during exercise, reaching 80–85% of cardiac output at maximal exercise, whereas blood flow to less active organs such as the liver and kidneys decreases [34].

Classical studies demonstrated that global and rCBF were relatively stable during exercise in both experimental animals [35–37] and humans [38]. Therefore, it was commonly believed that blood flow to the brain did not change during exercise because consistent blood flow to the brain is necessary for maintaining brain function even during exercise. However, in 1980, Gross and colleagues first revealed that rCBF in the brain regions associated with motor control (motor and sensory cortices, cerebellum) increased with treadmill exercise at moderate intensity in dogs [39]. In 2001, Delp and colleagues reported a study that extensively examined rCBF distributions in 23 brain regions during treadmill exercise at 70% and 100% of VdotO_2 max in miniature swine [40]. They reported a significant increase in rCBF in areas especially involved in motor and cardiorespiratory control, maintenance of equilibrium, and visual processing, while no changes were observed in regions related to auditory and olfactory functions [40]. These findings are further supported by the study that demonstrated rCBF increases in the neural circuits involving motor and sensory function (primary motor cortex, primary somatosensory cortex, dorsolateral striatum, ventrolateral thalamus, cerebellum, and visual cortex) in rats [41].

Thus, these studies have elucidated that exercise specifically increases rCBF in the brain regions associated with motor execution and control. It is noteworthy that the early 2000s marked a significant leap in our understanding of cerebral circulatory responses during exercise. Regarding the hippocampus, Delp and colleagues reported a tendency of increased blood flow in the hippocampus ($p = 0.056$) with exercise [40], but clear results were not reported yet. Furthermore, we noted that the aforementioned studies measured rCBF by radiolabeled microspheres or tracers. Although these methods have an advantage in examining the distribution of rCBF in the brain, dynamic and real-time change in rCBF during exercise was uncertain. In addition, with these methods unable to examine the molecular mechanisms regulating rCBF response during exercise. Therefore, it was unable to conclude that the increased rCBF was a result of neural activation triggered by exercise.

2.4 *Laser Doppler Flowmetry, a New Method for Real-Time Monitoring of rCBF in Conscious and Freely Moving Rats*

Laser Doppler Flowmetry (LDF) was first used to measure cutaneous blood flow [42], and in 1990s, garnered increasing attention in the field of neuroscience as a novel method for monitoring rCBF [43–45]. Theoretically, LDF measures rCBF by emitting laser light into the target tissue and analyzing the light returned to the photodetector through optical fibers. In brief, the laser light emitted from an optical fiber repeat scattering reflections and undergoes Doppler shift when it encounters moving red blood cells. As the Doppler shift depends on the speed of the moving red blood cells, the velocity of the red blood cells is calculated from the profile of frequency modulation. The volume of the red blood cells depends on the amount of the reflected light. Therefore, by calculating the velocity and volume of red blood cells, it is possible to determine the amount of blood flow of red blood cells, which can be considered as rCBF.

One of the notable advantages of LDF is its both spatial and temporal high resolution [43, 45]. The commercially available LDF is capable of measuring rCBF within a localized area (about a hemispherical space of 1 mm from the tip of the optical fiber). The use of laser light allows for a high temporal resolution within a few milliseconds. Furthermore, LDF enables continuous and real-time monitoring of rCBF, which is an extremely valuable feature for research in neurovascular coupling [46–48]. On the other hand, a disadvantage of LDF is its susceptibility to mechanical artifacts due to misalignment between the tip of the optical fiber and the target tissue [43]. Therefore, in most studies utilizing LDF, the experiments were performed with anesthetized animals.

Since the mid-1990s, although the number of literature was small, LDF studies measuring rCBF in conscious and free-moving rats have been documented [49–54]. A common aspect in these studies was the use of a guide cannula or a probe adaptor-connector device that was stereotaxically fixed to the skull in advance, providing secure fixation for the measuring optical fiber. Furthermore, it is crucial to acclimate the rats to the experimental environment beforehand, which helps keep the rats calm and contributes to the reduction of mechanical artifacts during the experiments [52]. These methodological refinements have enabled the real-time measurement of rCBF using LDF in conscious and freely moving rats. While it may not be possible to completely eliminate mechanical artifacts caused by vigorous movements, the spike-like artifacts can be easily discerned and removed through data analysis.

Using LDF, Osborne (1997) was the first to report that hippocampal cerebral blood flow (Hip-CBF) increases by approximately 20–35% during walking and running in rats [52], suggesting that exercise increases Hip-CBF. However, this study measured Hip-CBF in rats placed in a running wheel cage, and the duration of both walking and running was short (about 1 min). Nakajima et al. (2003) reported that Hip-CBF increased by about 4–5% during treadmill walking in a speed-dependent manner (1.2 ~ 4.8 m/min) in rats, but the duration of the treadmill walking was also short (only 30 s) [49]. Therefore, it was still unclear whether the increased Hip-CBF was sustained during prolonged exercise.

Nakajima et al. (2003) demonstrated that the increased Hip-CBF during treadmill walking was partially mediated by nicotinic acetylcholine receptors and the production of NO, suggesting that the increased Hip-CBF was regulated under neuronal activation [49]. However, in this study, the intravenous injection of either an antagonist of nicotinic acetylcholine receptors or an inhibitor of NO synthase altered mean arterial pressure, which made it difficult to conclude the regulating mechanisms of Hip-CBF during exercise. Continuing an extensive literature review, we found a unique experimental approach that combines LDF and microdialysis for examining molecular mechanisms regulating neurovascular coupling [55]. They placed the microdialysis probe adjacent to the location where blood flow was being measured by LDF. Importantly, local administration of NMDA into the hippocampus through microdialysis increased Hip-CBF, but no systemic effects such as changes in blood pressure or arterial gas partial pressure were observed. Therefore, we considered that the combined use of microdialysis and LDF was highly effective in investigating the regulatory mechanisms of Hip-CBF during exercise.

In summary, in the early 2000s, while interest in the enhancement of hippocampal neuroplasticity through habitual exercise received increasing attention, there was a lack of information for translating results obtained from animal experiments to humans. In particular, we focused on the question of whether hippocampal neural activity was activated by a single bout of exercise. From the perspective of exercise physiology, we strongly believed that the activation of hippocampal neurons with a single bout of exercise was a crucial trigger for leading adaptive changes, resulting in the enhancement of hippocampal neuroplasticity when exercise was repeated. A few studies suggested that exercise triggered hippocampal neural activation by c-Fos immunohistochemistry using postmortem rat brain [25, 26], but little was known about functional neuronal activity in the exercising rat hippocampus. Therefore, we endeavored to develop a unique experimental model in rats that enables continuous and real-time monitoring of Hip-CBF during prolonged treadmill running using LDF. Furthermore, local drug administration by microdialysis enabled us to examine whether the increased Hip-CBF seen during running was responsible for hippocampal neuronal activation and further molecular mechanisms (Fig. 1) [11, 12].

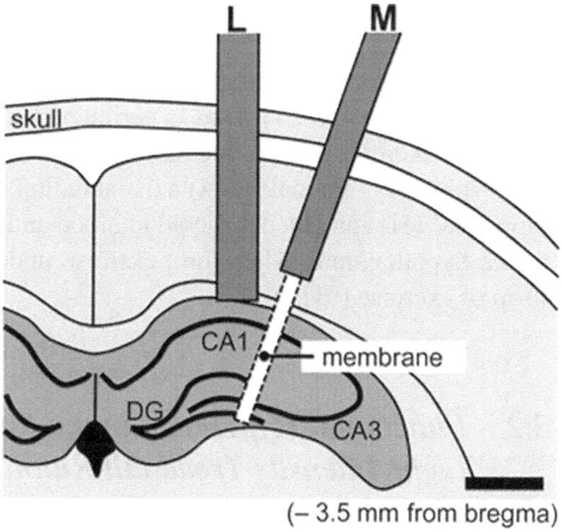

Fig. 1 Combined use of LDF and microdialysis to examine whether functional hyperemia occurs during light-intensity treadmill running. L: guide cannula for LDF; M: guide cannulae for microdialysis. The microdialysis membrane (2 mm) protrudes from the tip of the guide cannula. Scale bar represents 1 mm. (Adopted from Nishijima et al. [11])

3 Hippocampal Neural Activation During Light-Intensity Treadmill Running: Examination by Real-Time Monitoring of Hip-CBF

3.1 Exercise Intensity and Stress Responses

To translate the results obtained from animal studies to humans, we considered that exercise intensity was a crucial factor. Therefore, in our studies [11, 12], we set the speed of the treadmill running at a mild speed (10 m/min) and examined whether light-intensity treadmill running leads a functional hyperemia, an increase in rCBF in response to neuronal activation, in the rat hippocampus. The reason for choosing light-intensity exercise was due to some technical constraints. Indeed, as the running speed increases, there was a pronounced increase in head movement, making it more prone to mechanical artifacts. In addition, to investigate the molecular mechanisms regulating the increase in Hip-CBF, a sufficient amount of time was required for drug treatment via microdialysis. If the exercise intensity was high, there was a possibility that the physiological state of the running rats may change while waiting for the drugs to take effect, which could potentially impact Hip-CBF level. However, the most significant reason for choosing light-intensity exercise was the belief that if it became evident that light-intensity exercise enhances hippocampal neural activity, such insights would be highly relevant for application to humans (details will be discussed later).

During the same period in the early 2000s, colleagues in our laboratory were investigating whether exercise induces stress responses in rats, with a focus on exercise intensity [56–59]. In their studies, exercise intensity was determined based on the lactate threshold (LT) which refers to the point where the steady state of blood lactate accumulation breaks down and begins to increase rapidly during incremental exercise. Importantly, our colleagues confirmed the presence of the lactate threshold during incremental exercise both in rats [57, 59] and mice [60], like humans. Therefore, it becomes possible to generalize the impact of exercise intensity, i.e., below or above the LT, from experimental animals to humans. They found that supra-LT exercise induces a stress response in rats, i.e., activation of the hypothalamic–pituitary–adrenal (HPA) axis including increases in adrenocorticotropic hormone (ACTH) and glucocorticoid in blood and activation of paraventricular nucleus of the hypothalamus. Therefore, exercise under the LT is considered a stress-free form of exercise [61].

3.2 Functional Hyperemia in the Rat Hippocampus During Light-Intensity Treadmill Running

To examine whether functional hyperemia can occur in the rat hippocampus during light-intensity treadmill exercise, we combined LDF for continuous and real-time monitoring of Hip-CBF and microdialysis for local drug administration to the loci where Hip-CBF was monitored [12]. As described earlier, the disadvantage of LDF was its susceptibility to mechanical artifacts associated with head movement. Similar to the previous literature, we securely fixed the optical fiber with the guide cannula that had been stereotaxically fixed with the skull prior to the experiments performed. We also carried out a familiarization trial to ensure that the rats were fully habituated to the treadmill running with the optical fiber attached to the head. With these methodological considerations, we successfully demonstrated for the first time that light-intensity treadmill running increased Hip-CBF (Fig. 2a). Although measuring changes in Hip-CBF in a running rat seemed a reckless challenge, it was surprising for us that the LDF signals from the rats running at light intensity were more stable compared to those from freely moving rats because the rats kept their head position stable while running at a consistent speed.

The Hip-CBF increased during light-intensity treadmill running; however, significant questions remained whether the increased Hip-CBF was caused by the hippocampal neural activation. Indeed, exercise increases cardiac output and mean arterial pressure (MAP). Indeed, we found that the light-intensity treadmill running slightly increased MAP [12]. Therefore, the increase in Hip-CBF seen during light-intensity treadmill running may be a passive change in global CBF brought about by the increased MAP. Therefore, we locally administrated tetrodotoxin (TTX, a sodium channel blocker) via microdialysis to the loci where Hip-CBF was monitored. As expected, intrahippocampal infusion of arterial cerebrospinal fluid (aCSF) caused no statistical changes in Hip-CBF, but the infusion of TTX completely attenuated the increase in Hip-CBF (Fig. 3). With these results, we were able to demonstrate that light-intensity exercise activates hippocampal neurons, and subsequently causes functional hyperemia in the rat hippocampus.

We further investigated the molecular mechanisms by which exercise increased Hip-CBF, and found that the increased Hip-CBF during light-intensity treadmill running was attenuated by either NMDA receptor antagonists (MK-801) or NO synthase inhibitors (N(G)-nitro-l-arginine methyl ester, L-NAME) [11]. These results demonstrate that activation of the NMDA receptor/NO signal cascade is a potent mediator in functional hyperemia elicited by light-intensity treadmill running (Fig. 4). It was well known that NMDA receptor/NO signaling was a major regulatory mechanism in neurovascular coupling; however, the previous studies were only performed in anesthetized animals [62–65]. Therefore, we would like to note here that our results provide evidence for the first time that this signal cascade is also involved in the regulation of neurovascular coupling in a conscious physiological state, such as running. We also found that rCBF in the olfactory bulb was unchanged during light-intensity treadmill running. First, this result refuted the possibility that the increase in Hip-CBF was a nonspecific mechanical artifact. More importantly, the region-dependent changes in rCBF strongly supported our finding that Hip-CBF during running was regulated by local neuronal activity in the hippocampus.

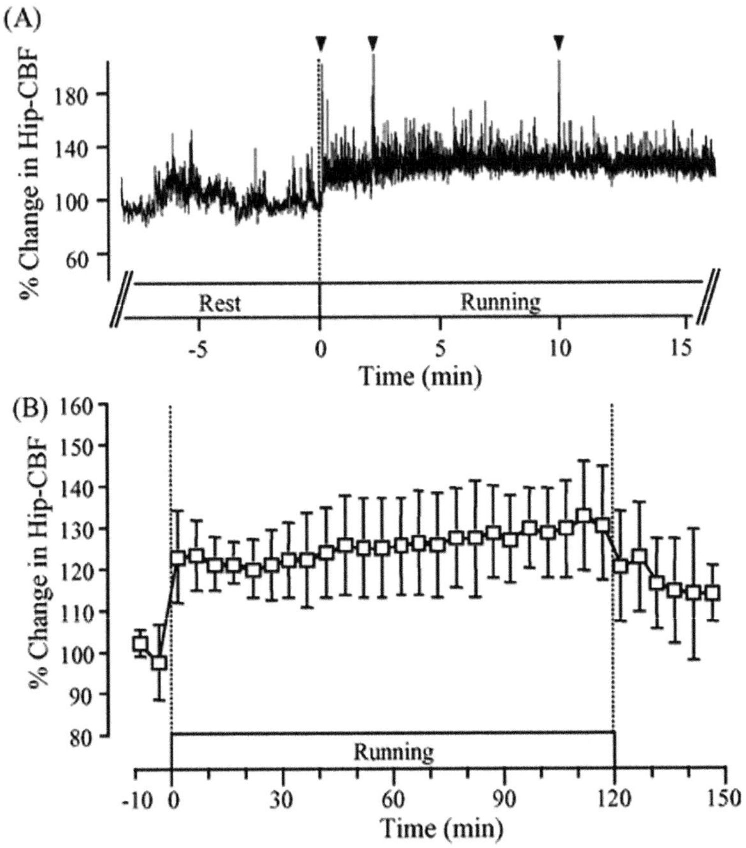

Fig. 2 Light-intensity treadmill running increases Hip-CBF. (**a**) Typical raw data of Hip-CBF monitored by LDF. Arrowheads represent movement-induced mechanical artifacts. (**b**) Changes in Hip-CBF vs. basal revel during light-intensity treadmill running. Data are expressed as mean ± SD ($n = 7$). (Adopted from Nishijima and Soya [12])

4 Light-Intensity Exercise Evokes Hippocampal Neuronal Activation: Significance for Translating Animal to Human Studies

As described earlier, by measuring Hip-CBF during exercise using LDF and further investigating its regulatory mechanisms, we revealed that even light-intensity exercise below the LT evokes neuronal activation in the rat hippocampus. In the field of exercise physiology, it was generally believed that the exercise intensity needed to be moderate or higher to enhance the function of skeletal muscles or the

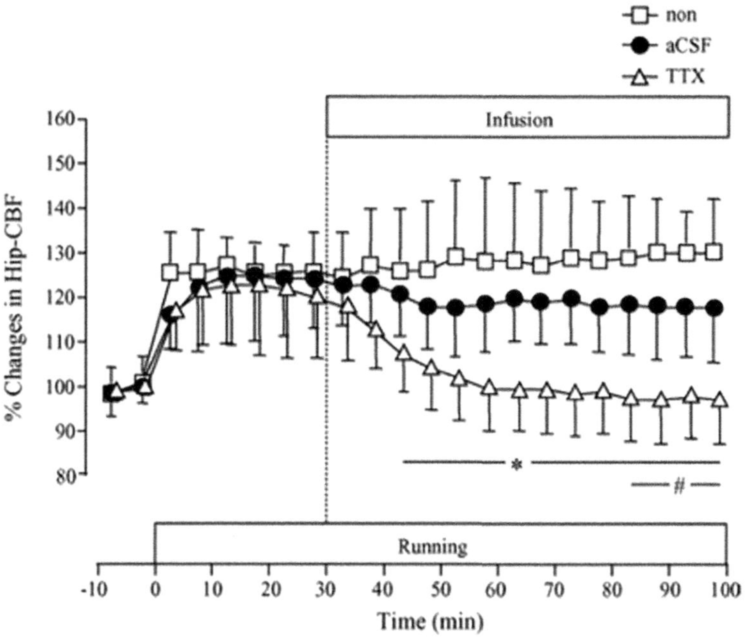

Fig. 3 The increase in Hip-CBF during light-intensity treadmill running was TTX-dependent. Data are expressed as mean ± SD (n = 6 in each group). *$p < 0.05$ vs. non-infusion group; # $p < 0.05$ vs. aCSF-infusion group. (Adopted from Nishijima and Soya [12])

respiratory–circulatory system. Therefore, the findings suggesting that even light-intensity exercise was sufficiently effective in enhancing hippocampal function encouraged us to further investigate the effects of low-intensity exercise on brain function. Indeed, we then found that habituated exercise at light intensity successfully enhanced BDNF expression [66], neurogenesis in the hippocampus [67–69], and hippocampus-dependent memory in rodents [70]. Furthermore, our colleagues demonstrated that transient light-intensity exercise also enhances not only the frontal cortex-dependent [71, 72] but also the hippocampus-dependent function in humans [73, 74].

In the early 2000s, while conducting these Hip-CBF studies, we and our colleagues simultaneously initiated an exercise intervention study targeting local elderly residents. For implementing exercise interventions in the elderly, we considered that the exercise program should be easy for the elderly to engage with and sustain for a long period. Therefore, the intervention program was designed with exercise at low-to-mild intensity that was easy for the elderly to implement even at home. The 2-year exercise intervention at light intensity improved cognitive performance and prevented the age-associated reduction of the prefrontal volume [75]. Thus, over the past two decades, we and our colleagues elucidated the positive

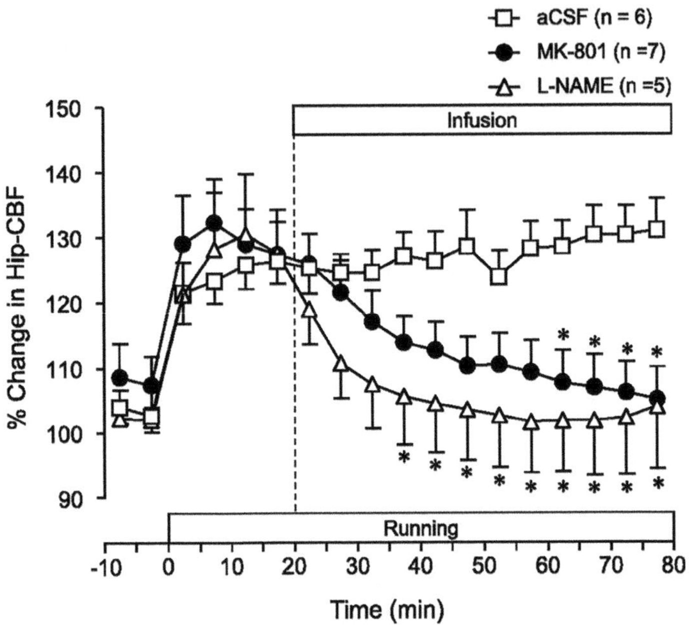

Fig. 4 The increase in Hip-CBF during light-intensity exercise was mediated by NMDA receptor/NO signaling. Data are expressed as mean ± SE. $*p < 0.05$ vs. aCSF infusion group. (Adopted from Nishijima et al. [11])

effects of light-intensity exercise on the brain. This was motivated and achieved by the recognition that investigating the impact of exercise intensity was crucial for translating findings from animal experiments to humans and our findings that light-intensity exercise can activate the hippocampus.

5 Further Perspectives: Beyond the Intensity of Exercise

In humans, it has been mentioned that the type of exercise is another critical factor influencing the effect of exercise on the brain, including depression [76, 77] and cognitive function [78, 79]. However, because of the difficulties in controlling exercise conditions in human cross-sectional studies, it was unclear whether the type of exercise truly influenced the effects of exercise, even when the intensity and duration were matched. Recently, we revealed that the impact of exercise on the hippocampus varies depending on the type of exercise in mice [80]. In brief, we compared the effects of an acute bout of exercise on hippocampal neuronal activation between two different types of exercise: treadmill and rotarod exercise. The exercise intensities were matched at just below the LT. While treadmill exercise activated the hippocampus, this effect was not observed with rotarod exercise. These results suggest that there may be a specific exercise type that enhances hippocampal function, or the intended benefits of exercise on the hippocampus may not be gained if the exercise type is not appropriate. So far, it was uncertain why treadmill exercise enhanced the hippocampal activity more than the rotarod exercise. If the reasons behind this question can be elucidated, it would be a new translational finding for considering exercise prescriptions in humans.

The effectiveness of exercise in promoting physical and mental health is widely acknowledged. However, in reality, initiating or maintaining regular exercise is exceedingly challenging, the number of individuals not meeting recommended levels of physical activity is increasing in both adult [81] and adolescent [82] population worldwide. Therefore, it is crucial not only to recommend exercise but also to increase physical activity levels, including daily life activities. In animal studies, it is well known that an enriched environment enhances brain functions, including the hippocampus [83]. Therefore, we tested a hypothesis that enriching the housing environment has the potential to increase physical activity levels, and subsequently enhance hippocampal neurogenesis in mice [84]. A body-implantable actimeter enabled us to measure the amount of physical activity in group-housed mice under a complex environment with a variety of objects. Surprisingly, contrary to our hypothesis, the enriched environment did not increase physical activity, but enhanced hippocampal neurogenesis. First, these findings may point out an issue that enriching the environment, i.e., enlarging housing space and increasing spatial complexity, may be insufficient to promote physical activity levels. Second, it is more intriguing for us that hippocampal neurogenesis was enhanced without increasing physical activity, which challenges the common belief that increasing the level of physical activity through exercise is a prerequisite for enhancing hippocampal neurogenesis. As the recent study demonstrated that contextual factors, such as location and type of activity, can affect the impact of exercise on mood, it may be feasible to consider the quality of physical activity performed [85], or environment as a contextual factor affecting the effects of physical activity on the hippocampus. Further studies are necessary to translate these results to humans.

6 Conclusion

In this chapter, we looked back at our studies in the early 2000s and summarized why we researched to answer the question of whether a bout of light-intensity exercise evoked neuronal activation in the rat hippocampus. Even in basic research using experimental animals, we prioritized obtaining insights applicable to humans, especially focusing on the intensity of exercise. Indeed, our finding that light-intensity exercise activates the rat hippocampus encouraged us and our colleagues to further investigate the effects of low-intensity exercise on brain function. Recently, we found that, in addition to the intensity of exercise, the type of exercise and the environment in which physical activity is performed may also influence the effects of exercise on the hippocampus. By continuing basic research aimed at translating to humans, we believe that we can contribute to the realization of well-being in the future.

References

1. Neeper SA, Gómez-Pinilla F, Choi J, Cotman C (1995) Exercise and brain neurotrophins. Nature 373:109
2. Wang CS, Kavalali ET, Monteggia LM (2022) BDNF signaling in context: from synaptic regulation to psychiatric disorders. Cell 185:62–76
3. van Praag H, Christie BR, Sejnowski TJ, Gage FH (1999) Running enhances neurogenesis, learning, and long-term potentiation in mice. Proc Natl Acad Sci USA 96:13427–13431
4. van Praag H, Kempermann G, Gage FH (1999) Running increases cell proliferation and neurogenesis in the adult mouse dentate gyrus. Nat Neurosci 2:266–270
5. Black JE, Isaacs KR, Anderson BJ, Alcantara AA, Greenough WT (1990) Learning causes synaptogenesis, whereas motor activity causes angiogenesis, in cerebellar cortex of adult rats. Proc Natl Acad Sci USA 87:5568–5572
6. Isaacs KR, Anderson BJ, Alcantara AA, Black JE, Greenough WT (1992) Exercise and the brain: angiogenesis in the adult rat cerebellum after vigorous physical activity and motor skill learning. J Cereb Blood Flow Metab 12:110–119
7. Petersen RC, Smith G, Kokmen E, Ivnik RJ, Tangalos EG (1992) Memory function in normal aging. Neurology 42:396–401
8. Mecocci P, Cherubini A, Mariani E, Ruggiero C, Senin U (2004) Depression in the elderly: new concepts and therapeutic approaches. Aging Clin Exp Res 16:176–189
9. Friedenberg RM (2003) Dementia: one of the greatest fears of aging. Radiology 229:632–635
10. Kramer AF, Hahn S, Cohen NJ et al (1999) Ageing, fitness and neurocognitive function. Nature 400:418–419
11. Nishijima T, Okamoto M, Matsui T, Kita I, Soya H (2012) Hippocampal functional hyperemia mediated by NMDA receptor/NO signaling in rats during mild exercise. J Appl Physiol 112:197
12. Nishijima T, Soya H (2006) Evidence of functional hyperemia in the rat hippocampus during mild treadmill running. Neurosci Res 54:186
13. Neeper SA, Gómez-Pinilla F, Choi J, Cotman CW (1996) Physical activity increases mRNA for brain-derived neurotrophic factor and nerve growth factor in rat brain. Brain Res 726:49–56
14. Gómez-Pinilla F, Dao L, So V (1997) Physical exercise induces FGF-2 and its mRNA in the hippocampus. Brain Res 764:1–8
15. Adlard PA, Cotman CW (2004) Voluntary exercise protects against stress-induced decreases in brain-derived neurotrophic factor protein expression. Neuroscience 124:985–992
16. Griesbach GS, Hovda DA, Molteni R, Wu A, Gomez-Pinilla F (2004) Voluntary exercise following traumatic brain injury: brain-derived neurotrophic factor upregulation and recovery of function. Neuroscience 125:129–139
17. Laurin D, Verreault R, Lindsay J, MacPherson K, Rockwood K (2001) Physical activity and risk of cognitive impairment and dementia in elderly persons. Arch Neurol 58:498–504
18. Blumenthal JA, Babyak MA, Moore KA et al (1999) Effects of exercise training on older patients with major depression. Arch Intern Med 159:2349–2356
19. Babyak M, Blumenthal JA, Herman S, Khatri P, Doraiswamy M, Moore K, Craighead WE, Baldewicz TT, Krishnan KR (2000) Exercise treatment for major depression: maintenance of therapeutic benefit at 10 months. Psychosom Med 62:633–638
20. Cotman CW, Berchtold NC (2002) Exercise: a behavioral intervention to enhance brain health and plasticity. Trends Neurosci 25:295–301

21. Hardingham GE, Fukunaga Y, Bading H (2002) Extrasynaptic NMDARs oppose synaptic NMDARs by triggering CREB shut-off and cell death pathways. Nat Neurosci 5:405–414
22. Arvidsson A, Kokaia Z, Lindvall O (2001) N-methyl-D-aspartate receptor-mediated increase of neurogenesis in adult rat dentate gyrus following stroke. Eur J Neurosci 14:10–18
23. Cameron HA, McEwen BS, Gould E (1995) Regulation of adult neurogenesis by excitatory input and NMDA receptor activation in the dentate gyrus. J Neurosci 15:4687–4692
24. Kitamura T, Mishina M, Sugiyama H (2003) Enhancement of neurogenesis by running wheel exercises is suppressed in mice lacking NMDA receptor epsilon 1 subunit. Neurosci Res 47:55–63
25. Lee T-H, Jang M-H, Shin M-C, Lim B-V, Kim Y-P, Kim H, Choi H-H, Lee K-S, Kim E-H, Kim C-J (2003) Dependence of rat hippocampal c-Fos expression on intensity and duration of exercise. Life Sci 72:1421–1436
26. Oladehin A, Waters RS (2001) Location and distribution of Fos protein expression in rat hippocampus following acute moderate aerobic exercise. Exp Brain Res 137:26–35
27. Sokoloff L (1981) Relationships among local functional activity, energy metabolism, and blood flow in the central nervous system. Fed Proc 40:2311–2316
28. Lauritzen M (2001) Relationship of spikes, synaptic activity, and local changes of cerebral blood flow. J Cereb Blood Flow Metab 21:1367–1383
29. Roy CS, Sherrington CS (1890) On the regulation of the blood-supply of the brain. J Physiol 11:85–158
30. Raichle ME (1998) Behind the scenes of functional brain imaging: a historical and physiological perspective. Proc Natl Acad Sci 95:765–772
31. Hosford PS, Gourine AV (2019) What is the key mediator of the neurovascular coupling response? Neurosci Biobehav Rev 96:174–181
32. Florence G, Seylaz J (1992) Rapid autoregulation of cerebral blood flow: a laser-Doppler Flowmetry study. J Cereb Blood Flow Metab 12:674–680
33. Hernåndez MJ, Brennan RW, Bowman GS (1978) Cerebral blood flow autoregulation in the rat. Stroke 9:150–154
34. Powers SK, Howley ET, Quindry J (2007) Exercise physiology: theory and application to fitness and performance, 10th edn. McGraw-Hill, New York
35. Hohimer AR, Hales JR, Rowell LB, Smith OA (1983) Regional distribution of blood flow during mild dynamic leg exercise in the baboon. J Appl Physiol 55:1173–1177
36. Globus M, Melamed E, Keren A, Tzivoni D, Granot C, Lavy S, Stern S (1983) Effect of exercise on cerebral circulation. J Cereb Blood Flow Metab 3:287–290
37. Foreman DL, Sanders M, Bloor CM (1976) Total and regional cerebral blood flow during moderate and severe exercise in miniature swine. J Appl Physiol 40:191–195
38. Zobl EG, Talmers FN, Christensen RC, Baer LJ (1965) Effect of exercise on the cerebral circulation and metabolism. J Appl Physiol 20:1289–1293
39. Gross PM, Marcus ML, Heistad DD (1980) Regional distribution of cerebral blood flow during exercise in dogs. J Appl Physiol 48:213–217
40. Delp MD, Armstrong RB, Godfrey DA, Laughlin MH, Ross CD, Wilkerson MK (2001) Exercise increases blood flow to locomotor, vestibular, cardiorespiratory and visual regions of the brain in miniature swine. J Physiol 533:849–859
41. Holschneider DP, Maarek J-MI, Yang J, Harimoto J, Scremin OU (2003) Functional brain mapping in freely moving rats during treadmill walking. J Cereb Blood Flow Metab 23:925–932
42. Holloway GA Jr, Watkins DW (1977) Laser Doppler measurement of cutaneous blood flow. J Invest Dermatol 69(3):306–309
43. Borgos J (1996) Laser Doppler monitoring of cerebral blood flow. Neurol Res 18:251–255
44. Frerichs KU, Feuerstein GZ (1990) Laser-doppler flowmetry. Mol Chem Neuropathol 12:55–70
45. Fukuda O, Endo S, Kuwayama N, Harada J, Takaku A (1995) The characteristics of laser-Doppler Flowmetry for the measurement of regional cerebral blood flow. Neurosurgery 36:358–364

46. Caesar K, Gold L, Lauritzen M (2003) Context sensitivity of activity-dependent increases in cerebral blood flow. Proc Natl Acad Sci 100:4239–4244
47. Jones M, Berwick J, Johnston D, Mayhew J (2001) Concurrent optical imaging spectroscopy and laser-Doppler Flowmetry: the relationship between blood flow, oxygenation, and volume in rodent barrel cortex. NeuroImage 13:1002–1015
48. Martin C, Martindale J, Berwick J, Mayhew J (2006) Investigating neural–hemodynamic coupling and the hemodynamic response function in the awake rat. NeuroImage 32:33–48
49. Nakajima K, Uchida S, Suzuki A, Hotta H, Aikawa Y (2003) The effect of walking on regional blood flow and acetylcholine in the hippocampus in conscious rats. Auton Neurosci 103:83–92
50. Gu W, Jiang W, Wester P (2003) Real-time cortical cerebral blood flow follow-up in conscious, freely moving rats by laser Doppler flowmetry. Methods 30:172–177
51. Gu W, Jiang W, Brännström T, Wester P (1999) Long-term cortical CBF recording by laser-Doppler flowmetry in awake freely moving rats subjected to reversible photothrombotic stroke. J Neurosci Methods 90:23–32
52. Osborne PG (1997) Hippocampal and striatal blood flow During behavior in rats: chronic laser Doppler Flowmetry study. Physiol Behav 61:485–492
53. Kimura A, Okada K, Sato A, Suzuki H (1994) Regional cerebral blood flow in the frontal, parietal and occipital cortices increases independently of systemic arterial pressure during slow walking in conscious rats. Neurosci Res 20:309–315
54. Sato A, Sato Y, Uchida S (1994) Blood flow in the sciatic nerve is regulated by vasoconstrictive and vasodilative nerve fibers originating from the ventral and dorsal roots of the spinal nerves. Neurosci Res 21:125–133
55. Kaiser MG, During MJ (1995) Combining laser doppler flowmetry with microdialysis: a novel approach to investigate the coupling of regional cerebral blood flow to neuronal activity. J Neurosci Methods 60:165–173
56. Ohiwa N, Chang H, Saito T, Onaka T, Fujikawa T, Soya H (2007) Possible inhibitory role of prolactin-releasing peptide for ACTH release associated with running stress. Am J Phys Regul Integr Comp Phys 292:R497–R504
57. Saito T, Soya H (2004) Delineation of responsive AVP-containing neurons to running stress in the hypothalamus. Am J Phys Regul Integr Comp Phys 286:R484–R490
58. Ohiwa N, Saito T, Chang H, Omori T, Fujikawa T, Asada T, Soya H (2006) Activation of A1 and A2 noradrenergic neurons in response to running in the rat. Neurosci Lett 395:46–50
59. Soya H, Mukai A, Deocaris CC, Ohiwa N, Chang H, Nishijima T, Fujikawa T, Togashi K, Saito T (2007) Threshold-like pattern of neuronal activation in the hypothalamus during treadmill running: establishment of a minimum running stress (MRS) rat model. Neurosci Res 58:341–348
60. Okamoto M, Yamamura Y, Liu Y-F et al (2015) Hormetic effects by exercise on hippocampal neurogenesis with glucocorticoid signaling. Brain Plast 1:149–158
61. Soya H, Okamoto M, Matsui T et al (2011) Brain activation via exercise: exercise conditions leading to neuronal activation & hippocampal neurogenesis. J Exerc Nutr Biochem 6:1
62. Yang G, Iadecola C (1998) Activation of cerebellar climbing fibers increases cerebellar blood flow. Stroke 29:499–508
63. Meng W, Tobin JR, Busija DW (1995) Glutamate-induced cerebral vasodilation is mediated by nitric oxide through *N*-methyl-D-aspartate receptors. Stroke 26:857–863
64. Chi O, Liu X, Weiss HR (2003) Effects of inhibition of neuronal nitric oxide synthase on NMDA-induced changes in cerebral blood flow and oxygen consumption. Exp Brain Res 148:256–260
65. Akgören N, Fabricius M, Lauritzen M (1994) Importance of nitric oxide for local increases of blood flow in rat cerebellar cortex during electrical stimulation. Proc Natl Acad Sci 91:5903–5907

66. Soya H, Nakamura T, Deocaris CC, Kimpara A, Iimura M, Fujikawa T, Chang H, McEwen BS, Nishijima T (2007) BDNF induction with mild exercise in the rat hippocampus. Biochem Biophys Res Commun 358:961
67. Inoue K, Okamoto M, Shibato J, Lee MC, Matsui T, Rakwal R, Soya H (2015) Long-term mild, rather than intense, exercise enhances adult hippocampal neurogenesis and greatly changes the transcriptomic profile of the hippocampus. PLoS One 10:e0128720
68. Okamoto M, Yamamura Y, Liu Y-F et al (2015) Hormetic effects by exercise on hippocampal neurogenesis with glucocorticoid signaling. Brain Plasticity 1:149–158
69. Okamoto M, Hojo Y, Inoue K, Matsui T, Kawato S, McEwen BS, Soya H (2012) Mild exercise increases dihydrotestosterone in hippocampus providing evidence for androgenic mediation of neurogenesis. Proc Natl Acad Sci 109:13100–13105
70. Inoue K, Hanaoka Y, Nishijima T, Okamoto M, Chang H, Saito T, Soya H (2014) Long-term mild exercise training enhances hippocampus-dependent memory in rats. Int J Sports Med 36:280–285
71. Yanagisawa H, Dan I, Tsuzuki D, Kato M, Okamoto M, Kyutoku Y, Soya H (2010) Acute moderate exercise elicits increased dorsolateral prefrontal activation and improves cognitive performance with Stroop test. NeuroImage 50:1702–1710
72. Hyodo K, Dan I, Suwabe K, Kyutoku Y, Yamada Y, Akahori M, Byun K, Kato M, Soya H (2012) Acute moderate exercise enhances compensatory brain activation in older adults. Neurobiol Aging 33:2621–2632
73. Suwabe K, Byun K, Hyodo K et al (2018) Rapid stimulation of human dentate gyrus function with acute mild exercise. Proc Natl Acad Sci 115:10487–10492
74. Suwabe K, Hyodo K, Byun K, Ochi G, Yassa MA, Soya H (2017) Acute moderate exercise improves mnemonic discrimination in young adults. Hippocampus 27:229–234
75. Tamura M, Nemoto K, Kawaguchi A, Kato M, Arai T, Kakuma T, Mizukami K, Matsuda H, Soya H, Asada T (2015) Long-term mild-intensity exercise regimen preserves prefrontal cortical volume against aging. Int J Geriatr Psychiatry 30:686–694
76. Chen P-J, Chen K-M, Hsu H-F, Belcastro F (2022) Types of exercise and training duration on depressive symptoms among older adults in long-term care facilities. Ageing Res Rev 77:101613
77. Kim K, Koo K-M (2022) Influence of different exercise types on health-related quality-of-life in men with depressive disorder in South Korea. Front Public Health 10
78. Gökçe E, Güneş E, Arı F, Hayme S, Nalçacı E (2021) Comparison of the effects of open- and closed-skill exercise on cognition and peripheral proteins: a cross-sectional study. PLoS One 16:e0251907
79. Gu Q, Zou L, Loprinzi PD, Quan M, Huang T (2019) Effects of open versus closed skill exercise on cognitive function: a systematic review. Front Psychol 10
80. Tsuchida R, Yamaguchi T, Funabashi D, Koumi Y, Kita I, Nishijima T (2022) Exercise type influences the effect of an acute bout of exercise on hippocampal neuronal activation in mice. Neurosci Lett 783:136707
81. Guthold R, Stevens GA, Riley LM, Bull FC (2018) Worldwide trends in insufficient physical activity from 2001 to 2016: a pooled analysis of 358 population-based surveys with 1·9 million participants. Lancet Glob Health 6:e1077–e1086
82. Guthold R, Stevens GA, Riley LM, Bull FC (2020) Global trends in insufficient physical activity among adolescents: a pooled analysis of 298 population-based surveys with 1·6 million participants. Lancet Child Adolesc Health 4:23–35
83. Rojas-Carvajal M, Sequeira-Cordero A, Brenes JC (2022) The environmental enrichment model revisited: a translatable paradigm to study the stress of our modern lifestyle. Eur J Neurosci 55:2359–2392

84. Funabashi D, Tsuchida R, Matsui T, Kita I, Nishijima T (2023) Enlarged housing space and increased spatial complexity enhance hippocampal neurogenesis but do not increase physical activity in mice. Front Sports Act Living 5
85. White RL, Ryan D, Young C, Elston R, Rossi T (2023) How does the context of physical activity influence perceived mood and wellbeing after exercise? Ment Health Phys Act 24:100504

Hippocampal Neurogenesis via Light-Intensity Running and Its Mechanism

Koshiro Inoue and Hideaki Soya

Abstract Adult hippocampal neurogenesis (AHN) is the process of generating new neurons in the adult hippocampal dentate gyrus (DG). Exercise promotes AHN and improves hippocampal function through neuroplastic enhancement. The underlying regulatory factors of this process are currently being vigorously studied. However, many previous studies have used a rodent wheel-running model, in which the exercise condition (e.g., volume, intensity, duration) cannot be controlled. In contrast, treadmill running (TR) allows the precise regulation of conditions such that animals can run according to specific experimental aims. Understanding the intensity-dependent effects of exercise on AHN and hippocampal functions, and the underlying mechanisms, is crucial for the development of exercise prescriptions for humans in diverse educational and clinical fields. Based on the lactate threshold (LT), an inflection point at which blood lactate accumulation drastically rises during incremental exercise, exercise can be defined as minimal-stress *light-intensity* exercise (below LT) and exercise-derived-stress *vigorous-intensity* exercise (above LT). This chapter begins with a brief overview of AHN, followed by a discussion of LT-based exercise effects on AHN and hippocampal function as they vary with exercise intensity, primarily following the findings from the TR models, and closing with the molecular factors involved in AHN and hippocampal function regulation.

K. Inoue (✉)
Center for Education in Liberal Arts and Sciences/School of Rehabilitation Sciences, Health Sciences University of Hokkaido, Hokkaido, Japan
e-mail: ikoshiro@hoku-iryo-u.ac.jp

H. Soya
Laboratory of Exercise Biochemistry and Neuroendocrinology, Institute of Health and Sport Sciences, University of Tsukuba, Tsukuba, Ibaraki, Japan

Sport Neuroscience Division, Kokoro Division, Advanced Research Initiative for Human High Performance (ARIHHP), Institute of Health and Sport Sciences, University of Tsukuba, Tsukuba, Ibaraki, Japan

Center for Cybernics Research, University of Tsukuba, Tsukuba, Ibaraki, Japan

H. Soya (ed.), *Exercise Brain Stimulation for Cognitive Function and Mental Health*, Advances in Neurobiology 44, https://doi.org/10.1007/978-981-95-0066-6_3

Keywords Treadmill running · Exercise intensity · Lactate threshold · Adult hippocampal neurogenesis · Hippocampal function · Hypothetical mechanisms

1 Introduction

In 1965, Altman and Das provided the first evidence that ongoing cell differentiation and neurogenesis occur near the granular cell layer of the dentate gyrus (DG) in the adult hippocampus by using the [^{3}H]-thymidine labeling method [1]. However, the scientific community did not acknowledge the discovery for about 20 years, as the dogma established by Ramon y Cajal that the adult brain cannot produce new cells (1894) was strongly believed. Subsequent advances in techniques for quantifying newly born cells, such as a thymidine analogue bromodeoxyuridine (BrdU) labeling method, have successively led to results that overturned this dogma, leading to the acceptance of the existence of the neurogenesis in the adult hippocampal DG [2]. This phenomenon is now referred to as adult hippocampal neurogenesis (AHN), and is also characterized as a stepwise cell-generation process (Fig. 1) [3, 4].

In the hippocampal DG, a large number of glia-like neural stem cells (NSCs, type-1 cells) capable of producing identical progeny through self-renewal reside in the subgranular zone (SGZ) in a quiescent and undifferentiated state. When activated, the NSCs self-renew or generate intermediate progenitor cells (IPCs, type-2 cells), which have the capacity to proliferate and differentiate into multiple lineage cell types but have no capacity to self-renew. The IPCs are highly proliferative, and thus the expansion of the pool of newborn cells is considered to mostly occur in this phase. The IPCs subsequently differentiate into neuroblasts (type 3 cells) and then into postmitotic immature neurons. During this phase, the proliferative activity gradually decreased, and radial migration toward their final destination begins. Only a small fraction of the proliferating immature cells survive and pass through the maturation phase, which is characterized by the development and elaboration of the axon and dendritic tree, and are finally integrated into the pre-existing functional hippocampal circuits. Since several molecular markers are specifically expressed during each generating process of AHN (Fig. 1), these markers, in combination with methods for labeling newborn cells (e.g., BrdU labeling), can be used to follow the time-course and fate specification of newborn cells within the adult hippocampal DG in a precise and detailed manner (for extensive reviews on this topic, *see* [5]).

The neurogenic process is strictly regulated by various factors, both intrinsic and extrinsic, within a specialized local microenvironment of the SGZ, also known as a neurogenic niche, which composes neuronal and non-neuronal cells such as astrocytes, microglia, and blood vessels. While the molecular players and signaling pathways are not yet fully characterized, numerous signaling factors (e.g., growth factors, neurotrophins, hormones, and neurotransmitters), transcription factors, and epigenetic regulators have been shown to be involved in regulating several targeted

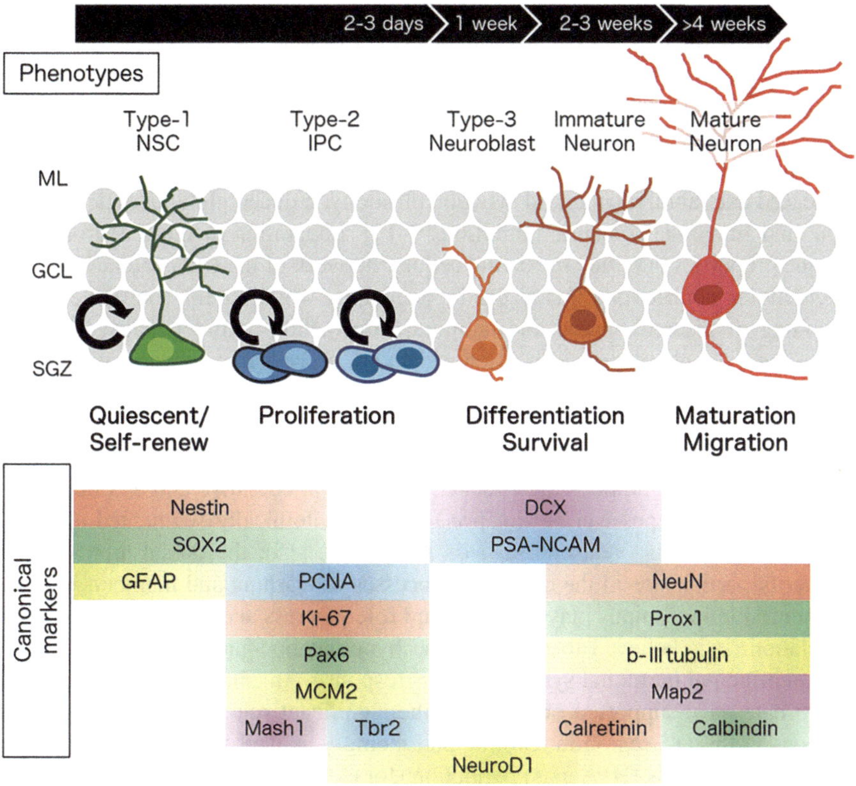

Fig. 1 Process of adult hippocampal neurogenesis and specific markers for each developmental stage. NSC neural stem cell, IPC intermediate progenitor cell, ML molecular layer, GCL granular cell layer, SGZ subgranular zone, GFAP glial fibrillary acidic protein, PCNA proliferating cell nuclear antigen, MCM2 minichromosome maintenance complex component 2, DCX doublecortin, PSA-NCAM polysialylated neuronal cell adhesion molecule, NeuN neuronal-nuclei, Map2 microtubule-associated protein 2

stages throughout the neurogenic process. AHN is a sequential process from proliferation to differentiation, maturation, survival, and functional integration, and the net effects on AHN are therefore thought to result from the combined effects of these factors on the respective stages of the neurogenic process. In the neurogenic niche, the multiple factors and the different cellular components interact to orchestrate the complex and precise development of newly born neurons, culminating in their correct integration into the preexisting hippocampal networks. AHN modulated via this cellular and molecular interactions is also highly sensitive to physiological and environmental stimuli, and to pathological status. For example, learning, physical activity, and an enriched environment are commonly known to have positive effects on AHN, while aging and neurodegenerative disorders, such as Alzheimer's disease, have negative effects on AHN. Moreover, stress, epilepsy, and stroke bi-directionally influence AHN depending on the timing, degree, specific paradigms, and so on (for extensive reviews on this topic, *see* [6, 7]).

Since the first paper relating to the role attributable to AHN in trace memory was reported [8], the contribution of newborn neurons to hippocampal function has become increasingly becoming clear. Neurons newly born in adulthood have intrinsic physiological properties that are not identical to mature granule cells such as increased excitability, enhanced synaptic plasticity, reduced glutamatergic innervation, and functional synaptic inhibition [9–12]. Integration of these neurons with unique properties into the preexisting cellular network of the DG generates specific intercellular connections inside the hippocampus which could contribute to the encoding of discrete spatial and temporal information as independent encoding units. For example, pattern separation, a network process that converts overlapping inputs into less similar outputs, is a core function of the DG and has been previously shown to be fine-tuned by the contribution of the newborn neurons [13, 14]. Higher neural network flexibility enhanced by newborn neurons also contributes to the regulation of some hippocampus-related functions at cognitive and emotional levels. The hippocampus can be functionally and anatomically segmented along its anteroposterior axis (ventrodorsal axis in rodents) [15]: the dorsal hippocampus primarily participates in the cognitive processes of learning and memory, whereas the ventral hippocampus plays a significant role in stress and emotional behavioral regulation. It is believed to modulate the hypothalamic–pituitary–adrenal (HPA) axis, a principal hormonal system in stress response. An extensive number of studies to date have supported or proposed the significant role of increased newborn neurons in the DG region in learning and memory as well as in mood (emotional behavior) and stress (HPA axis) regulation (for extensive reviews on this topic, *see* [16–18]).

2 Intensity-Dependent Impact of Treadmill Running on AHN and Hippocampal Function

Access to exercise (e.g., a voluntary running wheel in rodent models) is a well-proven environmental stimulus that improves hippocampal neural remodeling, primarily by promoting AHN in the DG. An early study reported the beneficial effects of enriched environment with nest, tunnels, toys, and running wheels (source of exercise) on AHN, especially on the survival of newborn cells [19]. Building upon the result, subsequent studies have shown that this type exercise, which was incorporated as one component of an enriched environment, could be the chief neurogenic, neurotrophic, and, further, pro-cognitive stimulus [20, 21]. In the two studies, significant increment of proliferating cells and newborn neurons, enhanced BDNF levels, and improved learning performance in the Morris water maze (MWM) were observed in conditions that combined enriched environments and running or running alone but not in the condition of environmental enrichment alone. To date, numerous studies, including the first study by van Praag et al. in 1999 [22], have reported that exercise enhances AHN and improves the related hippocampal functions, such as spatial memory, pattern separation, and alleviated anxiety-like behavior [23–26], supporting the idea that exercise is a useful strategy for maintaining healthy hippocampal function. However, these previous animal studies often used a running wheel, and, thus, lacked a precise assessment of the effects of exercise on AHN and related hippocampal functions in line with the experimental conditions.

The wheel running model places a lighter time burden on experimenters because it leaves the execution of exercise to the spontaneity of the experimental animals; however, it is difficult to adjust the factors (conditions), such as intensity, duration, and frequency, influencing the exercise's effectiveness. On the other hand, the treadmill running (TR) allows us to strictly control the factors (conditions) according to the purpose of experiment, although it requires more time and effort on the part of the experimenters because the animals are forced to exercise and must be monitored during exercise. In aerobic exercise, the combination of intensity, duration, and frequency determines the applied exercise load, and the optimization of these load components produces adaptive responses that improve general health and physical performance [27, 28]. Notably, the exercise intensity is considered a predominant component of overload induction to disrupt body homeostasis, as exercise over a minimum intensity, or threshold, is a foundation for the progress toward an individual's fitness or health, while the benefits of exercise may be attenuated or unobservable when exercise is performed below a minimum intensity or at an unadaptable high intensity [27, 29]. Given the importance of intensity in aerobic exercise, it is essential to examine the exercise-intensity-dependent changes in the AHN and hippocampal functions using TR models in order to develop exercise prescriptions that benefit the hippocampus and to accumulate and build upon such actionable knowledge. As stated earlier (for details, *see* chapter "Treadmill Running Model for Rodents as a Translational Approach Based on Human Physiological Responses"), the intensity of TR can be defined either as vigorous-intensity exercise with exercise stress above the lactate threshold (LT) or as stress-free light-intensity exercise below the LT, since exercise at intensities above the LT causes the activation of the hypothalamic paraventricular nucleus (PVN), a stress center, and also increases the blood levels of the adrenocorticotropic hormone (ACTH), one of the representative stress hormones [30, 31]. Our previous studies have confirmed in both rats and mice that the running speed at which the LT is reached is approximately 20 m/min, which corresponds to the moderate intensity range of 41.0–65.5% of maximal oxygen consumption [31–33]. Thus, the following section outlines the impact of TR at different LT-based intensities on AHN, according to the individual neurogenic process (i.e., cell proliferation, differentiation, survival, maturation, and integration), as well as the impact on hippocampal function, so as to provide useful information about intensity as it pertains to hippocampus-targeted exercise prescriptions.

2.1 Impact of TR on Cell Proliferation (NSCs and IPCs)

In 2001, Trejo et al. first reported the stimulatory effects of TR on AHN, and the role of circulating insulin-like growth factor 1 (IGF-1) on regulating its effects [34]. Their study assigned the rats to one of two groups: 2 weeks of light-intensity TR exercise (hereinafter, light-intensity TR) at below the LT (running speed: 17 m/min) and 2 weeks of sedentary control, and compared the number of proliferating and surviving newborn cells between the groups by using the BrdU-pulse labeling technique. The result showed that light-intensity TR enhanced the proliferation and survival of newborn cells, but did not modify the proportion of neurons to glia and other cell types for those cells. Since the publication of their results, numerous studies have not only confirmed the increase of cell proliferation with a single bout of or long-term TR, but have also observed intensity-dependent differences in the impact of TR.

Studies examining the effects of a single-bout of or relatively short-term (about a week) TR on cell proliferation have indicated that light-intensity TR below the LT has positive effects, while moderate to vigorous-intensity TR at or above the LT might have no effects [35–40]. For example, the BrdU-labeled proliferation cells increased with 3–7 days of TR at a running speed below the LT, but not with a speed above the LT. Also, interestingly, at running speeds below the LT, the increase in BrdU-labeled proliferating cells was more pronounced for slower running speeds, suggesting that the lighter treadmill exercise had greater enhancing effects on cell proliferation [35–37]. On the other hand, when training for longer than 2 weeks, both light- and moderate- or vigorous-intensity TR have been reported to increase cell proliferation (Fig. 2a) [41–45]. These studies have assessed cell proliferation by applying BrdU-labeling just before the end of treadmill training, or by using proliferation markers, such as Ki-67, and reported that the positive cells for BrdU or the markers increased with TR at running speeds below as well as above the LT. By contrast, there are a few results showing no impact for cell proliferation with long term, above moderate TR on cell proliferation [32, 46], probably due to differences in exercise intensity and duration, and further investigation from these perspectives is required. Together, these results imply that light-intensity TR is sufficient as a stimulus to increase cell proliferation and, moreover, that the stimulating effect is not temporary but persists over the long term. Most importantly, while none of the studies reported a decrease in cell proliferation with vigorous-intensity TR at above the LT, which may elicit stress response, other types of powerful or inescapable stress, such as exposing rats to a predator's odor [47] or to inescapable shock stimulus [48], cause decreased cell proliferation. The TR-derived stress is therefore inherently different from other types of stress, at least in that it does not become a negative factor inhibiting cell proliferation.

As seen above, TR, especially at light intensities, expands the proliferating population of NSCs and IPCs in the SGZ, but it has not yet been experimentally confirmed whether changes in cell populations are due to recruitment of quiescent NSCs to the active pool, to expansive divisions of NSCs and IPCs, or to a combination of these effects. A study has examined the changes in the number of NSCs and

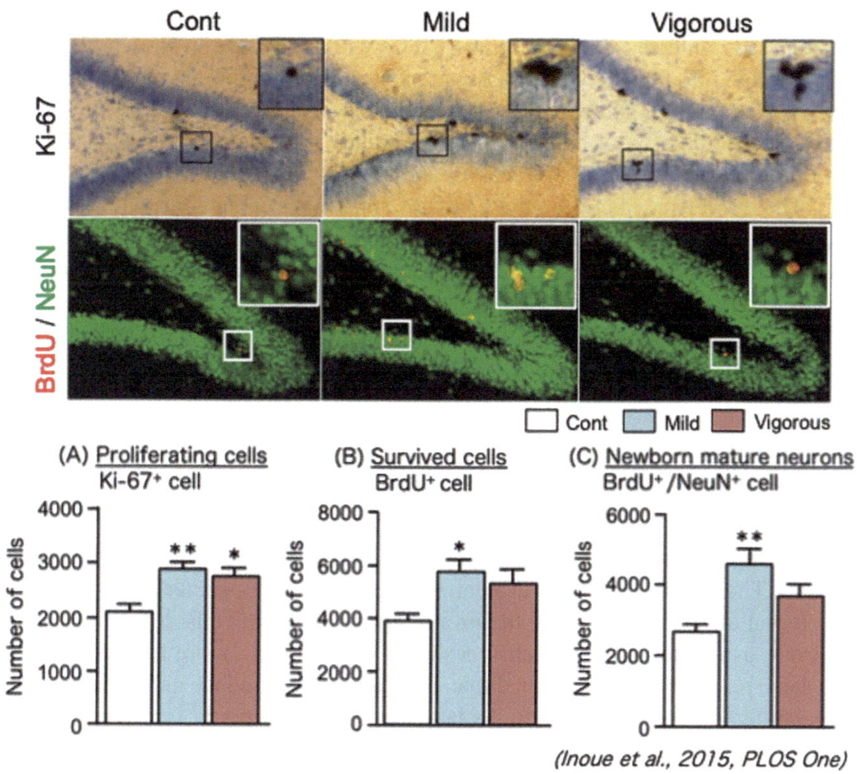

Fig. 2 Intensity-dependent changes in adult hippocampal neurogenesis induced by 6-week treadmill running (TR). Effects of light-intensity TR (< the lactate threshold, LT) or vigorous-intensity TR (> LT) on the number of proliferating cells (**a**), survived cells (**b**), and newborn mature neurons (**c**). Values are expressed as mean ± SE. * $p < 0.05$; ** $p < 0.01$ compared to sedentary control (Cont)

IPCs using the specific marker Nestin via a forced running wheel exercise model, which allows free and precise adjustment of running conditions similar to TR, at different running speeds (i.e., 7, 10, and 30 m/min) [41]. The study found that the Nestin^{+} cells increased in the exercise groups, and that the positive effects of exercise were greater with lighter intensities. Another study reported an increase in the number of cells positive for the cell proliferation marker, proliferating cell nuclear antigen (PCNA), the astrocyte marker, glial fibrillary acidic protein (GFAP), and cells positive for both markers, suggesting that the activation of NSCs or astrocytic proliferation due to moderate TR at the LT (15–22 m/min) [39]. These results suggest that light-intensity TR contributes to the activation of quiescent NSCs and to the expansion of NSCs and IPCs, but the evidence remains indirect and speculative. On the other hand, in the case of voluntary running, some studies have identified distinct cell populations during the proliferation phase by using various types of proliferation or neuronal activity markers and transgenic mice expressing green fluorescent protein and have quantified exercise-induced changes in or neuronal responses to a proliferating cell population. Early studies indicated that voluntary running led to increased proliferation of IPCs with no impact on NSC proliferation or activation [49, 50]. However, more recent studies have reported results that differ from earlier studies, showing that voluntary running recruits quiescent NSCs to an active proliferating pool, namely that NSCs enter into the cell cycle, not that the proliferation rate of IPCs is increased, and thereby neurogenesis is increased [51, 52]. The discrepancy between these studies applying a voluntary running model is thought to be a result of the difference in exercise conditions associated with the free conduction of exercise by experimental animals. Further in-depth research based on condition, especially on intensity, using a TR model is needed to solve this problem and to determine the actual contribution of exercise to cells such as NSCs and IPCs during the proliferating phase.

2.2 Impact of TR on Differentiation

Only a certain percentage of newly proliferated cells become neurons (fate specification). Though the first study evaluating changes in fate specification with TR did not reveal any modification of the proportion of newborn cells differentiated into neurons, astrocytes, or other cell types [34], a subsequent, more focused study by the same research team confirmed that TR does indeed produce an increase in neuronal differentiated cells [53]. In the latter study, the number of immature neurons, detected by markers such as doublecortin (DCX) and/or calretinin, was estimated using a new stereological cell counting method, and was compared between mice in two conditions: a 2-week resting control group and a 2-week TR (running below the LT at 12 m/min) group. Briefly, this cell counting method allows the estimation of the total number of immature neurons by multiplying the density of marker-positive cells by the total volume of the SGZ area, a method that may reflect a more holistic understanding of the neurogenic situation in the SGZ than was possible in the former study. The results of the latter study show more DCX and calretinin double-positive cells of the SGZ in the mice trained with light-intensity TR than in the sedentary control mice.

Two other studies using this cell counting method have shown enhanced neuronal differentiation with light-intensity TR, but not with vigorous-intensity TR. First one measured the changes in the number of DCX^+ cells in the SGZ of rats subjected to 2 weeks of TR at different intensities: rest, below the LT (13.5 m/min), or above the LT (28 m/min). The rats trained at below the LT had a significantly increased number of immature neurons (DCX^+ cells) compared to the untrained rats (rest), while there was no increase in the rats trained at above the LT. Second one reported similar results in a study using mice [32]. This work has shown that the number of newly born cells that differentiated into immature neurons ($BrdU^+/DCX^+$ cells) increased with 2 weeks of TR exercise at an intensity below the LT (15.0 m/min) compared to the rest condition, but that this increase did not occur with TR exercise at an intensity above the LT (30.0 m/min). The increase of DCX^+ and $BrdU^+/DCX^+$

cells with light-intensity TR below the LT has also been observed in both shorter (3–7 days) and longer (5–8 weeks) training periods [54, 55]. However, while these findings indicate that light-intensity TR leads to an increased number of immature neurons, it is not yet clear whether this is the result of an expansion of proliferating cell populations, a promotion of neuronal differentiation, or a combination of both. In a study investigating the effects of aging and mild TR (9–10 m/min) on AHN, exercise was reported to delay the aging-induced decrease not only in the number of immature neurons (DCX$^+$ or BrdU$^+$/DCX$^+$ cells) but also in the rates of neural cell specification, as estimated by dividing the number of BrdU$^+$/DCX$^+$ cells by the number of BrdU$^+$ cells [56]. These results suggest that light-intensity TR has the potential to increase the specification rates into neuron within newly born cells, and, if we also take into consideration the exercise-enhanced cell proliferation, which was discussed above, it is likely that mild TR promotes both cell proliferation and neuronal differentiation.

Astrocytes are another type of cell lineage from which newly born cells, that is NSCs, may differentiate. In a previous study [46], the exercise-intensity-dependent changes in astrocyte differentiation rates were investigated in the SGZ of mice subjected to 6 weeks of TR exercise at light-to-moderate (15 m/min, <LT) or vigorous (30 m/min, >LT) intensities. It is interesting to note that the ratios of BrdU$^+$/NeuN$^+$ cells to BrdU$^+$ cells increased in all mice that exercised regardless of the exercise intensity, while the ratios of BrdU$^+$/GFAP$^+$ cells to BrdU$^+$ cells decreased only in those who exercised at a light-to-moderate intensity, compared to resting control mice. This implies that mild-to-moderate TR shifts the balance between neurogenesis and gliogenesis in AHN toward increased neuron production. Conversely, others have reported astrocytic proliferation (PCNA$^+$/GFAP$^+$ cells) with moderate (at the LT) TR [39] as well as increased astrocytes (GFAP$^+$ cells) with short term (3–15 days) of light-intensity (below the LT) TR [55] and with 7 days of wheel running [57] in the DG. In the rat model of depression, 6 weeks of TR at the LT (20 m/min) has been reported to reverse the declined number of BrdU$^+$/GFAP$^+$ cells in the DG, suggesting that moderate TR induces an increase in newly generated astrocytes [58]. Thus, there are only a few studies on changes in the numbers of newborn astrocytes with TR, and those that do exist are inconsistent and do not provide a unified view—whether treadmill exercise leads to increases or decreases in newborn astrocytes. Astrocytes are a predominant component of the neurogenic niche, and a potential vital regulator of the differentiation and growth of newborn granule cells through the release of various trophic factors, gliotransmitters, and energetic source [59–61]. For this reason, it is important not only that changes in astrocyte numbers be explored, but also that the effects of TR on astrocyte morphology and activation be analyzed in future studies to fully understand the connection between AHN and astrocytes with regards to exercise. In fact, one study reported that 4 weeks of daily exercise using a combination of TR (5 days/weeks, 15 m/min at below the LT) and free access wheel running (7 days/weeks) neither changed the total number of astrocytes nor the distance between astrocytes, but did cause an expansion of the area projected by the astrocytes [62]. Moreover, we have reported that 2 weeks of TR at below the LT promotes the secretion of astrocyte-derived Wnt3, a factor triggering neuronal differentiation of hippocampal NSCs, which in turn increases the number of NeuroD1$^+$/BrdU$^+$ cells in the DG of young and aged mice [63]. Given that NeuroD1 is a marker transiently expressed in dividing progenitor cells and immature granule neurons [64], those results suggest that light-intensity TR promotes AHN via astrocyte activation.

2.3 *Impact of TR on Cell Survival, Maturation, and Integration*

From among newborn, immature neurons, only a small fraction is selected to survive, but not to apoptosis. The selected cells undergo a postmitotic maturation period, which is related to the axonal and dendritic outgrowth, synaptogenesis, and development of dendritic spines, and they eventually integrate into the existing functional hippocampal circuitry. After cell-cycle exit, the newborn neurons express postmitotic mature neuron markers, such as NeuN (RbFox3), calretinin, calbindin, and microtubule-associated protein 2 (MAP2). In early studies, aerobic exercise, such as voluntary wheel running, was thought to upregulate AHN, primarily by expanding the proliferation of progenitors in the DG, thereby making them available for further differentiation and maturation [22, 49]. Conversely, stimuli that were believed to more specifically induce hippocampal activity, such as environmental enrichment and learning complex tasks, actually promote the survival of adult-born granule cells [19, 65]. The differences and possible underling mechanisms are outlined by Olson et al. and Fabel and Kempermann [66, 67]. However, subsequent studies, including a meta-analysis, have indicated that the beneficial effects of aerobic exercise, especially wheel running, on AHN extend to postmitotic survival and maturation [68, 69].

Evidence showing the positive effects of TR on the postmitotic development period of AHN is limited but growing. In a series of experiments, we subjected rats injected with BrdU to 2 or 6 weeks of TR at different LT-based intensities: light (running speed < LT) and vigorous (running speed > LT). We found that the rats under the light-intensity TR condition had increased BrdU$^+$ cells in the DG, while such increases were not observed in the vigorous-intensity TR condition (Fig. 2b) [43, 45]. These rat studies analyzed the BrdU$^+$ cells either 2 or 6 weeks after its injection, and thus, the increased BrdU$^+$ cells reflected both the proliferation of NSCs/IPCs and the survival of the proliferated cells. On the other hand, in mice injected with BrdU at the 2-week point of a 6-week TR training, the light-intensity TR (15 m/min, <LT) led to an increased amount of surviving BrdU$^+$ cells at the end of training, with no impact on proliferating BrdU$^+$ cells at the 2-week stage of training [46]. The discrepancy between exercise-induced changes in proliferating and surviving BrdU$^+$ cells prompted us to speculate that NSC/IPC proliferation and the

survival rate of newborn cells could be affected separately via TR. The specific effects of TR at a running speed of 10–11 m/min on the survival of newborn cells in middle-aged mice were investigated by counting the BrdU$^+$ cells in the DG 2 days and 3 weeks after BrdU injection and by comparing the changes in their numbers [70]. In the study using middle-aged mice, the number of BrdU$^+$ cells that were detected 3 weeks after injection was lower than that detected 2 days after injection on both the TR and non-TR conditions. Compared to the number of BrdU$^+$ cells detected 2 days after injection, however, only around 30% of the cells in the control condition survived up to 3 weeks, whereas the rate was increased to 50% in the exercise condition. These results suggest that light-to-moderate TR can lead not only to increased cell proliferation, but also to increased newborn cell survival itself. However, a study investigating age-related changes in AHN and the effects of a mild TR intervention on these changes has reported that exercise does not lead to improvement of the survival rate of newborn cells [56]. Thus, further investigation is needed to determine the exact impact of light-intensity TR on the survival of newborn cells, especially in terms of changes in the survival rate of viable cells.

Our studies that we described above have also shown that 2 or 6 weeks of light-intensity, but not vigorous-intensity, TR induced increases in BrdU$^+$/NeuN$^+$ cells in the DG (Fig. 2c) [32, 43, 45]. Other studies have demonstrated that 4–5 weeks of TR at a speed of 10–11 m/min increases the number of BrdU$^+$/NeuN$^+$ or BrdU$^+$/MAP2$^+$ cells, along with the Ki-67$^+$ or BrdU$^+$ cells and the DCX$^+$ or DCX$^+$/BrdU$^+$ cells, in the DG of young-adult and middle-aged mice [70–72]. These results indicate that long-term light-intensity TR expands the cellular population of newborn mature neurons. In addition to these cellular population changes, TR impacts the morphological development of newborn neurons as well as migration toward their final destinations. Five weeks of mild TR increases the number of DCX$^+$ cells, whose dendrites project up to the granule cell layer (GCL) or across the GCL to the molecular layer and whose cell body axis is perpendicular to the SGZ in young-adult and middle-aged mice [70]. Also, the dendritic lengths of DCX$^+$ immature neurons quantitatively analyzed were enhanced with exercise [56, 70]. One study found that dendritic complexity quantified by the Sholl analysis and dendritic spine density are enhanced with 14 days of TR at a light intensity (8 m/min, <LT) without serum corticosterone elevation, but not at a moderate intensity (20 m/min, at LT) with the hormone elevation, in both ipsilesional and contralesional hippocampal DG of transient ischemic model rats [73]. Moreover, TR at a light intensity (15 m/min, <LT) decreases the percentage of newborn neurons located in the inner GCL, but increases that located in the middle and outer GCL [46]. Similar changes have been observed in vigorous-intensity TR (30 m/min, >LT). Hence, it is considered that TR facilitates the migration of newborn neurons in the radial direction of the GCL with no dependence on exercise intensity. As discussed above, TR, especially light-to-moderate TR, appears to enhance the survival and maturation of postmitotic neurons, and morphological development of dendrite.

3 Impact of TR on Hippocampal Functions

AHN is often described as an investment in the future, and the highly plastic hippocampal network, which includes an accumulation of new cells generated through exercise, has the potential to prime the hippocampus for upcoming relevant stimuli [67]. Supporting this idea, the increase in newborn cells/neurons induced by exercise and the subsequent improvement in hippocampus-associated functions, such as spatial memory, pattern separation, and anxiety- or depression-like behavior, are well synchronized [24–26, 74]. Furthermore, suppression of exercise-induced newborn cell generation by gamma irradiation inhibits the functional improvements of the hippocampus [23]. As a certain period is required for newborn cells to mature and to be incorporated into the existing hippocampal network [6], exercise-induced functional improvement via AHN could be achieved through long-term exercise.

Focusing on the impact of TR on AHN, 4–6 weeks of TR at a light intensity (8–15 m/min, <LT) has been reported to increase immature and mature newborn neurons in both the dorsal and ventral DG regions [45, 75, 76]. Given the anatomical segregation along the anteroposterior axis of the hippocampus [15], it is possible that AHN in the dorsal and ventral DG regions that is enhanced by light-intensity TR provides greater flexibility for the existing neural network in the hippocampus, thereby enhancing the dorsal region related to learning and memory capacity, but also improving the ventral region related to stress regulation and anxious/depressive behavior.

Indeed, using the MWM test, we have demonstrated that spatial memory ability is enhanced with 6 weeks of light-intensity TR (15 m/min, <LT), which is a condition under which mature newborn neurons in the DG increase [45, 77]. Specifically, in the MWM test, light-intensity TR did not affect memory acquisition performance during learning trials but did lead to improved memory retention during the probe trial (Fig. 3b, c). In contrast, these positive effects observed during MWM test performance were not present in the vigorous-intensity TR condition, which is a condition under which mature newborn neurons do not increase. Notably, substantial evidence pertaining to the link between acute or chronic manipulations of adult-born GCs in DG and hippocampal functions suggests that adult-born neurons may play a distinct role in learning and memory [78, 79]. Changes in adult-born GCs in the DG, especially their reduction in number, have little or no impact on the general learning stage of the MWM, while effect on memory formation differed depending on the age of the new neurons when being tested [79]. Regarding the age of adult-born neurons, it has been shown that immature new neurons of approximately 4–6 weeks old are preferentially recruited to hippocampal memory processes and contribute to form an essential component of hippocampal memory trace, which influences memory retrieval [80–82]. The term of functional maturation required for this memory formation coincides with the period of exercise in our 6-week model, so it is possible that the light-intensity TR caused the reinforcement of memory formation during the MWM via an increase in newborn mature neurons in the DG. However, some studies have shown improvements in both memory and

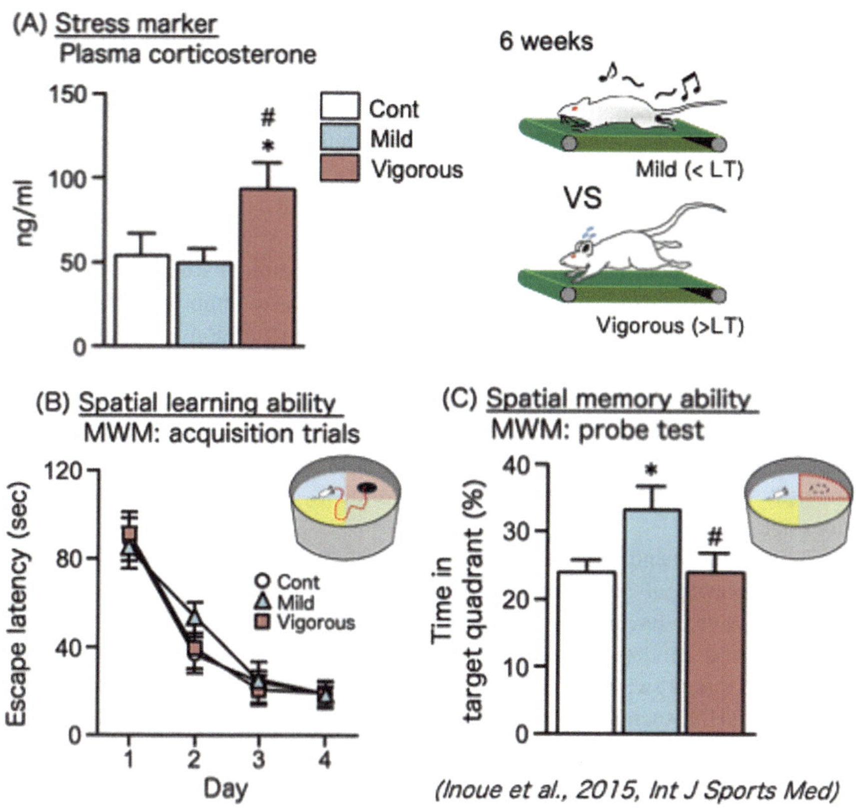

Fig. 3 Intensity-dependent changes in stress response and hippocampal function induced by 6-week treadmill running (TR). Effects of light-intensity TR (< the lactate threshold, LT) or vigorous-intensity TR (> LT) on plasma corticosterone levels (**a**), acquisition trials in the Morris water maze (MWM) (**b**), and probe test in the MWM (**c**). Values are expressed as mean ± SE. * $p < 0.05$ compared to sedentary control (Cont). # $p < 0.05$ compared to light-intensity TR

learning performance with the MWM test [71, 83, 84] as well as with other spatial tests such as the Y and radial maze tests [41, 85], following long-term TR exercise. Therefore, further careful consideration is needed to understand the impact of TR-induced AHN on hippocampal-dependent memory and learning ability.

In addition, pattern separation (spatial discrimination)—the ability to independently remember and represent similar overlapping stimuli or experiences—is one of the central characteristics of episodic memory. This ability is a core function of the DG, and studies over the past couple of decades have shown that this is accomplished with significant contributions from newborn neurons in the DG [13]. Wheel running exercise was reported to improve pattern separation capacity in mice, especially when the separation between stimuli was small rather than large, which is to say under the more difficult task condition [24]. Transgenic mice with increased AHN exhibited better differentiation ability between overlapping contextual representations, indicating an enhanced capacity for pattern separation [25]. The ability to make finer spatial distinctions with wheel running exercise may thus be due, at least in part, to the increased number of newborn neurons in the DG with this exercise [24, 86]. In contrast, few studies have investigated the effects of TR on this ability. One study examined the intensity-dependent effects of TR on spatial pattern separation ability [46] using Clelland et al.'s radial arm maze testing paradigm [13]. The results showed that 6 weeks of treadmill exercise at a light intensity (15 m/min, <LT) led to enhanced spatial pattern separation ability with a concurrent global improvement in neurogenic stage from cell proliferation to maturation. In contrast, vigorous-intensity exercise (30 m/min, >LT) caused partial improvement of neurogenesis in the adult DG, such as enhanced neuronal maturation and migration, but did not alter the pattern separation ability, a result which was probably due to the stress induced by the intensity of the exercise condition. To summarize, with regard to the effect of TR on spatial learning/memory and pattern separation, which are thought to be strongly related to the dorsal hippocampus, it appears that light-intensity TR effectively results in improvement, while vigorous-intensity TR does not.

On the other hand, there are some studies that have examined the effects of TR on the hippocampal ventral region, which is related to stress regulation and anxious and depressive behavior. For instance, 4 weeks of light-intensity TR (8–15 m/min, <LT) induced antidepressant and antianxiety activities in the tail suspension test (TST), the forced swim test (FST), and the open field test (OFT) accompanied with enhanced AHN, such as the promotion of proliferation and differentiation/survival of newborn cells and the increase of mature newborn neurons ($BrdU^+$/$NeuN^+$ cells) in the dorsal and ventral DG [76, 87]. Another study has confirmed similar antidepressant effect in the FST and increased DCX^+ immature cells in the ventral DG with 4 weeks of both light- (15 m/min, <LT) and vigorous-intensity TR (25 m/min, >LT) [75]. Moderate TR (20 m/min, at LT) for 3 weeks has also been shown to reduce depression-like behaviors in the FST and OFT with AHN improvement [88]. The study further examined tolerance to chronic unpredictable stress and found that exercise led to the recovery of the stress-induced reduction in hippocampal neurogenesis and improved aggravated depressive behaviors. These results suggest that long-term TR, at least under light-intensity TR condition, can improve depressive/anxious behaviors, probably through the regulation of AHN. Nevertheless, on its own, increasing adult hippocampal neurogenesis is not sufficient to produce an anti-anxiety- or antidepressant-like behavioral response [25]. There is no doubt that the ventral region of the hippocampus is an important area that contributes to the regulation of anxious and depressive behaviors. However, these behaviors function through connections with other brain structures that receive input from the ventral hippocampus, such as the amygdala and the nucleus accumbens [15]. Therefore, the role of AHN, which is enhanced by light-intensity TR, should also be examined more carefully in relation to these connections.

Today, even simple translations of results from rodent studies to humans are difficult because the role and characteristics of AHN in humans remains a topic of intense scientific debate [89, 90]. However, the evidence from rodent studies are promising, in that long term (4 weeks or longer) of light-intensity TR is useful for boosting both AHN and AHN-related functions such as memory formation and anxiety−/depression-like behaviors. Exercise is potently pro-neurogenic but also simultaneously causes numerous adaptations in the hippocampus and even in the whole brain. Thus, exercise-induced AHN does not seem to be the only factor for these positive behavioral outcomes. Indeed, it has been reported that short-term (2–3 weeks) mild TR, which is an insufficient timeframe for newborn neurons to mature, exerts antidepressant and antianxiety effects as observed behaviorally in the FST, the Elevated Plus Maze, and in the Novelty-Suppressed Feeding test [91, 92]. A more detailed examination that, for example, applies techniques such as the genetic manipulation of AHN, is essential in order to clarify whether the beneficial effects induced by light-intensity TR are predominantly brought by increased newborn neurons in the DG.

4 Mechanism by Which Light-Intensity Exercise Promotes Adult Neurogenesis

The mechanisms underlying exercise-induced AHN are highly complex and multifactorial. Indeed, exercise can exert beneficial effects on proliferation, neural differentiation, survival, and maturation of newborn cells by modulating various factors such as neurotrophic factors (e.g., brain-derived neurotrophic factor (BDNF) [93], vascular endothelial growth factor (VEGF) [94], and insulin-like growth factor I (IGF-1) [91]), modulators of inflammation (e.g., interleukin 6 and 10 [95], clusterin [96]), chemokines (e.g., CX_3CL1 [97]), neurotransmitters (e.g., β-endorphin [98], serotonin (5-HT) [99]), steroid hormones (e.g., androgen [43], glucocorticoid [32]), epigenetic regulators (e.g., MiR-135 [100]), and other signaling molecules (e.g., Wnt [63], bone morphogenetic protein (BMP) [101], ErbB4 [87]). Recently, it has been suggested that the effects of exercise are regulated not just locally in target tissues, but rather via systemic organ crosstalk through paracrine, autocrine, and/or endocrine mechanisms by key molecular signal carriers and drivers of inter-organ communication, called exerkines [102]. In the regulation of AHN by exercise, it has been suggested that peripherally derived factors that can pass through the blood–brain barrier (BBB), such as lactate [103], FNDC5/irisin [104], cathepsin B (CTPB) [105], adiponectin [106], and IGF-1 [107], contribute to exercise-induced AHN enhancement. As such, although it has become clear through intensive investigation that various factors are involved in the enhancement of AHN through exercise, previous studies have not adequately examined the regulatory mechanisms depending on exercise conditions, especially on exercise intensity, that are important in order to bridge to exercise prescriptions for humans. This section will focus mainly on studies using TR and will summarize the regulatory factors that may contribute to the modulation of exercise-intensity-specific effects on AHN (Table 1 and Fig. 4).

Table 1 Intensity-dependent effects of treadmill running on adult hippocampal neurogenesis and function, and associated regulatory factors

Animal		Exercise condition		AHN process			Behavioral assessment	Measured factors	References
Species	Treatment	Running speed	Duration	Pro	Diff/ Sur	Mat/ Mig			
Rat	Vehicle	17 m/min	2 W	+	+			Circulating IGF-1: Essential	[34]
	Anti-IGF-1			NC	NC				
Rat		2–8 m/min	3 D	++					[35]
		7–13 m/min		+					
		8–22 m/min		NC					
Rat		4–10 m/min	7 D	++					[36]
		10–16 m/ min		+					
Rat		2–8 m/min	3 D	++					[37]
		8–14 m/min		+					
		8–22 m/min		NC					
		2–8 m/min	1–28 D	+					
Rat		15–22 m/ min	7 D	+	+				[39]
Mouse		12 m/min	2 W		+				[53]
Rat		7 m/min	6 W	++			Y-maze: +		[41]
		10 m/min		+					
		13 m/min		+					
Rat		15–22 m/ min	7 D	+					[38]
Rat		5–11 m/min	1 W	+	+			*BDNF, Nmdar1, Flk-1*: +, *VEGF*: NC	[40]
		8–14 m/min		NC	NC			*Nmdar1: +, BDNF, Flk-1, VEGF: NC*	

Species	Treatment	Running speed	Duration	Pro	Diff/ Sur	Mat/ Mig			
		8–22 m/min		NC	NC			*BDNF, Nmdar1, Flk-3, VEGF: NC*	
Mouse	WT	12–m/min	3 W		+		NSFT, FST: +	Circulating IGF-1: Essential	[91]
	LID			–	–		MWM: –, NSFT: NC		
Mouse	Sham	10–11 m/ min	1–8 W		+			CORT, BNDF, TrkB: +	[54]
	ADX				+			CORT: –, BDNF, TrkB: NC	
	MR antagonist				+			CORT: +, BDNF, TrkB: NC	
Mouse	8-M old	10–11 m/ min	6 W	+	+	+ (dend)		CORT: NC, BDNF, TrkB: +	[70]
	12-M old			+	+	+ (dend)		CORT: NC, BDNF, TrkB: +	
Mouse	Vehicle	12 m/min	2 W	+		+ (spine)		Circulating IGF-1: Essential	[42]
	Anti-IGF-1			NC		+ (spine)			
Mouse	Vehicle	12 m/min	2 W	+	+		MWM, NOP: NC, EPM: +	Circulating IGF-1: Essential	[92]
	Anti-IGF-1			NC	+		MWM, NOP: –, EPM: NC		
Rat		10 m/min	3–15 D	+	+			BDNF: NC	[55]
Mouse	Young (12-W old)	15 m/min	2 W		+			Wnt-3a: +	[63]
	Aged (9-M old)	10 m/min	2 W		+				
Mouse		20 m/min	2 W	+			FST, OFT: +	VGF-Flk-1 signaling	[88]

(continued)

Table 1 (continued)

Animal		Exercise condition		AHN process			Behavioral assessment	Measured factors	References
Species	Treatment	Running speed	Duration	Pro	Diff/ Sur	Mat/ Mig			
Rat	Sham	13.5 m/min	2 W	+	+	+		HIP androgen: +	[43]
	AR antagonist			+	NC	NC			
	ORX or ER antagonist			+	+	+			
	ORX + AR antagonist			+	NC	NC			
	Naive	28 m/min		+	NC	NC			
Rat		13 m/min	Single	+				GR: NC	[171]
Rat		2–8 m/min	4 W	+	+			BDNF: +	[44]
Mouse	Regular ex	10 m/min	4 W	++	++	++	MWM: +	CORT: +	[71]
	Irregular ex			+	+	+	MWM: NC	CORT: +	
	Irregular time-of-day ex			+	+	+	MWM: NC	CORT: +	
Rat		15–22 m/min	7 D	+					[172]
Mouse	Vehicle	15 m/min	2 W	+	+	+		CORT signaling essential	[32]
	GR or MR antagonist			NC	NC	NC			
	Vehicle	30 m/min		NC	NC	NC			
	GR or MR antagonist			NC	NC	NC			
Rat		15 m/min	6 W	+	+	+		Microarray	[45]
		40 m/min		+	NC	NC			
Mouse		9–10 m/min	5 W	+	+	+ (dend)			[56]

Species	Treatment	Running speed	Duration	Pro	Diff/ Sur	Mat/ Mig			
Mouse		10–15 m/ min	6 W	+	+	+	RAM: +	CORT, BDNF, VEGF, IGF-1, EPO: +	[46]
		5–30 m/min		NC	NC	+	RAM: NC	CORT, BDNF, VEGF, IGF-1, EPO: NC	
Rat	Young (5-W old)	2–8 m/min	6 W	+	+		RAM: +		[85]
	Aged (20-W old)			+	+		RAM: NC		
Rat		15 m/min	4 W		+ d, v		FST: +		[75]
		25 m/min			+ v, NC d		FST: +		
Mouse		10 m/min	4 W	+	+	+			[72]
Mouse		8–14 m/min	4 W	+ d, v, t	+ d, v, t		FST, TST: +	FNDC5: +	[76]
Mouse	WT	10 m/min	4 W	+	+			CORT: Essential	[149]
	L/L			NC	NC				
Rat	Ex only	3–15 m/min	3 W	NC			MWM: NC		[173]
	Ex+EE			+			MWM: +		
Mouse	Ex only	15 m/min	10 D	+ d, NC v	+ t		MWM: +		[174]
	Ex+WMW				+ d, t				
Mouse	Vehicle	15–25 m/ min	5 W	+	+		Y-maze, RAM: +	FNDC5, BDNF, PGC-1a: +	[109]
	Lactate			+	+		Y-maze, RAM: ++	FNDC5, BDNF, PGC-1a: ++	
Mouse	Vehicle	5–8 m/min	4 W	+		+	MWM, T-maze, FC, OF: +	ErbB4 signaling	[87]

(continued)

Table 1 (continued)

Animal		Exercise condition		AHN process			Behavioral assessment	Measured factors	References
Species	Treatment	Running speed	Duration	Pro	Diff/ Sur	Mat/ Mig			
	ErbB4 inhibitor			NC		NC	MWM, T-maze, FC, OF: NC/+		

+ increase vs sedentary control, – decrease vs sedentary control, *NC* no change vs sedentary control, ++ increase vs other exercise group, *dend* dendrite outgrowth, *spine* spine density, *ADX* adrenalectomy, *MR* mineralocorticoid receptor, *AR* androgen receptor, *ORX* orchidectomy, *ER* estrogen receptor, *GR* glucocorticoid receptor, *L/L* Cyp11a1 promotor mutation mice (basal cortisol levels are normal but no fluctuations even after stress stimulation), *EE* enriched environment, *MWM* Morris water maze (spatial learning/memory), *D* day, *W* week, *NSFT* novelty-suppressed feeding test (anxiety), *FST* forced swim test (depression), *NOP* novel-object preference test (spatial memory), *EPM* elevated plus maze (anxiety), *OFT* open field test (anxiety), *RAM* radial arm maze (spatial learning/memory, pattern separation), *Y*- or T-maze, *Y*- or T-maze test (spatial learning/memory), *FC* contextual fear conditioning (non-spatial and reference memory), *IGF-1* insulin-like growth factor-1, *BDNF* brain-derived neurotrophic factor, *TrkB* tropomyosin-related kinase B (a receptor for BDNF), *NNDAR1* N-methyl-d-aspartate receptor 1, *VEGF* vascular endothelial growth factor, *Flk-1* receptor tyrosine kinase Flk-1 (a receptor for VEGF), *CORT* corticosterone, *EPO* erythropoietin, *FNDC5* fibronectin type III domain containing 5, *PGC-1a* peroxisome proliferator-activated receptor gamma, coactivator 1 alpha, *ErbB4* erb-b2 receptor tyrosine kinase 4. Italics indicate mRNA expression and capital letters indicate protein expression

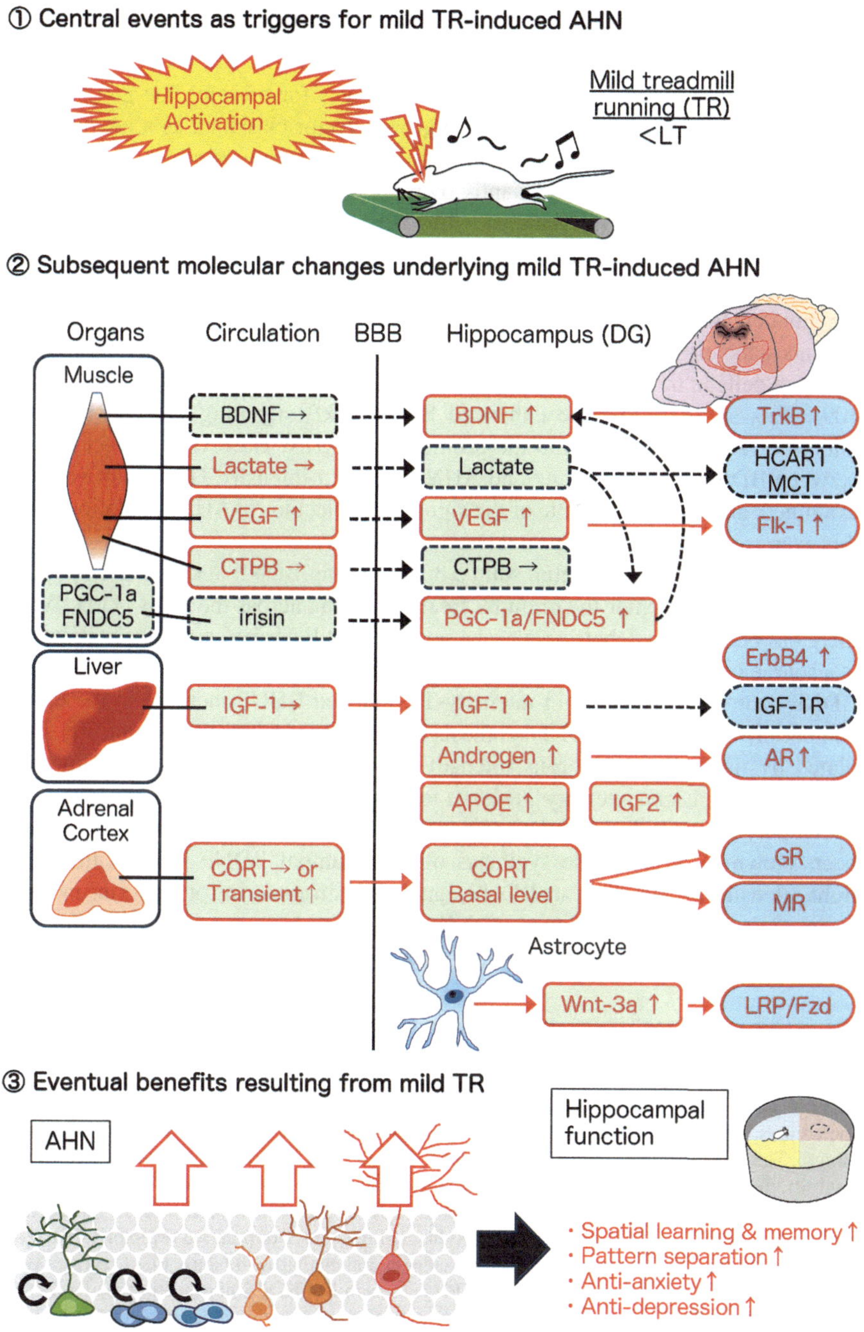

Fig. 4 Mechanisms underlying light-intensity treadmill running (TR)-induced adult hippocampal neurogenesis (AHN) and hippocampal functional development. Light-intensity TR at an intensity below the lactate threshold (LT) stimulates neural activity in the hippocampus, including the

4.1 BDNF and Potential Exerkines Affecting BDNF

One of the most enthusiastically studied candidates for bridging the gap between exercise and AHN is BDNF. BDNF is a growth factor classified as a neurotrophic factor family member, and functions by binding to tropomyosin receptor kinase B (TrkB), which is expressed abundantly by hippocampal neurons. After binding, the BDNF-TrkB complex is internalized within neurons and serves as a docking site for many signaling cascades, protein phosphorylation cascades, and secondary signaling systems [108]. TR at a light intensity, but not at or over a moderate intensity, upregulates the BDNF and TrkB protein expression in the hippocampus and enhances the proliferation, neuronal differentiation, survival, and maturation of newborn cells in the AHN, synchronizing with the upregulation [40, 44, 46, 70, 85, 109]. This suggests that the enhanced BDNF-TrkB signaling induced by light-intensity TR may contribute to the promotion of AHN. However, a direct linkage between BDNF-TrkB signaling and AHN with light-intensity TR has not yet been examined, and so the exact role of the signaling molecules in AHN regulation under this exercise condition remains unclear.

Skeletal muscle is a potential source of circulating BDNF during exercise [110], and the BDNF can enter the brain by freely diffusing across the BBB [111]. While there has been much debate about whether the BDNF derived from skeletal muscle is a source of hippocampal BDNF during exercise, it is unlikely that this peripheral BDNF contributes to the light-intensity-TR-induced BDNF increase in the hippocampus. This is because BDNF synthesis in skeletal muscle tissue is typically stimulated by moderate-to-vigorous intensity exercise, rather than by mild intensity exercise [112]. Light-intensity TR leads to an increase of BDNF mRNA in the hippocampus, particularly in its DG region [40, 113]. This type of exercise has also been shown to promote the synthesis of hippocampal BDNF protein; therefore, light-intensity TR may activate BDNF signaling in an autocrine or paracrine manner.

In addition, exercise-induced BDNF increase in the hippocampus is affect by exerkines released from various organs in response to exercise [110]. For example,

Fig. 4 (continued) dentate gyrus (DG) ①. This activity alters the behavior of various molecules associated with AHN, promoting the proliferation of newborn cells as well as the differentiation, survival, and maturation of newborn neurons ②. Consequently, it enhances hippocampal functions, including spatial learning and memory, pattern separation, and reductions in anxiety and depression ③. The parts shown in red or with red solid lines have been proven by previous research. In contrast, the parts shown in black or with black dotted lines are still open questions. BBB blood–brain barrier, BDNF brain-derived neurotrophic factor, TrkB tropomyosin-related kinase B (a receptor for BDNF), HCAR1 hydroxycarboxylic acid receptor 1 (a putative receptor for lactate), MCT monocarboxylate transporter, VEGF vascular endothelial growth factor, Flk-1 receptor tyrosine kinase Flk-1 (a receptor for VEGF), CTPB cathepsin B, FNDC5 fibronectin type III domain containing 5, PGC-1a peroxisome proliferator-activated receptor gamma, coactivator 1 alpha, ErbB4 erb-b2 receptor tyrosine kinase 4, IGF-1 insulin-like growth factor-1, IGF-1R IGF-1 receptor, AR androgen receptor, IGF-2 insulin-like growth factor-2, CORT corticosterone, GR glucocorticoid receptor, MR mineralocorticoid receptor, Wnt-3a wnt family member 3A, LRP low-density lipoprotein receptor-related proteins, Fzd frizzled receptors

myokines such as irisin [114], CTPB [105], and lactate [115, 116], which are secreted by skeletal muscles during exercise, may act on the hippocampus by crossing the BBB or via signal transduction at the BBB, resulting in an increase in BDNF and/or AHN.

Irisin, a cleaved from of fibronectin type III domain-containing protein 5 (FNDC5), is synthesized mainly in skeletal muscle tissue, secreted into the blood circulation during exercise, and regulated by the peroxisome proliferator-activated receptor γ coactivator-1α (PGC-1α) pathway. Exercise has been shown to increase hippocampal BDNF expression via the PGC-1α/FNDC5 pathway [114]. Furthermore, AHN and hippocampal function have been shown to improve with exercise through increased levels of both FNDC5 and BDNF in an Alzheimer's mouse model [117]. While these are findings from the wheel running model, similar effects can also be seen in the TR model. Indeed, a study indicated that 4 weeks of light-intensity TR increases cell proliferation (Ki-67$^+$ cells), neuronal differentiation (DCX$^+$ cells), and survival (BrdU$^+$ cells) in the DG, and enhances antidepression effects in the FST and TST, all of which is accompanied by elevated FNDC5/irisin expression in the hippocampus [76]. Such findings seem to indicate that light-intensity TR enhances hippocampal function and neurogenesis via a complex regulation of the hippocampal BDNF and FNDC5/irisin systems; however, there are still questions to be explored, such as the source of FNDC5/irisin and the relationship between the two systems.

The muscle secretory factor CTPB, a lysosomal cysteine protease, is a myokine, which may contribute to the improvement of AHN and hippocampus-dependent memory via exercise. A study demonstrated that voluntary running led to an increase in CTPB levels in the blood and gastrocnemius muscle, as well as enhanced CTPB expression in the mouse hippocampus [105]. The study further reported that the CTPB knockout (KO) mice showed no exercise-induced AHN or spatial memory improvement and that KO mice injected intravenously with CTPB had significantly increased CTPB levels in both blood and brain tissues. Interestingly, intravenous administration of recombinant CTPB to the KO mice increased DCX and BDNF protein levels in adult hippocampal progenitor cells. These findings suggest that the CTPB that increases with exercise may cross the BBB and contribute to promoting AHN and improving hippocampal function. However, since the exercise also enhances hippocampal CTPB mRNA expression, the effects of CTPB within the hippocampus could involve both direct and indirect mechanisms. In addition, focusing on exercise-intensity, exercise-induced CTPB and BDNF elevations appear to be caused by exercise at a vigorous intensity with blood lactate accumulation, but not at a light intensity [105, 118].

Lactate is produced in the brain and skeletal muscle as a metabolite during exercise, and peripherally derived lactate can cross the BBB through passive transport by monocarboxylate transporters (MCTs). Recently, it has been noted that central and peripheral lactate may not only function as an energy supplying substrate, but also as a signaling molecule to regulate neuronal plasticity in the central nervous system [119]. During TR, circulating lactate levels begin to increase from moderate-intensity exercise and above, but do not change during light-intensity exercise [31, 33]. A past study showed that intermittent vigorous-intensity TR at an intensity above the LT (≧20 m/min) performed until exhaustion causes blood lactate accumulation and the upregulation of PGC-1α, BDNF, MCT1, and MCT2 mRNA expression in the hippocampus [120]. However, the inhibition of peripheral lactate transport into the brain by the intraperitoneal administration of MCT inhibitors (AR-C155858 or UK5099) abolishes the vigorous-intensity-TR-induced upregulation of these gene expressions [115, 120]. Interestingly, another study reported that light-to-moderate-intensity TR-induced improvements in the performance of spatial memory tasks, as well as reference and retention tasks, are enhanced by exogenous lactate administration [109]. These findings suggest that peripheral blood lactate positively influences hippocampal function and the regulatory PGC-1α/BDNF pathway through the MCT system during TR, especially vigorous-intensity TR. However, since no improvement in AHN or hippocampal function has been observed under vigorous-intensity TR conditions [45, 77], it is highly likely that the positive effects of lactate are counteracted by the effects of stress hormones, etc., as discussed below (see Sect. 4.4). By contrast, it remains unclear how lactate produced by the hippocampus changes in response to exercise intensity and whether this hippocampus-derived lactate directly affects hippocampal neuroplasticity and function. In the future, the role of hippocampus-derived lactate during exercise, especially during light-intensity TR, should be carefully investigated in relation to AHN and hippocampal function, while also considering the dynamics of lactate derived from the skeletal muscle tissue.

4.2 IGF-1

Even though IGF-1 is produced in multiple tissue types, including skeletal muscles, kidneys, the spleen, and the brain, most IGF-1 is secreted by the liver and transported to other tissues, acting as an important mediator of metabolism and tissue growth and reconstruction [121, 122]. Circulating IGF-1 is predominantly bound to IGF-binding proteins (IGF-BPs), which are regulators of IGF-1 bioactivity with an affinity equal to or greater than that of the IGF receptor. At the target tissues, IGF-1 exerts growth-promoting effects via the IGF-1 receptor, a membrane-bound receptor tyrosine kinase, and activation of its downstream signaling pathways [121]. Peripheral IGF-1 can pass through the BBB by a mechanism called transcytosis, which is derived by neuronal activity [123, 124] (*see* chapter “Neuronal Activity in Rat Hippocampus During Light-Intensity Treadmill Running” for a more detailed explanation), and induces AHN [125]. As regards AHN regulation, IGF-1 signaling is known to be a potent regulator of NSC proliferation and survival through its pro-mitotic and antiapoptotic actions [126].

It appears that moderate-to-vigorous intensity aerobic or resistance exercise stimulates an acute increase in circulating concentrations of IGF-1, and increases the resting levels of IGF-1 via the exercise [127–129]. In contrast, it remains unclear whether light-intensity exercise has the same increase effects. For example, circulating IGF-1 levels have been reported to be unchanged after acute exercise below the LT in rats (17 m/min) [130], and decreased after 6 weeks of exercise at around the LT in humans [131]. Importantly, the study using rats reported that light-intensity TR increases the uptake of circulating IGF-I into the brain, including the hippocampus, even in the absence of elevated circulating IGF-1 levels [130]. Moreover, increased IGF-1 protein levels in the hippocampus have been confirmed in mice subjected to light-intensity TR (10–15 m/min, <LT), but not in those subjected to vigorous-intensity TR (30 m/min, >LT), for 6 weeks [46]. The uptake of IGF-1 into the brain is derived from neural activity [124], and light-intensity TR causes neural activation of the hippocampal DG [31, 113]. Taken together, these results indicate that light-intensity TR increases hippocampal DG IGF-1 levels in a neural-activity-dependent manner, regardless of changes in blood IGF-1, which may contribute to exercise-induced AHN.

Indeed, 2 weeks of light-intensity TR (17 m/min, <LT) enhanced NSC proliferation (assessed by determining the number of BrdU$^+$ cells at 24 h post-administration) and survival (assessed by determining the number of BrdU$^+$ cells at 2 weeks post-administration), but blocking peripheral IGF-1 from entering the brain by its antiserum prevented the TR-induced AHN [34]. Similar results were found in a study using liver-IGF-I deficient (LID) mice, a model of low-circulation IGF-I [91]. The mice showed impaired spatial learning together with reduced AHN at the maturation/survival phase. These functional and neurogenic deficits were not improved with 3–4 weeks of light-intensity TR (12 m/min, <LT). These studies suggest that peripheral IGF-1 is involved in the promotion of AHN through light-intensity TR. However, it remains unclear whether this promotion of AHN is solely due to the effects of IGF-1 on the hippocampus or whether it involves interactions with other neurotrophic factors, such as BDNF. Some studies have reported that circulatory or hippocampal IGF-1 influences BDNF expression. For example, IGF-1 injected into the carotid artery induced hippocampal BDNF mRNA expression [130]. Additionally, enhanced BDNF expression and cell survival signaling activation in the hippocampus with wheel running are abolished by blocking the hippocampal IGF-1 receptor [132] or by the systemic blockade of IGF-1 uptake using its antiserum [133]. The detailed role of IGF-1 as the regulator of exercise effects is also explained in chapter "Interoceptive Signaling by Circulating Insulin Like Growth Factor I and Neuroprotection by Exercise".

4.3 VEGF

A potent angiogenic factor, VEGF is involved in the crosstalk between endothelial cells and neuronal precursors in the neurovascular niche of the SGZ [134]. VEGF can promote AHN indirectly by enhancing the production of endothelial cells (angiogenesis), which secrete soluble factors with differentiation-accelerating action and by interacting with neurotrophic factors such as BDNF, and directly by stimulating NSC division or survival through Flk-1-mediated signaling [135, 136]. Past studies have shown that light-intensity, but not above moderate-intensity, TR upregulates hippocampal VEGF and Flk-1 mRNA/protein expression, and in synchrony with this change, stimulates the neurogenic processes of proliferation, survival, maturation, and migration [40, 46]. Since light-intensity TR also increases other neurogenesis regulators, such as BDNF, erythropoietin (EPO), and IGF-1 [46], in the hippocampus, this exercise-induced AHN is unlikely to be derived from the VEGF effect alone but may be modulated by interactions with these factors. In contrast, there are some conflicting reports on the expression of VEGF in the hippocampus, in which the increased VEGF mRNA and protein levels in the hippocampus are in response to TR at above a moderate intensity [137–139]. Although the appropriate exercise intensity necessary to induce hippocampal VEGF remains unclear, these findings suggest that TR promotes VEGF synthesis and secretion in the hippocampus, and this may well be possible even with light-intensity exercise.

Given that VEGF passes through the BBB [140], the exploration of exercise-induced regulation of AHN should consider the influence not only of hippocampal-derived VEGF, but also of peripheral VEGF originating in skeletal muscles [138]. Studies using wheel running have shown that peripheral VEGF, especially that derived from skeletal muscles, is essential for exercise-induced AHN enhancement. For instance, sequestering circulating VEGF with a soluble receptor inhibited wheel-running-induced increases in cell proliferation (Ki-67$^+$ cells) and neuronal differentiation (DCX$^+$ cells) in the DG, with no effect on their baseline levels [94]. Moreover, the increase in hippocampal NSC proliferation (Nestin$^+$/BrdU$^+$ cells) with wheel running was abolished in mice lacking the skeletal myofiber-specific VEGF gene [137]. In contrast, endothelial cell proliferation (CD31$^+$/BrdU$^+$ cells) in the hippocampus was not altered by exercise in mice with or without the ablation of skeletal myofiber VEGF [137], indicating that angiogenesis was not associated with the increased neurogenesis. These findings imply that during exercise, peripheral VEGF affects the hippocampus by crossing the BBB, playing a direct role in the neurogenic process. However, all of these studies were conducted using wheel-running exercise, and no studies have yet closely examined the intensity-dependent role of peripheral VEGF using a TR model. Interestingly, a rat study using TR revealed that the VEGF protein levels in the soleus and gastrocnemius muscles and in the hippocampus were significantly higher in their 6 weeks of light-intensity TR (8 m/min, <LT) group than in their sedentary control group, and that the increase observed in the exercise group was accompanied by improved learning and capacity during the MWM test [141]. Moreover, there were positive correlations between VEGF levels in the hippocampus and each muscle. While it is unclear to what extent the amount of hippocampal VEGF is affected by peripheral VEGF, we can see that even light-intensity TR may promote VEGF synthesis and secretion in skeletal muscles, which in turn regulates AHN.

4.4 Glucocorticoids

Glucocorticoids, such as cortisol in primates and corticosterone (CORT) in rodents, are stress-induced steroid hormones that are released from the adrenal cortex into the bloodstream, and play a pivotal role in the homeostatic regulation involved in metabolic, immune, and cardiovascular responses, among others [142]. Also, glucocorticoids easily enter neuronal target cells in all brain regions by passing through the BBB as they are small lipophilic components, and exert multimodal influences on neuronal plasticity and function through two types of nuclear receptors (i.e., mineralocorticoid receptor (MR) and glucocorticoid receptor (GR)), which each have a different affinity for glucocorticoids [143, 144]. Since the MR has higher affinity than the GR, the MR appears to be largely occupied even at basal and unstressed levels of glucocorticoids, whereas the low-affinity GR can be activated in response to increased levels of stress-induced glucocorticoids or at the peak of the circadian rhythm [142]. Among all brain regions, it is the hippocampus, particularly the DG, that expresses these receptors at a high density, resulting in being sensitive to the fluctuating levels of glucocorticoid [145]. Therefore, the balance between MR and GR occupancies is thought to play a central role in the regulation of hippocampal neuroplasticity, including AHN, even though indirect mechanisms independent of this pathway are also assumed.

It is generally agreed that stress and stress-induced glucocorticoids are injurious to AHN because they inhibit proliferation, neuronal differentiation, and survival; further, the negative effects of glucocorticoid on AHN are most likely mediated through the GRs. However, the relationships between stress and its inducible changes of granular cells or AHN are far more complex than a simple inhibitory impact. For example, acute immobilization stress for 3 h elevated plasma CORT concentrations and caused induction of cell proliferation in the DGs of rats [146]. Removal of circulating adrenal steroids, including CORT, by adrenalectomy (ADX) caused almost complete loss of hippocampal granule cells and a reduction in their electrophysiological responses, while CORT supplementation reversed the ADX-induced adverse effects [147]. Furthermore, several learning paradigms, enriched environments, and physical activity not only stimulate the proliferation, neuronal differentiation, and survival of newborn cells, but also activate the stress response, leading to elevated glucocorticoid levels (for extensive reviews on this topic, *see* [148]). Much of the data published to date have implied a biphasic (or an inverted U-shape like) relationship between the glucocorticoid effects and the developments of AHN: the effects of a proper level of glucocorticoid coincide with increased AHN, while the effects of lower or higher levels of glucocorticoid have a reducing or inhibiting impact on AHN [148].

Previous studies by our laboratory and others have examined the role of CORT and its receptors in AHN regulation during TR. The first example is a study by Okamoto et al. [32], which examines the role of glucocorticoid receptors (MR and GR) on exercise-induced AHN with 2 weeks of light- (15 m/min, <LT) or vigorous- (30 m/min, >LT) intensity TR in mice. This study found that the mild TR, which had no effect on CORT secretion, led to increased newborn cells (BrdU^{+} cells), newborn immature neurons (BrdU^{+}/DCX^{+} cells), and newborn mature neurons (BrdU^{+}/NeuN^{+} cells), and that the AHN-promoting effect via light-intensity TR was eliminated by the intraperitoneal injection of antagonists for MR, spironolactone, or GR, mifepristone. In contrast, vigorous-intensity TR, which caused a transient, but not chronic, elevation of blood CORT levels, eliminated the AHN-promoting effect seen in the light-intensity TR condition, and antagonists of the MR/GR did not reverse this elimination. These findings indicate that the CORT-receptor-dependent signaling is essential for the mild-TR-induced AHN promotion. The elimination of TR-induced AHN promotion has also been observed in mice with mutations in the Cyp11a1 promotor (termed L/L mice), in which basal cortisol levels are normal with no significant fluctuations even after stress stimulation, including bouts of exercise [149]. Interestingly, in the study, 4 weeks of TR at 10 m/min (<LT) causes a transient elevation of blood CORT levels and an increase of newborn immature neurons (BrdU^{+}/DCX^{+} cells) in wild-type mice, but not in L/L mice. Although it is unclear why CORT was secreted at speeds below the LT in the wild-type mice, this result shows that mild CORT stimulation is essential for exercise-induced AHN. Another study examined the influence of ADX on exercise-induced AHN in mice exposed to light-intensity TR (11 m/min, <LT) for 5 weeks [54]. The results revealed that ADX, which led to a dramatic reduction in circulating CORT levels, attenuated the TR-induced increase in immature neurons cells (BrdU^{+}/DCX^{+} cells) without impacting the basal numbers of its cells in sedentary mice. These studies suggest an indispensable and facilitative nature of the glucocorticoid signaling pathway in increasing AHN with light-intensity TR.

In contrast, the excess level of glucocorticoid derived from exercise may have a negative impact on AHN, resulting in the loss of exercise effectiveness for AHN and hippocampal functions. In addition to the 2-week TR model by Okamoto et al. [32] mentioned above, we also showed that, in a 6-week TR model, light-intensity TR (<LT) enhances the proliferation, neuronal differentiation, and maturation/survival of NSCs and spatial memory, while these beneficial effects, excluding the proliferation, were eliminated with vigorous-intensity TR (>LT) (Fig. 2) [45]. Importantly, the 6-week long-term period of vigorous-intensity TR at above the LT, but not light-intensity TR below the LT, resulted in an elevated resting basal level of CORT, indicating chronic activation of stress response systems (Fig. 3a). A similar result was observed in a study on voluntary exercise using spontaneously hypertensive rats, which are characterized by significantly more active running compared to other rats [150]. This study reported that the number of proliferating (BrdU^{+}) cells increased with 9 days of short-term wheel running with no change in basal blood CORT levels but decreased with 24 days of long-term running accompanied with elevated blood CORT levels in the rats. The suppression of AHN and the increase in CORT levels caused by the long-term running were prevented by restricting daily running distances (by 30–50%) for 24 days. These results indicate that the AHN-promoting effects of exercise should not be anticipated in situations where CORT action is chronically enhanced, that is when stress responses are chronically evoked, by vigorous-intensity TR above the LT intensity. Given the hormesis effects of CORT in light-intensity TR condition, the exercise-induced change in CORT levels might also produce an inverse U-shaped-like action with regard to AHN regulation. However, Okamoto et al.'s study, presented above [32], indicated that the elimination of exercise-induced AHN by vigorous-intensity TR, that is to say due to exercise-stress, may be regulated by an independent CORT receptor mechanism. Careful investigation is thus needed in order to determine whether changes in AHN with vigorous-intensity TR are resulting from an increase in CORT or from other negative factors, such as proinflammatory cytokines.

4.5 Androgen

Androgens, such as testosterone and 5α-dihydrotestosterone (DHT), are known to be sex steroid hormones with AHN-promoting properties, exerting their influence by interacting with androgen receptors (ARs) [151]. In general, brain androgens are thought to be supplied by circulating androgens derived from the testes; however, in addition to this system, the hippocampus possesses a complete sex hormone synthesis system capable of synthesizing androgens from cholesterol [152]. This hippocampal-synthesized androgen seems to play an important role in AHN regulation during exercise.

Okamoto et al. found that 2 weeks of light-intensity TR exercise (13.5 m/min, <LT) enhances the local dihydrotestosterone (DHT) synthesis in the hippocampus and increases AHN, especially the newly born immature (DCX^+) and mature ($BrdU^+/NeuN^+$) neurons, through androgenic mediation [43]. Indeed, light-intensity TR led to increased hippocampal DHT levels as well as the hippocampal mRNA expressions of DHT and 5α-reductases, a DHT synthase, in the orchiectomized (ORX) rats after exercise, suggesting exercise-induced androgen synthesis in the hippocampus. The subcutaneous injection of flutamide, an androgen receptor antagonist, suppressed the light-intensity-TR-induced increase in AHN, but the same suppression was not observed with the subcutaneous injection of tamoxifen, an estrogen receptor antagonist. Notably, the depletion of androgen from blood circulation due to ORX also did not suppress the exercise effect, but the combination of ORX and flutamide injection did. These results provide direct evidence of the involvement of hippocampus-synthesized androgens during light-intensity-TR-induced AHN regulation.

4.6 Omics Analysis

The underlying mechanisms of beneficial, exercise-induced effects on the hippocampus, including AHN, have been vigorously investigated, not only through research focusing on single or multiple factors, but also through attempts to systematically understand the entire regulatory mechanism based on comprehensive changes in multiple genes and proteins, using omics analysis techniques such as microarrays and proteomics. The early cDNA microarray studies [153, 154], which reported comprehensive changes in hippocampal gene expression following exercise, had a profound impact on subsequent research into the mechanisms linking exercise and hippocampal function. In these studies, the effects of relatively short-term (3 or 7 days) or long-term (3–4 weeks) of wheel running on hippocampal gene expression were examined and were reported as follows: (1) exercise up-regulates genes related to synaptic trafficking, signal transduction pathways, transcription regulators, and the glutamatergic system, while it down-regulates genes related to the gamma aminobutyric acid (GABA) system; (2) exercise up-regulates the BDNF gene at all time points; and (3) exercise up-regulates BDNF signaling-related genes, such as TrkB and CREB, at some but not all time points. Since these studies, much of the subsequent research has focused on BDNF to consider the mechanisms of exercise-induced benefits to the hippocampus. Meanwhile, global changes in protein expression in the hippocampus following exercise was reported for the first time using a proteomic approach that combined 2D gel electrophoresis and mass spectrometry in a 5-day wheel running model [155]. The results showed that exercise up-regulates the expression of proteins involved in energy metabolism, synaptic plasticity, cytoskeleton, and molecular chaperones, suggesting that exercise improves hippocampal neuroplasticity and function by regulating these processes.

Subsequent studies have focused on the DG, a brain region where neurogenesis persists into adulthood, to investigate comprehensive changes in gene and protein expression after wheel running by using techniques such as microarray analysis [156], RNA-seq analysis [157], and proteome analysis [158]. More interestingly, there is a study that comprehensively investigated the intrinsic molecular mechanisms of exercise-induced AHN by using an adult newborn neuron specific whole-genome gene expression analysis [159]. This study utilized a mouse line ($Rpl22^{HA}$:Nes-$CreER^{T2}$) engineered for specific ribosome tagging in adult new neurons. Upon tamoxifen (TAM) injection, ribosomes are tagged with hemagglutinin in NESTIN-expressing NSCs and their differentiated progeny. When the mice was subjected to 11, 18, and 28 days of wheel running after the TAM injection, gene expression changes in response to wheel running were observed in the adult new neurons. The number of genes altered in the neurons was greater than that in the whole DG, suggesting that newborn neurons may have a more responsive nature to exercise than mature cells. The authors also investigated the role of G protein signaling 6 (RGS6), which was most significantly upregulated with wheel running in the adult new neurons. They demonstrated that RGS6 knockdown abolished the wheel-running-induced neuronal maturation, as well as the improvements in spatial learning observed in the novel location test, contextual learning in the contextual fear conditioning test, and the antianxiety effect in the elevated plus maze test. In addition, the morphological and physiological maturation of adult born neurons resulting from the running was shown to be mimicked by overexpression of RGS6. These results indicate that the RGS6 intrinsic to adult new neurons may be a key regulator of exercise-enhanced AHN.

In this way, gene and protein expression profiles revealed through omics approaches provide valuable insights into identifying candidate factors and pathways involved in the exercise-induced improvement of hippocampal neuroplasticity and function. These approaches have been applied not only to healthy animals, but also to reveal hippocampal transcriptional profiles underlying the therapeutic effects of exercise in aging animals [160–162] and those with Alzheimer's disease (AD) [163]. However, as all of such studies have been performed using voluntary wheel running, the exercise conditions could not be strictly controlled, resulting in a lack of information about exercise intensity.

To address this gap in the research, we investigated the effects of 6 weeks of TR at different intensities based on the LT—light-intensity TR (15 m/min, <LT) and vigorous-intensity TR (40 m/min, >LT)—on the global transcriptional profile of the hippocampus using a microarray [45]. It is important to note that stress-free light-intensity TR is a condition that promotes AHN and hippocampal-dependent spatial memory, while vigorous-intensity TR is a condition that leads to a chronic increase in the stress response and abolishes the beneficial effects observed in the light-intensity TR condition. Among the genes upregulated or downregulated in each exercise group compared to the resting control group, there were no factors that had been considered important for hippocampal adaptation during exercise, such as BDNF and IGF-1. In addition, the total number of differentially expressed genes was greater in the light-intensity TR group (604 genes) than in the vigorous-intensity TR group (415 genes), and only 41 genes changed in both groups. Similar results were also shown in another study using a wheel running model [157]. That study examined the DG gene expression profile using RNA-seq in mice exposed to either a voluntary running wheel (RUN), a locked wheel (LD), or a LD plus social enrichment, which was a complex environment (CE), for 4 weeks. RUN mice showed an increase in the number of NeuroD^{+} neuroblasts in the DG, and a greater transcriptomic change than CE mice (RUN: 207 genes, CE: 110 genes), with only a few genes commonly changing in both mice (29 genes). RUN mice were further divided into low run (L-RUN, 10.65 ± 0.39 km/day) or high run (H-RUN, 17.00 ± 0.45 km/day) groups based on individual running distances. L-RUN and H-RUN mice showed surprisingly different patterns of transcriptional changes, with only a very small overlap between them. Also, gene expression change in L-RUN mice was almost four times greater than that in H-RUN mice (L-RUN: 424 genes, H-RUN: 112 genes). These results suggest that light-intensity and proper exercise induces greater transcriptional changes in the hippocampus and DG compared to vigorous-intensity and excessive exercise. This difference may underlie the distinct effects of exercise observed under each condition.

In our study [45] mentioned above, we further categorized and ranked the genes changed in the light- and vigorous-intensity TR groups. The results revealed that the genes influenced by mild TR were principally related to lipid metabolism (e.g., APOE), protein synthesis (e.g., IGF2, IRS1), and inflammatory response (e.g., IL1B, TNF), all of which are recognized as being associated with AHN. In contrast, genes influenced by vigorous-intensity TR were linked to an excessive inflammatory immune response (e.g., IL10, IL1R1), which is a negative regulator of hippocampal neuroadaptation. Although further research is needed to explore the direct relationship between the identified genes and AHN or hippocampal function, the gene inventory specific to light-intensity TR offers fundamental insights into why mild TR is more effective than vigorous-intensity TR in improving exercise-induced hippocampal adaptation. Also, as with previous studies using the wheel-running model [156–159], further progress in developing a more detailed understanding of the mechanisms behind light-intensity TR-induced AHN can expected through a comprehensive analysis of local gene and protein expressions in newborn neurons or the DG, where AHN occurs.

Previous studies have compared differences in gene and proteome expression in the hippocampus between TR and other exercise modes such as resistance exercise and wheel running. For example, a study examined changes in anxiety-like behaviors using the OFT, nest-building test (NBT), and FST, as well as hippocampal global transcriptional changes analyzed through RNA-seq in mice following 4 weeks of TR or weight-bearing ladder resistance training [164]. The intensity of the TR training was light (<LT; 16 m/min) but unknown for the resistance training. The results showed that Dcx gene expression in the hippocampus, which is related to AHN, increased with both forms of exercise training, whereas anxiety-like behaviors in the OFT and NBT were improved with TR training but not with resistance training. Moreover, bioinformatics pathway analysis of differentially expressed genes revealed that TR altered mitogen-activated protein kinase (MAPK) related genes and resistance training altered phosphatidylinositol-3-kinase (PI3K)/AKT and genes related to neurotrophic signaling pathways compared to resting control. Another study reported on hippocampal protein profiles of rats undertaking either wheel running or moderate-intensity TR (20 m/min, at LT) for 18 months using a proteomics approach [165]. Among the proteins that showed changes, the levels of heterogeneous nuclear ribonucleoprotein K, one protein variant of the heat shock cognate 71 kDa protein and BAG family molecular chaperone regulator 5 had significant differences between the groups. However, it cannot be simply concluded that these differences arose solely from variations in exercise modes, as exercise conditions, particularly intensity, cannot be strictly controlled with wheel running and resistance training. Nonetheless, the gene and protein inventories obtained from the light-to-moderate intensity TR groups may be useful information in elucidating the mechanisms underlying how TR promotes AHN and hippocampal function.

5 Conclusion

Treadmill running (TR) improves hippocampal neurogenesis and function, and these benefits can be gained through stress-free light-intensity exercise performed at an intensity below the lactate threshold (LT) (Table 1). Further detailed investigation is necessary to determine the extent to which light-intensity TR influences each stage of development of newborn neurons in the hippocampal dentate gyrus (DG), but the most important fact is that mild TR leads to an increase in the number of mature newborn neurons, which are the terminal product, resulting in the functional development of the hippocampus in memory tests and the anxiety/depression tests [45, 46, 75–77]. In contrast, although there is no consensus on the effects of vigorous-intensity TR at intensities above the LT, it is highly likely that the positive effects observed with light-intensity TR are diminished when vigorous-intensity TR induces chronic stress accompanied by elevated resting plasma corticosterone levels [45, 77]. Notably, uncontrollable chronic stress, such as social defeat—a paradigm in which a rodent is exposed to a dominant conspecific, inducing social subordination and stress—leads to reduced AHN and hippocampal dysfunction [166], while controllable exercise-induced stress does not cause similar exacerbation, even when the positive effects of exercise diminish.

Since stress-free light-intensity exercise is safe and easily accessible to everyone, it holds great potential for application across a wide range of populations, such as exercise beginners, individuals with low physical fitness levels, the elderly, and Alzheimer's patients, who often experience declines in AHN and hippocampal function. In fact, our and others' studies have reported improvements in AHN and hippocampal function in aged animals [63] and Alzheimer's disease model animals [167, 168] with light-intensity exercise. In addition, vigorous-intensity TR can enhance AHN and hippocampal function if carefully designed to prevent the induction of chronic stress. For instance, we have demonstrated that intermittent exercise, which combines short sessions of TR at intensities above the LT with rest, and resistance-like training through loaded wheel running, can efficiently increase the exercise load without raising resting corticosterone levels, thereby avoiding chronic stress and enhancing AHN and hippocampal function [169, 170].

Light-intensity TR likely promotes AHN and enhances hippocampal function through a complex network involving hippocampus-derived factors, such as BDNF and androgens, and peripheral factors, such as glucocorticoids (corticosterone), IGF-1, and VEGF (Fig. 4). While the key factors and pathways are not yet fully elucidated, we hypothesize that increased hippocampal neural activity induced by exercise, particularly light-intensity TR, acts as a central event that drives the activation of these hippocampal and peripheral factors. For instance, we have demonstrated that the uptake of peripheral IGF-1 into the brain is regulated in a neural activity-dependent manner [124]. Furthermore, we have shown that light-intensity TR enhances hippocampal neural activity [31] and significantly alters gene expression in the hippocampus [45]. These findings suggest that hippocampal neural activity triggered by light-intensity TR serves as a critical mechanism for mobilizing these factors, thereby enhancing hippocampal plasticity. To validate this hypothesis, it is essential to gain a comprehensive understanding of the issue by adopting not only targeted analysis of individual or multiple factors but also comprehensive approaches such as omics analysis using RNA-seq and microarrays in future research. Additionally, compared to studies using wheel running models, research investigating the mechanisms of exercise-induced AHN using TR models remains limited, posing a significant challenge. Moving forward, leveraging TR models—which enable precise control of exercise conditions—will be essential to uncover the mechanisms specific to different exercise conditions, especially intensity, and to generate insights that can inform the development of effective exercise prescriptions for humans. In closing, we offer the following. One of the biggest mysteries and fundamental points of interest is the question, why does exercise activate the hippocampus and enhance its plasticity and function? And, why is this activation necessary?

References

1. Altman J, Das GD (1965) Autoradiographic and histological evidence of postnatal hippocampal neurogenesis in rats. J Comp Neurol 124(3):319–335. https://doi.org/10.1002/cne.901240303
2. Taupin P (2007) BrdU immunohistochemistry for studying adult neurogenesis: paradigms, pitfalls, limitations, and validation. Brain Res Rev 53(1):198–214. https://doi.org/10.1016/j.brainresrev.2006.08.002
3. Ehninger D, Kempermann G (2008) Neurogenesis in the adult hippocampus. Cell Tissue Res 331(1):243–250. https://doi.org/10.1007/s00441-007-0478-3
4. Gonçalves JT, Schafer ST, Gage FH (2016) Adult neurogenesis in the hippocampus: from stem cells to behavior. Cell 167(4):897–914. https://doi.org/10.1016/j.cell.2016.10.021
5. Kempermann G, Song H, Gage FH (2015) Neurogenesis in the adult hippocampus. Cold Spring Harb Perspect Biol 7(9):a018812-a. https://doi.org/10.1101/cshperspect.a018812
6. Zhao C, Deng W, Gage FH (2008) Mechanisms and functional implications of adult neurogenesis. Cell 132(4):645–660. https://doi.org/10.1016/j.cell.2008.01.033
7. Ming GL, Song H (2011) Adult neurogenesis in the mammalian brain: significant answers and significant questions. Neuron 70(4):687–702. https://doi.org/10.1016/j.neuron.2011.05.001
8. Shors TJ, Miesegaes G, Beylin A, Zhao M, Rydel T, Gould E (2001) Neurogenesis in the adult is involved in the formation of trace memories. Nature 410(6826):372–376. https://doi.org/10.1038/35066584
9. Schmidt-Hieber C, Jones P, Bischofberger J (2004) Enhanced synaptic plasticity in newly generated granule cells of the adult hippocampus. Nature 429(6988):184–187. https://doi.org/10.1038/nature02553
10. Ge S, Yang CH, Hsu KS, Ming GL, Song H (2007) A critical period for enhanced synaptic plasticity in newly generated neurons of the adult brain. Neuron 54(4):559–566. https://doi.org/10.1016/j.neuron.2007.05.002
11. Mongiat LA, Esposito MS, Lombardi G, Schinder AF (2009) Reliable activation of immature neurons in the adult hippocampus. PLoS One 4(4):e5320. https://doi.org/10.1371/journal.pone.0005320
12. Dieni CV, Nietz AK, Panichi R, Wadiche JI, Overstreet-Wadiche L (2013) Distinct determinants of sparse activation during granule cell maturation. J Neurosci 33(49):19131–19142. https://doi.org/10.1523/JNEUROSCI.2289-13.2013
13. Clelland CD, Choi M, Romberg C, Clemenson GD Jr, Fragniere A, Tyers P et al (2009) A functional role for adult hippocampal neurogenesis in spatial pattern separation. Science 325(5937):210–213. https://doi.org/10.1126/science.1173215
14. Franca TFA, Bitencourt AM, Maximilla NR, Barros DM, Monserrat JM (2017) Hippocampal neurogenesis and pattern separation: a meta-analysis of behavioral data. Hippocampus 27(9):937–950. https://doi.org/10.1002/hipo.22746
15. Fanselow MS, Dong HW (2010) Are the dorsal and ventral hippocampus functionally distinct structures? Neuron 65(1):7–19. https://doi.org/10.1016/j.neuron.2009.11.031
16. Anacker C, Hen R (2017) Adult hippocampal neurogenesis and cognitive flexibility – linking memory and mood. Nat Rev Neurosci 18(6):335–346. https://doi.org/10.1038/nrn.2017.45
17. Baptista P, Andrade JP (2018) Adult hippocampal neurogenesis: regulation and possible functional and clinical correlates. Front Neuroanat 12(June):1–23. https://doi.org/10.3389/fnana.2018.00044
18. Kempermann G (2022) What is adult hippocampal neurogenesis good for? Front Neurosci 16(April):1–8. https://doi.org/10.3389/fnins.2022.852680
19. Kempermann G, Kuhn HG, Gage FH (1997) More hippocampal neurons in adult mice living in an enriched environment. Nature 386:493–495
20. Kobilo T, Liu QR, Gandhi K, Mughal M, Shaham Y, van Praag H (2011) Running is the neurogenic and neurotrophic stimulus in environmental enrichment. Learn Mem 18(9):605–609. https://doi.org/10.1101/lm.2283011

21. Mustroph ML, Chen S, Desai SC, Cay EB, DeYoung EK, Rhodes JS (2012) Aerobic exercise is the critical variable in an enriched environment that increases hippocampal neurogenesis and water maze learning in male C57BL/6J mice. Neuroscience 219:62–71. https://doi.org/10.1016/j.pharmthera.2010.05.003.Pharmacology
22. van Praag H, Kempermann G, Gage FH (1999) Running increases cell proliferation and neurogenesis in the adult mouse dentate gyrus. Nat Neurosci 2(3):266–270. https://doi.org/10.1038/6368
23. Clark PJ, Brzezinska WJ, Thomas MW, Ryzhenko NA, Toshkov SA, Rhodes JS (2008) Intact neurogenesis is required for benefits of exercise on spatial memory but not motor performance or contextual fear conditioning in C57BL/6J mice. Neuroscience 155(4):1048–1058. https://doi.org/10.1016/j.neuroscience.2008.06.051
24. Creer DJ, Romberg C, Saksida LM, van Praag H, Bussey TJ (2010) Running enhances spatial pattern separation in mice. PNAS 107(5):2367–2372. https://doi.org/10.1073/pnas.0911725107
25. Sahay A, Scobie KN, Hill AS, O'Carroll CM, Kheirbek MA, Burghardt NS et al (2011) Increasing adult hippocampal neurogenesis is sufficient to improve pattern separation. Nature 472(7344):466–470. https://doi.org/10.1038/nature09817
26. Fuss J, Abdallah NMB, Vogt MA, Touma C, Pacifici PG, Palme R et al (2010) Voluntary exercise induces anxiety-like behavior in adult C57BL/6J mice correlating with hippocampal neurogenesis. Hippocampus 376:364–376. https://doi.org/10.1002/hipo.20634
27. Liguori G (2021) ACSM'S guidelines for exercise testing and prescription, 11th edn. Lippincott Williams & Wilkins
28. Wenger HA, Bell GJ (1986) The interactions of intensity, frequency and duration of exercise training in altering cardiorespiratory fitness. Sports Med 3(5):346–356. https://doi.org/10.2165/00007256-198603050-00004
29. Swain DP, Franklin BA (2002) VO2 reserve and the minimal intensity for improving cardiorespiratory fitness. Med Sci Sports Exerc 34(1):152–157
30. Soya H (2001) Stress response to exercise and its hypothalamic regulation: role of arginine-vasopressin. In: Nose H, Gisolfi CV, Imaizumi K (eds) Exercise nutrition and environmental stress. Cooper Publishing, Traverse City, pp 21–37
31. Soya H, Mukai A, Deocaris CC, Ohiwa N, Chang H, Nishijima T et al (2007) Threshold-like pattern of neuronal activation in the hypothalamus during treadmill running: establishment of a minimum running stress (MRS) rat model. Neurosci Res 58(4):341–348. https://doi.org/10.1016/j.neures.2007.04.004
32. Okamoto M, Yamamura Y, Liu Y-F, Min-Chul L, Matsui T, Shima T et al (2015) Hormetic effects by exercise on hippocampal neurogenesis with glucocorticoid signaling. Brain Plast 1(1):149–158. https://doi.org/10.3233/bpl-150012
33. Inoue K, Soya H, Murakumo K, Araki Y, Hiraga T, Soya S et al (2024) Setting treadmill intensity for rat aerobic training using lactate and gas exchange thresholds. Med Sci Sports Exerc 57:434. https://doi.org/10.1249/MSS.0000000000003562
34. Trejo JL, Carro E, Torres-Aleman I (2001) Circulating insulin-like growth factor I mediates exercise-induced increases in the number of new neurons in the adult hippocampus. J Neurosci 21(5):1628–1634
35. Ra SM, Kim H, Jang MH, Shin MC, Lee TH, Lim BV et al (2002) Treadmill running and swimming increase cell proliferation in the hippocampal dentate gyrus of rats. Neurosci Lett 333(2):123–126. https://doi.org/10.1016/S0304-3940(02)01031-5
36. Kim SH, Kim HB, Jang MH, Lim BV, Kim YJ, Kim YP et al (2002) Treadmill exercise increases cell proliferation without altering of apoptosis in dentate gyrus of Sprague-Dawley rats. Life Sci 71(11):1331–1340. https://doi.org/10.1016/s0024-3205(02)01849-0
37. Kim YP, Kim HB, Jang MH, Lim BV, Kim YJ, Kim H et al (2003) Magnitude- and time-dependence of the effect of treadmill exercise on cell proliferation in the dentate gyrus of rats. Int J Sports Med 24(2):114–117. https://doi.org/10.1055/s-2003-38202

38. Uda M, Ishido M, Kami K (2007) Features and a possible role of Mash1-immunoreactive cells in the dentate gyrus of the hippocampus in the adult rat. Brain Res 1171(1):9–17. https://doi.org/10.1016/j.brainres.2007.06.099
39. Uda M, Ishido M, Kami K, Masuhara M (2006) Effects of chronic treadmill running on neurogenesis in the dentate gyrus of the hippocampus of adult rat. Brain Res 1104(1):64–72. https://doi.org/10.1016/j.brainres.2006.05.066
40. Lou S, Liu J, Chang H, Chen P (2008) Hippocampal neurogenesis and gene expression depend on exercise intensity in juvenile rats. Brain Res 1210:48–55. https://doi.org/10.1016/j.brainres.2008.02.080
41. Xu W-P, Shan L-D, Gong S, Chen L, Zhang Y-J, Yin Q-Z et al (2006) Forced running enhances neurogenesis in the hippocampal dentate gyrus of adult rats and improves learning ability. Acta Physiol Sinica 58(5):415–420
42. Glasper ER, Llorens-Martin MV, Leuner B, Gould E, Trejo JL (2010) Blockade of insulin-like growth factor-I has complex effects on structural plasticity in the hippocampus. Hippocampus 20(6):706–712. https://doi.org/10.1002/hipo.20672
43. Okamoto M, Hojo Y, Inoue K, Matsui T, Kawato S, McEwen BS et al (2012) Mild exercise increases dihydrotestosterone in hippocampus providing evidence for androgenic mediation of neurogenesis. Proc Natl Acad Sci 109(32):13100–13105. https://doi.org/10.1073/pnas.1210023109
44. Kwon SJ, Park JS, Park SY, Song KS, Jung ST, Jung SB et al (2013) Low-intensity treadmill exercise and/or bright light promote neurogenesis in adult rat brain. Neural Regen Res 8(10):922–929. https://doi.org/10.3969/j.issn.1673-5374.2013.10.007
45. Inoue K, Okamoto M, Shibato J, Lee MC, Matsui T, Rakwal R et al (2015) Long-term mild, rather than intense, exercise enhances adult hippocampal neurogenesis and greatly changes the transcriptomic profile of the hippocampus. PLoS One 10(6):e0128720. https://doi.org/10.1371/journal.pone.0128720
46. So JH, Huang C, Ge M, Cai G, Zhang L, Lu Y et al (2017) Intense exercise promotes adult hippocampal neurogenesis but not spatial discrimination. Front Cell Neurosci 11(January):1–12. https://doi.org/10.3389/fncel.2017.00013
47. Tanapat P, Hastings NB, Rydel TA, Galea LA, Gould E (2001) Exposure to fox odor inhibits cell proliferation in the hippocampus of adult rats via an adrenal hormone-dependent mechanism. J Comp Neurol 437(4):496–504. https://doi.org/10.1002/cne.1297
48. Malberg JE, Duman RS (2003) Cell proliferation in adult hippocampus is decreased by inescapable stress: reversal by fluoxetine treatment. Neuropsychopharmacology 28(9):1562–1571. https://doi.org/10.1038/sj.npp.1300234
49. Kronenberg G, Reuter K, Steiner B, Brandt MD, Jessberger S, Yamaguchi M et al (2003) Subpopulations of proliferating cells of the adult hippocampus respond differently to physiologic neurogenic stimuli. J Comp Neurol 467(4):455–463. https://doi.org/10.1002/cne.10945
50. Steiner B, Zurborg S, Hörster H, Fabel K, Kempermann G (2008) Differential 24 h responsiveness of Prox1-expressing precursor cells in adult hippocampal neurogenesis to physical activity, environmental enrichment, and kainic acid-induced seizures. Neuroscience 154(2):521–529. https://doi.org/10.1016/j.neuroscience.2008.04.023
51. Dong J, Pan YB, Wu XR, He LN, Liu XD, Feng DF et al (2019) A neuronal molecular switch through cell-cell contact that regulates quiescent neural stem cells. Sci Adv 5(2):eaav4416. https://doi.org/10.1126/sciadv.aav4416
52. Lugert S, Basak O, Knuckles P, Haussler U, Fabel K, Götz M et al (2010) Quiescent and active hippocampal neural stem cells with distinct morphologies respond selectively to physiological and pathological stimuli and aging. Cell Stem Cell 6(5):445–456. https://doi.org/10.1016/j.stem.2010.03.017
53. Llorens-Martín M, Torres-Alemán I, Trejo JL (2006) Pronounced individual variation in the response to the stimulatory action of exercise on immature hippocampal neurons. Hippocampus 16:480–490. https://doi.org/10.1002/hipo

54. Chang YT, Chen YC, Wu CW, Yu L, Chen HI, Jen CJ et al (2008) Glucocorticoid signaling and exercise-induced downregulation of the mineralocorticoid receptor in the induction of adult mouse dentate neurogenesis by treadmill running. Psychoneuroendocrinology 33(9):1173–1182. https://doi.org/10.1016/j.psyneuen.2008.05.014
55. Ferreira AFB, Real CC, Rodrigues AC, Alves AS, Britto LRG (2011) Short-term, moderate exercise is capable of inducing structural, bdnf-independent hippocampal plasticity. Brain Res 1425:111–122. https://doi.org/10.1016/j.brainres.2011.10.004
56. Yang TT, Lo CP, Tsai PS, Wu SY, Wang TF, Chen YW et al (2015) Aging and exercise affect hippocampal neurogenesis via different mechanisms. PLoS One 10(7). https://doi.org/10.1371/journal.pone.0132152
57. Komitova M, Zhao LR, Gido G, Johansson BB, Eriksson P (2005) Postischemic exercise attenuates whereas enriched environment has certain enhancing effects on lesion-induced subventricular zone activation in the adult rat. Eur J Neurosci 21(9):2397–2405. https://doi.org/10.1111/j.1460-9568.2005.04072.x
58. Li Y, Luo Y, Tang J, Liang X, Wang J, Xiao Q et al (2021) The positive effects of running exercise on hippocampal astrocytes in a rat model of depression. Translational. Psychiatry 11(1). https://doi.org/10.1038/s41398-021-01216-x
59. Alvarez Z, Hyrossova P, Perales JC, Alcantara S (2016) Neuronal progenitor maintenance requires lactate metabolism and PEPCK-M-directed cataplerosis. Cereb Cortex 26(3):1046–1058. https://doi.org/10.1093/cercor/bhu281
60. Quesseveur G, David DJ, Gaillard MC, Pla P, Wu MV, Nguyen HT et al (2013) BDNF overexpression in mouse hippocampal astrocytes promotes local neurogenesis and elicits anxiolytic-like activities. Transl Psychiatry 3(4):e253. https://doi.org/10.1038/tp.2013.30
61. Sultan S, Li L, Moss J, Petrelli F, Casse F, Gebara E et al (2015) Synaptic integration of adult-born hippocampal neurons is locally controlled by astrocytes. Neuron 88(5):957–972. https://doi.org/10.1016/j.neuron.2015.10.037
62. Fahimi A, Baktir MA, Moghadam S, Mojabi FS, Sumanth K, McNerney MW et al (2017) Physical exercise induces structural alterations in the hippocampal astrocytes: exploring the role of BDNF-TrkB signaling. Brain Struct Funct 222(4):1797–1808. https://doi.org/10.1007/s00429-016-1308-8
63. Okamoto M, Inoue K, Iwamura H, Terashima K, Soya H, Asashima M et al (2011) Reduction in paracrine Wnt3 factors during aging causes impaired adult neurogenesis. FASEB J 25(10):3570–3582. https://doi.org/10.1096/fj.11-184697
64. Von Bohlen Und Halbach O (2011) Immunohistological markers for proliferative events, gliogenesis, and neurogenesis within the adult hippocampus. Cell Tissue Res 345(1):1–19. https://doi.org/10.1007/s00441-011-1196-4
65. Gould E, Beylin A, Tanapat P, Reeves A, Shors TJ (1999) Learning enhances adult neurogenesis in the hippocampal formation. Nat Neurosci 2(3):260–265. https://doi.org/10.1038/6365
66. Olson AK, Eadie BD, Ernst C, Christie BR (2006) Environmental enrichment and voluntary exercise massively increase neurogenesis in the adult hippocampus via dissociable pathways. Hippocampus 16(3):250–260. https://doi.org/10.1002/hipo.20157
67. Fabel K, Kempermann G (2008) Physical activity and the regulation of neurogenesis in the adult and aging brain. NeuroMolecular Med 10(2):59–66. https://doi.org/10.1007/s12017-008-8031-4
68. Yildirim AB (2021) The effect of exercise on the total number of BrdU + cell counts in rats' hippocampal dentate gyrus: a meta-analysis study. Brain Res 1766:147512. https://doi.org/10.1016/j.brainres.2021.147512
69. Gao Y, Syed M, Zhao X (2023) Mechanisms underlying the effect of voluntary running on adult hippocampal neurogenesis. Hippocampus 33(February):1–18. https://doi.org/10.1002/hipo.23520
70. Wu C-WW, Chang Y-TT, Yu L, H-iI C, Jen CJ, Wu S-YY et al (2008) Exercise enhances the proliferation of neural stem cells and neurite growth and survival of neuronal progenitor cells in dentate gyrus of middle-aged mice. J Appl Physiol 105(5):1585–1594. https://doi.org/10.1152/japplphysiol.90775.2008

71. Li H, Liang A, Guan F, Fan R, Chi L, Yang B (2013) Regular treadmill running improves spatial learning and memory performance in young mice through increased hippocampal neurogenesis and decreased stress. Brain Res 1531:1–8. https://doi.org/10.1016/j.brainres.2013.07.041
72. Kim JW, Nam SM, Yoo DY, Jung HY, Kim IY, Hwang IK et al (2017) Comparison of adult hippocampal neurogenesis and susceptibility to treadmill exercise in nine mouse strains. Neural Plast 2017:1. https://doi.org/10.1155/2017/5863258
73. Shih PC, Yang YR, Wang RY (2013) Effects of exercise intensity on spatial memory performance and hippocampal synaptic plasticity in transient brain ischemic rats. PLoS One 8(10):e78163. https://doi.org/10.1371/journal.pone.0078163
74. van Praag H, Christie BR, Sejnowski TJ, Gage FH, Praag HV, Christie BR et al (1999) Running enhances neurogenesis, learning, and long-term potentiation in mice. Proc Natl Acad Sci USA 96(23):13427–13431. https://doi.org/10.1073/pnas.96.23.13427
75. Nishii A, Amemiya S, Kubota N, Nishijima T, Kita I (2017) Adaptive changes in the sensitivity of the dorsal raphe and hypothalamic paraventricular nuclei to acute exercise, and hippocampal neurogenesis may contribute to the antidepressant effect of regular treadmill running in rats. Front Behav Neurosci 11(November):1–13. https://doi.org/10.3389/fnbeh.2017.00235
76. Siteneski A, Olescowicz G, Pazini FL, Camargo A, Fraga DB, Brocardo PS et al (2020) Antidepressant-like and pro-neurogenic effects of physical exercise: the putative role of FNDC5/irisin pathway. J Neural Transm 127(3):355–370. https://doi.org/10.1007/s00702-020-02143-9
77. Inoue K, Hanaoka Y, Nishijima T, Okamoto M, Chang H, Saito T et al (2015) Long-term mild exercise training enhances hippocampus-dependent memory in rats. Int J Sports Med 36:280. https://doi.org/10.1055/s-0034-1390465
78. Miller SM, Sahay A (2019) Functions of adult-born neurons in hippocampal memory interference and indexing. Nat Neurosci 22(10):1565–1575. https://doi.org/10.1038/s41593-019-0484-2
79. Hernandez-Mercado K, Zepeda A (2021) Morris water maze and contextual fear conditioning tasks to evaluate cognitive functions associated with adult hippocampal neurogenesis. Front Neurosci 15:782947. https://doi.org/10.3389/fnins.2021.782947
80. Kee N, Teixeira CM, Wang AH, Frankland PW (2007) Preferential incorporation of adult-generated granule cells into spatial memory networks in the dentate gyrus. Nat Neurosci 10(3):355–362. https://doi.org/10.1038/nn1847
81. Farioli-Vecchioli S, Saraulli D, Costanzi M, Pacioni S, Cina I, Aceti M et al (2008) The timing of differentiation of adult hippocampal neurons is crucial for spatial memory. PLoS Biol 6(10):e246. https://doi.org/10.1371/journal.pbio.0060246
82. Gu Y, Arruda-Carvalho M, Wang J, Janoschka SR, Josselyn SA, Frankland PW et al (2012) Optical controlling reveals time-dependent roles for adult-born dentate granule cells. Nat Neurosci 15(12):1700–1706. https://doi.org/10.1038/nn.3260
83. Uysal N, Tugyan K, Kayatekin BM, Acikgoz O, Bagriyanik HA, Gonenc S et al (2005) The effects of regular aerobic exercise in adolescent period on hippocampal neuron density, apoptosis and spatial memory. Neurosci Lett 383(3):241–245. https://doi.org/10.1016/j.neulet.2005.04.054
84. Liu Y-F, Chen H-i, Wu C-L, Kuo Y-M, Yu L, Huang AM et al (2009) Differential effects of treadmill running and wheel running on spatial or aversive learning and memory: roles of amygdalar brain-derived neurotrophic factor and synaptotagmin I. J Physiol 587(13):3221–3231. https://doi.org/10.1113/jphysiol.2009.173088
85. Jin JJ, Ko IG, Kim SE, Hwang L, Lee MG, Kim DY et al (2017) Age-dependent differences of treadmill exercise on spatial learning ability between young- and adult-age rats. J Exerc Rehabil 13(4):381–386. https://doi.org/10.12965/jer.1735070.535
86. Bolz L, Heigele S, Bischofberger J (2015) Running improves pattern separation during novel object recognition. Brain Plast 1(1):129–141. https://doi.org/10.3233/BPL-150010

87. Yi Y, Zhang Y, Song Y, Lu Y (2024) Treadmill running regulates adult neurogenesis, spatial and non-spatial learning, Parvalbumin neuron activity by ErbB4 signaling. Cell Mol Neurobiol 44(1):17. https://doi.org/10.1007/s10571-023-01439-0
88. Kiuchi T, Lee H, Mikami T (2012) Regular exercise cures depression-like behavior via VEGF-Flk-1 signaling in chronically stressed mice. Neuroscience 207:208–217. https://doi.org/10.1016/j.neuroscience.2012.01.023
89. Kempermann G, Gage FH, Aigner L, Song H, Curtis MA, Thuret S et al (2018) Human adult neurogenesis: evidence and remaining questions. Cell Stem Cell 23(1):25–30. https://doi.org/10.1016/j.stem.2018.04.004
90. Moreno-Jimenez EP, Terreros-Roncal J, Flor-Garcia M, Rabano A, Llorens-Martin M (2021) Evidences for adult hippocampal neurogenesis in humans. J Neurosci 41(12):2541–2553. https://doi.org/10.1523/JNEUROSCI.0675-20.2020
91. Trejo JL, Llorens-Martín MV, Torres-Alemán I (2008) The effects of exercise on spatial learning and anxiety-like behavior are mediated by an IGF-I-dependent mechanism related to hippocampal neurogenesis. Mol Cell Neurosci 37(2):402–411. https://doi.org/10.1016/j.mcn.2007.10.016
92. Llorens-Martin M, Torres-Aleman I, Trejo JL (2010) Exercise modulates insulin-like growth factor 1-dependent and -independent effects on adult hippocampal neurogenesis and behaviour. Mol Cell Neurosci 44(2):109–117. https://doi.org/10.1016/j.mcn.2010.02.006
93. Liu PZ, Nusslock R (2018) Exercise-mediated neurogenesis in the hippocampus via BDNF. Front Neurosci 12:52. https://doi.org/10.3389/fnins.2018.00052
94. Fabel K, Fabel K, Tam B, Kaufer D, Baiker A, Simmons N et al (2003) VEGF is necessary for exercise-induced adult hippocampal neurogenesis. Eur J Neurosci 18(10):2803–2812. https://doi.org/10.1111/j.1460-9568.2003.03041.x
95. Connolly MG, Bruce SR, Kohman RA (2022) Exercise duration differentially effects age-related Neuroinflammation and hippocampal neurogenesis. Neuroscience 490:275–286. https://doi.org/10.1016/j.neuroscience.2022.03.022
96. De Miguel Z, Khoury N, Betley MJ, Lehallier B, Willoughby D, Olsson N et al (2021) Exercise plasma boosts memory and dampens brain inflammation via clusterin. Nature 600(7889):494–499. https://doi.org/10.1038/s41586-021-04183-x
97. Vukovic J, Colditz MJ, Blackmore DG, Ruitenberg MJ, Bartlett PF (2012) Microglia modulate hippocampal neural precursor activity in response to exercise and aging. J Neurosci 32(19):6435–6443. https://doi.org/10.1523/JNEUROSCI.5925-11.2012
98. Koehl M, Meerlo P, Gonzales D, Rontal A, Turek FW, Abrous DN (2008) Exercise-induced promotion of hippocampal cell proliferation requires beta-endorphin. FASEB J 22(7):2253–2262. https://doi.org/10.1096/fj.07-099101
99. Kondo M, Nakamura Y, Ishida Y, Shimada S (2015) The 5-HT3 receptor is essential for exercise-induced hippocampal neurogenesis and antidepressant effects. Mol Psychiatry 20(11):1428–1437. https://doi.org/10.1038/mp.2014.153
100. Pons-Espinal M, Gasperini C, Marzi MJ, Braccia C, Armirotti A, Potzsch A et al (2019) MiR-135a-5p is critical for exercise-induced adult neurogenesis. Stem Cell Reports 12(6):1298–1312. https://doi.org/10.1016/j.stemcr.2019.04.020
101. Gobeske KT, Das S, Bonaguidi MA, Weiss C, Radulovic J, Disterhoft JF et al (2009) BMP signaling mediates effects of exercise on hippocampal neurogenesis and cognition in mice. PLoS One 4(10):e7506. https://doi.org/10.1371/journal.pone.0007506
102. Chow LS, Gerszten RE, Taylor JM, Pedersen BK, van Praag H, Trappe S et al (2022) Exerkines in health, resilience and disease. Nat Rev Endocrinol 18(5):273–289. https://doi.org/10.1038/s41574-022-00641-2
103. Muller P, Duderstadt Y, Lessmann V, Muller NG (2020) Lactate and BDNF: key mediators of exercise induced neuroplasticity? J Clin Med 9(4). https://doi.org/10.3390/jcm9041136
104. Sadier NS, El Hajjar F, Al Sabouri AAK, Abou-Abbas L, Siomava N, Almutary AG et al (2024) Irisin: an unveiled bridge between physical exercise and a healthy brain. Life Sci 339:122393. https://doi.org/10.1016/j.lfs.2023.122393

105. Moon HY, Becke A, Berron D, Becker B, Sah N, Benoni G et al (2016) Running-induced systemic Cathepsin B secretion is associated with memory function. Cell Metab 24(2):332–340. https://doi.org/10.1016/j.cmet.2016.05.025
106. Yau SY, Li A, Hoo RL, Ching YP, Christie BR, Lee TM et al (2014) Physical exercise-induced hippocampal neurogenesis and antidepressant effects are mediated by the adipocyte hormone adiponectin. Proc Natl Acad Sci USA 111(44):15810–15815. https://doi.org/10.1073/pnas.1415219111
107. Llorens-Martín M, Torres-Alemán I, Trejo JL (2008) Growth factors as mediators of exercise actions on the brain. NeuroMolecular Med 10(2):99–107. https://doi.org/10.1007/s12017-008-8026-1
108. Phillips C, Baktir MA, Srivatsan M, Salehi A (2014) Neuroprotective effects of physical activity on the brain: a closer look at trophic factor signaling. Front Cell Neurosci 8:170. https://doi.org/10.3389/fncel.2014.00170
109. Hwang D, Kim J, Kyun S, Jang I, Kim T, Park HY et al (2023) Exogenous lactate augments exercise-induced improvement in memory but not in hippocampal neurogenesis. Sci Rep 13(1):5838. https://doi.org/10.1038/s41598-023-33017-1
110. Cefis M, Chaney R, Wirtz J, Meloux A, Quirie A, Leger C et al (2023) Molecular mechanisms underlying physical exercise-induced brain BDNF overproduction. Front Mol Neurosci 16:1275924. https://doi.org/10.3389/fnmol.2023.1275924
111. Pan W, Banks WA, Fasold MB, Bluth J, Kastin AJ (1998) Transport of brain-derived neurotrophic factor across the blood-brain barrier. Neuropharmacology 37(12):1553–1561. https://doi.org/10.1016/s0028-3908(98)00141-5
112. Renteria I, Garcia-Suarez PC, Fry AC, Moncada-Jimenez J, Machado-Parra JP, Antunes BM et al (2022) The molecular effects of BDNF synthesis on skeletal muscle: a mini-review. Front Physiol 13:934714. https://doi.org/10.3389/fphys.2022.934714
113. Soya H, Nakamura T, Deocaris CC, Kimpara A, Iimura M, Fujikawa T et al (2007) BDNF induction with mild exercise in the rat hippocampus. Biochem Biophys Res Commun 358(4):961–967. https://doi.org/10.1016/j.bbrc.2007.04.173
114. Wrann CD, White JP, Salogiannnis J, Laznik-Bogoslavski D, Wu J, Ma D et al (2013) Exercise induces hippocampal BDNF through a PGC-1alpha/FNDC5 pathway. Cell Metab 18(5):649–659. https://doi.org/10.1016/j.cmet.2013.09.008
115. El Hayek L, Khalifeh M, Zibara V, Abi Assaad R, Emmanuel N, Karnib N et al (2019) Lactate mediates the effects of exercise on learning and memory through SIRT1-dependent activation of hippocampal brain-derived neurotrophic factor (BDNF). J Neurosci 39(13):2369–2382. https://doi.org/10.1523/JNEUROSCI.1661-18.2019
116. Lev-Vachnish Y, Cadury S, Rotter-Maskowitz A, Feldman N, Roichman A, Illouz T et al (2019) L-lactate promotes adult hippocampal neurogenesis. Front Neurosci 13:403. https://doi.org/10.3389/fnins.2019.00403
117. Choi SH, Bylykbashi E, Chatila ZK, Lee SW, Pulli B, Clemenson GD et al (2018) Combined adult neurogenesis and BDNF mimic exercise effects on cognition in an Alzheimer's mouse model. Science 361(6406). https://doi.org/10.1126/science.aan8821
118. Ji M, Cho C, Lee S (2024) Acute effect of exercise intensity on circulating FGF-21, FSTL-1, cathepsin B, and BDNF in young men. J Exerc Sci Fit 22(1):51–58. https://doi.org/10.1016/j.jesf.2023.11.002
119. Xue X, Liu B, Hu J, Bian X, Lou S (2022) The potential mechanisms of lactate in mediating exercise-enhanced cognitive function: a dual role as an energy supply substrate and a signaling molecule. Nutr Metab (Lond) 19(1):52. https://doi.org/10.1186/s12986-022-00687-z
120. Park J, Kim J, Mikami T (2021) Exercise-induced lactate release mediates mitochondrial biogenesis in the hippocampus of mice via Monocarboxylate transporters. Front Physiol 12:736905. https://doi.org/10.3389/fphys.2021.736905
121. Laron Z (2001) Insulin-like growth factor 1 (IGF-1): a growth hormone. Mol Pathol 54(5):311–316. https://doi.org/10.1136/mp.54.5.311

122. Yakar S, Liu JL, Stannard B, Butler A, Accili D, Sauer B et al (1999) Normal growth and development in the absence of hepatic insulin-like growth factor I. Proc Natl Acad Sci USA 96(13):7324–7329. https://doi.org/10.1073/pnas.96.13.7324
123. Pan W, Kastin AJ (2000) Interactions of IGF-1 with the blood-brain barrier in vivo and in situ. Neuroendocrinology 72(3):171–178. https://doi.org/10.1159/000054584
124. Nishijima T, Piriz J, Duflot S, Fernandez AM, Gaitan G, Gomez-Pinedo U et al (2010) Neuronal activity drives localized blood-brain-barrier transport of serum insulin-like growth factor-I into the CNS. Neuron 67(5):834–846. https://doi.org/10.1016/j.neuron.2010.08.007
125. Aberg MA, Aberg ND, Hedbacker H, Oscarsson J, Eriksson PS (2000) Peripheral infusion of IGF-I selectively induces neurogenesis in the adult rat hippocampus. J Neurosci 20(8):2896–2903. https://doi.org/10.1523/JNEUROSCI.20-08-02896.2000
126. O'Kusky J, Ye P (2012) Neurodevelopmental effects of insulin-like growth factor signaling. Front Neuroendocrinol 33(3):230–251. https://doi.org/10.1016/j.yfrne.2012.06.002
127. Kraemer RR, Durand RJ, Acevedo EO, Johnson LG, Kraemer GR, Hebert EP et al (2004) Rigorous running increases growth hormone and insulin-like growth factor-I without altering ghrelin. Exp Biol Med (Maywood) 229(3):240–246. https://doi.org/10.1177/153537020422900304
128. Kraemer RR, Kraemer BR (2023) The effects of peripheral hormone responses to exercise on adult hippocampal neurogenesis. Front Endocrinol (Lausanne) 14:1202349. https://doi.org/10.3389/fendo.2023.1202349
129. Zouhal H, Jayavel A, Parasuraman K, Hayes LD, Tourny C, Rhibi F et al (2022) Effects of exercise training on anabolic and catabolic hormones with advanced age: a systematic review. Sports Med 52(6):1353–1368. https://doi.org/10.1007/s40279-021-01612-9
130. Carro E, Nunez A, Busiguina S, Torres-Aleman I (2000) Circulating insulin-like growth factor I mediates effects of exercise on the brain. J Neurosci 20(8):2926–2933. https://doi.org/10.1523/JNEUROSCI.20-08-02926.2000
131. Nishida Y, Matsubara T, Tobina T, Shindo M, Tokuyama K, Tanaka K et al (2010) Effect of low-intensity aerobic exercise on insulin-like growth factor-I and insulin-like growth factor-binding proteins in healthy men. Int J Endocrinol 2010:1. https://doi.org/10.1155/2010/452820
132. Ding Q, Vaynman S, Akhavan M, Ying Z, Gomez-Pinilla F (2006) Insulin-like growth factor I interfaces with brain-derived neurotrophic factor-mediated synaptic plasticity to modulate aspects of exercise-induced cognitive function. Neuroscience 140(3):823–833. https://doi.org/10.1016/j.neuroscience.2006.02.084
133. Chen MJ, Russo-Neustadt AA (2007) Running exercise- and antidepressant-induced increases in growth and survival-associated signaling molecules are IGF-dependent. Growth Factors 25(2):118–131. https://doi.org/10.1080/08977190701602329
134. Palmer TD, Willhoite AR, Gage FH (2000) Vascular niche for adult hippocampal neurogenesis. J Comp Neurol 425(4):479–494
135. Louissaint A, Rao S, Leventhal C, Goldman SA (2002) Coordinated interaction of neurogenesis and angiogenesis in the adult songbird brain. Neuron 34(6):945–960. https://doi.org/10.1016/S0896-6273(02)00722-5
136. Jin K, Zhu Y, Sun Y, Mao XO, Xie L, Greenberg DA (2002) Vascular endothelial growth factor (VEGF) stimulates neurogenesis in vitro and in vivo. Proc Natl Acad Sci USA 99(18):11946–11950. https://doi.org/10.1073/pnas.182296499
137. Rich B, Scadeng M, Yamaguchi M, Wagner PD, Breen EC (2017) Skeletal myofiber vascular endothelial growth factor is required for the exercise training-induced increase in dentate gyrus neuronal precursor cells. J Physiol 595(17):5931–5943. https://doi.org/10.1113/JP273994
138. Tang K, Xia FC, Wagner PD, Breen EC (2010) Exercise-induced VEGF transcriptional activation in brain, lung and skeletal muscle. Respir Physiol Neurobiol 170(1):16–22. https://doi.org/10.1016/j.resp.2009.10.007

139. Elfving B, Christensen T, Ratner C, Wienecke J, Klein AB (2013) Transient activation of mTOR following forced treadmill exercise in rats. Synapse 67(9):620–625. https://doi.org/10.1002/syn.21668
140. Cotman CW, Berchtold NC, Christie LA (2007) Exercise builds brain health: key roles of growth factor cascades and inflammation. Trends Neurosci 30(9):464–472. https://doi.org/10.1016/j.tins.2007.06.011
141. Karakilic A, Yuksel O, Kizildag S, Hosgorler F, Topcugil B, Ilgin R et al (2021) Regular aerobic exercise increased VEGF levels in both soleus and gastrocnemius muscles correlated with hippocampal learning and VEGF levels. Acta Neurobiol Exp (Wars) 81(1):1–9. https://doi.org/10.21307/ane-2021-001
142. De Kloet ER, Vreugdenhil E, Oitzl MS, Joels M (1998) Brain corticosteroid receptor balance in health and disease. Endocr Rev 19(3):269–301. https://doi.org/10.1210/edrv.19.3.0331
143. Reul JM, de Kloet ER (1985) Two receptor systems for corticosterone in rat brain: microdistribution and differential occupation. Endocrinology 117(6):2505–2511. https://doi.org/10.1210/endo-117-6-2505
144. Arriza JL, Weinberger C, Cerelli G, Glaser TM, Handelin BL, Housman DE et al (1987) Cloning of human mineralocorticoid receptor complementary DNA: structural and functional kinship with the glucocorticoid receptor. Science 237(4812):268–275. https://doi.org/10.1126/science.3037703
145. McEwen BS (1999) Stress and hippocampal plasticity. Annu Rev Neurosci 22:105–122. https://doi.org/10.1146/annurev.neuro.22.1.105
146. Kirby ED, Muroy SE, Sun WG, Covarrubias D, Leong MJ, Barchas LA et al (2013) Acute stress enhances adult rat hippocampal neurogenesis and activation of newborn neurons via secreted astrocytic FGF2. elife 2:e00362. https://doi.org/10.7554/eLife.00362
147. Sloviter RS, Valiquette G, Abrams GM, Ronk EC, Sollas AL, Paul LA et al (1989) Selective loss of hippocampal granule cells in the mature rat brain after adrenalectomy. Science 243(4890):535–538. https://doi.org/10.1126/science.2911756
148. Saaltink DJ, Vreugdenhil E (2014) Stress, glucocorticoid receptors, and adult neurogenesis: a balance between excitation and inhibition? Cell Mol Life Sci 71(13):2499–2515. https://doi.org/10.1007/s00018-014-1568-5
149. Tzu-Feng Wang TF, Tsai SF, Zhao ZW, Shih M, Wang CY, Yang TT et al (2021) Exercise-induced increases of corticosterone contribute to exercise-enhanced adult hippocampal neurogenesis in mice. Chin J Physiol 64(4):186–193. https://doi.org/10.4103/cjp.cjp_39_21
150. Naylor AS (2005) Extended voluntary running inhibits exercise-induced adult hippocampal progenitor proliferation in the spontaneously hypertensive rat. J Neurophysiol 93(5):2406–2414. https://doi.org/10.1152/jn.01085.2004
151. Blankers SA, Galea LAM (2021) Androgens and adult neurogenesis in the hippocampus. Androg Clin Res Ther 2(1):203–215. https://doi.org/10.1089/andro.2021.0016
152. Hojo Y, Kawato S (2018) Neurosteroids in adult hippocampus of male and female rodents: biosynthesis and actions of sex steroids. Front Endocrinol 9(APR):1–8. https://doi.org/10.3389/fendo.2018.00183
153. Tong L, Shen H, Perreau VM, Balazs R, Cotman CW (2001) Effects of exercise on gene-expression profile in the rat hippocampus. Neurobiol Dis 8(6):1046–1056. https://doi.org/10.1006/nbdi.2001.0427
154. Molteni R, Ying Z, Gómez-Pinilla F (2002) Differential effects of acute and chronic exercise on plasticity-related genes in the rat hippocampus revealed by microarray. Eur J Neurosci 16(6):1107–1116. https://doi.org/10.1046/j.1460-9568.2002.02158.x
155. Ding Q, Vaynman S, Souda P, Whitelegge JP, Gomez-Pinilla F (2006) Exercise affects energy metabolism and neural plasticity-related proteins in the hippocampus as revealed by proteomic analysis. Eur J Neurosci 24(5):1265–1276. https://doi.org/10.1111/j.1460-9568.2006.05026.x
156. Guerrieri D, van Praag H (2015) Exercise-mimetic AICAR transiently benefits brain function. Oncotarget 6(21):18293–18313. https://doi.org/10.18632/oncotarget.4715

157. Gregoire CA, Tobin S, Goldenstein BL, Samarut E, Leclerc A, Aumont A et al (2018) RNA-sequencing reveals unique transcriptional signatures of running and running-independent environmental enrichment in the adult mouse dentate gyrus. Front Mol Neurosci 11:126. https://doi.org/10.3389/fnmol.2018.00126
158. Frey S, Schieweck R, Forne I, Imhof A, Straub T, Popper B et al (2020) Physical activity dynamically regulates the hippocampal proteome along the Dorso-ventral Axis. Int J Mol Sci 21(10). https://doi.org/10.3390/ijms21103501
159. Gao Y, Shen M, Gonzalez JC, Dong Q, Kannan S, Hoang JT et al (2020) RGS6 mediates effects of voluntary running on adult hippocampal neurogenesis. Cell Rep 32(5):107997. https://doi.org/10.1016/j.celrep.2020.107997
160. Stranahan AM, Lee K, Becker KG, Zhang Y, Maudsley S, Martin B et al (2010) Hippocampal gene expression patterns underlying the enhancement of memory by running in aged mice. Neurobiol Aging 31(11):1937–1949. https://doi.org/10.1016/j.neurobiolaging.2008.10.016
161. Kohman RA, Rodriguez-Zas SL, Southey BR, Kelley KW, Dantzer R, Rhodes JS (2011) Voluntary wheel running reverses age-induced changes in hippocampal gene expression. PLoS One 6(8). https://doi.org/10.1371/journal.pone.0022654
162. Alvarez-López MJ, Castro-Freire M et al (2013) Long-term exercise modulates hippocampal gene expression in senescent female mice. J Alzheimer's Dis 33(4):1177–1190. https://doi.org/10.3233/JAD-121264
163. Zhang SS, Zhu L, Peng Y, Zhang L, Chao FL, Jiang L et al (2022) Long-term running exercise improves cognitive function and promotes microglial glucose metabolism and morphological plasticity in the hippocampus of APP/PS1 mice. J Neuroinflammation 19(1):34. https://doi.org/10.1186/s12974-022-02401-5
164. Yuk KH, Lee SM, Bae WR, Park JY, Woo SW, Song P et al (2023) Distinct effect of exercise modes on mood-related behavior in mice. Biochem Biophys Res Commun 646:36–43. https://doi.org/10.1016/j.bbrc.2023.01.047
165. Chen WQ, Diao WF, Viidik A, Skalicky M, Höger H, Lubec G (2008) Modulation of the hippocampal protein machinery in voluntary and treadmill exercising rats. Biochim et Biophys Acta Prot Proteomics 1784(3):555–562. https://doi.org/10.1016/j.bbapap.2008.01.002
166. Lehmann ML, Brachman RA, Martinowich K, Schloesser RJ, Herkenham M (2013) Glucocorticoids orchestrate divergent effects on mood through adult neurogenesis. J Neurosci 33(7):2961–2972. https://doi.org/10.1523/JNEUROSCI.3878-12.2013
167. Yu H, Zhang C, Xia J, Xu B (2021) Treadmill exercise ameliorates adult hippocampal neurogenesis possibly by adjusting the app proteolytic pathway in app/ps1 transgenic mice. Int J Mol Sci 22(17). https://doi.org/10.3390/ijms22179570
168. Kim BK, Shin MS, Kim CJ, Baek SB, Ko YC, Kim YP (2014) Treadmill exercise improves short-term memory by enhancing neurogenesis in amyloid beta-induced Alzheimer disease rats. J Exerc Rehabil 10(1):2–8. https://doi.org/10.12965/jer.140086
169. Okamoto M, Mizuuchi D, Omura K, Lee M, Oharazawa A, Yook JS et al (2021) High-intensity intermittent training enhances spatial memory and hippocampal neurogenesis associated with BDNF signaling in rats. Cereb Cortex 31(9):4386–4397. https://doi.org/10.1093/cercor/bhab093
170. Lee MC, Okamoto M, Liu YF, Inoue K, Matsui T, Nogami H et al (1985) Voluntary resistance running with short distance enhances spatial memory related to hippocampal BDNF signaling. J Appl Physiol 113(8):1260–1266. https://doi.org/10.1152/japplphysiol.00869.2012
171. Lee S-W, Kim Y-S, Jun T-W, Seo J-H, Kim K, Shin M-S et al (2013) The impact of duration of one bout treadmill exercise on cell proliferation and central fatigue in rats. J Exerc Rehabil 9(5):463–469. https://doi.org/10.12965/jer.130069
172. Mukuda T, Koyama Y, Hamasaki S, Kaidoh T, Furukawa Y (2014) Systemic angiotensin II and exercise-induced neurogenesis in adult rat hippocampus. Brain Res 1588:92–103. https://doi.org/10.1016/j.brainres.2014.09.019

173. Khodadadegan MA, Negah SS, Saheb M, Gholami J, Arabi MH, Hajali V (2021) Combination effect of exercise and environmental enrichment on cognitive functions and hippocampal neurogenesis markers of rat. Neuroreport 32:1234–1240. https://doi.org/10.1097/WNR.0000000000001713
174. Avila-Gamiz F, Perez-Cano AM, Perez-Berlanga JM, Mullor-Vigo RM, Zambrana-Infantes EN, Santin LJ et al (2023) Sequential treadmill exercise and cognitive training synergistically increase adult hippocampal neurogenesis in mice. Physiol Behav 266:114184. https://doi.org/10.1016/j.physbeh.2023.114184

Minimum Exercise Model in Humans Based on Oxygen Uptake and Physiological Parameters

Morimasa Kato, Hyukki Chang, and Hideaki Soya

Abstract Understanding exercise intensity is essential for optimizing training outcomes and minimizing health risks. This chapter introduces key physiological and subjective parameters used to assess exercise intensity, including heart rate reserve (HRR), oxygen uptake reserve (VO2R), maximal oxygen consumption (VO2max), and ratings of perceived exertion (RPE). Standardized classifications from organizations such as the American College of Sports Medicine (ACSM) are presented, alongside practical methods like the Talk Test for field applications. Incremental exercise testing is highlighted for identifying physiological thresholds, including lactate and ventilatory thresholds, which serve as critical markers for personalized training. Additionally, recent advances in neuroimaging—including electroencephalography (EEG), near-infrared spectroscopy (NIRS), and functional magnetic resonance imaging (fMRI)—are reviewed to explore how different exercise intensities affect brain activity. Evidence suggests that even low to moderate-intensity exercise can positively influence cognitive function and cerebral blood flow. The integration of wearable technologies has further enabled real-time monitoring of both physiological and neurocognitive responses. Overall, this chapter underscores the impor-

M. Kato (✉)
Department of Health and Nutrition, Yamagata Prefectural Yonezawa University of Nutrition Sciences, Yamagata, Japan
e-mail: m-kato@yone.ac.jp

H. Chang
Department of Sport and Exercise Science, Seoul Women's University, Seoul, Republic of Korea

H. Soya
Laboratory of Exercise Biochemistry and Neuroendocrinology, Institute of Health and Sport Sciences, University of Tsukuba, Tsukuba, Ibaraki, Japan

Sport Neuroscience Division, Kokoro Division, Advanced Research Initiative for Human High Performance (ARIHHP), Institute of Health and Sport Sciences, University of Tsukuba, Tsukuba, Ibaraki, Japan

Center for Cybernics Research, University of Tsukuba, Tsukuba, Ibaraki, Japan

H. Soya (ed.), *Exercise Brain Stimulation for Cognitive Function and Mental Health*, Advances in Neurobiology 44, https://doi.org/10.1007/978-981-95-0066-6_4

tance of individualized, evidence-based approaches in exercise prescription and highlights emerging methods for linking exercise intensity with brain function.

Keywords Exercise intensity · Heart Rate · Oxygen uptake · Ratings of perceived Exertion (RPE) · Lactate threshold (LT) · Ventilatory threshold (VT)

1 Physiological Parameters for Assessing Exercise Intensity

To understand how exercise affects the human body, it is necessary to accurately assess the stimuli that provoke such responses. Exercise is composed of several key variables, including type, duration, frequency, and intensity. Among these, exercise intensity plays a particularly crucial role, as it reflects the level of effort exerted by the individual. It is commonly expressed through various physiological parameters, including the amount of energy expended per unit of time during physical activity. Leading organizations such as the American College of Sports Medicine (ACSM) [1], the Canadian Society for Exercise Physiology (CSEP) [2], and Exercise & Sports Science Australia (ESSA) [3] have established standardized classifications of exercise intensity. These classifications utilize a range of physiological indicators, including heart rate reserve (HRR), oxygen uptake reserve (VO2R), maximal heart rate (HRmax), maximal oxygen uptake (VO_{2max}), rating of perceived exertion (RPE), and metabolic equivalents (METs). The ACSM categorizes exercise intensity into five levels—very light, light, moderate, vigorous, and near-maximal—based on both relative and absolute values (Tables 1a and 1b).

These classifications are detailed in Tables 1a and 1b, which outline both relative and absolute intensity levels across various age groups using physiological and perceptual measures.

Table 1a Classification of relative exercise intensity

	Relative intensity			
Intensity	HRR or VO_2R (%)	HRmax (%)	VO_2max (%)	Perceived exertion (Rating on 6–20 RPE score)
Very light	<30	<57	<37	Very light (RPE <9)
Light	30–39	57–63	37–45	Very light to fairly light (RPE 9-11)
Moderate	40–59	64–76	46–63	Fairly light to somewhat hard (RPE 12-13)
Vigorous	60–89	77–95	64–90	Somewhat hard to very hard (RPE 14-17)
Near-maximal to maximal	≥ 90	≥ 96	≥ 91	≥ Very hard (RPE ≥ 18)

Note: *HRR* heart rate reserve, *VO2R* oxygen uptake reserve, *HRmax* maximal heart rate, *V̇O2max* maximal oxygen uptake

Table 1b Classification of absolute exercise intensity by age group (in METs)

	Absolute intensity	Absolute intensity (MET) by age		
		Young		Older
Intensity	METs	(20–39 years)	Middle age (40–64 years)	(≥ 65 years)
Very light	<2.0	<2.4	<2.0	<1.6
Light	2.0–2.9	2.4–4.7	2.0–3.9	1.6–3.1
Moderate	3.0–5.9	4.8–7.1	4.0–5.9	3.2–4.7
Vigorous	6.0–8.7	7.2–10.1	6.0–8.4	4.8–6.7
Near-maximal to maximal	≥ 8.8	≥ 10.2	≥ 8.5	≥ 6.8

Note: *MET* metabolic equivalent of task. MET values vary by age to reflect differences in functional capacity

1.1 Individualized Indicators of Relative Exercise Intensity

Among the various indicators used to assess relative exercise intensity on an individual basis, heart rate is generally considered the most accessible and practical. There are two common methods for estimating exercise intensity using heart rate: the heart rate reserve (HRR) method, also known as the Karvonen method, and the maximal heart rate (HRmax) method. The HRR method estimates target exercise intensity by calculating heart rate reserve using resting heart rate (HRrest) and age- or sex-estimated maximal heart rate (HRmax), and then deriving a target heart rate range. In contrast, the HRmax method estimates exercise intensity as a percentage of the maximal heart rate alone. While heart rate is widely used in both sports settings and laboratory environments, it may not be suitable for individuals with arrhythmias or those using pacemakers [4, 5]. When estimating maximal heart rate, the formula "220 – age" is most used (as seen in the Karvonen method), but alternative equations also exist, such as "226 – age" and "208 – (0.7 × age)". The choice of formula should consider factors such as age, sex, training status, and testing protocol. Additionally, although not identical, pulse rate is closely aligned with heart rate. As such, pulse readings obtained via wearable devices are increasingly being used to estimate exercise intensity, based on these same principles (Table 2).

In addition to heart rate, oxygen uptake during exercise—measured via respiratory gas analysis—is another key indicator for assessing relative exercise intensity. Two commonly used methods are the percentage of oxygen uptake reserve (%VO2R) method and the %VO2max method.

The %VO2R method is calculated using the difference between resting and maximal oxygen uptake, using the formula: %VO2R = (submaximal VO2 – resting VO2)/VO2 reserve × 100. Research indicates that exercise intensity calculated using the %VO2R method closely matches that estimated using the HRR method. Meanwhile, the %VO2max method expresses exercise intensity as a percentage of

Table 2 Commonly used equations for predicting maximal heart rate

Author	Equation	Population
Fox et al. [6]	220—Age	Small group of men and women
Warburon et al. [2]	226—Age	For women
Tanaka et al. [7]	208—(0.7 × age)	Healthy men and women

maximal oxygen uptake. Both methods are valuable for tailoring exercise prescriptions to individual physiological capacities.

Adding brief explanations about VO2max and VO2peak. VO2max (maximal oxygen uptake) refers to the maximum amount of oxygen the body can utilize during intense, whole-body exercise. It is considered the gold standard measure of cardiorespiratory fitness and reflects the aerobic physical capacity of an individual. VO2max is typically measured in milliliters of oxygen consumed per minute per kilogram of body weight (ml/kg/min). VO2peak, on the other hand, refers to the highest oxygen uptake value recorded during an exercise test, regardless of whether a true physiological maximum has been achieved. It is often used in clinical or submaximal testing situations where a subject may not reach their actual VO2max due to factors such as early fatigue, motivational limitations, and medical conditions. The main difference between VO2max and VO2peak lies in whether a true physiological maximum was reached. VO2max requires specific criteria to be met, such as a plateau in oxygen uptake despite increasing workload, while VO2peak is simply the highest value observed during the test.

1.2 Subjective Indicators of Exercise Intensity: The Ratings of Perceived Exertion (RPE)

In addition to physiological parameters, a widely used subjective indicator for assessing exercise intensity is the ratings of perceived exertion (RPE) [8]. Developed by Swedish researcher Gunnar Borg, the RPE is a psychophysiological scale that quantifies how hard an individual feels they are working during physical activity. The original Borg RPE Scale is a 15-point category scale ranging from 6 to 20, with each numerical value roughly corresponding to heart rate when multiplied by 10—a design intended to approximate physiological effort. For example, an RPE of 13 would suggest a heart rate around 130 beats per minute. Later, to better capture the proportional nature of perceived exertion, Borg introduced the Category-Ratio 10 (CR10) Scale, which ranges from 0 to 10 [9]. This scale allows for finer gradations in perceived intensity, including the use of decimal values (e.g., 2.5 or 7.3) to reflect subtle differences. Importantly, the CR10 scale also accommodates values exceeding 10 when individuals experience exertion levels beyond their previously known maximum. This feature helps avoid ceiling effects in high-intensity conditions, such as maximal effort or pain perception.

Table 3 provides a comparative overview of the Borg Rating of Perceived Exertion (RPE) scale and the Category-Ratio 10 (CR10) scale, both widely used

Table 3 Comparison of the Borg RPE Scale and the CR10 Scale Borg RPE scale descriptor

Borg RPE scale		Borg CR10 scale	
6	No exertion at all	0	Nothing at all
7	Extremely light	0.3	
8		0.5	Extremely weak
9	Very light	1	Very weak
10		1.5	
11	Light	2	Weak
12		2.5	
13	Somewhat hard	3	Moderate
14		4	
15	Hard (heavy)	5	Strong
16		6	
17	Very hard	7	Very strong
18		8	
19	Extremely hard	9	
20	Maximal exertion	10	Extremely strong

tools for assessing subjective exercise intensity. The RPE scale ranges from 6 to 20, originally designed so that multiplying the rating by 10 approximates heart rate (e.g., an RPE of 13 suggests a heart rate around 130 bpm). This scale is particularly effective for tracking exertion during aerobic activities such as running or cycling. The CR10 scale, on the other hand, ranges from 0 to 10 and was developed to better reflect the proportional nature of perceived exertion. Unlike the 6–20 scale, the CR10 allows for more precise reporting, including decimal values (e.g., 2.5 or 7.3), and even values greater than 10 in cases where perceived effort exceeds prior maximal experience. This makes the CR10 especially useful in clinical, high-performance, or pain-related assessments, where subtle changes in perceived intensity are important.

By comparing the two scales side-by-side, the table demonstrates how similar descriptive terms (e.g., "very light," "somewhat hard," "extremely hard") are represented across different numerical systems. This facilitates flexibility in selecting a scale depending on the population, setting, or purpose of the evaluation. Ultimately, both scales offer valuable insight into an individual's perceived effort and can serve as practical, non-invasive tools to guide exercise intensity in both clinical and athletic environments.

2 Physiological Parameters for Assessing Exercise Intensity in Incremental Load Tests

One widely used approach for determining exercise intensity involves conducting *incremental* load tests using devices such as treadmills or cycle ergometers. During these tests, physiological thresholds such as the lactate threshold (LT) and the

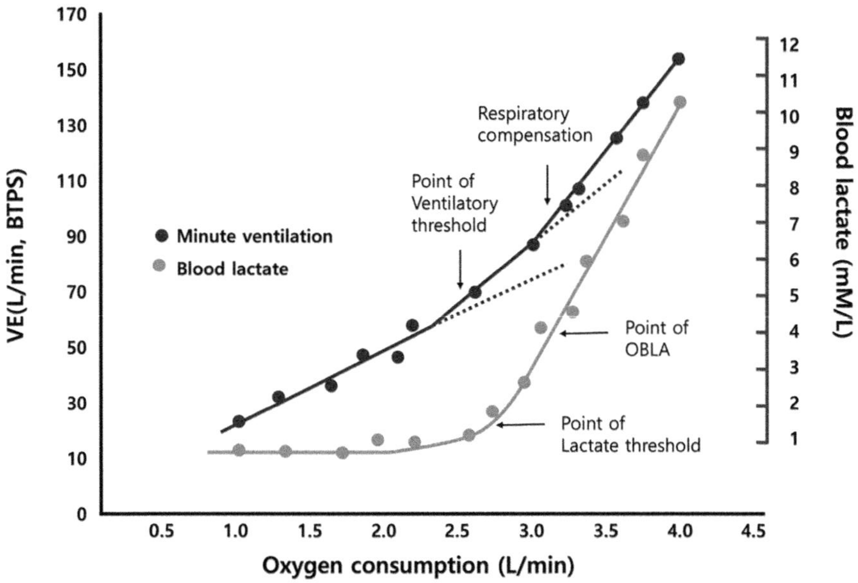

Fig. 1 Lactate and ventilatory thresholds

ventilatory threshold (VT) are assessed by measuring blood lactate levels and ventilatory responses as exercise intensity progressively increases (see Fig. 1). These thresholds reflect substrate metabolism in response to energy demand in the working muscles during exercise. In the early stages of exercise, energy production primarily relies on the aerobic system, utilizing free fatty acids and other lipids to synthesize adenosine triphosphate (ATP). However, as intensity increases, the aerobic system alone becomes insufficient, prompting the recruitment of the anaerobic glycolytic system for additional ATP supply. As glycolysis accelerates, lactate production in the active muscles exceeds its clearance, resulting in lactate accumulation in the blood and the onset of metabolic acidosis. In response, the body increases carbon dioxide (CO2) output through ventilation, which leads to a non-linear rise in both ventilation rate and CO2 exhalation. These physiological responses mark the ventilatory and lactate thresholds. These thresholds are closely linked to endurance performance and are known to shift with training and aging [10, 11]. Setting exercise intensity based on individual physiological responses not only helps minimize the risk of injury and fatigue but also enables the design of personalized and optimal training plans. Such an approach can significantly enhance exercise prescription accuracy, maximize athletic adaptation, and improve overall performance [12, 13].

2.1 *Talk Test: A Subjective Alternative for Assessing Ventilatory Thresholds*

In addition to the RPE scale, another self-reported subjective method used to estimate exercise intensity is the Talk Test [14, 15], which has been applied in contexts like the lactate threshold and ventilatory threshold. The Talk Test involves having individuals read aloud a passage during exercise, with the degree of breathlessness while speaking serving as an indirect indicator of exercise intensity. This method is based on the observation that speech production is closely tied to ventilation and oxygen consumption [16]. Comfortable speech requires controlled breathing frequency and ventilation volume. As such, the point at which speech becomes noticeably difficult or disrupted is thought to correspond with the ventilatory threshold. Research has increasingly explored this association, suggesting that the Talk Test may serve as a non-invasive and practical surrogate for physiological markers such as the ventilatory threshold, which typically require specialized equipment and laboratory settings to measure [14, 17, 18]. However, despite its practicality and accessibility, the Talk Test has limitations—most notably, its lack of precision for fine-grained intensity control. It is therefore best suited for general guidance in field settings or with populations where more invasive or technical methods are not feasible.

3 Exercise Intensity and Brain Function

As with physiological responses such as respiratory and circulatory changes during exercise, examining how brain function reflects exercise intensity is of growing interest. Various neuroimaging techniques are available to assess human brain activity, including electroencephalography (EEG), magnetoencephalography (MEG), functional magnetic resonance imaging (fMRI), near-infrared spectroscopy (NIRS), positron emission tomography (PET), and transcranial Doppler (TCD).

Each method has its own advantages and limitations in terms of equipment constraints, temporal resolution, and amount of information obtained, requiring researchers to choose the appropriate tool depending on the experimental setting. Techniques such as PET, fMRI, and MEG can analyze structural plasticity, neural responses to stimuli, functional brain networks, regional or global cerebral blood flow and metabolism, as well as neurotransmitter release and receptor binding, offering detailed insights into brain activity during exercise [19–21]. PET allows the measurement of metabolic dynamics during exercise but has drawbacks such as radiation exposure and high cost. Although fMRI and MEG are non-invasive, they must be conducted in magnetically shielded environments, which limits compatibility with traditional metallic exercise equipment. However, recent advancements such as MRI-compatible cycle ergometers have enabled brain activity measurements during exercise within the MRI suite. If stable protocols are established, this approach could allow the investigation of brain responses to varying exercise intensities (see Fig. 2).

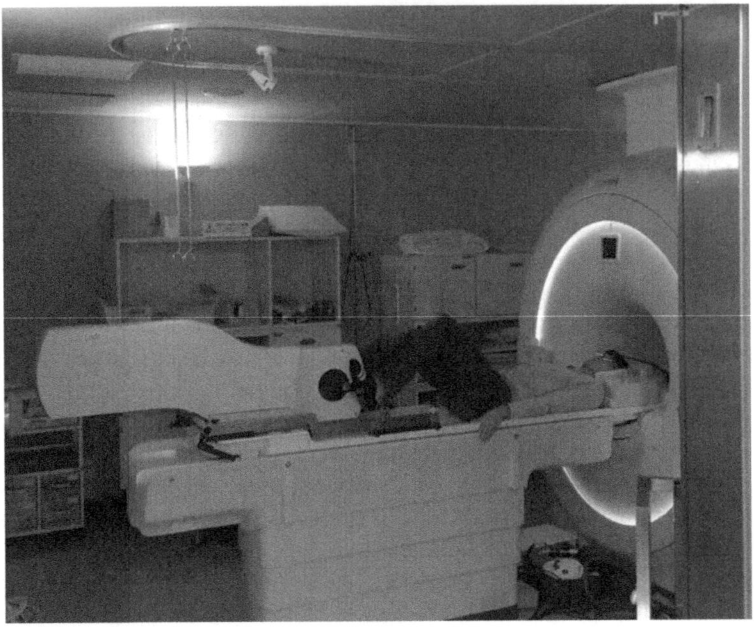

Fig. 2 Non-magnetic MRI-compatible ergometer

3.1 Portable Brain Monitoring: EEG and NIRS

Unlike the above methods, EEG and NIRS are portable, non-invasive, and relatively resistant to movement artifacts. While EEG was once considered unsuitable for use during exercise due to motion-induced noise, recent technological advances have made it highly effective for recording dynamic brain activity during physical tasks [22, 23]. Studies using EEG—including those analyzing event-related potentials (ERPs)—have shown that exercise intensity modulates cortical activity and activation patterns. However, findings remain inconsistent: some studies suggest a dose–response relationship with intensity, while others propose an inverted U-shaped curve, where moderate intensity yields optimal brain activation [24–26]. These discrepancies may be due to variables such as hormonal responses (e.g., adrenaline), participant training status, and experimental design [24, 26, 27]. NIRS uses near-infrared light, which penetrates the skull and measures changes in oxygenated and deoxygenated hemoglobin in the cortical surface. Based on the principle of neurovascular coupling, in which increased neural activity leads to elevated cerebral blood flow via glial modulation, NIRS provides a useful proxy for brain activity. Although it has lower temporal resolution than EEG and MEG, NIRS enables real-time assessment of brain function during exercise on a second-by-second basis.

Interpretation of NIRS data during exercise must consider physiological confounders, such as changes in arterial CO2 (PaCO2) and arterial O2 (PaO2) levels, which can affect vascular tone. Nevertheless, research suggests that cerebral

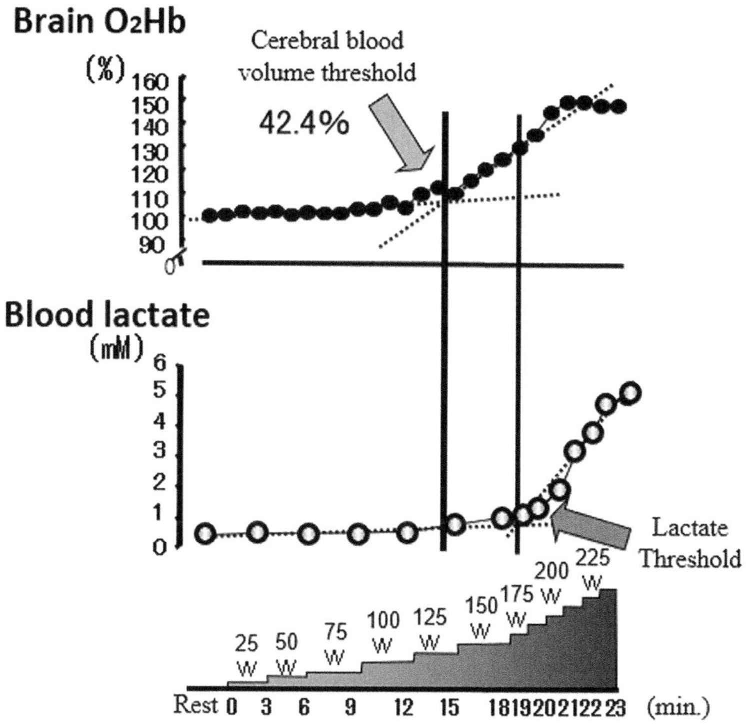

Fig. 3 Frontal O_2Hb and blood lactate during incremental cycling exercise. (Modified from Timinkul et al. [28])

hemodynamic responses during graded exercise tests may differ from trends seen in ventilation and blood lactate [28]. Additionally, during all-out exercise, oxygenated hemoglobin (HbO2) concentrations in the prefrontal cortex (PFC) may decline before exercise termination, indicating intensity-dependent changes [28, 29]. Recent developments in multi-channel EEG and NIRS systems have expanded the possibilities for mapping cortical network dynamics and functional connectivity during exercise. Future studies are expected to explore how varying levels of exercise intensity influence specific cortical regions and network interactions (Fig. 3).

4 Conclusion

In summary, while further investigation is needed to identify brain-derived indicators that reliably reflect exercise intensity, accumulating evidence from single- and multi-channel NIRS and intervention studies suggests that even light-to-moderate-intensity exercise may play a beneficial role in enhancing human brain function [30–32].

References

1. American College of Sports Medicine (2021) Guidelines for exercise testing and prescription, 11th edn. Wolters Kluwer Health, Philadelphia
2. Warburton DE, Nicol CW, Bredin SS (2006) Prescribing exercise as preventive therapy. CMAJ 174:961–974
3. Norton K, Norton L, Sadgrove D (2010) Position statement on physical activity and exercise intensity terminology. J Sci Med Sport 13:496–502
4. Achten J, Jeukendrup AE (2003) Heart rate monitoring: applications and limitations. Sports Med 33:517–538
5. Gajda R, Gajda J, Czuba M, Knechtle B, Drygas W (2024) Sports heart monitors as reliable diagnostic tools for training control and detecting arrhythmias in professional and leisure-time endurance athletes: an expert consensus statement. Sports Med 54:1–21
6. Fox SM, Naughton JP, Haskell WL (1971) Physical activity and the prevention of coronary heart disease. Ann Clin Res 3:404–432
7. Tanaka H, Monahan KD, Seals DR (2001) Age-predicted maximal heart rate revisited. JACC 37:153–156
8. Borg GA (1973) Perceived exertion: a note on history and methods. Med Sci Sports 5:90–93
9. Hareendran A, Leidy NK, Monz BU, Winnette R, Becker K, Mahler DA (2012) Proposing a standardized method for evaluating patient report of the intensity of dyspnea during exercise testing in COPD. Int J Chron Obstruct Pulmon Dis 7:345–355
10. Londeree BR (1997) Effect of training on lactate/ventilatory thresholds: a meta-analysis. Med Sci Sports Exerc 29:837–843
11. Lenti M, De Vito G, Scotto di Palumbo A, Sbriccoli P, Quattrini FM, Sacchetti M (2011) Effects of aging and training status on ventilatory response during incremental cycling exercise. J Strength Cond Res 25:1326–1332
12. Scharhag-Rosenberger F, Walitzek S, Kindermann W, Meyer T (2012) Differences in adaptations to 1 year of aerobic endurance training: individual patterns of nonresponse. Scand J Med Sci Sports 22:113–118
13. Stöggl TL, Sperlich B (2015) The training intensity distribution among well-trained and elite endurance athletes. Front Physiol 6:295
14. Reed JL, Pipe AL (2014) The talk test: a useful tool for prescribing and monitoring exercise intensity. Curr Opin Cardiol 29:475–480
15. Bok D, Rakovac M, Foster C (2022) An examination and critique of subjective methods to determine exercise intensity: the talk test, feeling scale, and rating of perceived exertion. Sports Med 52:2085–2109
16. Meckel Y, Rotstein A, Inbar O (2002) The effects of speech production on physiologic responses during submaximal exercise. Med Sci Sports Exerc 34:1337–1343
17. Jeanes EM, Foster C, Porcari JP, Gibson M, Doberstein S (2011) Translation of exercise testing to exercise prescription using the talk test. J Strength Cond Res 25:590–596
18. Fletcher GF, Ades PA, Kligfield P et al (2013) Exercise standards for testing and training: a scientific statement from the American Heart Association. Circulation 128:873–934
19. Boecker H, Drzezga A (2016) A perspective on the future role of brain PET imaging in exercise science. NeuroImage 131:73–80
20. Zhang K, Jan YK, Liu Y, Zhao T, Zhang L, Liu R, Liu J, Cao C (2022) Exercise intensity and brain plasticity: what's the difference of brain structural and functional plasticity characteristics between elite aerobic and anaerobic athletes? Front Hum Neurosci 16:757522
21. Ligeza TS, Vens MJ, Bluemer T, Junghofer M (2023) Acute aerobic exercise benefits allocation of neural resources related to selective attention. Sci Rep 13:8624
22. Lin MA, Meng LF, Ouyang Y, Chan HL, Chang YJ, Chen SW, Liaw JW (2021) Resistance-induced brain activity changes during cycle ergometer exercises. BMC Sports Sci Med Rehabil 13:27

23. Bigliassi M, Cabral DF, Kotler S, Mannino M, Mavrantza AM, Oparina E, Gomes-Osman J (2024) Electroencephalography spectral coherence analysis during cycle ergometry in low- and high-tolerant individuals. Psychophysiology 61:e14437
24. Brümmer V, Schneider S, Abel T, Vogt T, Strüder HK (2011) Brain cortical activity is influenced by exercise mode and intensity. Med Sci Sports Exerc 43:1863–1772
25. Ligeza TS, Maciejczyk M, Kałamała P, Szygula Z, Wyczesany M (2018) Moderate-intensity exercise boosts the N2 neural inhibition marker: a randomized and counterbalanced ERP study with precisely controlled exercise intensity. Biol Psychol 135:170–179
26. Hosang L, Mouchlianitis E, Guérin SMR, Karageorghis CI (2024) Effects of exercise on electroencephalography-recorded neural oscillations: a systematic review. Int Rev Sport Exerc Psychol 17:926–979
27. Stock C, Baum M, Rosskopf P, Schober F, Weiss M, Liesen H (1996) Electroencephalogram activity, catecholamines, and lymphocyte subpopulations after resistance exercise and during regeneration. Eur J Appl Physiol Occup Physiol 72:235–241
28. Timinkul A, Kato M, Omori T, Deocaris CC, Ito A, Kizuka T, Sakairi Y, Nishijima T, Asada T, Soya H (2008) Enhancing effect of cerebral blood volume by mild exercise in healthy young men: a near-infrared spectroscopy study. Neurosci Res 61:242–248
29. Song Q, Cheng X, Zheng R, Yang J, Wu H (2022) Effects of different exercise intensities of race-walking on brain functional connectivity as assessed by functional near-infrared spectroscopy. Front Hum Neurosci 16:1002793
30. Yanagisawa H, Dan I, Tsuzuki D, Kato M, Okamoto M, Kyutoku Y, Soya H (2010) Acute moderate exercise elicits increased dorsolateral prefrontal activation and improves cognitive performance with Stroop test. NeuroImage 50:1702–1710
31. Tamura M, Nemoto K, Kawaguchi A, Kato M, Arai T, Kakuma T, Mizukami K, Matsuda H, Soya H, Asada T (2015) Long-term mild-intensity exercise regimen preserves prefrontal cortical volume against aging. Int J Geriatr Psychiatry 30:686–694
32. Soya H, Kato M, Sakamaki Y, Motoyama T, Asada T (2005) Enhanced memory function of elderly people by exercise intervention with enjoyable and mild intensity: tone project. Alzheimers Dement 1:S98

Exercise Benefits Dentate Gyrus Function: Bridging from Animal Models to Human High-Resolution Functional MRI of Pattern Separation

Kazuya Suwabe, Michael A. Yassa, Kenji Suzuki, Yoshiyuki Sankai, and Hideaki Soya

Abstract Animal studies have demonstrated that physical exercise enhances neural plasticity, particularly in the hippocampal dentate gyrus (DG), and improves learning and memory. Notably, these effects are more pronounced with light-intensity exercise compared to higher-intensity exercise. Recent advancements have enabled the investigation of hippocampal memory in humans, especially DG-dependent pattern separation, providing a bridge between animal and human research. In this chapter, we outline the methodological approaches for assessing hippocampal

K. Suwabe
Faculty of Health and Sport Sciences, Ryutsu Keizai University, Ibaraki, Japan

Laboratory of Exercise Biochemistry and Neuroendocrinology, Institute of Health and Sport Sciences, University of Tsukuba, Tsukuba, Ibaraki, Japan

M. A. Yassa (✉)
Department of Neurobiology and Behavior, University of California, Irvine, CA, USA

Center for the Neurobiology of Learning and Memory, University of California, Irvine, CA, USA

Sport Neuroscience Division, Kokoro Division, Advanced Research Initiative for Human High Performance (ARIHHP), Institute of Health and Sport Sciences, University of Tsukuba, Tsukuba, Ibaraki, Japan

Center for Cybernics Research, University of Tsukuba, Tsukuba, Ibaraki, Japan
e-mail: myassa@uci.edu

K. Suzuki · Y. Sankai
Center for Cybernics Research, University of Tsukuba, Tsukuba, Ibaraki, Japan

H. Soya
Laboratory of Exercise Biochemistry and Neuroendocrinology, Institute of Health and Sport Sciences, University of Tsukuba, Tsukuba, Ibaraki, Japan

Sport Neuroscience Division, Kokoro Division, Advanced Research Initiative for Human High Performance (ARIHHP), Institute of Health and Sport Sciences, University of Tsukuba, Tsukuba, Ibaraki, Japan

Center for Cybernics Research, University of Tsukuba, Tsukuba, Ibaraki, Japan

H. Soya (ed.), *Exercise Brain Stimulation for Cognitive Function and Mental Health*, Advances in Neurobiology 44, https://doi.org/10.1007/978-981-95-0066-6_5

pattern separation using mnemonic discrimination tasks and high-resolution MRI, and we review key studies examining the effects of aerobic exercise on human DG function. Evidence from cross-sectional, acute, and chronic exercise intervention studies supports the positive impact of exercise and physical activity, with some findings aligning with the exercise hormesis observed in animals. However, further research is needed to determine the optimal exercise conditions for enhancing hippocampal memory and to elucidate the underlying neural mechanisms.

Keywords Pattern separation · Hippocampus · Episodic memory · Aerobic fitness

1 Introduction

Animal and human studies have consistently demonstrated that physical exercise enhances neuronal plasticity and cognitive function [1–3]. Following the pioneering work of van Praag et al., the dentate gyrus (DG) in the hippocampus has been identified as a key site where exercise facilitates brain plasticity, particularly neurogenesis, which contributes to improved learning and memory [4, 5]. However, studies using rodent voluntary wheel-running models have limitations, as they cannot control exercise parameters such as intensity, duration, or frequency. To address this, Soya and colleagues developed a treadmill-based exercise model for animals, allowing precise control of exercise intensity based on physiological stress responses. Using this model, they demonstrated that even light-intensity exercise below the lactate threshold (LT) is sufficient to activate neurons in the hippocampal DG [6]. Furthermore, long-term light-intensity exercise training (lasting 2–6 weeks) effectively increases neurogenesis in the DG and enhances learning and memory [7–10]. Interestingly, the benefits of light-intensity exercise were diminished or completely reversed with high-intensity exercise, highlighting a potential hormetic effect of exercise [7–10]. The negative impact of high-intensity exercise on hippocampal plasticity may be mediated by stress-related pathways, including the hypothalamic–pituitary–adrenal (HPA) axis [11, 12]. These findings underscore the importance of light-intensity exercise in promoting hippocampal function, offering valuable insights for promoting exercise adherence and implementation.

Conversely, while most human studies on the effects of exercise on cognitive function have primarily focused on the prefrontal cortex, relatively few have explored its impact on hippocampal memory function. This gap may stem from methodological challenges in assessing memory performance specific to the hippocampus. However, recent advancements in neuroimaging techniques and cognitive tasks now allow for the evaluation of hippocampal memory function at the level of individual subregions. Among these, pattern separation—a critical feature of episodic memory—has been identified as a function highly dependent on the DG. This has garnered significant attention as a potential bridge between animal and human research [2]. In this chapter, we highlight key studies examining the effects of exercise on the human hippocampus, with a particular focus on DG-mediated pattern separation.

2 Pattern Separation in the Dentate Gyrus

The hippocampus, located at the core of the medial temporal lobe (MTL), plays a critical role in episodic memory by interacting with other MTL regions such as the entorhinal cortex (EC), perirhinal cortex (PRC), and parahippocampal cortex (PHC). Key features of episodic memory include the ability to recall related items sequentially from a single cue and to preserve distinct representations of similar events. Computational studies suggest that two processes, pattern completion and pattern separation, underlie these abilities [13]. Pattern completion involves reconstructing a complete memory by filling in missing details, while pattern separation encodes similar activity patterns as distinct, non-overlapping representations, creating a unique index for each (Fig. 1a) [14]. Impairments in pattern separation can lead to interference from past, similar information, ultimately contributing to memory deficits.

The brain mechanisms underlying pattern completion and pattern separation have been attributed to specific hippocampal circuitry. The recurrent collaterals of the CA3 region are thought to mediate pattern completion, while mossy fibers from the granule cell layer of the DG are responsible for pattern separation [13, 15, 16]. These roles have been empirically confirmed through electrophysiological and molecular techniques. Place cells in the hippocampus display consistent firing patterns within the same environment, but this activity undergoes "remapping" when the environment changes. In the CA1 region, remapping occurs in a linear fashion, with firing pattern changes proportional to the degree of change in input

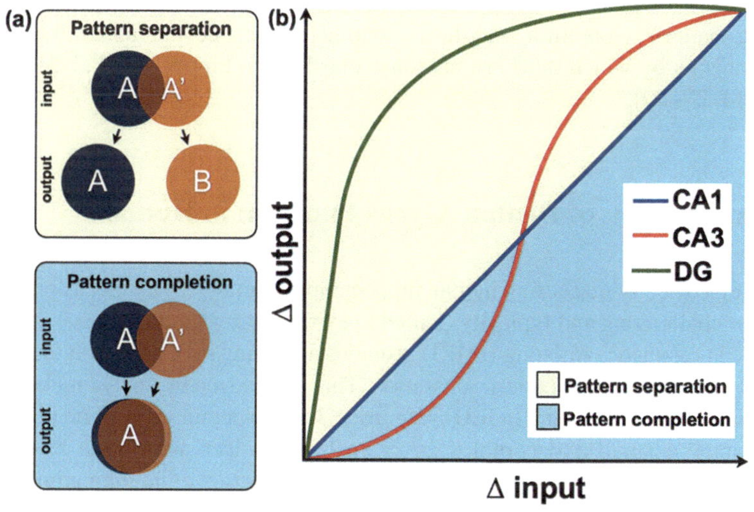

Fig. 1 (**a**) A conceptual representation of pattern separation and pattern completion. (**b**) Relationship between mnemonic interference level of the input signal and difference in neural responses, in each subregion of the hippocampus

information. When remapping in CA1 shows minimal changes despite differences in input, it reflects a pattern completion function, as similar activity patterns are preserved. Conversely, substantial firing pattern changes in response to small differences in input signify pattern separation, amplifying distinctions between inputs [14, 17].

In the CA3 region, the remapping rate varies based on sensory input differences. For small differences, the rate is lower than in CA1, favoring pattern completion. For larger differences, the rate exceeds that of CA1, supporting pattern separation [18–20]. The DG, however, consistently exhibits highly distinct activity patterns in response to even small input differences, emphasizing its critical role in pattern separation (Fig. 1b) [21]. Behavioral studies corroborate these findings; DG-specific damage or selective knockout of NMDA receptors in DG granule cells impairs the ability to discriminate slight differences while preserving discrimination of larger differences [22, 23].

A growing body of evidence suggests that newborn cells generated through neurogenesis play a critical role in pattern separation within the DG. Creer et al. demonstrated a strong correlation between exercise-induced increases in the number of newborn neurons and improved spatial pattern separation, as assessed by discrimination performance at small distances using a touch screen task [24]. In another study, enhancing neurogenesis through genetic modification selectively improved performance on tasks requiring discrimination between two similar contexts, while performance on tasks that did not involve distinguishing similar experiences, such as the Morris water maze and novel object recognition tests, remained unchanged [25]. Furthermore, the suppression of neurogenesis through X-irradiation was shown to impair performance on spatial learning tasks requiring fine spatial discrimination [26]. These findings underscore the importance of neurogenesis in facilitating pattern separation and have spurred further debate about the specific mechanisms by which newborn neurons contribute to hippocampal circuitry function [14, 27–29].

3 Assessment of Dentate Gyrus Function in Humans

Although directly measuring human hippocampal neural activity electrophysiologically is challenging and typically limited to extraordinary circumstances, functional magnetic resonance imaging (fMRI) provides a non-invasive approach to assessing brain activity related to pattern separation. This is achieved by analyzing blood oxygenation level-dependent (BOLD) responses during specific tasks. The human hippocampus, a small and complex structure located deep within the brain, poses unique challenges for imaging, including the segmentation of its subregions. Several of these challenges have been addressed including increasing resolution to a level that is sufficient to parse hippocampal subfields, using an acquisition that reduces distortions and minimizes signal dropout, and using a regional alignment technique that maximizes inter-individual overlap of hippocampal subfields [30, 31].

One phenomenon leveraged in fMRI studies is repetition suppression, whereby the BOLD response to a repeated stimulus is reduced, perhaps due to a form of adaptation [32]. Bakker et al. (2008) applied this principle to investigate pattern completion and pattern separation. They hypothesized that reduced activity for a similar stimulus indicates the stimuli are treated as identical (pattern completion), whereas comparable activity for similar stimuli suggests they are treated as distinct (pattern separation). Their results showed that activity in response to similar lure stimuli was reduced in the medial temporal lobe (MTL) region, including CA1, compared to novel stimuli. In contrast, the DG/CA3 region exhibited comparable activity levels for both lure and novel stimuli, indicative of pattern-separated representation [30].

A follow-up study examined neural responses to varying levels of similarity between stimuli. CA1 activity displayed a linear increase, with a decreasing repetition-suppression effect as similarity diminished. Conversely, DG/CA3 regions showed a natural logarithmic response, demonstrating heightened sensitivity to highly similar stimuli [33].

Although no memory tests were conducted in the two experiments mentioned earlier, other studies have analyzed brain activity based on recognition responses. These studies found that when participants correctly rejected a similar lure, DG/CA3 activity tended to be higher compared to instances where participants incorrectly identified the lure as "old" [34]. Additionally, the lure discrimination index (LDI) has been shown to correlate with DG/CA3 activity in response to changes in similarity. Specifically, lower discrimination performance was associated with a steeper DG/CA3 activity slope, resembling the response pattern of CA1 [35]. These findings suggest that hippocampal pattern separation is closely tied to behavioral mnemonic discrimination responses.

Based on findings from animal studies and human neuroimaging, it is theoretically feasible to assess pattern separation in the hippocampus through mnemonic discrimination performance. The mnemonic discrimination task was specifically designed to characterize the behavioral outcomes of neural pattern separation [36]. In this task, participants are typically instructed to view stimuli sequentially and determine whether each stimulus is exactly the "same" as one previously shown, "similar" to a prior stimulus, or entirely "new." A key feature of this task is that stimuli are designed to vary in mnemonic interference, ranging from identical (target) to similar (lure) to entirely different (foil). By categorizing lures according to mnemonic similarity—rather than purely physical similarity—researchers can parametrize interference levels, enabling more nuanced analysis (Fig. 2).

The lure discrimination index (LDI) is calculated as the proportion of correct "similar" responses to lures minus the proportion of incorrect "similar" responses to foils, accounting for any response biases. The LDI is closely linked to functional activity in the DG/CA3 [34, 35], making it a valuable marker for neurobiological changes associated with aging and neurological disorders [37, 38], Importantly, the LDI is significantly impaired by selective hippocampal lesions, while target recognition remains unaffected [39]. Thus, mnemonic discrimination performance serves as a robust behavioral index of pattern separation.

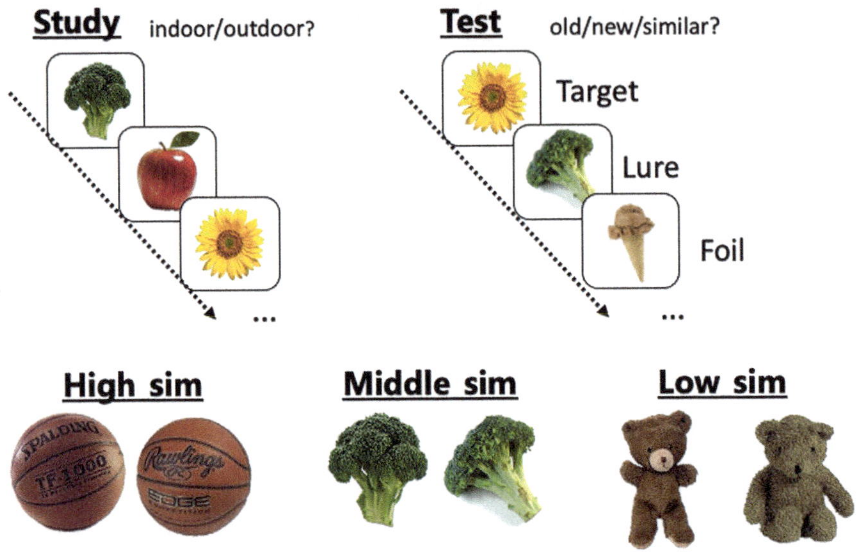

Fig. 2 Outline of the study/test version of mnemonic discrimination task. Encoding (study) phase was an incidental indoor–outdoor judgment task. Recognition (test) was tested using an old–similar–new judgment task using targets, foils, and similar lures

4 Acute Effects of Exercise on Hippocampal Pattern Separation

Over the past decade, significant research has focused on exploring the effects of exercise on hippocampal memory in humans, utilizing mnemonic discrimination tasks and high-resolution fMRI. This chapter examines both the acute and chronic impacts of exercise on hippocampal function.

When examining the effects of exercise on brain health, it is crucial to control exercise conditions, particularly exercise intensity, which should be tailored to an individual's cardiorespiratory fitness level. The gold standard measure of cardiorespiratory fitness is maximal oxygen uptake (VO_{2max}) [40]. Measuring VO_{2max} requires a graded exercise test to exhaustion, during which oxygen uptake is assessed at maximal effort. While heart rate and the rating of perceived exertion (RPE) are also useful indicators of exercise intensity, they should ideally be calibrated against VO_{2max} in carefully controlled laboratory settings. Soya and colleagues developed an acute-exercise model based on individual peak oxygen uptake (VO_{2peak}) and demonstrated its positive effects on human cognitive function [41, 42].

Suwabe et al. utilized this experimental model to investigate the effects of exercise on hippocampal memory. The study involved 21 young adults who participated under two conditions: an exercise condition, where participants completed a memory test following 10 min of pedaling exercise, and a rest condition, where they remained seated on the ergometer without exercising. The exercise intensity was individualized to achieve 50% of each participant's VO_{2peak}, as determined by a prior exercise test. The mnemonic discrimination task was employed for the memory assessment. During the retrieval phase, following the encoding phase, participants were asked to categorize stimuli as "old," "similar," or "new," and the lure discrimination index (LDI) was calculated across three levels of similarity.

The results revealed a decrease in "same" responses and an increase in "similar" responses to lures in the exercise condition, with the LDI for highly similar stimuli being significantly higher compared to the rest condition [43]. Since DG/CA3 regions are particularly sensitive to small differences in sensory input [33], these findings suggest that the observed improvement in discrimination performance for high-similarity items is attributable to enhanced pattern separation in the DG/CA3.

Follow-up studies have supported the findings of Suwabe et al. For instance, Acevedo-Triana et al. demonstrated that moderate-intensity exercise (50% VO_{2peak}) on a treadmill improves discrimination performance, with a greater effect observed after 20 min of exercise compared to 10 min [44]. However, no effects were observed in retrieval tests conducted 1 day, 2 days, or 1 week after encoding. The lack of effects over longer retention periods remains unclear but may be attributed to increased task difficulty leading to a floor effect or potential confusion caused by the repeated presentation of stimuli.

Bernstein et al. explored the transient effects of exercise on discrimination performance in relation to individual characteristics such as depressive symptoms and exercise habits in healthy young and middle-aged adults. Their study found that 30 min of moderate-intensity cycling (60–70% estimated HR_{max}) improved the lure discrimination index (LDI) compared to stretching, particularly in participants with mild or no symptoms of depression. This effect was more pronounced among regular exercisers [45].

Animal studies suggest that even light-intensity exercise may be sufficient to stimulate the hippocampus and enhance memory [7–10]. Building on this hypothesis, Suwabe et al. examined the effects of very light-intensity exercise (30% VO_{2peak}) using the same paradigm as previous moderate-intensity experiments. They found that improvements in the lure discrimination index (LDI) with very light-intensity exercise were comparable to those observed at moderate intensity [46]. To explore the neural basis of this effect, a follow-up experiment employed high-resolution fMRI. Sixteen participants performed 10 min of very light-intensity pedaling exercise (30% VO_{2peak}) or rest, followed by a continuous version of the mnemonic discrimination task in an MRI scanner 5 min later. The exercise effects were replicated, with the LDI being higher in the exercise condition.

Neural activity related to pattern separation, assessed through the critical contrast (lure correct rejections minus lure false alarms) [47], was significantly greater after exercise across hippocampal subfields, including the DG/CA3, CA1, and subiculum, as well as surrounding neocortical regions such as the entorhinal cortex (EC) and parahippocampal cortex (PHC). Additionally, psychophysiological interaction (PPI) analysis revealed increased functional connectivity (FC) between the DG/CA3 and cortical regions, including the PHC, angular gyrus, and fusiform gyrus. Importantly, the magnitude of this enhanced connectivity predicted the degree of improvement in discrimination performance at an individual level [46].

These findings suggest that visual information, initially processed in the visual cortex, is transmitted to association areas like the angular and fusiform gyri before converging in the EC via the PHC and perirhinal cortex (PRC), and ultimately reaching the hippocampus. The results indicate that light exercise enhances DG function by facilitating communication within this neocortical–hippocampal network (Fig. 3).

Thus far, we have discussed studies of exercise and hippocampal function in young adults. Effects on older adults have only recently been explored. Aging is associated with reduced pattern separation in the DG/CA3, particularly at higher interference levels, leading to impaired mnemonic discrimination compared to younger adults [38]. As a result, it has been unclear whether exercise could yield similar benefits for older adults. To address this question, Suwabe et al. conducted experiments involving 21 healthy older adults (aged 66–81 years) using the same experimental paradigm as their 2018 study, but without fMRI scanning. Exercise intensity was individualized to be light, corresponding to 60% of each participant's ventilatory threshold (VT). The results demonstrated that 10 min of light exercise significantly improved LDI scores for high-similarity lures, indicating better discrimination performance [48]. Notably, pupillometry data suggested that the ascending arousal system might play a role in the transient improvement of discrimination performance observed with light-intensity exercise. These findings underscore the potential for light exercise to enhance hippocampal function in older adults and highlight the need for future studies to investigate the underlying neural substrates using fMRI.

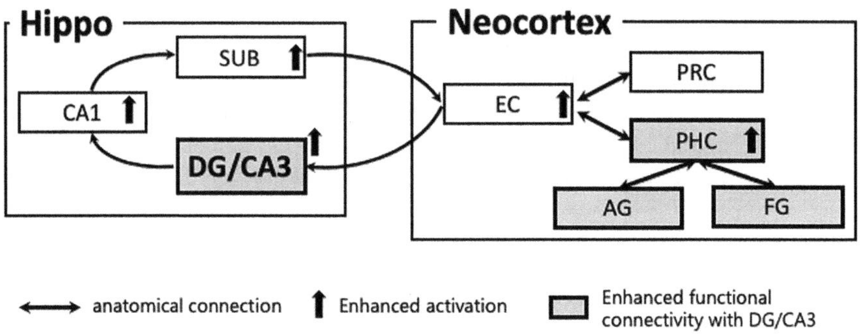

Fig. 3 Potential neural basis of transient improvement of mnemonic discrimination with very light-intensity exercise measured by high-resolution fMRI. Hippo hippocampus, SUB subiculum, DG dentate gyrus, EC entorhinal cortex, PRC perirhinal cortex, PHC parahippocampus cortex, AG angular gyrus, FG fusiform gyrus

In contrast, a series of studies conducted by Callow et al. investigated the effects of moderate-to-vigorous intensity exercise on mnemonic discrimination in older adults. In the first experiment, which did not involve fMRI, 31 healthy older adults (aged 60–89 years) participated in both exercise and resting conditions. During the exercise condition, participants completed 30 min of moderate-to-vigorous intensity exercise (78% HR_{max}), including a 5-min warm-up and cool-down. In the resting condition, participants engaged in 30 min of seated rest. The mnemonic discrimination task was administered twice during each session: before and after exercise or rest. The results showed that while the LDI decreased following seated rest, this decline was preserved in the exercise condition, suggesting a beneficial effect of exercise in maintaining discrimination performance [49].

In the second experiment, fMRI was used to explore the underlying mechanisms. Using the same within-subject crossover design, 36 healthy older adults (aged 60–89 years) completed 30 min of either moderate-to-vigorous intensity exercise (82% HR_{max}) or seated rest, followed by a mnemonic discrimination task performed in the scanner. Unlike the first experiment, the results indicated that higher intensity exercise had adverse effects on discrimination performance and pattern separation in the DG. The LDI was higher in the resting condition compared to the exercise condition, and mnemonic discrimination-related activity in the DG/CA3 (assessed via lure correct rejections minus lure false alarms) was reduced in the exercise condition [50].

These findings suggest that moderate-to-vigorous intensity exercise may negatively impact hippocampal function, potentially through a stress response mediated by the hypothalamic–pituitary–adrenal (HPA) axis. While the reasons for the discrepancy between the two experiments remain unclear, the authors speculated that the time-restricted and forced-choice nature of the continuous task performed in the scanner, combined with the stress of exercise, could have contributed to poorer performance during exercise conditions [50]. Further research is needed to clarify the effects of higher intensity exercise on hippocampal mnemonic discrimination. However, given the lack of improvement in LDI following higher intensity exercise [49] and the observed relationship between greater perceived effort and poorer LDI performance [50], negative effects are more likely with rigorous exercise.

In summary, acute exercise impacts discrimination performance, with functional modulation of hippocampal networks—particularly in the DG/CA3—likely playing a key role in the underlying mechanisms. Findings from studies in older adults highlight exercise intensity as a critical factor influencing these transient effects on hippocampal function. Low-intensity exercise appears to have beneficial effects, while high-intensity exercise may exert toxic effects [48–50]. This pattern is particularly intriguing as it mirrors the concept of exercise hormesis observed in animal studies [7–10].

5 The Impacts of Chronic Exercise on Mnemonic Discrimination

Next, we review the long-term effects of exercise on hippocampal memory function, starting with cross-sectional investigations. While previous studies have established a positive association between aerobic fitness levels [51–55] and physical activity levels [54] with hippocampal memory using relational memory paradigms, the link between these factors and mnemonic discrimination under varying levels of interference was less clear. Suwabe et al. hypothesized that individuals with higher aerobic capacity and regular exercise habits would exhibit "memory flexibility"—the opposite of "memory rigidity," a selective impairment in processing similar experiences often seen in older adults (Fig. 4a) [37].

To test this hypothesis, the study assessed aerobic fitness, physical activity, and mnemonic discrimination performance in 75 healthy young adults. VO_{2peak}, measured through a graded exercise test to exhaustion, served as a marker of aerobic fitness. Participants were divided into high- and low-fitness groups based on a median split. Results showed that while the Low-Fit group demonstrated a linear relationship between interference levels and discrimination performance, the High-Fit group excelled in discriminating middle similarity lures (Fig. 4b). Mediation analysis further supported the causal hypothesis, indicating that higher fitness levels were driven by greater physical activity in daily life [56].

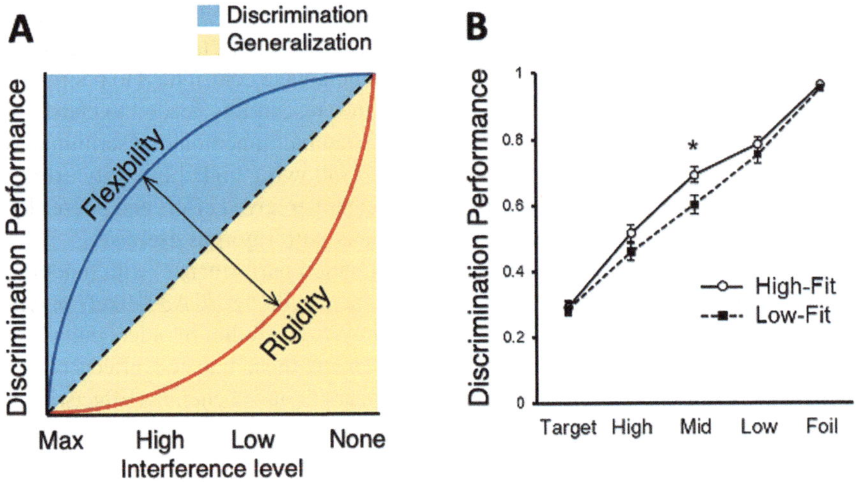

Fig. 4 (**a**) Conceptual model of memory rigidity and flexibility. Memory rigidity is observed with loss of DG function and loss of pattern separation as seen in aging and in loss of neurogenesis and other models of pathology. Conversely, memory flexibility is expected with enhancement of DG function and enhanced pattern separation. (**b**) Mnemonic discrimination performance differences between High-Fit and Low-Fit as a function of interference levels. Discrimination performance = 1 − *p*("old"|Type). Values are mean ± SE, $*p < 0.05$

Follow-up studies suggest that memory flexibility linked to higher fitness in young adults may also extend to older adults. One study involving older adults (aged 60–88 years) found a positive correlation between aerobic fitness and performance on high-interference mnemonic discrimination tasks [57]. Another study examined the relationship between exercise habits, assessed using the George Non-Exercise Test, and discrimination performance. The results showed that more regular exercise predicted better LDI scores, with this trend being especially pronounced in older participants [45].

In addition to cross-sectional studies showing that individuals with high levels of exercise, physical activity, and aerobic capacity excel in discriminating similar experiences, long-term exercise intervention studies have also demonstrated that increased aerobic capacity correlates with improved mnemonic discrimination. In a single-arm trial, 13 young adults without regular exercise habits underwent 6 weeks of high-intensity interval training (HIIT), and a positive correlation was observed between changes in aerobic fitness and the LDI before and after training [58]. Similarly, in another study involving 12 weeks of moderate-to-vigorous continuous aerobic training (80% HR_{max}) and resistance training (as an active control) in young adults, increased fitness was positively associated with improved discrimination performance in a delayed-matching-to-sample task using novel face stimuli. This effect was particularly pronounced in participants with lower baseline fitness levels [59].

Although these studies did not directly compare groups, other randomized controlled trials (RCTs) have evaluated the effectiveness of various exercise conditions for enhancing mnemonic discrimination, suggesting that HIIT may be particularly promising. One study in older adults compared the effects of 12 weeks of HIIT with moderate continuous training and found that the LDI increased exclusively in the HIIT group compared to a resting control group. A combined analysis across all groups revealed a relationship between increased aerobic capacity and improved LDI [60]. Similarly, a study in adolescents compared 12 weeks of HIIT (30-s high-intensity exercises, such as sprints and jumps, followed by 30-s resistance exercises, such as push-ups and squats) to regular moderate intensity physical education classes and found that the LDI improved only in the HIIT group [61].

The benefits of HIIT may stem from its efficiency in improving aerobic capacity [62] and the psychological advantage of being more enjoyable compared to continuous exercise [63, 64]. Taken together, increasing aerobic fitness through exercise appears to be a key factor in enhancing mnemonic discrimination, with HIIT emerging as a particularly effective exercise mode for improving both fitness and memory. However, the causal relationship between aerobic capacity improvements and mnemonic discrimination remains unclear, as comparisons between exercise conditions with and without aerobic capacity gains have not been thoroughly explored. Additionally, future studies should investigate optimal exercise conditions, including light-intensity exercise, which has shown effectiveness in transient studies [46, 48].

6 Potential Neural Basis for the Long-Term Effects of Exercise on Mnemonic Discrimination

Given the limited research on the neural mechanisms underlying the long-term effects of exercise on mnemonic discrimination, we review prominent neuroimaging studies investigating the impact of chronic exercise on the hippocampus. Erickson and colleagues were among the first to demonstrate that a one-year moderate-to-vigorous intensity exercise intervention (60–75% heart rate reserve, HRR) increased anterior hippocampal volume and improved spatial memory in older adults [65]. Similar effects have been reported in young to middle-aged adults, individuals with mild cognitive impairment (MCI), and those with multiple sclerosis [66–70].

Several studies have further analyzed hippocampal volume segmented by subfields. Nauer et al. found that fitness improvements with moderate-intensity exercise training were associated with increased volume, particularly in the left DG/CA3 head, in young adults [59]. Rosano et al. observed significant increases in CA1–3 volume and a trend toward increased DG/CA4 volume following 24 months of moderate exercise in older adults [71]. Additionally, daily walking activity, as measured objectively, was cross-sectionally associated with greater hippocampal volume in older women, particularly in the subiculum [72, 73]. Aerobic fitness has also been shown to predict bilateral subiculum volume [74].

Interestingly, evidence suggests that exercise effects extend beyond the DG, implicating the subiculum as well. The dorsal subiculum-to-mammillary bodies circuit, which is involved in stress hormone regulation, may also benefit from exercise-induced changes [75].

In addition to structural changes in the hippocampus, aerobic exercise may enhance functional connectivity with other brain regions. A cross-sectional study in youths found that higher fitness levels correlated not only with increased anterior hippocampal volume but also with stronger resting-state functional connectivity (FC) between the anterior hippocampus and several regions, including the prefrontal cortex [76]. Another study explored the neural basis of mnemonic discrimination improvement and suggested that increases in hippocampal volume and connectivity with other brain regions contribute to this enhancement. The study compared four groups undergoing 16 weeks of interventions that combined moderate aerobic exercise (70% HR_{max}) and/or video game training. Results showed that aerobic exercise increased left hippocampal volume, with the most significant improvement in LDI observed in the group that combined exercise and video games. Moreover, gains in aerobic fitness predicted hippocampal volume increases, which, in turn, predicted LDI improvements among participants who engaged in exercise. Additionally, the positive effects of video game training on FC, including connectivity with the hippocampus, were associated with better LDI performance in participants who played video games [77].

It is still a matter of debate what structural changes are responsible for exercise-induced increases in hippocampal volume. Pioneering work by Pereira et al. utilized gadolinium-based perfusion imaging to measure regional cerebral blood volume (rCBV) and examine the effects of exercise. Their study demonstrated in animals that rCBV is coupled with neurogenesis in the DG and could serve as an indirect marker of neurogenesis in humans. In a sample of 11 adults (aged 21–45 years) undergoing 12 weeks of aerobic exercise training, they observed increased rCBV in the DG, which may also be driven by angiogenesis [78]. Similarly, Maass et al. reported that hippocampal head volume changes in older adults after 12 weeks of moderate-intensity exercise (65–75% HR_{max}) were associated with increased hippocampal rCBV, suggesting that both neurogenesis and angiogenesis contribute to exercise-induced hippocampal hypertrophy [79]. Finally, Thomas et al. found that 6 weeks of moderate-to-vigorous exercise (55–85% HR_{max}) increased anterior hippocampal volume in young to middle-aged adults. However, this volumetric change was more strongly associated with myelination rather than vascular changes [67]. Based on these studies, it is likely that exercise-induced changes in hippocampal volume have several drivers, including possible contributions from neurogenesis, angiogenesis, and myelination. While difficult to measure directly in humans, it is also likely that synaptogenesis and dendrogenesis are contributors to these effects.

7 Conclusion

This chapter reviewed key studies examining the effects of aerobic exercise on human DG function, assessed through mnemonic discrimination performance and high-resolution MRI. The impact of exercise on the hippocampus appears to depend on exercise intensity. Research on acute exercise effects indicates that very light-to-moderate intensity exercise positively influences discrimination performance within a short timeframe, potentially through functional plasticity in hippocampal memory networks. Conversely, moderate-to-vigorous intensity acute exercise has been shown to suppress hippocampal activity and impair discrimination performance. With respect to chronic exercise effects, cross-sectional and intervention studies suggest a positive association between aerobic fitness and both mnemonic discrimination and hippocampal DG function. These findings imply that exercises aimed at improving aerobic capacity may enhance hippocampal function, though no randomized controlled trials (RCTs) have yet demonstrated a direct causal relationship. Additionally, the long-term effects of light-intensity exercise training, which has shown promise in transient studies, remain unexplored. Future research should focus on identifying optimal exercise conditions for improving hippocampal memory and uncovering the underlying neural mechanisms. Such efforts will benefit from employing a combination of advanced neuroimaging techniques and sensitive memory testing to provide a more comprehensive understanding of the effects of exercise on the hippocampus.

References

1. Voss MW, Vivar C, Kramer AF, van Praag H (2013) Bridging animal and human models of exercise-induced brain plasticity. Trends Cogn Sci 17:525–544
2. Voss MW, Soto C, Yoo S, Sodoma M, Vivar C, van Praag H (2019) Exercise and hippocampal memory systems. Trends Cogn Sci 23:318–333
3. Raichlen DA, Alexander GE (2017) Adaptive capacity: an evolutionary neuroscience model linking exercise, cognition, and brain health. Trends Neurosci 40:408–421
4. van Praag H, Kempermann G, Gage FH (1999) Running increases cell proliferation and neurogenesis in the adult mouse dentate gyrus. Nat Neurosci 2:266–270
5. van Praag H, Christie BR, Sejnowski TJ, Gage FH (1999) Running enhances neurogenesis, learning, and long-term potentiation in mice. Proc Natl Acad Sci USA 96:13427–13431
6. Soya H, Nakamura T, Deocaris CC, Kimpara A, Iimura M, Fujikawa T, Chang H, McEwen BS, Nishijima T (2007) BDNF induction with mild exercise in the rat hippocampus. Biochem Biophys Res Commun 358:961–967
7. Inoue K, Okamoto M, Shibato J, Lee MC, Matsui T, Rakwal R, Soya H (2015) Long-term mild, rather than intense, exercise enhances adult hippocampal neurogenesis and greatly changes the transcriptomic profile of the hippocampus. PLoS One 10:e0128720
8. Okamoto M, Hojo Y, Inoue K, Matsui T, Kawato S, McEwen BS, Soya H (2012) Mild exercise increases dihydrotestosterone in hippocampus providing evidence for androgenic mediation of neurogenesis. Proc Natl Acad Sci USA 109:13100–13105
9. Okamoto M, Yamamura Y, Liu Y, Min-chul L, Matsui T (2015) Hormetic effects by exercise on hippocampal neurogenesis with glucocorticoid signaling. Brain Plast 1:149–158
10. Inoue K, Hanaoka Y, Nishijima T, Okamoto M, Chang H, Saito T, Soya H (2015) Long-term mild exercise training enhances hippocampus-dependent memory in rats. Int J Sports Med 36:280–285
11. McEwen BS (1999) Stress and hippocampal plasticity. Annu Rev Neurosci 22:105–122
12. McEwen BS, Bowles NP, Gray JD, Hill MN, Hunter RG, Karatsoreos IN, Nasca C (2015) Mechanisms of stress in the brain. Nat Neurosci 18:1353–1363
13. Marr D (1971) Simple memory: a theory for archicortex. Philos Trans R Soc Lond Ser B Biol Sci 262:23–81
14. Yassa MA, Stark CEL (2011) Pattern separation in the hippocampus. Trends Neurosci 34:515–525
15. McNaughton BL, Morris RGM (1987) Hippocampal synaptic enhancement and information storage within a distributed memory system. Trends Neurosci 10:408–415
16. Rolls ET (1987) Information representation, processin, and storage in the brain: analysis at the single neuron level. Neural Mol Bases Learn:503–539
17. Guzowski J, Knierim J, Moser E (2004) Ensemble dynamics of hippocampal regions CA3 and CA1. Neuron 44:581–584
18. Lee I, Yoganarasimha D, Rao G, Knierim JJ (2004) Comparison of population coherence of place cells in hippocampal subfields CA1 and CA3. Nature 430:456–459
19. Leutgeb S, Leutgeb JK, Treves A, Moser M-B, Moser EI (2004) Distinct ensemble codes in hippocampal areas CA3 and CA1. Science 305:1295–1298
20. Vazdarjanova A, Guzowski JF (2004) Differences in hippocampal neuronal population responses to modifications of an environmental context: evidence for distinct, yet complementary, functions of CA3 and CA1 ensembles. J Neurosci 24:6489–6496
21. Leutgeb JK, Leutgeb S, Moser M-B, Moser EI (2007) Pattern separation in the dentate gyrus and CA3 of the hippocampus. Science 315:961–966
22. Hunsaker MR, Rosenberg JS, Kesner RP (2008) The role of the dentate gyrus, CA3a,b, and CA3c for detecting spatial and environmental novelty. Hippocampus 18:1064–1073
23. McHugh TJ, Jones MW, Quinn JJ, Balthasar N, Coppari R, Elmquist JK, Lowell BB, Fanselow MS, Wilson MA, Tonegawa S (2007) Dentate gyrus NMDA receptors mediate rapid pattern separation in the hippocampal network. Science 317:94–99

24. Creer DJ, Romberg C, Saksida LM, van Praag H, Bussey TJ (2010) Running enhances spatial pattern separation in mice. Proc Natl Acad Sci USA 107:2367–2372
25. Sahay A, Scobie KN, Hill AS, O'Carroll CM, Kheirbek M et al (2011) Increasing adult hippocampal neurogenesis is sufficient to improve pattern separation. Nature 472:466–470
26. Clelland CD, Choi M, Romberg C et al (2009) A functional role for adult hippocampal neurogenesis in spatial pattern separation. Science 325:210–203
27. Aimone JB, Deng W, Gage FH (2011) Resolving new memories: a critical look at the dentate gyrus, adult neurogenesis, and pattern separation. Neuron 70:589–596
28. Sahay A, Wilson DA, Hen R (2011) Pattern separation: a common function for new neurons in hippocampus and olfactory bulb. Neuron 70:582–588
29. Deng W, Aimone JB, Gage FH (2010) New neurons and new memories: how does adult hippocampal neurogenesis affect learning and memory. Nat Rev Neurosci 11:339–350
30. Bakker A, Kirwan CB, Miller M, Stark CEL (2008) Pattern separation in the human hippocampal CA3 and dentate gyrus. Science 319:1640–1642
31. Kirwan CB, Jones CK, Miller MI, Stark CEL (2007) High-resolution fMRI investigation of the medial temporal lobe. Hum Brain Mapp 28:959–966
32. Krekelberg B, Boynton GM, van Wezel RJA (2006) Adaptation: from single cells to BOLD signals. Trends Neurosci 29:250–256
33. Lacy JW, Yassa MA, Stark SM, Muftuler LT, Stark CEL (2011) Distinct pattern separation related transfer functions in human CA3/dentate and CA1 revealed using high-resolution fMRI and variable mnemonic similarity. Learn Mem 18:15–18
34. Kirwan CB, Stark CEL (2007) Overcoming interference: an fMRI investigation of pattern separation in the medial temporal lobe. Learn Mem 14:625–633
35. Yassa MA, Mattfeld AT, Stark SM, Stark CEL (2011) Age-related memory deficits linked to circuit-specific disruptions in the hippocampus. Proc Natl Acad Sci USA 108:8873–8878
36. Stark SM, Yassa MA, Lacy JW, Stark CEL (2013) A task to assess behavioral pattern separation (BPS) in humans: data from healthy aging and mild cognitive impairment. Neuropsychologia 51:2442–2449
37. Leal SL, Yassa MA (2015) Neurocognitive aging and the hippocampus across species. Trends Neurosci 38:800–812
38. Leal SL, Yassa MA (2018) Integrating new findings and examining clinical applications of pattern separation. Nat Neurosci 21:163–173
39. Kirwan B et al (2012) Pattern separation deficits following damage to the hippocampus. Neuropsychologia 50:2408–2414
40. Suwabe K, Byun K, Hyodo K et al (2018) Reply to Gronwald et al.: exercise intensity does indeed matter; maximal oxygen uptake is the gold-standard indicator. Proc Natl Acad Sci USA 115:E11892–E11893
41. Yanagisawa H, Dan I, Tsuzuki D, Kato M, Okamoto M, Kyutoku Y, Soya H (2010) Acute moderate exercise elicits increased dorsolateral prefrontal activation and improves cognitive performance with Stroop test. NeuroImage 50:1702–1710
42. Byun K, Hyodo K, Suwabe K, Ochi G, Sakairi Y, Kato M, Dan I, Soya H (2014) Positive effect of acute mild exercise on executive function via arousal-related prefrontal activations: an fNIRS study. NeuroImage 98:336–345
43. Suwabe K, Hyodo K, Byun K, Ochi G, Yassa MA, Soya H (2017) Acute moderate exercise improves mnemonic discrimination in young adults. Hippocampus 27:229–234
44. Acevedo-Triana C, Córdoba-Patiño D, Muñoz JF, Cifuentes J, Pinto VM, Oscar Rodríguez W, Hurtado-Parrado C (2021) Comparing the short and long-term effects of acute moderate-intensity exercise on mnemonic similarity and emotional memory tasks. Acta Neuropsychol 19:33–61
45. Bernstein EE, McNally RJ (2019) Examining the effects of exercise on pattern separation and the moderating effects of mood symptoms. Behav Ther 50:582–593
46. Suwabe K, Byun K, Hyodo K et al (2018) Rapid stimulation of human dentate gyrus function with acute mild exercise. Proc Natl Acad Sci USA 115:10487–10492

47. Yassa MA, Stark SM, Bakker A, Albert MS, Gallagher M, Stark CEL (2010) High-resolution structural and functional MRI of hippocampal CA3 and dentate gyrus in patients with amnestic mild cognitive impairment. NeuroImage 51:1242–1252
48. Suwabe K, Kuwamizu R, Hyodo K, Yoshikawa T, Otsuki T, Zempo-Miyaki A, Yassa MA, Soya H (2024) Improvement of mnemonic discrimination with acute light exercise is mediated by pupil-linked arousal in healthy older adults. Neurobiol Aging 133:107–114
49. Callow DD, Pena GS, Stark CEL, Smith JC (2023) Effects of acute aerobic exercise on mnemonic discrimination performance in older adults. J Int Neuropsychol Soc 29:519–528
50. Callow DD, Kommula Y, Stark CEL, Smith JC (2023) Acute cycling exercise and hippocampal subfield function and microstructure in healthy older adults. Hippocampus 33:1123–1138
51. Chaddock L, Erickson KI, Prakash RS et al (2010) A neuroimaging investigation of the association between aerobic fitness, hippocampal volume, and memory performance in preadolescent children. Brain Res 1358:172–183
52. Erickson KI, Prakash RS, Voss MW, Chaddock L, Hu L, Morris KS, White SM, Wójcicki TR, McAuley E, Kramer AF (2009) Aerobic fitness is associated with hippocampal volume in elderly humans. Hippocampus 19:1030–1039
53. Chaddock L, Hillman CH, Buck SM, Cohen NJ (2010) Aerobic fitness and executive control of relational memory in preadolescent children. Med Sci Sport Exerc 43:344–349
54. Hayes SM, Alosco ML, Hayes JP, Cadden M, Peterson KM, Allsup K, Forman DE, Sperling RA, Verfaellie M (2015) Physical activity is positively associated with episodic memory in aging. J Int Neuropsychol Soc 21:780–790
55. Baym CL, Khan NA, Pence A, Raine LB, Hillman CH, Cohen NJ (2014) Aerobic fitness predicts relational memory but not item memory performance in healthy young adults. J Cogn Neurosci 26:2645–2652
56. Suwabe K, Hyodo K, Byun K, Ochi G, Fukuie T, Shimizu T, Kato M, Yassa MA, Soya H (2017) Aerobic fitness associates with mnemonic discrimination as a mediator of physical activity effects: evidence for memory flexibility in young adults. Sci Rep 7:5140
57. Bullock AM, Mizzi AL, Kovacevic A, Heisz JJ (2018) The association of aging and aerobic fitness with memory. Front Aging Neurosci 10:1–7
58. Déry N, Pilgrim M, Gibala M, Gillen J, Wojtowicz JM, Macqueen G, Becker S (2013) Adult hippocampal neurogenesis reduces memory interference in humans: opposing effects of aerobic exercise and depression. Front Neurosci 7:66
59. Nauer RK, Dunne MF, Stern CE, Storer TW, Schon K (2020) Improving fitness increases dentate gyrus/CA3 volume in the hippocampal head and enhances memory in young adults. Hippocampus 30:488–504
60. Kovacevic A, Fenesi B, Paolucci E, Heisz JJ (2020) The effects of aerobic exercise intensity on memory in older adults. Appl Physiol Nutr Metab 45:591–600
61. Ben-Zeev T, Hirsh T, Weiss I, Gornstein M, Okun E (2020) The effects of high-intensity functional training (HIFT) on spatial learning, visual pattern separation and attention span in adolescents. Front Behav Neurosci 14:1–11
62. Gibala MJ, Little JP, Macdonald MJ, Hawley JA (2012) Physiological adaptations to low-volume, high-intensity interval training in health and disease. J Physiol 590:1077–1084
63. Kujach S, Byun K, Hyodo K, Suwabe K, Fukuie T, Laskowski R, Dan I, Soya H (2018) A transferable high-intensity intermittent exercise improves executive performance in association with dorsolateral prefrontal activation in young adults. NeuroImage 169:117–125
64. Bartlett JD, Close GL, MacLaren DPM, Gregson W, Drust B, Morton JP (2011) High-intensity interval running is perceived to be more enjoyable than moderate-intensity continuous exercise: implications for exercise adherence. J Sports Sci 29:547–553
65. Erickson KI, Voss MW, Prakash RS et al (2011) Exercise training increases size of hippocampus and improves memory. Proc Natl Acad Sci USA 108:3017–3022
66. Brinke T, Bolandzadeh N, Nagamatsu LS, Hsu CL, Davis JC, Miran-Khan K, Liu-Ambrose T (2014) Aerobic exercise increases hippocampal volume in older women with probable mild cognitive impairment: a 6-month randomised controlled trial. Br J Sports Med 1:1–8

67. Thomas AG, Dennis A, Rawlings NB et al (2016) Multi-modal characterization of rapid anterior hippocampal volume increase associated with aerobic exercise. NeuroImage 131:162–170
68. Leavitt VM, Cirnigliaro C, Cohen A, Farag A, Brooks M, Wecht JM, Wylie GR, Chiaravalloti ND, DeLuca J, Sumowski JF (2014) Aerobic exercise increases hippocampal volume and improves memory in multiple sclerosis: preliminary findings. Neurocase 20:695–697
69. Tarumi T, Rossetti H, Thomas BP et al (2019) Exercise training in amnestic mild cognitive impairment: a one-year randomized controlled trial. J Alzheimer's Dis 71:421–433
70. Wilckens KA, Stillman CM, Waiwood AM, Kang C, Leckie RL, Peven JC, Foust JE, Fraundorf SH, Erickson KI (2021) Exercise interventions preserve hippocampal volume: a meta-analysis. Hippocampus 31:335–347
71. Rosano C, Guralnik J, Pahor M et al (2017) Hippocampal response to a 24-month physical activity intervention in sedentary older adults. Am J Geriatr Psychiatry 25:209–217
72. Varma VR, Chuang Y-F, Harris GC, Tan EJ, Carlson MC (2015) Low-intensity daily walking activity is associated with hippocampal volume in older adults. Hippocampus 25:605–615
73. Varma VR, Tang X, Carlson MC (2016) Hippocampal sub-regional shape and physical activity in older adults. Hippocampus 1060:1051–1060
74. Kern KL, Storer TW, Schon K (2021) Cardiorespiratory fitness, hippocampal subfield volumes, and mnemonic discrimination task performance in aging. Hum Brain Mapp 42:871–892
75. Roy DS, Kitamura T, Okuyama T, Ogawa SK, Sun C, Obata Y, Yoshiki A, Tonegawa S (2017) Distinct neural circuits for the formation and retrieval of episodic memories. Cell 170:1000–1012.e19
76. Stillman CM, Uyar F, Huang H, Grove GA, Watt JC, Wollam ME, Erickson KI (2018) Cardiorespiratory fitness is associated with enhanced hippocampal functional connectivity in healthy young adults. Hippocampus 28:239–247
77. Cui X, Gui W, Miao J et al (2023) A combined intervention of aerobic exercise and video game in older adults: the efficacy and neural basis on improving mnemonic discrimination. J Gerontol A Biol Sci Med Sci 78:1436–1444
78. Pereira AC, Huddleston DE, Brickman AM, Sosunov AA, Hen R, McKhann GM, Sloan R, Gage FH, Brown TR, Small SA (2007) An in vivo correlate of exercise-induced neurogenesis in the adult dentate gyrus. Proc Natl Acad Sci USA 104:5638–5643
79. Maass A, Düzel S, Goerke M et al (2015) Vascular hippocampal plasticity after aerobic exercise in older adults. Mol Psychiatry 20:585–593

Pupil Dynamics Predict Exercise Brain Stimulation: An Overview of Exercise Pupillometry

Ryuta Kuwamizu, Yudai Yamazaki, Kazuya Suwabe, Kenji Suzuki, Yoshiyuki Sankai, and Hideaki Soya

Abstract Proper physical activity, even at a very light intensity such as walking or slow running, improves brain health related to prefrontal executive function and hippocampal memory. However, the neural mechanism behind the cognitive enhancement that occurs during dynamic aerobic exercise is elusive and remains unclear in humans. Recently, pupillometry has been attracting attention as a kind of readout of the brain's ascending arousal mechanism, especially for brain

R. Kuwamizu (✉)
Sport Neuroscience Division, Kokoro Division, Advanced Research Initiative for Human High Performance (ARIHHP), Institute of Health and Sport Sciences, University of Tsukuba, Tsukuba, Ibaraki, Japan

Laboratory of Exercise Biochemistry and Neuroendocrinology, Institute of Health and Sport Sciences, University of Tsukuba, Tsukuba, Ibaraki, Japan
e-mail: kuwamizu.ryuta.ft@u.tsukuba.ac.jp

Y. Yamazaki
Laboratory of Exercise Biochemistry and Neuroendocrinology, Institute of Health and Sport Sciences, University of Tsukuba, Tsukuba, Ibaraki, Japan

K. Suwabe
Laboratory of Exercise Biochemistry and Neuroendocrinology, Institute of Health and Sport Sciences, University of Tsukuba, Tsukuba, Ibaraki, Japan

Faculty of Health and Sport Sciences, Ryutsu Keizai University, Ryugasaki, Ibaraki, Japan

Center for Cybernics Research, University of Tsukuba, Tsukuba, Ibaraki, Japan

K. Suzuki · Y. Sankai
Center for Cybernics Research, University of Tsukuba, Tsukuba, Ibaraki, Japan

H. Soya
Sport Neuroscience Division, Kokoro Division, Advanced Research Initiative for Human High Performance (ARIHHP), Institute of Health and Sport Sciences, University of Tsukuba, Tsukuba, Ibaraki, Japan

Laboratory of Exercise Biochemistry and Neuroendocrinology, Institute of Health and Sport Sciences, University of Tsukuba, Tsukuba, Ibaraki, Japan

Center for Cybernics Research, University of Tsukuba, Tsukuba, Ibaraki, Japan

H. Soya (ed.), *Exercise Brain Stimulation for Cognitive Function and Mental Health*, Advances in Neurobiology 44, https://doi.org/10.1007/978-981-95-0066-6_6

noradrenergic and cholinergic system activation. Thus, to identify the neural mechanism behind the effects of very-light-intensity exercise, our recent work has focused on pupillometry during aerobic exercise, and we have successfully shown the efficacy of pupil dilation as a biological marker, even during very-light-/light-intensity exercise (below the ventilatory threshold). Interestingly, neuromelanin-MRI contrast in the LC, a marker of LC integrity, predicted the magnitude of exercise-induced pupil dilation and psychological arousal changes at the individual level. In addition, we have found that pupil dilation during exercise predicted the positive impact of acute very-light-/light-intensity exercise on prefrontal executive performance and hippocampal memory performance. The series of exercise pupillometry studies we will discuss here provides essential insights into the neural substrates of the advantages of exercise-induced brain stimulation in humans.

Keywords Pupil diameter · Eye tracking · Executive function · Memory · Arousal · Catecholamine

1 Introduction

While we perform physical activity, it is not only the skeletomuscular system, but also the brain that is transformed from a resting state to an active state. The spillover effects of single bouts of aerobic physical exercise on cognitive performance have been well documented [1, 2]. Recent translational studies from rodents to humans indicate that physical exercise, even at a very light/light (i.e., mild) intensity (below the ventilatory threshold and the lactate threshold), improved brain function related to higher order cognitive performance, possibly by enhancing prefrontal and hippocampal reactivity [3, 4]. However, the full potential of exercise remains unfulfilled due to the lack of a comprehensive understanding of neural mechanisms triggered during exercise and the absence of real-time readout tools for assessing exercise efficacy in humans.

Brain activation during exercise is a possible trigger for post-exercise cognitive enhancement and changes in neural responses during tasks. In rodents, exercise activates cognition-related brain region even at stress-free very light/light intensities below the lactate threshold (LT) [5, 6]. The ascending arousal system, which includes catecholaminergic pathways originating in the brainstem, has been proposed as a critical factor in regulating emotion-related and cognitive functions, such as executive function, learning, and memory [7, 8]. Recent research with rodents has suggested that even light-intensity exercise induces arousal promotion, specifically the catecholaminergic neurons from the locus coeruleus (LC) [9]. This is a potential neural substrate for the stimulation of the hippocampus and the prefrontal cortex (PFC) during very-light-/light-intensity exercise. However, it is important to note that this remains a hypothesis and lacks conclusive evidence in humans. One major reason for this is that in human studies, the challenges associated with body movement and cardiovascular responses during dynamic exercise make it difficult to employ neuroimaging techniques such as fMRI.

Recent technical advancements in neuroscience have revitalized interest in a non-invasive index for activity in the brain's arousal system, mainly the catecholaminergic system, through pupillometry [10, 11]. Pupil dilation, which is controlled by the brainstem, has long been employed in neurology to evaluate brainstem functions (e.g., absence of pupillary reflex as a sign of death). Recent research has established that pupil dynamics are a dependable marker for LC activity, and the cortical state is strongly influenced by the release of acetylcholine and noradrenaline [11, 12], but see [13].

Thus, pupillometry may be key to unlocking the neural circuit mechanism underlying physical exercise-induced brain stimulation. This chapter draws upon past studies, both our own and those of others in this field, to outline the neural substrate of pupil dynamics and discusses how they serve as a neurobiological marker of the benefit of exercise to human cognition.

2 Pupil Dynamics as a Real-Time Readout for the Brain's Arousal States

The pupil, a small aperture in the iris, serves as a gateway for light to reach the retina, channeling all visual information. The primary role of the pupil is to regulate the amount of light projected onto the retina by adjusting the pupil size. Pupil size regulation is carried out by two opposing neuromuscular circuits: pupil dilation is mediated by the sympathetic output to the dilator pupillae muscle, and pupil constriction is mediated by the parasympathetic output to the sphincter pupillae muscle [14] (Fig. 1). Additionally, it is well known that the pupil dilates due to activation of the sympathetic nervous system and inactivation of the parasympathetic nervous system when an animal is exposed to a stressor and is in the fight or flight state.

When not being influenced by lighting conditions or the distance of objects from the eye and high-cognitive control, pupil size seems to mainly reflect the brain's arousal state [10, 15]. In recent years, pupil dynamics have gained prominence in the field of neuroscience as an indirect indicator of the arousal-related neuromodulatory system: arousal-related neural circuits either directly or indirectly induce pupil dilation. In particular, there has been strong evidence of a causal or correlative relationship between LC activation and pupil size. The LC is a small nucleus located deep in the brainstem and is the initiating nucleus of the brain catecholaminergic system [8]. LC activity inhibits parasympathetic nerves via $\alpha 2$-adrenergic receptors and excites sympathetic nerves via $\alpha 1$-adrenergic receptors, resulting in pupil dilation upon LC activation (Fig. 1). In monkeys, there is a clear synchrony between the electrophysiological activity of the LC and pupillary fluctuations [16]. Recently, a remarkable pupil dilation effect, which was due to selective chemogenetic activation of the LC with noradrenaline increase in the cortex and hippocampus, was observed in rodents [17]. Optogenetic stimulation of the LC leads to activity-dependent pupil dilation, while suppression causes pupil constriction [18, 19]. Pupil

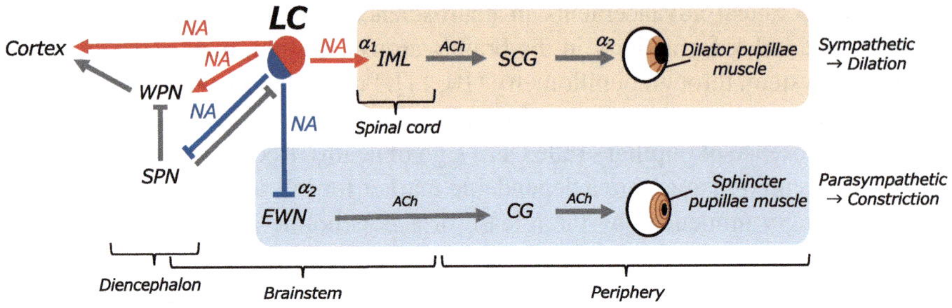

Fig. 1 Pathways controlling pupil diameter via LC
This was created based on previous review articles [14, 23]. Note: ACh acetylcholine, α1 α1 adrenergic receptor, α2 α2 adrenergic receptor, CG ciliary ganglion, EWN Edinger–Westphal nucleus (midbrain), IML intermediolateral cell column, LC locus coeruleus, NA noradrenaline, SCG superior cervical sympathetic ganglion, SPN sleep-promoting nuclei (ventrolateral preoptic nucleus of hypothalamus, basal forebrain), WPN wake-promoting nuclei (basal forebrain, thalamus, pedunculopontine tegmental nucleus, tuberomammillary nucleus, ventral tegmental area, dorsal raphe nucleus)

dynamics tracked the activation of LC axons in the PFC and real-time cortical excitability during locomotion [20, 21]. The administration of clonidine, which is well known as an LC activity inhibitor, completely eliminates pupil dilation during walking in mice [22]. The causal mechanisms of LC activation on pupil dilation have been well documented in both literature and technical reviews [11, 23].

Pupil dilation is driven not only by LC activity but also by the involvement of the serotonergic raphe nuclei and the cholinergic system [24, 25]. Reimer et al. [24] posit that the dilation of pupils associated with locomotion is not solely contingent upon noradrenergic activation but is also intricately linked to cholinergic activation within the cortex. Recently, there have been indications that the serotonergic system may play a role in pupil size modulation with uncertainty but not locomotion [25]. The orexinergic system influences LC-mediated pupil dilation [26]. Therefore, the background of pupil dilation is linked with the activation of various arousal-related neuromodulatory systems, although the LC is primarily causally related to pupil dilation [11, 27, 28] (Fig. 1).

3 Exercise-Induced Pupil Dilation

3.1 Exercise-Intensity-Dependent Pupil Dilation

Pupillometry has the potential to emerge as a valuable brain-state biomarker during exercise, operating independently of conventional neuroimaging tools such as functional magnetic resonance imaging (fMRI), functional near-infrared spectroscopy (fNIRS), or electroencephalogram (EEG). As mentioned above, recent animal studies have uncovered a link between pupil dynamics and arousal-related neural activity during locomotion, providing high time resolution insights [20–22, 24, 29]. Pupil dynamics allow the tracking of changes in cortical noradrenaline and acetylcholine levels as well as cortical electrophysiological signature changes [21, 24]. Intriguingly, clonidine, a known LC activity inhibitor, disrupts pupil dilation induced by locomotion [22]. These recent research has suggested the substantial utility of changes in pupil diameter as a marker of neuromodulatory system activation with physical activity.

Human studies have reported that pupils dilate during physical activity. Hayashi and colleagues observed that pupil diameter increases with dynamic exercise intensity [30]. To extend these findings, we examined pupil dilation patterns across graded exercise intensities, ranging from resting to all-out, to comprehensively understand the activation of the pupil-linked arousal system as it pertains to exercise intensity [31]. First, it was found that pupil dilation intensifies in an exercise-intensity-dependent manner but that the pupil dilation pattern is not linear. The pupil prominently dilates with exercise at a moderate intensity (>60% Vo_{2peak}) or greater, and the increase in dilation is exponential in an exercise-intensity-dependent manner, as is the ventilation increase pattern. These results show that, as with ventilation increase, pupil dilation has an inflection point rather than a linear trajectory; we named this point the pupil dilation threshold (PDT). This dilation pattern also resembles previous observations of increased blood catecholamine concentrations [32, 33]. Thus, PDT may be an anaerobic threshold sign as are lactate threshold (LT), ventilatory threshold (VT), blood catecholamine threshold (BCT), and heart rate (HR) variability threshold. Therefore, a rapid increase in LC activation can be expected beyond the VT and LT, where stress responses occur. This is also supported by the activation of blood noradrenaline concentrations, which correlates with VT and LT thresholds [34, 35] because increased blood noradrenaline concentrations may indirectly promote LC activity and noradrenergic activation in the brain [36, 37].

The study categorized exercise intensities based on the gold standard of % Vo_{2peak} index proposed by the American College of Sports Medicine (ACSM). Interestingly, the pupil dilation pattern is, in one important way, substantially different from other physiological parameters, such as blood lactate and ventilation, with thresholds: pupil diameter increases with very-light-intensity exercise (<37% Vo_{2peak}) compared with resting. The increase in pupil dilation from rest to very-light-intensity exercise is more substantial than the changes from very light-to-light intensity exercise (37~44% Vo_{2peak}) and light-to–moderate intensity exercise (46~63% Vo_{2peak}). This finding demonstrates that pupil dilation increases robustly even with very-light-intensity exercise, as demonstrated in Fig. 2. Recent studies have demonstrated that even very-light-intensity exercise (below 37% Vo_{2peak}) enhances arousal-related mood states and improves executive function with prefrontal activation and hippocampal memory function [3, 4], suggesting that the brain's arousal system may

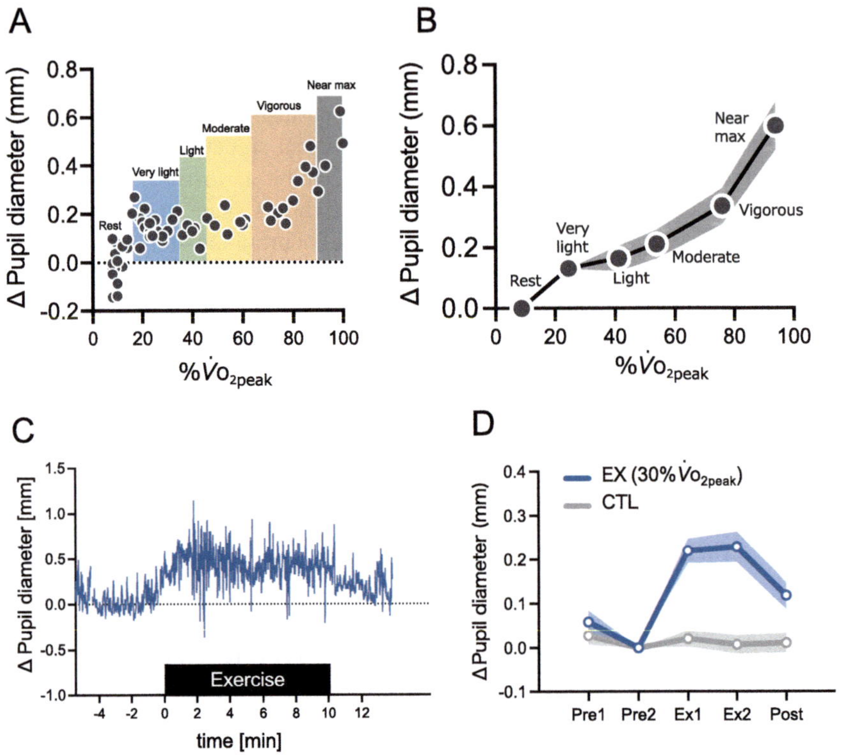

Fig. 2 Pupil dilation during exercise
(**a**) Typical example of the change in pupil diameter during graded exercise [31]. (**b**) Change in pupil diameter compared to rest during graded exercise [31]. The data for each intensity are plotted for % Vo_{2peak}. Exercise intensity was categorized exercise intensities based on the gold standard of %Vo_{2peak} index proposed by the American College of Sports Medicine (ACSM). (**c**) Typical example of the change in pupil diameter during 10 min constant load (30% $Vo_{2peak,}$ very light) intensity exercise [44]. (**d**) Pupil diameter in both resting control and 10 min constant load (30% Vo_{2peak}, very light intensity) conditions. The difference from baseline (just before exercise: Pre2) is indicated [44]

sufficiently activate with exercise intensities below the VT and LT. The study using graded exercise described above observed a positive correlation between interindividual variations in pupil changes and heightened differences in arousal-related mood states changes [31]. In other words, individuals exhibiting greater pupil dilation during very-light-intensity exercise also experienced amplified psychological arousal. This psychological examination indirectly corroborates our hypothesis that exercise-induced pupil dilation is a physiological indicator signaling arousal activation (Fig. 3).

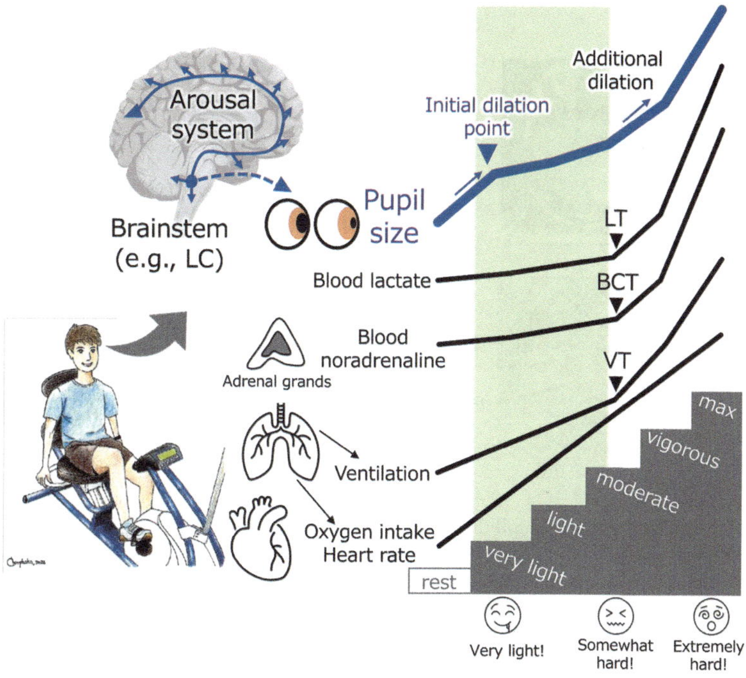

Fig. 3 Schematic of intensity-dependent pupil and other physiological/biochemical responses While other physiological and biochemical parameters show significant increases at the LT, VT, and BCT points, it is notable that pupil dilation does not occur only around these points or at higher exercise intensities (although at these points additional dilation does occur), but also during very-light-intensity exercise (i.e., initial dilation point). This pattern suggests the early activation of the ascending arousal system, providing a potential explanation for the positive impact of exercise intensities that fall below the LT, VT, and BCT points (within the region of the vertical green bar) on arousal states [31]

3.2 *Exercise-Induced Pupil Dilation and Locus Coeruleus Integrity*

Based on rodent locomotion studies, we speculate that the neural mechanism of locomotion-induced pupil dilation is due to activation of the ascending arousal system in the brain. We attempted to obtain clear evidence of this in humans by performing analyses of structural brain images [38].

In the ascending arousal system of the brain, LC activity is the most dominant neural basis for exercise-induced pupil dilation because there is a direct neural pathway between the LC and pupil dilation. Regrettably, the effect of exercise-induced LC activity on pupil dilation remains uncertain in humans. Rodent studies confirm the activation of LC neurons during running exercise, even at a light intensity below the LT [9]. However, measuring LC activity in humans during exercise poses challenges due to its small size, its anatomical position, and body movement. Therefore,

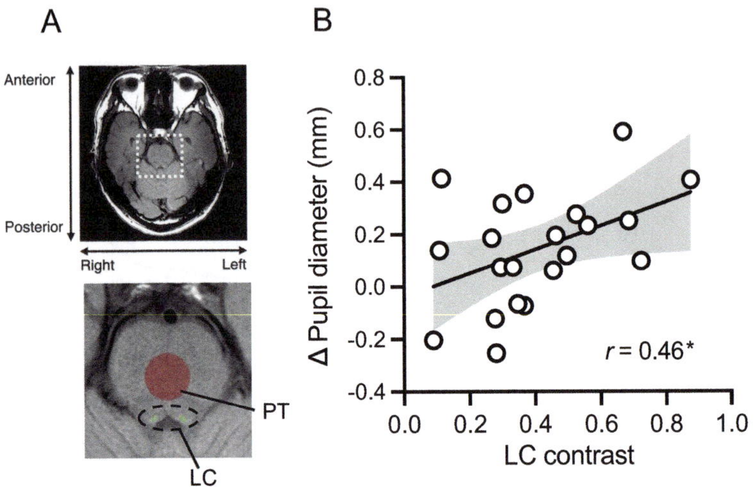

Fig. 4 Correlation between pupil diameter and LC contrast
(**a**) Individual examples of LC and position. The top image shows a representative complete slide from one participant measured using neuromelanin-sensitive weighted MRI. The bottom image shows an enlarged view of the brainstem area. Within this, the voxels of interest (VOI) outlined by the dotted line indicates the LC (green), and the large red zone above it represents the PT. (**b**) Correlations between changes in pupil diameter and LC contrast. The change (Δ) in pupil diameter is calculated as the difference between the pupil diameter during exercise conditions and the pupil diameter before exercise, from which the corresponding change in pupil diameter during the rest condition is subtracted

we introduced a neuromelanin-sensitive magnetic resonance imaging (MRI) system to address this issue. This specialized MRI technique generates T1-weighted images highlighting a hyperintense signal called LC contrast. This contrast within the LC is linked to the distribution of catecholaminergic neurons and neuromelanin concentration, indicating its reflection of neuronal integrity and density of these neurons. It must be noted that this measurement does not serve to directly measure the reactivity of the LC. The signal intensities from both sides were averaged to calculate the LC contrast. The pontine tegmentum (PT) was used as a reference region for LC contrast calculations (Fig. 4a).

We tested the role of the LC in modulating pupil dilation during very-light-intensity exercise, employing a cross-sectional examination between the individual differences in pupil dilation during exercise and LC integrity in 21 young male individuals [38]. MRI scans and exercise/rest conditions were conducted on separate days. Participants engaged in a 10-min, very-light-intensity bout of pedaling exercise, allowing us to assess pupil diameter and psychological arousal changes. The results revealed that very light intensity exercise causes increases in pupil diameter that correlate with changes in psychological arousal, which is consistent with previous findings [31]. Additionally, it was shown that the marker of LC integrity, LC contrast, predicts magnitudes of both pupil dilation and increases in psychological arousal induced by exercise (Fig. 4b). Although it should be noted that LC contrast is only indicative of the structural characteristics of the individual, these associations indicate the potential role of LC activation as a mechanism underlying arousal linked to pupil dilation during very-light-intensity exercise.

4 Pupil Dilation During Very-Light-/Light-Intensity Exercise Predicts Benefit to Cognitive Performance

4.1 *Prefrontal-Related Executive Function*

Growing research, including our own, has examined how exercise facilitates executive function performance [4, 39]. Over the past decade, there has been a gradual increase in the number of studies challenging this issue using neuroimaging. Byun et al. [4] used fNIRS to demonstrate that acute exercise, even very light intensity, improves cognitive conflict processing (i.e., Stroop interference) with the enhancement of the dorsolateral prefrontal cortex (DLPFC) activation in post-exercise [4]. Yet, the underlying trigger behind this phenomenon remains a pivotal but unresolved query. Neural dynamics, inducing prefrontal activation, during exercise, possibly works as a trigger mechanism. However, in human studies, the brain activity during exercise is yet to be comprehensively unraveled. The DLPFC plays a pivotal role in cognitive conflict processing [40, 41] and serves as a target region for neuromodulatory systems associated with pupil dilation, such as the catecholaminergic and cholinergic systems [7]. Moderate activity in the brain's arousal system enhances the signaling efficiency of the DLPFC and improves executive function [7]. These neuromodulatory systems are essential for understanding prefrontal cognitive processes [7, 42]. Thus, it is plausible that the facilitation of executive function with very-light-intensity exercise involves this neural circuit.

If pupil dilation induced by exercise reflects the activity of the brain's arousal system while exercising, then the magnitude of pupil dilation during exercise may predict the degree of subsequent improvement in executive function and prefrontal activation post-exercise. To test this hypothesis, we integrated pupillometry during exercise into well-established experimental models using the color-word Stroop task and fNIRS [2, 4, 39, 43]. We instructed 24 healthy young adults to engage in 10 min of very-light-intensity exercise or to remain seated at rest [44]. Each subject performed an executive function task—the color-word Stroop task—both before and after the exercise (Fig. 5a). The results revealed a reduction in Stroop interference, indicating enhanced executive function, in the exercise condition compared to the resting condition (Fig. 5b). Additionally, when we explored the association between pupil dilation during exercise and the exercise-induced enhancement of executive function through correlation and mediation analyses, we discovered that individuals exhibiting greater pupil dilation had a more significant reduction in Stroop interference (i.e., improvement of executive function as measured by the Stroop task) (Fig. 5d, e). These findings support that the activation of neural substrates linked to pupil dynamics, possibly the activity of the LC, is a crucial factor in the beneficial effects of very-light-intensity exercise on executive function.

Furthermore, when we examined PFC activity using fNIRS, we observed increased activity in the left DLPFC (l-DLPFC) (Fig. 5c), a region known to play a key role in executive function [44]. The activation of catecholaminergic neurons in the LC heightens cortical activity by modulating glutamatergic neurotransmission, a key determinant in the link between increased cortical activity and changes in hemodynamics [45]. Activation of the cholinergic system also has the effect of increasing regional blood flow in the rodent cortex during slow walking [46–48]. Very-light-intensity exercise seems to upregulate the neuromodulatory system, such as the catecholaminergic and cholinergic systems, in a manner linked to pupil response, potentially influencing oxy-Hb signal changes in l-DLPFC-associated Stroop interference. In this way, exercise-induced activation of the pupil-linked neuromodulatory system might contribute to elevated activity in the DLPFC, resulting in improved executive function. While the directional shift in l-DLPFC activation coincided with both pupil dilation and decreased Stroop interference, our correlation analysis, focusing on individual differences, did not reveal a significant linear association between pupil dilation during exercise and l-DLPFC activity. In the future, it would be beneficial to explore whether experimentally inhibiting the catecholaminergic arousal system, within ethical boundaries, correspondingly reduces pupil-linked arousal during exercise and task performance and prefrontal activation (Fig. 5).

(**a**) Illustration of the experimental design. (**b**) Differences in Stroop interference performance (inverse efficiency score, IES) between the rest and exercise (Ex) conditions. Reaction time (RT) was adjusted for accuracy using the formula: IES = RT [ms]/accuracy [%] × 100. (**c**) Brain regions showing exercise-enhanced cortical activation related to Stroop interference. The t-map highlights a very-light-intensity exercise effect on cortical activity. (**d**) Participants who exhibited greater pupil dilation during very-light-intensity exercise showed more pronounced improvements in executive function task performance, as indicated by reduced Stroop interference. (**e**) A mediation model illustrating the statistical examination of whether pupil diameter changes during exercise (mediator variable) explain the relationship between exercise (independent variable) and executive function (dependent variable). The total effect of exercise on executive function can be divided into a direct effect (not mediated by pupil diameter changes) and an indirect effect (mediated by pupil diameter changes). The results indicated that the mediation effect was statistically significant

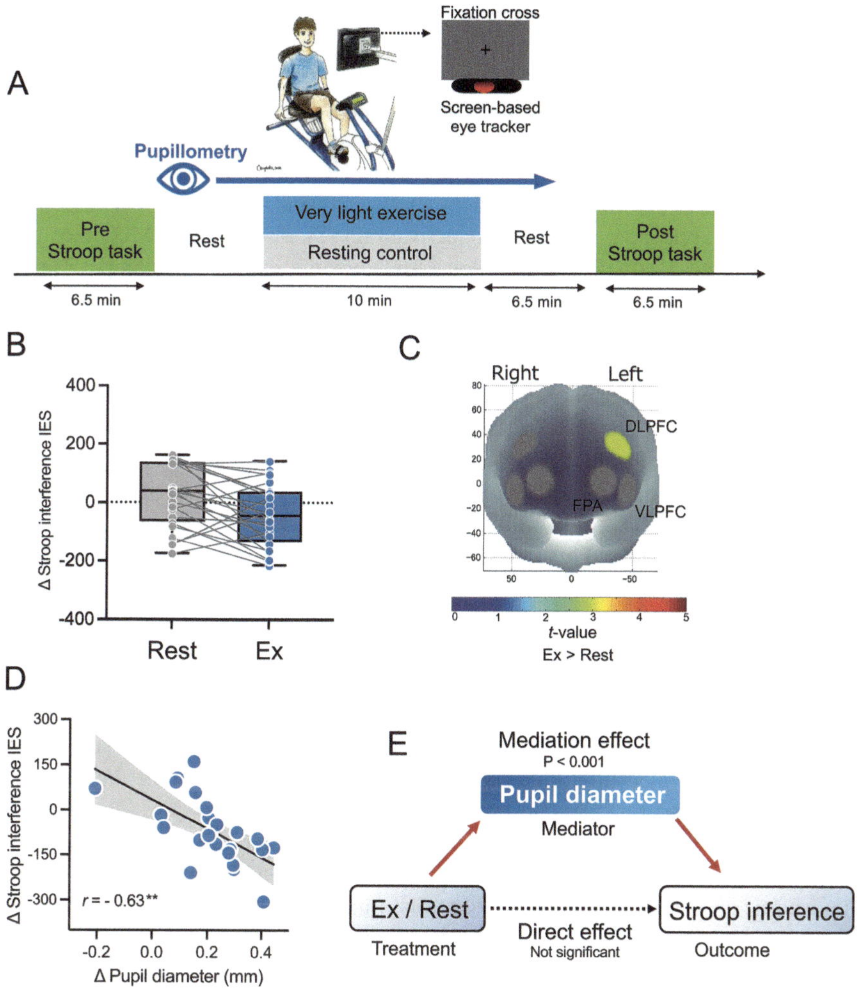

Fig. 5 The effect of very-light-intensity exercise on pupil size and prefrontal executive function

4.2 Hippocampal-Related Memory Function

Aging often involves hippocampal memory decline, partly attributed to reduced catecholaminergic pathways from the LC to the hippocampus [49, 50]. Physical exercise profoundly impacts hippocampal structure and function, positively influencing hippocampal memory, especially during age-related decline [51–53].

Past studies in young adults have highlighted how not only moderate but also very-light-intensity exercise can enhance memory, potentially by boosting hippocampus-neocortical connectivity [3, 54]. Considering the significant role of the neuromodulatory systems, including the catecholaminergic and cholinergic systems, in hippocampal learning and memory, the activation of arousal systems from the LC may contribute to exercise's positive impact on the hippocampus. A recent rodent study demonstrated that both vigorous-intensity exercise, triggering a stress response, and light-intensity exercise below the LT elevate catecholamine levels in the hippocampus [9]. Moreover, rodent studies suggest that slower paced walking can increase hippocampal cholinergic activation [47]. Furthermore, McGinley et al. [21] discovered a correlation between pupil dynamics and hippocampal electrophysiological activity in mice during various arousal states, suggesting that pupil-linked arousal may enhance hippocampal activity: a mechanism that is potentially applicable to humans through pupillometry [21]. Hence, we hypothesized that pupil dilation during light-intensity exercise might improve human hippocampal memory performance, akin to the effect seen in prefrontal-related executive function following exercise-induced pupil dilation.

A study involving 21 healthy older adults (66–81 years old) without dementia examined the impact of a 10-min bout of light-intensity exercise (at 60% of each individual's VT) followed by a memory test [55] (Fig. 6a). Pupil diameter was measured during the session. We employed a mnemonic discrimination test, which is an indicator of hippocampal memory function [56]. A recognition memory test required participants to categorize each item as either "old" (targets, i.e., previously seen items), "similar" (lures resembling but not identical to previously viewed images), or "new" (foils, i.e., entirely new items). This metric, termed the "lure discrimination index" (LDI), examines discrimination performance, which is closely associated with age-related alterations in input to the dentate gyrus and the CA3 and functional signals in the dentate gyrus/CA3 [57–60, 76].

The results revealed enhanced memory performance (i.e., LDI) after exercise for high-similarity lures, which is closely associated with the dentate gyrus of the hippocampus compared to the resting controls. Pupil diameter increased during light-intensity exercise compared to seated resting comparable to that found in young adult studies. Additionally, greater pupil dilation during exercise correlated with exercise-induced LDI improvement (Fig. 6b). An additional causal mediation analysis supported that pupil dilation during exercise mediated the enhancement of LDI performance with exercise (Fig. 6c). These results indicate that pupil-linked neuromodulatory systems, such as catecholaminergic and cholinergic systems in the brain, mediated the exercise-induced hippocampal memory enhancement. These findings suggest that pupil-linked neural substrates play a role in enhancing hippocampal memory following light-intensity exercise, as they do in enhanced prefrontal executive function. Insights from these pupillometry findings will aid in shedding light on neurobiological mechanisms for exercise-based hippocampus stimulation and in designing exercise regimens to improve hippocampal memory with aging.

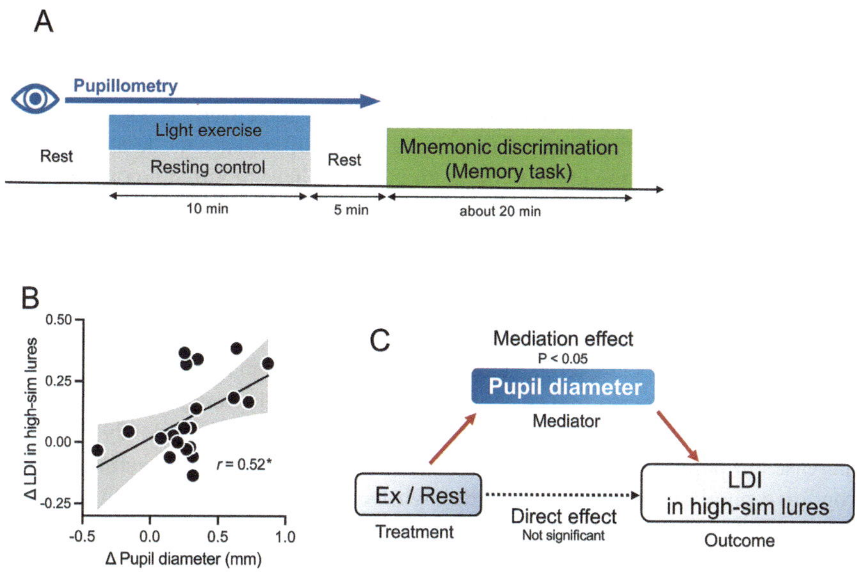

Fig. 6 Memory improvement mediated by pupil dilation during light-intensity exercise (**a**) Illustration of the experimental paradigm. (**b**) Participants who exhibited greater pupil dilation during light-intensity exercise showed more pronounced improvements in memory task performance, as indicated by the improved lure discrimination index scores in high similarity lures (LDI in high-sim lures). (**c**) A mediation model illustrating the statistical examination of whether pupil diameter changes during exercise (mediator variable) explain the relationship between exercise (independent variable) and memory function (dependent variable). The results indicate that the mediation effect was statistically significant

5 Open Question and Future Direction

5.1 *Duration of Pupil Dilation Remaining After a Single Bout of Exercise*

Our studies, which we introduced above, focused on pupil measurements only during exercise [31, 38, 44, 55]. This timeframe is consistent with our original hypothesis that pupil-linked arousal during exercise is a mechanism for triggering the beneficial effects of exercise on cognition. Activation of the LC catecholaminergic system increases the network of cortical excitatory and inhibitory neurons resulting in increased cortical activity [45] and this effect may carry over to post-activation due to dopamine and noradrenaline remaining in extracellular space [9, 34, 61]. Thus, activation of the ascending arousal system during exercise might modulate the availability, production, absorption, and clearance of neurotransmitters in the projected brain area, including the PFC and the hippocampus, in a post-exercise behavioral task.

There is no consistent consensus regarding the extent to which and the duration of which residual pupil dilation remains after exercise. One study reported a higher baseline pupil diameter during a flanker task 15 min after acute moderate-intensity exercise than after resting control [62]. Other studies have reported no difference in pupil diameter after exercise and active control [63]. We examined the sustained effect of pupil dilation after very-light-intensity exercise (30% Vo_{2peak}) compared to rest [64]. That study followed previous protocols, with neutral visual stimuli (fixation cross) maintained post-exercise. Five minutes after 10 min of very-light-intensity exercise, pupil size returned to resting levels [64]. These preliminary results suggest that post-exercise baseline pupil diameter may not directly contribute to the cognitive performance enhancement observed in previous studies [44].

Earlier human research has demonstrated that cognitive task demands can cause increases in pupil dilation when pupillometry is employed during such tasks (i.e., cognitive pupillometry; [65, 66]). Although we did not focus on cognitive pupillometry in this chapter, some recent studies have aimed to test the effect of acute exercise on cognitive pupillometry [62, 63, 67]. Recent cross-over design studies have tested the effect of acute exercise on task-evoked pupillary reactivity with a post-exercise executive function task [62, 63]. However, their results demonstrate that aerobic exercise did not modulate task-evoked pupillary reactivity, although both data set show exercise-induced improvement in executive function task performance. The link between exercise pupillometry and cognitive pupillometry would be an interesting topic.

5.2 Generalization Across Intensities and Types of Exercise

The studies discussed earlier mainly focused on very-light-/light-intensity exercise. Their findings could be interpreted as follows: the more the pupils dilate during exercise, the more improved the cognitive performance [44, 55]. Given the results of the studies showing pupil dilation in an exercise-intensity-dependent manner [31], we might assume that higher intensity exercise would induce linearly higher positive effects. However, this is a limited hypothesis, given that the effect from very light to vigorous-intensity exercise is comparable or an inversed U-shape [68, 69], which aligns with the knowledge that the prefrontal cortex and the hippocampus are the brain regions most sensitive to the detrimental effects of stress exposure [7, 70]. Moderate-intensity exercise under hypoxic conditions decreases activity in the DLPFC and impairs executive function [43]. It is interesting to study whether pupillometry can predict cognitive fatigue caused by high-intensity exercise or hypoxic exercise accompanied by a stress response.

5.3 *Can Pupillometry Be a Real-World Biomarker of Exercise Brain Stimulation?*

As wristwatch-type device technology has advanced, tracking heart rate and daily physical activity has become commonplace. Some readers might use these devices to monitor their exercise and sleep routines. Extending this idea, eyeglass-type devices can potentially measure pupil size, acting as biomarkers to indicate how exercise stimulates the brain. However, pupil sizes can be affected by various cognitive aspects and are particularly responsive to visual stimuli like light. The research discussed here primarily focused on brain mechanisms and implemented control factors by taking measurements in a room with consistent lighting and with static black and gray images. Consequently, it remains uncertain how applicable these findings would be in environments rich with complex visual information. To validate the real-world applicability of exercise pupillometry, further research in ecologically diverse settings is necessary. For example, further methodological improvements might be needed to measure pupil dilation more accurately for exercise types that involve greater head movement than cycling, such as running [71].

In addition to pupil dynamics, other promising eye-based measures exist. One such marker is spontaneous eye blink rate, a noninvasive biomarker that can be measured on video without the need for an eye tracker. Accumulating research suggests the involvement of the brain dopaminergic system in the neural basis of spontaneous eye blink rate [72]. Our recent findings in young men indicate a correlation between spontaneous eye blink rate, aerobic-based physical fitness, and prefrontal executive function performance [73]. Spontaneous blink rate may also serves as a biomarker that elucidate the positive effects of physical activity on cognitive function [73, 74]. Other eye-based measures, such as fixation, retinal vessel analysis, saccadic eye movements, have been reviewed elsewhere (e.g., Zou et al. [75]). Although the mechanisms of these eye-based markers are very complex, involving a variety of cognitive and physiological factors from lower to higher order levels, the combination of pupil dynamics and these other eye-based markers may have more predictive power for the effect of exercise-related brain stimulation with a real-world application.

6 Conclusion

In this chapter, we reviewed the evidence we have recently uncovered that exercise-induced pupil dilation predicts exercise-induced activation of the brain's arousal system and cognitive performance related to the PFC and the hippocampus. These results suggest that even light-intensity exercise upregulates the brain's arousal system via the LC, triggering beneficial effects on psychological and cognitive aspects (Fig. 7). In addition, these findings suggest that exercise pupillometry could be a new methodological approach and serve as an a readout of exercise brain stimulation.

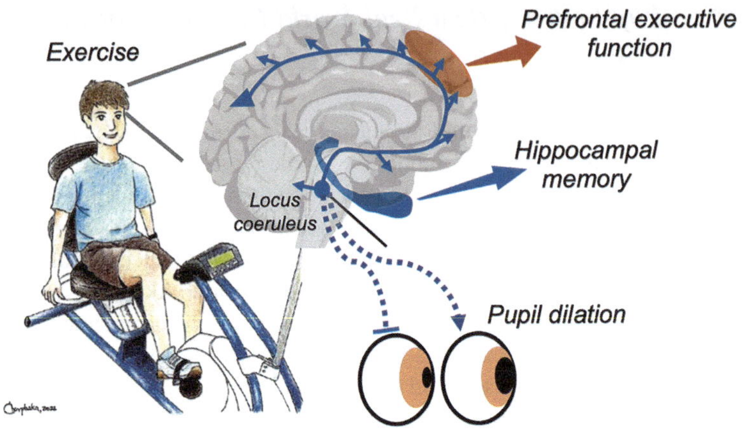

Fig. 7 Schematic of exercise-enhanced cognitive function and upregulation of the pupil-linked arousal system
Pupil dilation during very-light-/light-intensity exercise predicts the positive post-exercise effects of exercise on prefrontal executive function and hippocampal memory. These results suggest that even very light/light intensity exercise causes up-regulation of the brain's pupil-linked arousal system, mainly the LC activation, triggering the mechanism for the beneficial effects of such exercise on the brain

References

1. Pontifex MB, McGowan AL, Chandler MC, Gwizdala KL, Parks AC, Fenn K, Kamijo K (2019) A primer on investigating the after effects of acute bouts of physical activity on cognition. Psychol Sport Exerc 40:1–22
2. Damrongthai C, Kuwamizu R, Suwabe K, Ochi G, Yamazaki Y, Fukuie T, Adachi K, Yassa MA, Churdchomjan W, Soya H (2021) Benefit of human moderate running boosting mood and executive function coinciding with bilateral prefrontal activation. Sci Rep 11:22657
3. Suwabe K, Byun K, Hyodo K et al (2018) Rapid stimulation of human dentate gyrus function with acute mild exercise. Proc Natl Acad Sci USA 115:10487–10492
4. Byun K, Hyodo K, Suwabe K, Ochi G, Sakairi Y, Kato M, Dan I, Soya H (2014) Positive effect of acute mild exercise on executive function via arousal-related prefrontal activations: an fNIRS study. NeuroImage 98:336–345
5. Nishijima T, Soya H (2006) Evidence of functional hyperemia in the rat hippocampus during mild treadmill running. Neurosci Res 54:186–191
6. Soya H, Nakamura T, Deocaris CC, Kimpara A, Iimura M, Fujikawa T, Chang H, McEwen BS, Nishijima T (2007) BDNF induction with mild exercise in the rat hippocampus. Biochem Biophys Res Commun 358:961–967
7. Cools R, Arnsten AFT (2022) Neuromodulation of prefrontal cortex cognitive function in primates: the powerful roles of monoamines and acetylcholine. Neuropsychopharmacology 47:309–328
8. Poe GR, Foote S, Eschenko O et al (2020) Locus coeruleus: a new look at the blue spot. Nat Rev Neurosci 21:644–659
9. Hiraga T, Hata T, Soya S, Shimoda R, Takahashi K, Soya M, Inoue K, Johansen JP, Okamoto M, Soya H (2024) Light-exercise-induced dopaminergic and noradrenergic stimulation in the dorsal hippocampus: using a rat physiological exercise model. FASEB J 38:e70215

10. Viglione A, Mazziotti R, Pizzorusso T (2023) From pupil to the brain: new insights for studying cortical plasticity through pupillometry. Front Neural Circuits 17:1151847
11. Privitera M, Ferrari KD, von Ziegler LM et al (2020) A complete pupillometry toolbox for real-time monitoring of locus coeruleus activity in rodents. Nat Protoc 15:2301–2320
12. Schwalm M, Rosales Jubal E (2017) Back to pupillometry: how cortical network state fluctuations tracked by pupil dynamics could explain neural signal variability in human cognitive neuroscience. eNeuro 4: e0293-16
13. Megemont M, McBurney-Lin J, Yang H (2022) Pupil diameter is not an accurate real-time readout of locus coeruleus activity. Elife 11:e70510
14. Szabadi E (2018) Functional organization of the sympathetic pathways controlling the pupil: light-inhibited and light-stimulated pathways. Front Neurol 9:1069
15. Strauch C, Wang C-A, Einhäuser W, Van der Stigchel S, Naber M (2022) Pupillometry as an integrated readout of distinct attentional networks. Trends Neurosci 45:635–647
16. Aston-Jones G, Cohen JD (2005) An integrative theory of locus coeruleus-norepinephrine function: adaptive gain and optimal performance. Annu Rev Neurosci 28:403–450
17. Zerbi V, Floriou-Servou A, Markicevic M et al (2019) Rapid reconfiguration of the functional connectome after chemogenetic locus coeruleus activation. Neuron 103:702–718.e5
18. Hayat H, Regev N, Matosevich N et al (2020) Locus coeruleus norepinephrine activity mediates sensory-evoked awakenings from sleep. Sci Adv 6:eaaz4232
19. Breton-Provencher V, Drummond GT, Feng J, Li Y, Sur M (2022) Spatiotemporal dynamics of noradrenaline during learned behaviour. Nature 606:732–738
20. Breton-Provencher V, Sur M (2019) Active control of arousal by a locus coeruleus GABAergic circuit. Nat Neurosci 22:218–228
21. McGinley MJ, David SV, McCormick DA (2015) Cortical membrane potential signature of optimal states for sensory signal detection. Neuron 87:179–192
22. Lu L, Ren Y, Yu T et al (2020) Control of locomotor speed, arousal, and hippocampal theta rhythms by the nucleus incertus. Nat Commun 11:262
23. Joshi S, Gold JI (2020) Pupil size as a window on neural substrates of cognition. Trends Cogn Sci 24:466–480
24. Reimer J, McGinley MJ, Liu Y, Rodenkirch C, Wang Q, McCormick DA, Tolias AS (2016) Pupil fluctuations track rapid changes in adrenergic and cholinergic activity in cortex. Nat Commun 7:13289
25. Cazettes F, Reato D, Morais JP, Renart A, Mainen ZF (2021) Phasic activation of dorsal raphe serotonergic neurons increases pupil size. Curr Biol 31:192–197.e4
26. Grujic N, Tesmer A, Bracey E, Peleg-Raibstein D, Burdakov D (2023) Control and coding of pupil size by hypothalamic orexin neurons. Nat Neurosci 26:1160–1164
27. Grujic N, Polania R, Burdakov D (2024) Neurobehavioral meaning of pupil size. Neuron 112(20):3381–3395
28. Joshi S, Li Y, Kalwani R, m., Gold J i. (2016) Relationships between pupil diameter and neuronal activity in the locus coeruleus, colliculi, and cingulate cortex. Neuron 89:221–234
29. Vinck M, Batista-Brito R, Knoblich U, Cardin JA (2015) Arousal and locomotion make distinct contributions to cortical activity patterns and visual encoding. Neuron 86:740–754
30. Hayashi N, Someya N, Fukuba Y (2010) Effect of intensity of dynamic exercise on pupil diameter in humans. J Physiol Anthropol 29:119–122
31. Kuwamizu R, Yamazaki Y, Aoike N, Ochi G, Suwabe K, Soya H (2022) Pupil-linked arousal with very light exercise: pattern of pupil dilation during graded exercise. J Physiol Sci 72:23
32. Chmura J, Nazar K, Kaciuba-Uściłko H (1994) Choice reaction time during graded exercise in relation to blood lactate and plasma catecholamine thresholds. Int J Sports Med 15:172–176
33. Sales MM, Sousa CV, da Silva AS, Knechtle B, Nikolaidis PT, Alves PM, Simões HG (2019) An integrative perspective of the anaerobic threshold. Physiol Behav 205:29–32
34. Pagliari R, Peyrin L (1995) Norepinephrine release in the rat frontal cortex under treadmill exercise: a study with microdialysis. J Appl Physiol 78:2121–2130

35. McMorris T, Turner A, Hale BJ, Sproule J (2016) Beyond the catecholamines hypothesis for an acute exercise–cognition interaction: a neurochemical perspective. In: T MM (ed) Exercise-cognition interaction: neuroscience perspectives. Elsevier Academic Press, San Diego, pp 65–103
36. Miyashita T, Williams CL (2006) Epinephrine administration increases neural impulses propagated along the vagus nerve: role of peripheral beta-adrenergic receptors. Neurobiol Learn Mem 85:116–124
37. Mravec B (2006) Possible involvement of the vagus nerve in monitoring plasma catecholamine levels. Neurobiol Learn Mem 86:353–355. author reply 356-7
38. Yamazaki Y, Suwabe K, Nagano-Saito A et al (2023) A possible contribution of the locus coeruleus to arousal enhancement with mild exercise: evidence from pupillometry and neuromelanin imaging. Cereb Cortex Commun 4:tgad010
39. Yanagisawa H, Dan I, Tsuzuki D, Kato M, Okamoto M, Kyutoku Y, Soya H (2010) Acute moderate exercise elicits increased dorsolateral prefrontal activation and improves cognitive performance with Stroop test. NeuroImage 50:1702–1710
40. Frings C, Brinkmann T, Friehs MA, van Lipzig T (2018) Single session tDCS over the left DLPFC disrupts interference processing. Brain Cogn 120:1–7
41. MacDonald AW 3rd, Cohen JD, Stenger VA, Carter CS (2000) Dissociating the role of the dorsolateral prefrontal and anterior cingulate cortex in cognitive control. Science 288:1835–1838
42. Arnsten AFT, Li B-M (2005) Neurobiology of executive functions: catecholamine influences on prefrontal cortical functions. Biol Psychiatry 57:1377–1384
43. Ochi G, Kuwamizu R, Suwabe K, Fukuie T, Hyodo K, Soya H (2022) Cognitive fatigue due to exercise under normobaric hypoxia is related to hypoxemia during exercise. Sci Rep 12:9835
44. Kuwamizu R, Yamazaki Y, Aoike N, Hiraga T, Hata T, Yassa MA, Soya H (2023) Pupil dynamics during very light exercise predict benefits to prefrontal cognition. NeuroImage 277:120244
45. Toussay X, Basu K, Lacoste B, Hamel E (2013) Locus coeruleus stimulation recruits a broad cortical neuronal network and increases cortical perfusion. J Neurosci 33:3390–3401
46. Sato A, Sato Y, Uchida S (2004) Activation of the intracerebral cholinergic nerve fibers originating in the basal forebrain increases regional cerebral blood flow in the rat's cortex and hippocampus. Neurosci Lett 361:90–93
47. Nakajima K, Uchida S, Suzuki A, Hotta H, Aikawa Y (2003) The effect of walking on regional blood flow and acetylcholine in the hippocampus in conscious rats. Auton Neurosci 103:83–92
48. Kurosawa M, Okada K, Sato A, Uchida S (1993) Extracellular release of acetylcholine, noradrenaline and serotonin increases in the cerebral cortex during walking in conscious rats. Neurosci Lett 161:73–76
49. Leal SL, Yassa MA (2015) Neurocognitive aging and the hippocampus across species. Trends Neurosci 38(12):800–812
50. Ferguson LA, Leal SL (2022) Interactions of emotion and memory in the aging brain: neural and psychological correlates. Curr Behav Neurosci Rep 9:47–57
51. Erickson KI, Voss MW, Prakash RS, et al (2011) Exercise training increases size of hippocampus and improves memory. Proc Natl Acad Sci USA 108:3017–3022
52. van Praag H, Shubert T, Zhao C, et al (2005) Exercise enhances learning and hippocampal neurogenesis in aged mice. J Neurosci 25:8680–8685
53. Voss MW, Soto C, Yoo S, et al (2019) Exercise and hippocampal memory systems. Trends Cogn Sci 23:318–333
54. Suwabe K, Hyodo K, Byun K, Ochi G, Yassa MA, Soya H (2017) Acute moderate exercise improves mnemonic discrimination in young adults. Hippocampus 27:229–234
55. Suwabe K, Kuwamizu R, Hyodo K, Yoshikawa T, Otsuki T, Zempo-Miyaki A, Yassa MA, Soya H (2024) Improvement of mnemonic discrimination with acute light exercise is mediated by pupil-linked arousal in healthy older adults. Neurobiol Aging 133:107–114
56. Stark SM, Yassa MA, Lacy JW, Stark CEL (2013) A task to assess behavioral pattern separation (BPS) in humans: data from healthy aging and mild cognitive impairment. Neuropsychologia 51:2442–2449

57. Berron D, Cardenas-Blanco A, Bittner D, et al (2019) Higher CSF tau levels are related to hippocampal hyperactivity and object mnemonic discrimination in older adults. J Neurosci 39:8788–8797
58. Maass A, Berron D, Harrison TM, et al (2019) Alzheimer's pathology targets distinct memory networks in the ageing brain. Brain 142:2492–2509
59. Trelle AN, Carr VA, Wilson EN, et al (2021) Association of CSF biomarkers with hippocampal-dependent memory in preclinical Alzheimer disease. Neurology 96:e1470–e1481
60. Webb CE, Foster CM, Horn MM, et al (2020) Beta-amyloid burden predicts poorer mnemonic discrimination in cognitively normal older adults. Neuroimage 221:117199
61. Totah NKB, Logothetis NK, Eschenko O (2019) Noradrenergic ensemble-based modulation of cognition over multiple timescales. Brain Res 1709:50–66
62. Shigeta TT, Morris TP, Henry DH, Kucyi A, Bex P, Kramer AF, Hillman CH (2021) Acute exercise effects on inhibitory control and the pupillary response in young adults. Int J Psychophysiol 170:218-228
63. McGowan AL, Chandler MC, Brascamp JW, Pontifex MB (2019) Pupillometric indices of locus-coeruleus activation are not modulated following single bouts of exercise. Int J Psychophysiol 140:41–52
64. Kuwamizu R, Yamazaki Y, Aoike N, et al (2025) Limited potential for fNIRS photobiomodulation in exercise-induced pupil dynamics and improved executive function. Neuroimage 313:121233
65. van der Wel P, van Steenbergen H (2018) Pupil dilation as an index of effort in cognitive control tasks: a review. Psychon Bull Rev 25:2005–2015
66. Mathôt S, Vilotijević A (2023) Methods in cognitive pupillometry: design, preprocessing, and statistical analysis. Behav Res Methods 55:3055–3077
67. Ayala N, Heath M (2021) Pupillometry reveals the role of arousal in a postexercise benefit to executive function. Brain Sci 11:1048
68. Tari B, Shirzad M, Behboodpour N, Belfry GR, Heath M (2021) Exercise intensity-specific changes to cerebral blood velocity do not modulate a postexercise executive function benefit. Neuropsychologia 161:108018
69. Kashihara K, Maruyama T, Murota M, Nakahara Y (2009) Positive effects of acute and moderate physical exercise on cognitive function. J Physiol Anthropol 28:155–164
70. McEwen BS, Nasca C, Gray JD (2016) Stress effects on neuronal structure: hippocampus, amygdala, and prefrontal cortex. Neuropsychopharmacology 41:3–23
71. Damrongthai C, Kuwamizu R, Yamazaki Y et al (2024) Slow running benefits: boosts in mood and facilitation of prefrontal cognition even at very light intensity. bioRxiv. https://doi.org/10.1101/2024.01.29.575971
72. Jongkees BJ, Colzato LS (2016) Spontaneous eye blink rate as predictor of dopamine-related cognitive function-a review. Neurosci Biobehav Rev 71:58–82
73. Kuwamizu R, Suwabe K, Damrongthai C, Fukuie T, Ochi G, Hyodo K, Hiraga T, Nagano-Saito A, Soya H (2021) Spontaneous eye blink rate connects missing link between aerobic fitness and cognition. Med Sci Sports Exerc 53:1425–1433
74. Kuwamizu R, Yamazaki Y, Aoike N, et al (2025) Resting-state blink rate does not increase following very-light-intensity exercise but individual variation predicts executive function enhancement levels. J Physiol Anthropol 44:10
75. Zou L, Herold F, Ludyga S et al (2023) Look into my eyes: what can eye-based measures tell us about the relationship between physical activity and cognitive performance? J Sport Health Sci 12:568–591
76. Marks SM, Lockhart SN, Baker SL, et al (2017) Tau and β-amyloid are associated with medial temporal lobe structure, function, and memory encoding in normal aging. J Neurosci 37:3192–3201

Physical Exercise Benefits Cognition: A Narrative Review of Evidence and Possible Mechanisms

Tzu-Feng Wang, Onanong Mee-inta, and Yu-Min Kuo

Abstract This review aims to elucidate the positive effects of exercise on cognitive function and explore the underlying mechanisms. Extensive evidence supports the assertion that exercise positively influences neuroplasticity, learning and memory, and mitigates cognitive decline. Nevertheless, comprehending the intricate factors influencing the efficacy of exercise in cognitive improvement remains challenging. Further investigations are imperative to determine the optimal personalized exercise regimen, including the frequency, intensity, type, dosage, and duration, as a non-pharmacological, safe, and cost-effective approach to maximize cognitive benefits. This pursuit holds significant promise for advancing our understanding of exercise as a practical intervention to promote cognitive well-being.

Keywords Cognition · Exercise · Adult hippocampal neurogenesis · Neural plasticity · Neurotrophic factors

T.-F. Wang · Y.-M. Kuo (✉)
Department of Cell Biology and Anatomy and Institute of Basic Medical Sciences, College of Medicine, National Cheng Kung University, Tainan, Taiwan
e-mail: kuoym@mail.ncku.edu.tw

O. Mee-inta
Department of Cell Biology and Anatomy and Institute of Basic Medical Sciences, College of Medicine, National Cheng Kung University, Tainan, Taiwan

Chakri Naruebodindra Medical Institute, Faculty of Medicine Ramathibodi Hospital, Mahidol University, Samut Prakan, Thailand

H. Soya (ed.), *Exercise Brain Stimulation for Cognitive Function and Mental Health*, Advances in Neurobiology 44, https://doi.org/10.1007/978-981-95-0066-6_7

1 Introduction

The 2020 guidelines from the World Health Organization (WHO) recommend a weekly minimum duration of physical activity for adults: advocating for 150–300 min of moderate-intensity or 75–150 min of vigorous intensity exercise [1]. Engaging in physical activity serves not only to sustain physical fitness but also plays a role in enhancing mental well-being. Previous studies reported that physical activity alleviates symptoms related anxiety and depression. Furthermore, it plays a vital role in enhancing cognition, leading to an improved quality of life, and better sleep patterns [2–7]. It also reduces a risk for cardiovascular diseases (e.g., hypertension, heart disease, and stroke) [8], type 2 diabetes [9], and cancers [10].

Physical exercise is recognized for its ability to enhance brain function, mental well-being, cognition (particularly in attention and working memory domains), executive function, and memory. It is considered a non-pharmacological therapeutic approach for various brain diseases, given its potential advantages in preventing cognitive decline and promoting physical fitness. This review offers an in-depth examination of the relationship between exercise and cognition. The primary emphasis is then placed on the fundamental biological mechanisms responsible for the cognitive enhancements associated with exercise, specifically highlighting the promotion of neuroplasticity and heightened expression of growth factors.

2 Beneficial Effects of Exercise on Cognition

2.1 Behavior Level

2.1.1 Animal Models

The beneficial effect of exercise on learning and memory, particularly those dependent on the hippocampus, has been extensively validated through studies conducted on rodents. These studies employ a range of tests—novel object recognition [11, 12], object displacement [12, 13], Morris water maze [11, 14–19], radial arm maze [20–22], Y-maze [23], passive avoidance [24, 25], and contextual fear conditioning [26–28]—to evaluate performance in learning and memory. Studies conducted on young animals have underscored the significant influence of exercise on memory, specifically in terms of acquisition and retention.

Regular physical exercise is also known to delay age-related cognitive impairment in aged rodents [29–32]. For example, it has been demonstrated that chronic moderate intensity aerobic exercise enhances hippocampal neuroplasticity and memory performance in mice across various age groups, even in instances where memory impairment had advanced to a significant stage [32]. Additionally, engaging in moderate intensity aerobic exercise not only prevents age-related cognitive decline but also mitigates anxiety in middle-aged and elderly rats [33, 34].

Furthermore, a daily 30 min of treadmill exercise for 6 weeks improved short-term and spatial memories in 24-month-old rats [35]. It has been demonstrated that a duration of 20 min of treadmill running over a period of 2 weeks is adequate for 20-month-old rats to enhance their performance in the inhibitory avoidance test [36, 37]. Different exercise interventions, such as aerobic, acrobatic, resistance, and combined exercise, with a duration of 20 min, 3 times a week, for 12 weeks, can inhibit fear-associated memory impairment in 22-month-old Wistar rats [38].

2.1.2 Human Studies

Young Adults

It is recognized that physical activity improves cognitive and mental health [39] during adolescence, a stage characterized by dynamic neurodevelopment [40–42]. A systematic review spanning from 1990 to 2014, which investigated the relationship between physical activity and cognitive function in children aged 5–13 years, revealed a favorable impact of physical activity not only on cognition but also on brain structure and function [43]. Most studies suggest that the effectiveness of exercise is influenced significantly by its duration. For example, engaging in a single bout of physical activity shows a positive impact on attention. Nevertheless, sustained, consistent physical activity interventions spanning several weeks appear to hold greater promise for enhancing executive functions and academic performance in both preadolescent children (aged 6–12 years) [44] and adolescents (aged 13–17 years) [45]. Furthermore, research indicates that adolescents (aged 12–14 years) exhibit improved inhibitory control in cognition following a 12-week aerobic exercise program conducted 5 days a week [46].

Both acute and chronic interventions of physical activity show significant associations with memory function improvements in adolescents and young adults (12–30 years) [47, 48], as well as in young to middle-aged adults (18–50 years) [49]. A meta-analytic review of 29 randomized controlled trials focusing on individuals with a mean age of ≥18 years demonstrated that aerobic exercise training enhances attention, processing speed, and executive and memory functions [50]. Acute high-intensity resistance exercise has the potential to enhance episodic memory, provided there is a longer recovery period (more than 10 min) [51]. A single bout of resistance exercise [52], high-intensity functional training (HIFT), or a continuous session of moderate-intensity training has a benefit on global cognitive function, mainly in executive function and selective attention in healthy adults [53]. However, a 12-week program of mild-intensity aerobic exercise does not result in improved spatial learning and memory in young, healthy adults aged 18–30 years [54].

Exercise enhances pattern separation in humans, a process that eliminates redundancy from similar inputs to facilitate the separation of events and minimize interference [55]. In young adults, acute moderate-intensity aerobic exercise has been shown to improve pattern separation [56]. Aerobic fitness is closely linked to mnemonic discrimination ability, as evidenced by the superior performance of young female individuals with moderate interference levels compared to those with lower fitness levels in the pattern separation test [57]. Additionally, chronic physical activity has been found to enhance working memory performance in individuals who are physically and cognitively healthy across all age groups [58].

Elderly Adults

Age-related cognitive decline can be reduced by physical exercise [59–63]. Increasing exercise levels in an already ongoing fitness regimen for middle-aged, moderately fit adults brings benefits to both psychological and cognitive health [63]. Sessions lasting 45–60 min of moderate-intensity exercise improve cognitive function in individuals aged over 50 years, irrespective of their cognitive status [64]. A meta-analysis including 42 studies (from 1966 to 2010) on aerobic and extended cognitive practice training interventions for healthy adults aged 55 years and older demonstrated enhanced cognitive performance [65]. Examining the acute effects of different aerobic activities through a 5-day trial, wherein participants engaged in walking of equal duration at a self-selected pace (such as 18-hole golf, Nordic walking, and regular walking), revealed enhancements in cognitive functions among healthy older individuals [66]. Notably, Nordic walking and regular walking enhanced executive functions.

In a 12-week aerobic exercise program for healthy adults aged 50–70 years, improvements were observed in executive function and attention-speed, although combining it with computerized cognitive training did not yield additional benefits [67]. A 3-month physical exercise intervention enhanced dual-task performance, processing speed, and cardiorespiratory fitness in adults aged over 70 years [68]. A systematic review involving 33 studies with individuals aged 60 years and over, both with and without cognitive impairment, indicated that exercise holds the potential to enhance cognitive function in adults aged 55–75 years, particularly in executive functioning [69]. Another systematic review and meta-analysis, examining data from healthy middle-aged and older adults (45 years and older), revealed that a combination of aerobics and resistance training yielded greater benefits for global cognitive function, episodic memory, and word fluency [70].

In a randomized controlled trial studying the relationship between exercise and cognitive function in 120 healthy participants aged over 65 years, 6 months of exercise reversed age-related loss in hippocampus volume and improved spatial memory performance [71]. A 6-month moderate-intensity cycling regimen reduced the decline in global cognition among older adults with mild-to-moderate Alzheimer's disease and dementia [72]. Systematic reviews concluded that exercise in multiple moderate-intensity short bouts may yield the best cognitive results for older adults with cognitive impairments [73, 74]. These findings suggest that exercise not only enhances learning and memory in the young but also improves executive functioning and mitigates age-related cognitive decline in older individuals.

2.2 *Cellular Level*

Exercise-related structural and functional changes in the brain have been reported in both animal and human studies. Studies conducted on animal models indicate that aerobic training leads to an increase in cortical capillary supplies, synaptic connections, and the generation of new neurons. These changes have undergone thorough examination in previous reviews [75–78]. In this context, we offer a brief overview of these changes, with a specific focus on neurons and glial cells.

2.2.1 Neurons

Physical exercise induces structural changes in neurons across various brain regions, including the dentate gyrus [79, 80] and the cornu ammonis (CA) [27] areas of the hippocampus [81–83], medial prefrontal cortex [84], limbic structures [85], striatum [34], and cerebellum [86]. These changes are indicative of neuroplasticity, which refers to the brain's ability to modify neural networks through growth and reorganization, mediating existing structural and functional adaptations of synapses.

In studies involving humans, acute aerobic exercise augments the neuroplasticity of the motor cortex in both healthy male and female individuals aged 18–65 years [87]. Aerobic exercise training demonstrates effectiveness in preserving both gray and white matter regions [88, 89], with a notable impact on hippocampal volume loss [71, 90–92] in late adulthood. Regular exercise in the elderly improves physical fitness and white matter microstructural integrity, irrespective of the exercise type—whether walking, stretching, or practicing Tai Chi [93, 94]. Additionally, three trials of treadmill-based aerobic exercise are associated with enhanced effective connectivity within extra-hippocampal brain networks crucial for memory performance in a cohort of healthy young participants [95]. Furthermore, compared to healthy sedentary older adults, individuals engaged in lifelong exercise, particularly master athletes, exhibits higher concentrations of gray and white matter clusters in regions such as the subgyral, cuneus, and precuneus, which are associated with visuospatial function, motor control, and working memory [96].

Aerobic training enhances the functional connectivity of higher order cognitive areas, fostering communication between the frontal, posterior, and temporal cortices within the Default Mode Network and Frontal Executive Network of young adults [95] and aged individuals with mild cognitive impairment [97–99]. Additionally, findings from a functional MRI study suggest that individuals with cardiovascular fitness or those engaged in aerobic training exhibit increased activity in regions of the prefrontal and parietal cortices during cognitively challenging tasks [100]. Furthermore, a 12-week moderate walking exercise intervention (30 min per session, 4 sessions per week) has been shown to elevate hippocampal functional connectivity in older adults with mild cognitive impairment [101]. These suggest that physical activity can enhance "brain rejuvenation" in areas related cognitive and executive efficiency [102].

2.2.2 Astrocytes

Cognitive functions and the integrity of neural network are intimately regulated by astrocytes [103, 104]. Astrocytes, the major type of glial cells in the brain, regulate neurotransmission by clearing neurotransmitters and controlling the transfer of brain energy from the vasculature to the neurons [105–107]. Astrocytes have two functional states, A1 and A2, which represent to the neurotoxic phenotype and neuroprotective phenotype, respectively [108].

Aerobic exercise is known to activate astrocytes through distinct mRNA transcription processes, ultimately leading to structural and functional changes in astrocytes [109]. Consequently, this activation prompts the release of various factors, including neurotransmitters and growth factors [108], which further stimulate neuronal growth and differentiation, ultimately enhancing cognitive performance [78]. It has been shown that treadmill exercise leads to astrocyte morphological changes and increased cell numbers in the CA1 area [110, 111], prefrontal cortex, striatum [112], and entorhinal cortex [113]. Exercise also changes the orientation of astrocytic projections toward dentate gyrus, the region of significant neurogenesis augmentation after exercise [111]. Rats that engaged in 12 days of voluntary wheel running showed improvements in synaptic, dendritic, and astrocytic alterations within various brain regions associated with cognition, particularly in the prefrontal cortex [114]. On the other hand, aged mice with a sedentary lifestyle showed age-related cognitive decline and morphological alteration of astrocytes [115]. Interestingly, long-term voluntary physical exercise can counter cognitive impairment by increasing the number of new astrocytes, as well as the numbers of primary branches and the soma area in 7-month-old transgenic Alzheimer's disease (5xFAD) mouse hippocampi [116]. Many of these astrocytes are plaque associated, contributing to improvements in learning and memory.

Additionally, astrocytes play a crucial role in regulating the reuptake of neurotransmitters, influencing the equilibrium between excitatory and inhibitory signals, as well as neuronal activation. The key neurotransmitter glutamate is predominantly regulated by astrocytic excitatory amino acid transporters, which uptake glutamate from the synaptic cleft [107]. Research has shown that engaging in a combination of caloric restriction and moderate treadmill exercise for a 3-month period results in an increase in the reuptake of glutamate from the extracellular space [117]. This discovery provides insight into a cellular mechanism wherein exercise, by enhancing astrocytic function, can effectively prevent excitotoxic damage, serving as a protective measure against neuronal death and cognitive impairment.

Significantly, in aged mice, voluntary wheel running exercise can elevate the expression of the astrocytic water channel aquaporin 4 (AQP4) [109]. AQP4 is a key component of the glymphatic system, facilitating the exchange of cerebrospinal fluid and interstitial fluid through paravascular pathways and brain waste clearance. In the study using a streptozotocin-induced sporadic Alzheimer's disease rat model, it has been demonstrated that high-intensity interval training can improve the drainage of tau and amyloid-β from the cortex and the hippocampus via the glymphatic system [118]. This drainage is associated with the astrocytic polarization from A1 toward A2 phenotype, which expresses more AQP4 distribution, resulting in an amelioration of learning and memory deficits [118]. Additionally, AQP4 knockout in an animal model of Alzheimer's disease overexpressing amyloid-β inhibits the exercise effects in reducing of amyloid-β accumulation in the brain [119]. These findings indicate that physical exercise has potential to increase glymphatic activity by enhancing the polarized AQP4 expression in astrocytes, which eventually leads to cognitive improvement. However, the interaction between physical exercise and another clearance pathway (meningeal lymphatic system)-associated cognition need to be further explored.

2.2.3 Oligodendrocytes

Oligodendrocytes form myelin sheaths around axons facilitating rapid transmission of neuronal excitation and maintaining the integrity of axons [120]. Impairment or dysfunction of oligodendrocytes can interfere with the myelination process, potentially resulting in cognitive impairment [121, 122]. Voluntary physical exercise enhances the proliferation and differentiation of oligodendrocyte precursor cells, contributing to improved cognition [123, 124]. Additionally, in 6-month-old Alzheimer's disease transgenic mice (APP/PS1), a 4-month period of running exercise led to increased white matter volume, myelinated fiber volume, and axon volume compared to sedentary mice [125]. In another Alzheimer's disease mouse model, 8-month-old 3xTg mice displayed oligodendrocyte abnormalities, characterized by a high percentage of loosely structured and granulated myelin sheaths around axons [126]. However, 6 months of aerobic exercise preserved myelin sheath components and oligodendrocyte precursor cells while reducing the proportion of loosely structured and granulated myelin sheaths, rescuing cognitive decline.

Interestingly, studies have reported severe demyelination, decreased sheath thickness, and oligodendrocyte apoptosis in aged mice [127]. Nevertheless, aerobic exercise promotes remyelination and enhances myelin sheath thickness by upregulating the ROCK signaling pathway, an essential mechanism for promoting regeneration and preventing apoptosis [127]. This ultimately leads to an improvement in cognitive impairment caused by age-related demyelination.

2.2.4 Microglia

Microglia, the resident macrophage of the central nervous system, inspect synaptic activity and modify synapses by phagocytosis or by the release of cytokines or neurotrophic factors. In homeostatic status, microglia survey the surrounding and phagocytic of debris, support neural differentiation, and refine synaptic pruning, whereas inflammatory signaling can polarize microglia to inhibit neurogenesis and participate in excessive synaptic engulfment [128, 129].

Exercise has an impact on both the structure and function of microglia [130, 131]. Mice undergoing a 10-day running wheel exercise displayed increased microglial proliferation in the superficial cortical layers [132], leading to the promotion of a ramified and surveilling state of microglia in the hippocampus [133]. Following 8 weeks of wheel running, young mice exhibited an increase in the number of new microglia in the cortex [132] and the hippocampus [134]. In contrast, aged mice experienced a reduction in microglial proliferation in the hippocampus [135]. A recent investigation revealed a positive connection between cognitive resilience across the lifespan and the extent of physical activity in the later stages of life, as gauged by actigraphy monitoring in humans [136]. Additionally, the levels of physical activity were negatively correlated with microglial activation in post-mortem human brain tissue [136]. Endurance exercise can effectively modulate the proliferation, activation, and function of microglia in a manner linked to enhanced functionality.

The dysregulation of microglia plays a role in the pathogenesis of neurodegenerative diseases [137], while prophylactic exercise offers an adaptive advantage, conferring resilience against subsequent perturbations, whether infectious or non-infectious, to brain homeostasis in both young [138] and aged rodents [139]. Aging increases the gene expression levels of microglial activation markers such as CD86, CD68, MHCII, and CD11b [140]. However, voluntary wheel running exercise decreases the aging-induced activation and proliferation of microglia and preserves neurogenesis in the hippocampus [134, 135, 139, 141]. Also, voluntary wheel running exercise for 2 months inhibits microglial activation in aged mice [142]. Moreover, exercise reduces the potentiated inflammatory response of aged hippocampal microglia to pro-inflammatory stimulation [143]. Daily exercise decreases pro-inflammatory cytokines (IL-1β [144] and IL-6 [145]), but increases anti-inflammatory cytokine (IL-10) [145] in the hippocampus. It has been demonstrated that 3 months of mild-intensity treadmill running upregulate BDNF-TrkB-Erk-CREB signaling pathway [146] and inhibit aging-related microglial activation and nigral dopaminergic neuron loss in mice [147].

2.3 *Molecular Level*

2.3.1 Brain-Derived Neurotrophic Factor

The molecular mechanisms driving exercise-induced neuroplasticity encompass various neurotrophic and growth factors. Brain-derived neurotrophic factor (BDNF) holds a pivotal role in neurogenesis [148], neuronal development [149], plasticity [150, 151], and survival [152]. Evidence from both animal and human studies indicates that exercise upregulates the expression BDNF, which plays a crucial role in the induction of exercise-induced neuroplasticity [153–156]. The levels of BDNF increase in the hippocampus of both young and aged animals following different types of exercise, such as voluntary wheel-running and treadmill running [16, 25, 27, 157, 158].

BDNF and its receptor, tropomyosin receptor kinase B (TrkB), regulate intracellular signaling cascades, including the Ras/mitogen-activated protein kinase (MAPK)-ERK pathway [159, 160], the phosphatidylinositol-3-kinase (PI3K)-protein kinase B (Akt) pathways [161, 162], and phospholipase C-γ (PLCγ). Subsequently, this signaling amplifies the phosphorylation levels of cAMP response-element-binding (CREB) and promotes the translation of proteins related to synaptic plasticity [16, 27, 163–166], ultimately enhancing neuroplasticity.

Furthermore, TrkB activation can also increase the expression of peroxisome proliferator-activated receptor γ co-activator α (PGC1α), which in turn elevates BDNF expression in neurons through the PGC1α/FNDC5/BDNF pathway [167, 168]. Mice exhibiting higher levels of BDNF and TrkB receptors in the hippocampus demonstrate superior memory performance [20, 25]. Conversely, blocking the TrkB receptor with a TrkB receptor-IgG chimera disrupts the cognitive benefits of exercise through inhibiting exercise-induced upregulations of synaptic plasticity-related [163, 164]. Results from animal behavioral tests utilizing the Morris water maze [169], passive avoidance [25], and contextual fear condition [27] also suggest that blocking TrkB receptors diminishes the learning and memory improvements associated with exercise. Furthermore, a randomized controlled trial revealed that an increased BDNF level after exercise is linked to an augmented hippocampal volume in elderly individuals [71].

Six weeks of either daily forced swimming or voluntary running exercises elevate hippocampal BDNF levels and enhance short-term and long-term memories in male Wister rats [170]. Additionally, 5 weeks of mild-intensity treadmill running is enough to reverse the age-related decline in BDNF and TrkB and promote adult hippocampal neurogenesis (AHN) in middle-aged mice [171]. Moreover, the same exercise regimen can improve the adverse impact of lipopolysaccharide on AHN and learning and memory performance, potentially by restoring the BDNF signaling pathway compromised by lipopolysaccharide [172]. The BDNF induced by exercise contributes to improved cognitive and brain function through the BDNF pathway.

2.3.2 Insulin-Like Growth Factor-1

Insulin-like growth factor-1 (IGF-1) is a peptide secreted by the liver and various tissues, playing a role in neuronal functions related to synaptic plasticity, synapse density, neurotransmission, neurogenesis, and neuron differentiation in the central nervous system [173–177]. Additionally, there is a positive correlation between levels of IGF-1 and changes in hippocampal volume, as well as memory function, in aged adults [178]. Chronic exercise enhances IGF-1 signaling in the hippocampus, which influences AHN, as well as learning and memory performance [179–182].

The level of IGF-1 increases in both the hippocampus and blood of adolescent rats engaged in voluntary and mandatory physical exercise, positively correlated with their learning and memory abilities [183]. It has been found that the levels of IGF-1 in the brain increase after exercise, caused by the uptake of IGF-1 from the circulation [182, 184]. A single bout of 30 min of moderate-intensity aerobic exercise increases serum levels of BDNF and IGF-1, whereas acute resistance exercise only increases serum IGF-1 levels [185]. Nevertheless, the elevated levels of IGF-1 and other molecular markers from exercise return to baseline approximately 20 min after exercise in elderly individuals with mild cognitive impairment [185].

Neutralization of the hippocampal IGF-1 receptor inhibits the exercise-induced uptake of IGF-1 to the brain, subsequently impacting AHN and diminishing the effects of exercise on memory retention [176, 180, 182]. The exercise-induced upregulation of BDNF may be influenced by the IGF-1 pathway, as blocking the IGF-1 receptor during exercise hinders the exercise effects on the augmentation of pro-BDNF and BDNF in the hippocampus [176].

2.3.3 Nerve Growth Factor

Nerve growth factor NGF is essential for normal development of the central nervous system, the survival of neural progenitors, and the promotion of neuronal function [186], particularly in mature cholinergic neurons within the basal forebrain [187, 188]. The activity of forebrain cholinergic neurons is crucial for enhancing neurotransmitter function and improving learning and memory processes [189, 190].

A study involving 96 Italian pre-Olympic athletes revealed that prolonged physical exercise significantly increased NGF serum levels compared to the control group [191]. In a randomized controlled clinical trial with 60 multiple sclerosis patients engaged in daily moderate cycling exercises for 30 min over 3 weeks, either on a cycle ergometer or an aquatic bike, cardiorespiratory values significantly improved in both groups over time. Interestingly, BDNF concentration increased after water cycling but not after land cycling. However, NGF concentration did not change after either form of exercise [192].

In rats, engagement in wheel-running exercise leads to elevated levels of both BDNF and NGF mRNA in the hippocampus [193, 194]. Additionally, exercise can restore deficits in BDNF and NGF in rats affected by amnesia due to thiamine deficiency [195]. The expressions of hippocampal NGF and its receptor, tropomyosin receptor kinase A (TrkA), increase after 8 weeks of treadmill running in rodents. It has been shown that continuous infusion of NGF enhances the survival of new neurons in the granule cell layer of the dentate gyrus in young adult rats, but not in aged rats. NGF, potentially mediated through increased cholinergic tone, promotes AHN, reflecting the neurotropic action of NGF [186]. Similar to BDNF/TrkB signaling pathway, the NGF/TrkA signaling pathway stimulates the downstream transcription factor CREB and triggers various gene transcriptions related to cell survival and neuroplasticity [196].

2.3.4 Vascular Endothelial Growth Factor

Vascular endothelial growth factor (VEGF), an angiogenic protein, has been demonstrated to play an important role in regulating hippocampal progenitor proliferation [197] and exerting neuroprotective functions [198, 199]. VEGF primarily acts during the early (proliferative) phase of neurogenesis [200]. The synthesis and release of VEGF can occur in various tissues, including peripheral vascular endothelial cells, skeletal myofiber, and brain cells such as astrocytes, ependymal cells, and neuronal stem cells [201]. Administration of VEGF or the gene transfer of VEGF in the hippocampus increases number of BrdU-labeled cells in the subventricular zone and the subgranular zone of the dentate gyrus, suggesting a strong association between angiogenesis and neurogenesis [202, 203].

Regular treadmill exercise at an intensity sufficient to elevate hippocampal VEGF levels can directly or indirectly regulate both hippocampal blood flow and neurogenesis [204]. Exercise training has been shown to enhance cognitive function in Aβ-injected rats, possibly by augmenting VEGF signaling and reducing angiostatin [205].

Voluntary exercise increases the expressions of VEGF receptors, specifically fms-like tyrosine kinase (Flt-1/VEGF1) in vessels and fetal liver kinase (Flk-1 or VEGFR2) in astrocytes, in the hippocampus [206]. The activation of the intracellular tyrosine kinase domains of VEGF receptors triggers downstream signaling pathways, enhancing the proliferation of neuron precursors within an angiogenic niche [207, 208]. Peripheral VEGF appears to play an essential role in exercise-induced AHN. Blockade of VEGF signaling, achieved through intravenous injection of adenoviruses carrying the chimeric VEGFR1 receptor or conditional knockdown of skeletal myofiber-specific VEGF gene, abolishes the exercise-induced AHN [209, 210]. Decreases in VEGF expression induced by stress are associated with a reduction in the cell proliferation of progenitors closely located to the vasculature [211]. Blocking VEGF signaling eliminates the exercise-induced increase in BrdU-positive cell number (proliferation/short-term survival) [209] and hinders the augmentation in progenitor proliferation and behavioral effects of antidepressants such as desipramine [212].

2.3.5 Glucocorticoids

Glucocorticoids (GCs) are the endogenous adrenal steroids in most mammals, including humans [213, 214], with corticosterone (CORT) being the primary form of GCs in laboratory rodents [215, 216]. Stress activates the hypothalamic–pituitary–adrenal axis, leading to the release of stress hormones, such as CORT, from the adrenal cortex into circulation. These hormones are capable of crossing the blood–brain barrier. Administration of high levels of exogenous CORT diminishes AHN, while depleting CORT through bilateral adrenalectomy enhances AHN [217–220]. AHN is known to play crucial roles in the formation, exhibition, and/or extinction of specific types of hippocampus-dependent memory [218, 221].

Prolonged exposure to chronic stress adversely affect AHN and memory performance, which is attributed to increased levels of CORT and decreased levels of BDNF [222].

It is noteworthy that engaging in mild-to-moderate intensity running increases AHN, despite the concurrent rise in the activity of the hypothalamic–pituitary–adrenal axis and plasma GC levels during exercise [172, 218, 220]. This exercise–GC paradox might be elucidated by the more substantial positive effects on AHN, such as IGF-1 and BDNF, compared to the negative effects induced by exercise, including GCs and oxidative stress. Another possible explanation is the existence of an inverted-U-shaped relationship between GC levels and AHN, similar to the relationship described for GC levels and spatial memory [223]. It is plausible that the elevated GC levels prompted by exercise fall within an optimal range along the inverted-U curve. Support for this hypothesis comes from observations that the depletion of GCs by adrenalectomy decreased the mild/moderate running-enhanced AHN [220]. However, mice that received a daily injection of CORT to evoke serum CORT levels similar to those during exercise for 4 weeks did not affect AHN [218]. To dissect the underlying mechanism, the authors employed a unique mouse model, featuring normal basal levels of CORT but a blunted CORT response to exercise, demonstrated that exercise-induced CORT contributes to the exercise-enhanced AHN [218]. While CORT alone is inadequate to stimulate AHN, elevated levels of CORT can impede the process.

Finally, a prolonged aerobic exercise intervention has the potential to induce adaptations in stress pathways [224]. It has been shown that in rats with alimentary obesity, the stress response to moderate-intensity exercise results in heightened testosterone levels, mitigating the catabolic effects of CORT and promoting an increase in the weight of reproductive organs [225].

3 Conclusion

Exercise proves to be a valuable approach for enhancing cognitive functions and promoting brain health across all age groups, particularly in the elderly who are more susceptible to neurodegenerative disorders. The cognitive protective mechanisms of exercise training may involve improved cardiorespiratory fitness, trophic/growth factors, and enhanced function of glial cells. Despite ongoing debates about the minimum effective dosage of exercise to mitigate cognitive impairment, consistently engaging in mild-to-moderate intensity aerobic activities proves beneficial for enhancing neurocognitive performance in adults.

References

1. Bull FC, Al-Ansari SS, Biddle S, Borodulin K, Buman MP, Cardon G, Carty C, Chaput JP, Chastin S, Chou R et al (2020) World Health Organization 2020 guidelines on physical activity and sedentary behaviour. Br J Sports Med 54:1451–1462. https://doi.org/10.1136/bjsports-2020-102955
2. Lubans D, Richards J, Hillman C, Faulkner G, Beauchamp M, Nilsson M, Kelly P, Smith J, Raine L, Biddle S (2016) Physical activity for cognitive and mental health in youth: a systematic review of mechanisms. Pediatrics 138. https://doi.org/10.1542/peds.2016-1642
3. Jakicic JM, Kraus WE, Powell KE, Campbell WW, Janz KF, Troiano RP, Sprow K, Torres A, Piercy KL, Physical Activity Guidelines Advisory, C (2019) Association between bout duration of physical activity and health: systematic review. Med Sci Sports Exerc 51:1213–1219. https://doi.org/10.1249/MSS.0000000000001933
4. White RL, Babic MJ, Parker PD, Lubans DR, Astell-Burt T, Lonsdale C (2017) Domain-Specific physical activity and mental health: a meta-analysis. Am J Prev Med 52:653–666. https://doi.org/10.1016/j.amepre.2016.12.008
5. Jakicic JM, Powell KE, Campbell WW, Dipietro L, Pate RR, Pescatello LS, Collins KA, Bloodgood B, Piercy KL, Physical Activity Guidelines Advisory, C (2019) Physical activity and the prevention of weight gain in adults: a systematic review. Med Sci Sports Exerc 51:1262–1269. https://doi.org/10.1249/MSS.0000000000001938
6. Schuch FB, Vancampfort D (2021) Physical activity, exercise, and mental disorders: it is time to move on. Trends Psychiatry Psychother 43:177–184. https://doi.org/10.47626/2237-6089-2021-0237
7. Ekelund U, Tarp J, Steene-Johannessen J, Hansen BH, Jefferis B, Fagerland MW, Whincup P, Diaz KM, Hooker SP, Chernofsky A et al (2019) Dose-response associations between accelerometry measured physical activity and sedentary time and all cause mortality: systematic review and harmonised meta-analysis. BMJ 366:l4570. https://doi.org/10.1136/bmj.l4570
8. Araki E, Tanaka A, Inagaki N, Ito H, Ueki K, Murohara T, Imai K, Sata M, Sugiyama T, Ishii H et al (2021) Diagnosis, prevention, and treatment of cardiovascular diseases in people with type 2 diabetes and prediabetes: a consensus statement jointly from the japanese circulation society and the japan diabetes society. Diabetol Int 12:1–51. https://doi.org/10.1007/s13340-020-00471-5
9. Colberg SR, Sigal RJ, Fernhall B, Regensteiner JG, Blissmer BJ, Rubin RR, Chasan-Taber L, Albright AL, Braun B, American College of Sports, M, American Diabetes, A (2010) Exercise and type 2 diabetes: the American College of Sports Medicine and the American Diabetes Association: joint position statement. Diabetes Care 33:e147–e167. https://doi.org/10.2337/dc10-9990
10. McTiernan A, Friedenreich CM, Katzmarzyk PT, Powell KE, Macko R, Buchner D, Pescatello LS, Bloodgood B, Tennant B, Vaux-Bjerke A et al (2019) Physical activity in cancer prevention and survival: a systematic review. Med Sci Sports Exerc 51:1252–1261. https://doi.org/10.1249/MSS.0000000000001937
11. O'Callaghan RM, Ohle R, Kelly AM (2007) The effects of forced exercise on hippocampal plasticity in the rat: a comparison of LTP, spatial- and non-spatial learning. Behav Brain Res 176:362–366. https://doi.org/10.1016/j.bbr.2006.10.018
12. Griffin EW, Bechara RG, Birch AM, Kelly AM (2009) Exercise enhances hippocampal-dependent learning in the rat: evidence for a BDNF-related mechanism. Hippocampus 19:973–980. https://doi.org/10.1002/hipo.20631
13. Bechara RG, Lyne R, Kelly AM (2014) BDNF-stimulated intracellular signalling mechanisms underlie exercise-induced improvement in spatial memory in the male Wistar rat. Behav Brain Res 275:297–306. https://doi.org/10.1016/j.bbr.2013.11.015
14. Ang ET, Dawe GS, Wong PT, Moochhala S, Ng YK (2006) Alterations in spatial learning and memory after forced exercise. Brain Res 1113:186–193. https://doi.org/10.1016/j.brainres.2006.07.023

15. van Praag H, Christie BR, Sejnowski TJ, Gage FH (1999) Running enhances neurogenesis, learning, and long-term potentiation in mice. Proc Natl Acad Sci USA 96:13427–13431. https://doi.org/10.1073/pnas.96.23.13427
16. Liu YF, Chen HI, Wu CL, Kuo YM, Yu L, Huang AM, Wu FS, Chuang JI, Jen CJ (2009) Differential effects of treadmill running and wheel running on spatial or aversive learning and memory: roles of amygdalar brain-derived neurotrophic factor and synaptotagmin I. J Physiol 587:3221–3231. https://doi.org/10.1113/jphysiol.2009.173088
17. Taati M, Barzegar PEF, Raisi A (2022) Exercise improves spatial learning and memory performance through the central GLP-1 receptors. Behav Neurol 2022:2900628. https://doi.org/10.1155/2022/2900628
18. Inoue K, Hanaoka Y, Nishijima T, Okamoto M, Chang H, Saito T, Soya H (2015) Long-term mild exercise training enhances hippocampus-dependent memory in rats. Int J Sports Med 36:280–285. https://doi.org/10.1055/s-0034-1390465
19. Cassilhas RC, Lee KS, Fernandes J, Oliveira MG, Tufik S, Meeusen R, de Mello MT (2012) Spatial memory is improved by aerobic and resistance exercise through divergent molecular mechanisms. Neuroscience 202:309–317. https://doi.org/10.1016/j.neuroscience.2011.11.029
20. Berchtold NC, Castello N, Cotman CW (2010) Exercise and time-dependent benefits to learning and memory. Neuroscience 167:588–597. https://doi.org/10.1016/j.neuroscience.2010.02.050
21. Alomari MA, Alzoubi KH, Khabour OF (2021) Swimming exercise improves short- and long-term memories: time-course changes. Physiol Rep 9:e14851. https://doi.org/10.14814/phy2.14851
22. Dao AT, Zagaar MA, Levine AT, Salim S, Eriksen JL, Alkadhi KA (2013) Treadmill exercise prevents learning and memory impairment in Alzheimer's disease-like pathology. Curr Alzheimer Res 10:507–515. https://doi.org/10.2174/1567205011310050006
23. Van der Borght K, Havekes R, Bos T, Eggen BJ, Van der Zee EA (2007) Exercise improves memory acquisition and retrieval in the Y-maze task: relationship with hippocampal neurogenesis. Behav Neurosci 121:324–334. https://doi.org/10.1037/0735-7044.121.2.324
24. Chen HI, Lin LC, Yu L, Liu YF, Kuo YM, Huang AM, Chuang JI, Wu FS, Liao PC, Jen CJ (2008) Treadmill exercise enhances passive avoidance learning in rats: the role of downregulated serotonin system in the limbic system. Neurobiol Learn Mem 89:489–496. https://doi.org/10.1016/j.nlm.2007.08.004
25. Liu YF, Chen HI, Yu L, Kuo YM, Wu FS, Chuang JI, Liao PC, Jen CJ (2008) Upregulation of hippocampal TrkB and synaptotagmin is involved in treadmill exercise-enhanced aversive memory in mice. Neurobiol Learn Mem 90:81–89. https://doi.org/10.1016/j.nlm.2008.02.005
26. Baruch DE, Swain RA, Helmstetter FJ (2004) Effects of exercise on pavlovian fear conditioning. Behav Neurosci 118:1123–1127. https://doi.org/10.1037/0735-7044.118.5.1123
27. Lin TW, Chen SJ, Huang TY, Chang CY, Chuang JI, Wu FS, Kuo YM, Jen CJ (2012) Different types of exercise induce differential effects on neuronal adaptations and memory performance. Neurobiol Learn Mem 97:140–147. https://doi.org/10.1016/j.nlm.2011.10.006
28. Greenwood BN, Strong PV, Foley TE, Fleshner M (2009) A behavioral analysis of the impact of voluntary physical activity on hippocampus-dependent contextual conditioning. Hippocampus 19:988–1001. https://doi.org/10.1002/hipo.20534
29. El-Domiaty HF, El-Roghy ES, Salem HR (2022) Combination of magnesium supplementation with treadmill exercise improves memory deficit in aged rats by enhancing hippocampal neurogenesis and plasticity: a functional and histological study. Appl Physiol Nutr Metab 47:296–308. https://doi.org/10.1139/apnm-2021-0133
30. Vanzella C, Sanches EF, Odorcyk FK, Nicola F, Kolling J, Longoni A, Dos Santos TM, Wyse ATS, Netto CA (2017) Forced treadmill exercise prevents spatial memory deficits in aged rats probably through the activation of Na(+), K(+)-ATPase in the hippocampus. Neurochem Res 42:1422–1429. https://doi.org/10.1007/s11064-017-2196-x

31. Albeck DS, Sano K, Prewitt GE, Dalton L (2006) Mild forced treadmill exercise enhances spatial learning in the aged rat. Behav Brain Res 168:345–348. https://doi.org/10.1016/j.bbr.2005.11.008
32. Tsai SF, Ku NW, Wang TF, Yang YH, Shih YH, Wu SY, Lee CW, Yu M, Yang TT, Kuo YM (2018) Long-term moderate exercise rescues age-related decline in hippocampal neuronal complexity and memory. Gerontology 64:551–561. https://doi.org/10.1159/000488589
33. Pietrelli A, Lopez-Costa J, Goni R, Brusco A, Basso N (2012) Aerobic exercise prevents age-dependent cognitive decline and reduces anxiety-related behaviors in middle-aged and old rats. Neuroscience 202:252–266. https://doi.org/10.1016/j.neuroscience.2011.11.054
34. Pietrelli A, Lopez-Costa JJ, Goni R, Lopez EM, Brusco A, Basso N (2011) Effects of moderate and chronic exercise on the nitrergic system and behavioral parameters in rats. Brain Res 1389:71–82. https://doi.org/10.1016/j.brainres.2011.03.005
35. Kim SE, Ko IG, Kim BK, Shin MS, Cho S, Kim CJ, Kim SH, Baek SS, Lee EK, Jee YS (2010) Treadmill exercise prevents aging-induced failure of memory through an increase in neurogenesis and suppression of apoptosis in rat hippocampus. Exp Gerontol 45:357–365. https://doi.org/10.1016/j.exger.2010.02.005
36. Lovatel GA, Elsner VR, Bertoldi K, Vanzella C, Moyses Fdos S, Vizuete A, Spindler C, Cechinel LR, Netto CA, Muotri AR, Siqueira IR (2013) Treadmill exercise induces age-related changes in aversive memory, neuroinflammatory and epigenetic processes in the rat hippocampus. Neurobiol Learn Mem 101:94–102. https://doi.org/10.1016/j.nlm.2013.01.007
37. de Meireles LC, Bertoldi K, Cechinel LR, Schallenberger BL, da Silva VK, Schroder N, Siqueira IR (2016) Treadmill exercise induces selective changes in hippocampal histone acetylation during the aging process in rats. Neurosci Lett 634:19–24. https://doi.org/10.1016/j.neulet.2016.10.008
38. de Meireles LCF, Galvao F Jr, Walker DM, Cechinel LR, de Souza Grefenhagen AI, Andrade G, Palazzo RP, Lovatel GA, Basso CG, Nestler EJ, Siqueira IR (2019) Exercise modalities improve aversive memory and survival rate in aged rats: role of hippocampal epigenetic modifications. Mol Neurobiol 56:8408–8419. https://doi.org/10.1007/s12035-019-01675-w
39. Belcher BR, Zink J, Azad A, Campbell CE, Chakravartti SP, Herting MM (2021) The roles of physical activity, exercise, and fitness in promoting resilience during adolescence: effects on mental well-being and brain development. Biol Psychiatry Cogn Neurosci Neuroimaging 6:225–237. https://doi.org/10.1016/j.bpsc.2020.08.005
40. Herting MM, Chu X (2017) Exercise, cognition, and the adolescent brain. Birth Defects Res 109:1672–1679. https://doi.org/10.1002/bdr2.1178
41. Muntaner-Mas A, Mazzoli E, Abbott G, Mavilidi MF, Galmes-Panades AM (2022) Do physical fitness and executive function mediate the relationship between physical activity and academic achievement? An examination using structural equation modelling. Children (Basel):9. https://doi.org/10.3390/children9060823
42. Christiansen L, Beck MM, Bilenberg N, Wienecke J, Astrup A, Lundbye-Jensen J (2019) Effects of exercise on cognitive performance in children and adolescents with ADHD: potential mechanisms and evidence-based recommendations. J Clin Med 8. https://doi.org/10.3390/jcm8060841
43. Donnelly JE, Hillman CH, Castelli D, Etnier JL, Lee S, Tomporowski P, Lambourne K, Szabo-Reed AN (2016) Physical activity, fitness, cognitive function, and academic achievement in children: a systematic review. Med Sci Sports Exerc 48:1197–1222. https://doi.org/10.1249/MSS.0000000000000901
44. de Greeff JW, Bosker RJ, Oosterlaan J, Visscher C, Hartman E (2018) Effects of physical activity on executive functions, attention and academic performance in preadolescent children: a meta-analysis. J Sci Med Sport 21:501–507. https://doi.org/10.1016/j.jsams.2017.09.595
45. Verburgh L, Konigs M, Scherder EJ, Oosterlaan J (2014) Physical exercise and executive functions in preadolescent children, adolescents and young adults: a meta-analysis. Br J Sports Med 48:973–979. https://doi.org/10.1136/bjsports-2012-091441

46. Li X, Wang Z, Wang Y, Li X, Li D (2022) Effect of exercise on inhibitory control is dose-dependent for adolescents. Sports Med Health Sci 4:54–60. https://doi.org/10.1016/j.smhs.2021.10.005
47. Haverkamp BF, Wiersma R, Vertessen K, van Ewijk H, Oosterlaan J, Hartman E (2020) Effects of physical activity interventions on cognitive outcomes and academic performance in adolescents and young adults: a meta-analysis. J Sports Sci 38:2637–2660. https://doi.org/10.1080/02640414.2020.1794763
48. Loprinzi PD, Roig M, Tomporowski PD, Javadi AH, Kelemen WL (2022) Effects of acute exercise on memory: considerations of exercise intensity, post-exercise recovery period and aerobic endurance. Mem Cogn:1–16. https://doi.org/10.3758/s13421-022-01373-4
49. Loprinzi PD, Frith E, Edwards MK, Sng E, Ashpole N (2018) The effects of exercise on memory function among young to middle-aged adults: systematic review and recommendations for future research. Am J Health Promot 32:691–704. https://doi.org/10.1177/0890117117737409
50. Smith PJ, Blumenthal JA, Hoffman BM, Cooper H, Strauman TA, Welsh-Bohmer K, Browndyke JN, Sherwood A (2010) Aerobic exercise and neurocognitive performance: a meta-analytic review of randomized controlled trials. Psychosom Med 72:239–252. https://doi.org/10.1097/PSY.0b013e3181d14633
51. Loprinzi PD, Green D, Wages S, Cheke LG, Jones T (2020) Experimental effects of acute high-intensity resistance exercise on episodic memory function: consideration for post-exercise recovery period. J Lifestyle Med 10:7–20. https://doi.org/10.15280/jlm.2020.10.1.7
52. Wilke J, Giesche F, Klier K, Vogt L, Herrmann E, Banzer W (2019) Acute effects of resistance exercise on cognitive function in healthy adults: a systematic review with multilevel meta-analysis. Sports Med 49:905–916. https://doi.org/10.1007/s40279-019-01085-x
53. de Diego-Moreno M, Alvarez-Salvago F, Martinez-Amat A, Boquete-Pumar C, Orihuela-Espejo A, Aibar-Almazan A, Jimenez-Garcia JD (2022) Acute effects of high-intensity functional training and moderate-intensity continuous training on cognitive functions in young adults. Int J Environ Res Public Health 19. https://doi.org/10.3390/ijerph191710608
54. Ben-Zeev T, Weiss I, Ashri S, Heled Y, Ketko I, Yanovich R, Okun E (2020) Mild physical activity does not improve spatial learning in a virtual environment. Front Behav Neurosci 14:584052. https://doi.org/10.3389/fnbeh.2020.584052
55. Kesner RP (2013) Role of the hippocampus in mediating interference as measured by pattern separation processes. Behav Process 93:148–154. https://doi.org/10.1016/j.beproc.2012.09.018
56. Suwabe K, Hyodo K, Byun K, Ochi G, Yassa MA, Soya H (2017) Acute moderate exercise improves mnemonic discrimination in young adults. Hippocampus 27:229–234. https://doi.org/10.1002/hipo.22695
57. Suwabe K, Hyodo K, Byun K, Ochi G, Fukuie T, Shimizu T, Kato M, Yassa MA, Soya H (2017) Aerobic fitness associates with mnemonic discrimination as a mediator of physical activity effects: evidence for memory flexibility in young adults. Sci Rep 7:5140. https://doi.org/10.1038/s41598-017-04850-y
58. Rathore A, Lom B (2017) The effects of chronic and acute physical activity on working memory performance in healthy participants: a systematic review with meta-analysis of randomized controlled trials. Syst Rev 6:124. https://doi.org/10.1186/s13643-017-0514-7
59. Bherer L (2015) Cognitive plasticity in older adults: effects of cognitive training and physical exercise. Ann N Y Acad Sci 1337:1–6. https://doi.org/10.1111/nyas.12682
60. Angevaren M, Aufdemkampe G, Verhaar HJ, Aleman A, Vanhees L (2008) Physical activity and enhanced fitness to improve cognitive function in older people without known cognitive impairment. Cochrane Database Syst Rev CD005381. https://doi.org/10.1002/14651858.CD005381.pub2
61. Dascal JB, Sanders LMJ, Filho E, Hortobagyi T (2019) Dose-response effects of years of self-reported physical activity on old females' motor and cognitive function. Braz J Phys Ther 23:48–55. https://doi.org/10.1016/j.bjpt.2018.06.002

62. Buchman AS, Yu L, Wilson RS, Lim A, Dawe RJ, Gaiteri C, Leurgans SE, Schneider JA, Bennett DA (2019) Physical activity, common brain pathologies, and cognition in community-dwelling older adults. Neurology 92:e811–e822. https://doi.org/10.1212/WNL.0000000000006954
63. Boujelbane MA, Trabelsi K, Jahrami HA, Masmoudi L, Ammar A, Khacharem A, Boukhris O, Puce L, Garbarino S, Scoditti E et al (2022) Time-restricted feeding and cognitive function in sedentary and physically active elderly individuals: Ramadan diurnal intermittent fasting as a model. Front Nutr 9:1041216. https://doi.org/10.3389/fnut.2022.1041216
64. Northey JM, Cherbuin N, Pumpa KL, Smee DJ, Rattray B (2018) Exercise interventions for cognitive function in adults older than 50: a systematic review with meta-analysis. Br J Sports Med 52:154–160. https://doi.org/10.1136/bjsports-2016-096587
65. Hindin SB, Zelinski EM (2012) Extended practice and aerobic exercise interventions benefit untrained cognitive outcomes in older adults: a meta-analysis. J Am Geriatr Soc 60:136–141. https://doi.org/10.1111/j.1532-5415.2011.03761.x
66. Kettinen J, Tikkanen H, Hiltunen M, Murray A, Horn N, Taylor WR, Venojarvi M (2023) Cognitive and biomarker responses in healthy older adults to a 18-hole golf round and different walking types: a randomised cross-over study. BMJ Open Sport Exerc Med 9:e001629. https://doi.org/10.1136/bmjsem-2023-001629
67. Roig-Coll F, Castells-Sanchez A, Lamonja-Vicente N, Toran-Monserrat P, Pera G, Garcia-Molina A, Tormos JM, Montero-Alia P, Alzamora MT, Dacosta-Aguayo R et al (2020) Effects of aerobic exercise, cognitive and combined training on cognition in physically inactive healthy late-middle-aged adults: the projecte moviment randomized controlled trial. Front Aging Neurosci 12:590168. https://doi.org/10.3389/fnagi.2020.590168
68. Bherer L, Langeard A, Kaushal N, Vrinceanu T, Desjardins-Crepeau L, Langlois F, Kramer AF (2021) Physical exercise training effect and mediation through cardiorespiratory fitness on dual-task performances differ in younger-old and older-old adults. J Gerontol B Psychol Sci Soc Sci 76:219–228. https://doi.org/10.1093/geronb/gbz066
69. Chen FT, Etnier JL, Chan KH, Chiu PK, Hung TM, Chang YK (2020) Effects of exercise training interventions on executive function in older adults: a systematic review and meta-analysis. Sports Med 50:1451–1467. https://doi.org/10.1007/s40279-020-01292-x
70. Barha CK, Davis JC, Falck RS, Nagamatsu LS, Liu-Ambrose T (2017) Sex differences in exercise efficacy to improve cognition: a systematic review and meta-analysis of randomized controlled trials in older humans. Front Neuroendocrinol 46:71–85. https://doi.org/10.1016/j.yfrne.2017.04.002
71. Erickson KI, Voss MW, Prakash RS, Basak C, Szabo A, Chaddock L, Kim JS, Heo S, Alves H, White SM et al (2011) Exercise training increases size of hippocampus and improves memory. Proc Natl Acad Sci USA 108:3017–3022. https://doi.org/10.1073/pnas.1015950108
72. Yu F, Vock DM, Zhang L, Salisbury D, Nelson NW, Chow LS, Smith G, Barclay TR, Dysken M, Wyman JF (2021) Cognitive effects of aerobic exercise in Alzheimer's disease: a pilot randomized controlled trial. J Alzheimer's Dis 80:233–244. https://doi.org/10.3233/JAD-201100
73. Biazus-Sehn LF, Schuch FB, Firth J, Stigger FS (2020) Effects of physical exercise on cognitive function of older adults with mild cognitive impairment: a systematic review and meta-analysis. Arch Gerontol Geriatr 89:104048. https://doi.org/10.1016/j.archger.2020.104048
74. Sanders LMJ, Hortobagyi T, la Bastide-van Gemert S, van der Zee EA, van Heuvelen MJG (2019) Dose-response relationship between exercise and cognitive function in older adults with and without cognitive impairment: a systematic review and meta-analysis. PLoS One 14:e0210036. https://doi.org/10.1371/journal.pone.0210036
75. Voss MW, Vivar C, Kramer AF, van Praag H (2013) Bridging animal and human models of exercise-induced brain plasticity. Trends Cogn Sci 17:525–544. https://doi.org/10.1016/j.tics.2013.08.001
76. Chen J, Zhou R, Feng Y, Cheng L (2022) Molecular mechanisms of exercise contributing to tissue regeneration. Signal Transduct Target Ther 7:383. https://doi.org/10.1038/s41392-022-01233-2

77. Ryan SM, Kelly AM (2016) Exercise as a pro-cognitive, pro-neurogenic and anti-inflammatory intervention in transgenic mouse models of Alzheimer's disease. Ageing Res Rev 27:77–92. https://doi.org/10.1016/j.arr.2016.03.007
78. Zhou B, Zuo YX, Jiang RT (2019) Astrocyte morphology: diversity, plasticity, and role in neurological diseases. CNS Neurosci Ther 25:665–673. https://doi.org/10.1111/cns.13123
79. Redila VA, Christie BR (2006) Exercise-induced changes in dendritic structure and complexity in the adult hippocampal dentate gyrus. Neuroscience 137:1299–1307. https://doi.org/10.1016/j.neuroscience.2005.10.050
80. Lattanzi D, Savelli D, Pagliarini M, Cuppini R, Ambrogini P (2022) Short-term, voluntary exercise affects morpho-functional maturation of adult-generated neurons in rat hippocampus. Int J Mol Sci 23. https://doi.org/10.3390/ijms23126866
81. Cariati I, Bonanni R, Pallone G, Scimeca M, Frank C, Tancredi V, D'Arcangelo G (2021) Hippocampal adaptations to continuous aerobic training: a functional and ultrastructural evaluation in a young murine model. J Funct Morphol Kinesiol 6. https://doi.org/10.3390/jfmk6040101
82. Dong TN, Kramar EA, Beardwood JH, Al-Shammari A, Wood MA, Keiser AA (2022) Temporal endurance of exercise-induced benefits on hippocampus-dependent memory and synaptic plasticity in female mice. Neurobiol Learn Mem 194:107658. https://doi.org/10.1016/j.nlm.2022.107658
83. Yang TT, Lo CP, Tsai PS, Wu SY, Wang TF, Chen YW, Jiang-Shieh YF, Kuo YM (2015) Aging and exercise affect hippocampal neurogenesis via different mechanisms. PLoS One 10:e0132152. https://doi.org/10.1371/journal.pone.0132152
84. Eddy MC, Green JT (2017) Running wheel exercise reduces renewal of extinguished instrumental behavior and alters medial prefrontal cortex neurons in adolescent, but not adult, rats. Behav Neurosci 131:460–469. https://doi.org/10.1037/bne0000218
85. Watanasriyakul WT, Normann MC, Akinbo OI, Colburn W, Dagner A, Grippo AJ (2019) Protective neuroendocrine effects of environmental enrichment and voluntary exercise against social isolation: evidence for mediation by limbic structures. Stress 22:603–618. https://doi.org/10.1080/10253890.2019.1617691
86. Huang TY, Lin LS, Cho KC, Chen SJ, Kuo YM, Yu L, Wu FS, Chuang JI, Chen HI, Jen CJ (1985) Chronic treadmill exercise in rats delicately alters the Purkinje cell structure to improve motor performance and toxin resistance in the cerebellum. J Appl Physiol 2012(113):889–895. https://doi.org/10.1152/japplphysiol.01363.2011
87. Mellow ML, Goldsworthy MR, Coussens S, Smith AE (2020) Acute aerobic exercise and neuroplasticity of the motor cortex: a systematic review. J Sci Med Sport 23:408–414. https://doi.org/10.1016/j.jsams.2019.10.015
88. Colcombe SJ, Erickson KI, Scalf PE, Kim JS, Prakash R, McAuley E, Elavsky S, Marquez DX, Hu L, Kramer AF (2006) Aerobic exercise training increases brain volume in aging humans. J Gerontol A Biol Sci Med Sci 61:1166–1170. https://doi.org/10.1093/gerona/61.11.1166
89. Colcombe SJ, Erickson KI, Raz N, Webb AG, Cohen NJ, McAuley E, Kramer AF (2003) Aerobic fitness reduces brain tissue loss in aging humans. J Gerontol A Biol Sci Med Sci 58:176–180. https://doi.org/10.1093/gerona/58.2.m176
90. Erickson KI, Prakash RS, Voss MW, Chaddock L, Hu L, Morris KS, White SM, Wojcicki TR, McAuley E, Kramer AF (2009) Aerobic fitness is associated with hippocampal volume in elderly humans. Hippocampus 19:1030–1039. https://doi.org/10.1002/hipo.20547
91. Ahlskog JE, Geda YE, Graff-Radford NR, Petersen RC (2011) Physical exercise as a preventive or disease-modifying treatment of dementia and brain aging. Mayo Clin Proc 86:876–884. https://doi.org/10.4065/mcp.2011.0252
92. Demnitz N, Stathi A, Withall J, Stainer C, Seager P, De Koning J, Esser P, Wassenaar T, Dawes H, Brooks J et al (2022) Hippocampal maintenance after a 12-month physical activity intervention in older adults: the REACT MRI study. Neuroimage Clin 35:102762. https://doi.org/10.1016/j.nicl.2021.102762

93. Voss MW, Heo S, Prakash RS, Erickson KI, Alves H, Chaddock L, Szabo AN, Mailey EL, Wojcicki TR, White SM et al (2013) The influence of aerobic fitness on cerebral white matter integrity and cognitive function in older adults: results of a one-year exercise intervention. Hum Brain Mapp 34:2972–2985. https://doi.org/10.1002/hbm.22119
94. Chen FT, Soya H, Yassa MA, Li RH, Chu CH, Chen AG, Hung CL, Chang YK (2022) Effects of exercise types on white matter microstructure in late midlife adults: preliminary results from a diffusion tensor imaging study. Front Aging Neurosci 14:943992. https://doi.org/10.3389/fnagi.2022.943992
95. Kronman CA, Kern KL, Nauer RK, Dunne MF, Storer TW, Schon K (2020) Cardiorespiratory fitness predicts effective connectivity between the hippocampus and default mode network nodes in young adults. Hippocampus 30:526–541. https://doi.org/10.1002/hipo.23169
96. Tseng BY, Uh J, Rossetti HC, Cullum CM, Diaz-Arrastia RF, Levine BD, Lu H, Zhang R (2013) Masters athletes exhibit larger regional brain volume and better cognitive performance than sedentary older adults. J Magn Reson Imaging 38:1169–1176. https://doi.org/10.1002/jmri.24085
97. Voss MW, Prakash RS, Erickson KI, Basak C, Chaddock L, Kim JS, Alves H, Heo S, Szabo AN, White SM et al (2010) Plasticity of brain networks in a randomized intervention trial of exercise training in older adults. Front Aging Neurosci 2. https://doi.org/10.3389/fnagi.2010.00032
98. Yogev-Seligmann G, Eisenstein T, Ash E, Giladi N, Sharon H, Nachman S, Bregman N, Kodesh E, Hendler T, Lerner Y (2021) Neurocognitive plasticity is associated with cardiorespiratory fitness following physical exercise in older adults with amnestic mild cognitive impairment. J Alzheimer's Dis 81:91–112. https://doi.org/10.3233/JAD-201429
99. Farhani F, Shahrbanian S, Auais M, Hekmatikar AHA, Suzuki K (2022) Effects of aerobic training on brain plasticity in patients with mild cognitive impairment: a systematic review of randomized controlled trials. Brain Sci 12. https://doi.org/10.3390/brainsci12060732
100. Colcombe SJ, Kramer AF, Erickson KI, Scalf P, McAuley E, Cohen NJ, Webb A, Jerome GJ, Marquez DX, Elavsky S (2004) Cardiovascular fitness, cortical plasticity, and aging. Proc Natl Acad Sci USA 101:3316–3321. https://doi.org/10.1073/pnas.0400266101
101. Won J, Callow DD, Pena GS, Jordan LS, Arnold-Nedimala NA, Nielson KA, Smith JC (2021) Hippocampal functional connectivity and memory performance after exercise intervention in older adults With mild cognitive impairment. J Alzheimer's Dis 82:1015–1031. https://doi.org/10.3233/JAD-210051
102. Maugeri G, D'Agata V, Magri B, Roggio F, Castorina A, Ravalli S, Di Rosa M, Musumeci G (2021) Neuroprotective effects of physical activity via the adaptation of astrocytes. Cells 10. https://doi.org/10.3390/cells10061542
103. von Bartheld CS, Bahney J, Herculano-Houzel S (2016) The search for true numbers of neurons and glial cells in the human brain: a review of 150 years of cell counting. J Comp Neurol 524:3865–3895. https://doi.org/10.1002/cne.24040
104. Santello M, Toni N, Volterra A (2019) Astrocyte function from information processing to cognition and cognitive impairment. Nat Neurosci 22:154–166. https://doi.org/10.1038/s41593-018-0325-8
105. Ullian EM, Sapperstein SK, Christopherson KS, Barres BA (2001) Control of synapse number by glia. Science 291:657–661. https://doi.org/10.1126/science.291.5504.657
106. Sancho L, Contreras M, Allen NJ (2021) Glia as sculptors of synaptic plasticity. Neurosci Res 167:17–29. https://doi.org/10.1016/j.neures.2020.11.005
107. Tsai SF, Chen PC, Calkins MJ, Wu SY, Kuo YM (2016) Exercise counteracts aging-related memory impairment: a potential role for the astrocytic metabolic shuttle. Front Aging Neurosci 8:57. https://doi.org/10.3389/fnagi.2016.00057
108. Huang B, Chen K, Li Y (2023) Aerobic exercise, an effective prevention and treatment for mild cognitive impairment. Front Aging Neurosci 15:1194559. https://doi.org/10.3389/fnagi.2023.1194559

109. He XF, Liu DX, Zhang Q, Liang FY, Dai GY, Zeng JS, Pei Z, Xu GQ, Lan Y (2017) Voluntary exercise promotes glymphatic clearance of amyloid beta and reduces the activation of astrocytes and microglia in aged mice. Front Mol Neurosci 10:144. https://doi.org/10.3389/fnmol.2017.00144
110. Saur L, Baptista PP, de Senna PN, Paim MF, do Nascimento P, Ilha J, Bagatini PB, Achaval M, Xavier LL (2014) Physical exercise increases GFAP expression and induces morphological changes in hippocampal astrocytes. Brain Struct Funct 219:293–302. https://doi.org/10.1007/s00429-012-0500-8
111. Fahimi A, Baktir MA, Moghadam S, Mojabi FS, Sumanth K, McNerney MW, Ponnusamy R, Salehi A (2017) Physical exercise induces structural alterations in the hippocampal astrocytes: exploring the role of BDNF-TrkB signaling. Brain Struct Funct 222:1797–1808. https://doi.org/10.1007/s00429-016-1308-8
112. de Senna PN, Bagatini PB, Galland F, Bobermin L, do Nascimento PS, Nardin P, Tramontina AC, Goncalves CA, Achaval M, Xavier LL (2017) Physical exercise reverses spatial memory deficit and induces hippocampal astrocyte plasticity in diabetic rats. Brain Res 1655:242–251. https://doi.org/10.1016/j.brainres.2016.10.024
113. Lundquist AJ, Parizher J, Petzinger GM, Jakowec MW (2019) Exercise induces region-specific remodeling of astrocyte morphology and reactive astrocyte gene expression patterns in male mice. J Neurosci Res 97:1081–1094. https://doi.org/10.1002/jnr.24430
114. Brockett AT, LaMarca EA, Gould E (2015) Physical exercise enhances cognitive flexibility as well as astrocytic and synaptic markers in the medial prefrontal cortex. PLoS One 10:e0124859. https://doi.org/10.1371/journal.pone.0124859
115. Siqueira Mendes FCC, Almeida MNF, Falsoni M, Andrade MLF, Felicio APG, Paixao L, Junior F, Anthony DC, Brites D, Diniz CWP, Sosthenes MCK (2022) The sedentary lifestyle and masticatory dysfunction: time to review the contribution to age-associated cognitive decline and astrocyte morphotypes in the dentate gyrus. Int J Mol Sci 23. https://doi.org/10.3390/ijms23116342
116. Belaya I, Ivanova M, Sorvari A, Ilicic M, Loppi S, Koivisto H, Varricchio A, Tikkanen H, Walker FR, Atalay M et al (2020) Astrocyte remodeling in the beneficial effects of long-term voluntary exercise in Alzheimer's disease. J Neuroinflammation 17:271. https://doi.org/10.1186/s12974-020-01935-w
117. Santin K, da Rocha RF, Cechetti F, Quincozes-Santos A, de Souza DF, Nardin P, Rodrigues L, Leite MC, Moreira JC, Salbego CG, Goncalves CA (2011) Moderate exercise training and chronic caloric restriction modulate redox status in rat hippocampus. Brain Res 1421:1–10. https://doi.org/10.1016/j.brainres.2011.08.003
118. Feng S, Wu C, Zou P, Deng Q, Chen Z, Li M, Zhu L, Li F, Liu TC, Duan R, Yang L (2023) High-intensity interval training ameliorates Alzheimer's disease-like pathology by regulating astrocyte phenotype-associated AQP4 polarization. Theranostics 13:3434–3450. https://doi.org/10.7150/thno.81951
119. Liu Y, Hu PP, Zhai S, Feng WX, Zhang R, Li Q, Marshall C, Xiao M, Wu T (2022) Aquaporin 4 deficiency eliminates the beneficial effects of voluntary exercise in a mouse model of Alzheimer's disease. Neural Regen Res 17:2079–2088. https://doi.org/10.4103/1673-5374.335169
120. Funfschilling U, Supplie LM, Mahad D, Boretius S, Saab AS, Edgar J, Brinkmann BG, Kassmann CM, Tzvetanova ID, Mobius W et al (2012) Glycolytic oligodendrocytes maintain myelin and long-term axonal integrity. Nature 485:517–521. https://doi.org/10.1038/nature11007
121. Fields RD (2008) White matter in learning, cognition and psychiatric disorders. Trends Neurosci 31:361–370. https://doi.org/10.1016/j.tins.2008.04.001
122. Falkai P, Rossner MJ, Raabe FJ, Wagner E, Keeser D, Maurus I, Roell L, Chang E, Seitz-Holland J, Schulze TG, Schmitt A (2023) Disturbed oligodendroglial maturation causes cognitive dysfunction in Schizophrenia: a new hypothesis. Schizophr Bull 49:1614–1624. https://doi.org/10.1093/schbul/sbad065

123. Ohtomo R, Kinoshita K, Ohtomo G, Takase H, Hamanaka G, Washida K, Islam MR, Wrann CD, Katsuki H, Iwata A et al (2020) Treadmill exercise suppresses cognitive decline and increases white matter oligodendrocyte precursor cells in a mouse model of prolonged cerebral hypoperfusion. Transl Stroke Res 11:496–502. https://doi.org/10.1007/s12975-019-00734-7
124. Jiang T, Zhang L, Pan X, Zheng H, Chen X, Li L, Luo J, Hu X (2017) physical exercise improves cognitive function together with microglia phenotype modulation and remyelination in chronic cerebral hypoperfusion. Front Cell Neurosci 11:404. https://doi.org/10.3389/fncel.2017.00404
125. Zhang L, Chao FL, Luo YM, Xiao Q, Jiang L, Zhou CN, Zhang Y, Lv FL, He Q, Ma J, Tang Y (2017) exercise prevents cognitive function decline and demyelination in the white matter of APP/PS1 transgenic AD mice. Curr Alzheimer Res 14:645–655. https://doi.org/10.2174/1567205014666161213121353
126. Qiu D, Zhou S, Donnelly J, Xia D, Zhao L (2023) Aerobic exercise attenuates abnormal myelination and oligodendrocyte differentiation in 3xTg-AD mice. Exp Gerontol 182:112293. https://doi.org/10.1016/j.exger.2023.112293
127. Bao C, He C, Shu B, Meng T, Cai Q, Li B, Wu G, Wu B, Li H (2021) Aerobic exercise training decreases cognitive impairment caused by demyelination by regulating ROCK signaling pathway in aging mice. Brain Res Bull 168:52–62. https://doi.org/10.1016/j.brainresbull.2020.12.010
128. Hammond TR, Robinton D, Stevens B (2018) Microglia and the brain: complementary partners in development and disease. Annu Rev Cell Dev Biol 34:523–544. https://doi.org/10.1146/annurev-cellbio-100616-060509
129. Li Q, Barres BA (2018) Microglia and macrophages in brain homeostasis and disease. Nat Rev Immunol 18:225–242. https://doi.org/10.1038/nri.2017.125
130. Valero J, Paris I, Sierra A (2016) Lifestyle shapes the dialogue between environment, microglia, and adult neurogenesis. ACS Chem Neurosci 7:442–453. https://doi.org/10.1021/acschemneuro.6b00009
131. Mee-Inta O, Zhao ZW, Kuo YM (2019) Physical exercise inhibits inflammation and microglial activation. Cells 8. https://doi.org/10.3390/cells8070691
132. Ehninger D, Kempermann G (2003) Regional effects of wheel running and environmental enrichment on cell genesis and microglia proliferation in the adult murine neocortex. Cereb Cortex 13:845–851. https://doi.org/10.1093/cercor/13.8.845
133. Olah M, Ping G, De Haas AH, Brouwer N, Meerlo P, Van Der Zee EA, Biber K, Boddeke HW (2009) Enhanced hippocampal neurogenesis in the absence of microglia T cell interaction and microglia activation in the murine running wheel model. Glia 57:1046–1061. https://doi.org/10.1002/glia.20828
134. Vukovic J, Colditz MJ, Blackmore DG, Ruitenberg MJ, Bartlett PF (2012) Microglia modulate hippocampal neural precursor activity in response to exercise and aging. J Neurosci 32:6435–6443. https://doi.org/10.1523/JNEUROSCI.5925-11.2012
135. Kohman RA, DeYoung EK, Bhattacharya TK, Peterson LN, Rhodes JS (2012) Wheel running attenuates microglia proliferation and increases expression of a proneurogenic phenotype in the hippocampus of aged mice. Brain Behav Immun 26:803–810. https://doi.org/10.1016/j.bbi.2011.10.006
136. Casaletto KB, Lindbergh CA, VandeBunte A, Neuhaus J, Schneider JA, Buchman AS, Honer WG, Bennett DA (2022) Microglial correlates of late life physical activity: relationship with synaptic and cognitive aging in older adults. J Neurosci 42:288–298. https://doi.org/10.1523/JNEUROSCI.1483-21.2021
137. Silvin A, Qian J, Ginhoux F (2023) Brain macrophage development, diversity and dysregulation in health and disease. Cell Mol Immunol 20:1277–1289. https://doi.org/10.1038/s41423-023-01053-6
138. Mota BC, Kelly AM (2020) Exercise alters LPS-induced glial activation in the mouse brain. Neuronal Signal 4:NS20200003. https://doi.org/10.1042/NS20200003

139. Littlefield AM, Setti SE, Priester C, Kohman RA (2015) Voluntary exercise attenuates LPS-induced reductions in neurogenesis and increases microglia expression of a proneurogenic phenotype in aged mice. J Neuroinflammation 12:138. https://doi.org/10.1186/s12974-015-0362-0
140. Norden DM, Godbout JP (2013) Review: microglia of the aged brain: primed to be activated and resistant to regulation. Neuropathol Appl Neurobiol 39:19–34. https://doi.org/10.1111/j.1365-2990.2012.01306.x
141. Kohman RA, Bhattacharya TK, Wojcik E, Rhodes JS (2013) Exercise reduces activation of microglia isolated from hippocampus and brain of aged mice. J Neuroinflammation 10:114. https://doi.org/10.1186/1742-2094-10-114
142. Giorgetti E, Panesar M, Zhang Y, Joller S, Ronco M, Obrecht M, Lambert C, Accart N, Beckmann N, Doelemeyer A et al (2019) Modulation of microglia by voluntary exercise or CSF1R inhibition prevents age-related loss of functional motor units. Cell Rep 29:1539–1554.e1537. https://doi.org/10.1016/j.celrep.2019.10.003
143. Barrientos RM, Frank MG, Crysdale NY, Chapman TR, Ahrendsen JT, Day HE, Campeau S, Watkins LR, Patterson SL, Maier SF (2011) Little exercise, big effects: reversing aging and infection-induced memory deficits, and underlying processes. J Neurosci 31:11578–11586. https://doi.org/10.1523/JNEUROSCI.2266-11.2011
144. Speisman RB, Kumar A, Rani A, Foster TC, Ormerod BK (2013) Daily exercise improves memory, stimulates hippocampal neurogenesis and modulates immune and neuroimmune cytokines in aging rats. Brain Behav Immun 28:25–43. https://doi.org/10.1016/j.bbi.2012.09.013
145. Connolly MG, Bruce SR, Kohman RA (2022) Exercise duration differentially effects age-related neuroinflammation and hippocampal neurogenesis. Neuroscience 490:275–286. https://doi.org/10.1016/j.neuroscience.2022.03.022
146. Wu SY, Pan BS, Tsai SF, Chiang YT, Huang BM, Mo FE, Kuo YM (2020) BDNF reverses aging-related microglial activation. J Neuroinflammation 17:210. https://doi.org/10.1186/s12974-020-01887-1
147. Wang TF, Wu SY, Pan BS, Tsai SF, Kuo YM (2022) Inhibition of nigral microglial activation reduces age-related loss of dopaminergic neurons and motor deficits. Cells 11. https://doi.org/10.3390/cells11030481
148. Lee J, Duan W, Mattson MP (2002) Evidence that brain-derived neurotrophic factor is required for basal neurogenesis and mediates, in part, the enhancement of neurogenesis by dietary restriction in the hippocampus of adult mice. J Neurochem 82:1367–1375. https://doi.org/10.1046/j.1471-4159.2002.01085.x
149. Zigova T, Pencea V, Wiegand SJ, Luskin MB (1998) Intraventricular administration of BDNF increases the number of newly generated neurons in the adult olfactory bulb. Mol Cell Neurosci 11:234–245. https://doi.org/10.1006/mcne.1998.0684
150. Bramham CR, Messaoudi E (2005) BDNF function in adult synaptic plasticity: the synaptic consolidation hypothesis. Prog Neurobiol 76:99–125. https://doi.org/10.1016/j.pneurobio.2005.06.003
151. Lu B, Chow A (1999) Neurotrophins and hippocampal synaptic transmission and plasticity. J Neurosci Res 58:76–87
152. Sairanen M, Lucas G, Ernfors P, Castren M, Castren E (2005) Brain-derived neurotrophic factor and antidepressant drugs have different but coordinated effects on neuronal turnover, proliferation, and survival in the adult dentate gyrus. J Neurosci 25:1089–1094. https://doi.org/10.1523/JNEUROSCI.3741-04.2005
153. Erickson KI, Miller DL, Roecklein KA (2012) The aging hippocampus: interactions between exercise, depression, and BDNF. Neuroscientist 18:82–97. https://doi.org/10.1177/1073858410397054
154. Szuhany KL, Bugatti M, Otto MW (2015) A meta-analytic review of the effects of exercise on brain-derived neurotrophic factor. J Psychiatr Res 60:56–64. https://doi.org/10.1016/j.jpsychires.2014.10.003

155. Cotman CW, Berchtold NC (2002) Exercise: a behavioral intervention to enhance brain health and plasticity. Trends Neurosci 25:295–301. https://doi.org/10.1016/s0166-2236(02)02143-4
156. Lee MC, Okamoto M, Liu YF, Inoue K, Matsui T, Nogami H, Soya H (1985) Voluntary resistance running with short distance enhances spatial memory related to hippocampal BDNF signaling. J Appl Physiol 2012(113):1260–1266. https://doi.org/10.1152/japplphysiol.00869.2012
157. Lee M, Soya H (2017) Effects of acute voluntary loaded wheel running on BDNF expression in the rat hippocampus. J Exerc Nutr Biochem 21:52–57. https://doi.org/10.20463/jenb.2017.0034
158. Venezia AC, Guth LM, Sapp RM, Spangenburg EE, Roth SM (2016) Sex-dependent and independent effects of long-term voluntary wheel running on Bdnf mRNA and protein expression. Physiol Behav 156:8–15. https://doi.org/10.1016/j.physbeh.2015.12.026
159. Dimitropoulou A, Bixby JL (2000) Regulation of retinal neurite growth by alterations in MAPK/ERK kinase (MEK) activity. Brain Res 858:205–214. https://doi.org/10.1016/s0006-8993(00)01946-6
160. Jiang P, Zhu T, Xia Z, Gao F, Gu W, Chen X, Yuan T, Yu H (2015) Inhibition of MAPK/ERK signaling blocks hippocampal neurogenesis and impairs cognitive performance in prenatally infected neonatal rats. Eur Arch Psychiatry Clin Neurosci 265:497–509. https://doi.org/10.1007/s00406-015-0588-y
161. Huang EJ, Reichardt LF (2001) Neurotrophins: roles in neuronal development and function. Annu Rev Neurosci 24:677–736. https://doi.org/10.1146/annurev.neuro.24.1.677
162. Numakawa T, Suzuki S, Kumamaru E, Adachi N, Richards M, Kunugi H (2010) BDNF function and intracellular signaling in neurons. Histol Histopathol 25:237–258. https://doi.org/10.14670/HH-25.237
163. Vaynman S, Ying Z, Gomez-Pinilla F (2003) Interplay between brain-derived neurotrophic factor and signal transduction modulators in the regulation of the effects of exercise on synaptic-plasticity. Neuroscience 122:647–657. https://doi.org/10.1016/j.neuroscience.2003.08.001
164. Gomez-Pinilla F, So V, Kesslak JP (2001) Spatial learning induces neurotrophin receptor and synapsin I in the hippocampus. Brain Res 904:13–19. https://doi.org/10.1016/s0006-8993(01)02394-0
165. Ferreira AF, Real CC, Rodrigues AC, Alves AS, Britto LR (2011) Short-term, moderate exercise is capable of inducing structural, BDNF-independent hippocampal plasticity. Brain Res 1425:111–122. https://doi.org/10.1016/j.brainres.2011.10.004
166. Nakagawa S, Kim JE, Lee R, Malberg JE, Chen J, Steffen C, Zhang YJ, Nestler EJ, Duman RS (2002) Regulation of neurogenesis in adult mouse hippocampus by cAMP and the cAMP response element-binding protein. J Neurosci 22:3673–3682. https://doi.org/10.1523/JNEUROSCI.22-09-03673.2002
167. Jodeiri Farshbaf M, Ghaedi K, Megraw TL, Curtiss J, Shirani Faradonbeh M, Vaziri P, Nasr-Esfahani MH (2016) Does PGC1alpha/FNDC5/BDNF elicit the beneficial effects of exercise on neurodegenerative disorders? Neuromolecular Med 18:1–15. https://doi.org/10.1007/s12017-015-8370-x
168. Wrann CD, White JP, Salogiannnis J, Laznik-Bogoslavski D, Wu J, Ma D, Lin JD, Greenberg ME, Spiegelman BM (2013) Exercise induces hippocampal BDNF through a PGC-1alpha/FNDC5 pathway. Cell Metab 18:649–659. https://doi.org/10.1016/j.cmet.2013.09.008
169. Vaynman S, Ying Z, Gomez-Pinilla F (2004) Hippocampal BDNF mediates the efficacy of exercise on synaptic plasticity and cognition. Eur J Neurosci 20:2580–2590. https://doi.org/10.1111/j.1460-9568.2004.03720.x
170. Alomari MA, Khabour OF, Alzoubi KH, Alzubi MA (2013) Forced and voluntary exercises equally improve spatial learning and memory and hippocampal BDNF levels. Behav Brain Res 247:34–39. https://doi.org/10.1016/j.bbr.2013.03.007
171. Wu CW, Chang YT, Yu L, Chen HI, Jen CJ, Wu SY, Lo CP, Kuo YM (1985) Exercise enhances the proliferation of neural stem cells and neurite growth and survival of neuronal progenitor

cells in dentate gyrus of middle-aged mice. J Appl Physiol 2008(105):1585–1594. https://doi.org/10.1152/japplphysiol.90775.2008
172. Wu CW, Chen YC, Yu L, Chen HI, Jen CJ, Huang AM, Tsai HJ, Chang YT, Kuo YM (2007) Treadmill exercise counteracts the suppressive effects of peripheral lipopolysaccharide on hippocampal neurogenesis and learning and memory. J Neurochem 103:2471–2481. https://doi.org/10.1111/j.1471-4159.2007.04987.x
173. Fernandez AM, Torres-Aleman I (2012) The many faces of insulin-like peptide signalling in the brain. Nat Rev Neurosci 13:225–239. https://doi.org/10.1038/nrn3209
174. Brooker GJ, Kalloniatis M, Russo VC, Murphy M, Werther GA, Bartlett PF (2000) Endogenous IGF-1 regulates the neuronal differentiation of adult stem cells. J Neurosci Res 59:332–341. https://doi.org/10.1002/(sici)1097-4547(20000201)59:3<332::aid-jnr6>3.0.co;2-2
175. Aleman A, Torres-Aleman I (2009) Circulating insulin-like growth factor I and cognitive function: neuromodulation throughout the lifespan. Prog Neurobiol 89:256–265. https://doi.org/10.1016/j.pneurobio.2009.07.008
176. Ding Q, Vaynman S, Akhavan M, Ying Z, Gomez-Pinilla F (2006) Insulin-like growth factor I interfaces with brain-derived neurotrophic factor-mediated synaptic plasticity to modulate aspects of exercise-induced cognitive function. Neuroscience 140:823–833. https://doi.org/10.1016/j.neuroscience.2006.02.084
177. Lista I, Sorrentino G (2010) Biological mechanisms of physical activity in preventing cognitive decline. Cell Mol Neurobiol 30:493–503. https://doi.org/10.1007/s10571-009-9488-x
178. Maass A, Duzel S, Brigadski T, Goerke M, Becke A, Sobieray U, Neumann K, Lovden M, Lindenberger U, Backman L et al (2016) Relationships of peripheral IGF-1, VEGF and BDNF levels to exercise-related changes in memory, hippocampal perfusion and volumes in older adults. Neuroimage 131:142–154. https://doi.org/10.1016/j.neuroimage.2015.10.084
179. Malberg JE, Platt B, Rizzo SJ, Ring RH, Lucki I, Schechter LE, Rosenzweig-Lipson S (2007) Increasing the levels of insulin-like growth factor-I by an IGF binding protein inhibitor produces anxiolytic and antidepressant-like effects. Neuropsychopharmacology 32:2360–2368. https://doi.org/10.1038/sj.npp.1301358
180. Llorens-Martin M, Torres-Aleman I, Trejo JL (2010) Exercise modulates insulin-like growth factor 1-dependent and -independent effects on adult hippocampal neurogenesis and behaviour. Mol Cell Neurosci 44:109–117. https://doi.org/10.1016/j.mcn.2010.02.006
181. Trejo JL, Llorens-Martin MV, Torres-Aleman I (2008) The effects of exercise on spatial learning and anxiety-like behavior are mediated by an IGF-I-dependent mechanism related to hippocampal neurogenesis. Mol Cell Neurosci 37:402–411. https://doi.org/10.1016/j.mcn.2007.10.016
182. Trejo JL, Carro E, Torres-Aleman I (2001) Circulating insulin-like growth factor I mediates exercise-induced increases in the number of new neurons in the adult hippocampus. J Neurosci 21:1628–1634. https://doi.org/10.1523/JNEUROSCI.21-05-01628.2001
183. Cetinkaya C, Sisman AR, Kiray M, Camsari UM, Gencoglu C, Baykara B, Aksu I, Uysal N (2013) Positive effects of aerobic exercise on learning and memory functioning, which correlate with hippocampal IGF-1 increase in adolescent rats. Neurosci Lett 549:177–181. https://doi.org/10.1016/j.neulet.2013.06.012
184. Cassilhas RC, Lee KS, Venancio DP, Oliveira MG, Tufik S, de Mello MT (2012) Resistance exercise improves hippocampus-dependent memory. Braz J Med Biol Res 45:1215–1220. https://doi.org/10.1590/s0100-879x2012007500138
185. Tsai CL, Ukropec J, Ukropcova B, Pai MC (2018) An acute bout of aerobic or strength exercise specifically modifies circulating exerkine levels and neurocognitive functions in elderly individuals with mild cognitive impairment. Neuroimage Clin 17:272–284. https://doi.org/10.1016/j.nicl.2017.10.028
186. Frielingsdorf H, Simpson DR, Thal LJ, Pizzo DP (2007) Nerve growth factor promotes survival of new neurons in the adult hippocampus. Neurobiol Dis 26:47–55. https://doi.org/10.1016/j.nbd.2006.11.015

187. Gustilo MC, Markowska AL, Breckler SJ, Fleischman CA, Price DL, Koliatsos VE (1999) Evidence that nerve growth factor influences recent memory through structural changes in septohippocampal cholinergic neurons. J Comp Neurol 405:491–507. https://doi.org/10.1002/(sici)1096-9861(19990322)405:4<491::aid-cne4>3.0.co;2-n
188. Tuszynski MH, Yang JH, Barba D, Hoi-Sang U, Bakay RA, Pay MM, Masliah E, Conner JM, Kobalka P, Roy S, Nagahara AH (2015) nerve growth factor gene therapy: activation of neuronal responses in Alzheimer disease. JAMA Neurol 72:1139–1147. https://doi.org/10.1001/jamaneurol.2015.1807
189. Easton A, Douchamps V, Eacott M, Lever C (2012) A specific role for septohippocampal acetylcholine in memory? Neuropsychologia 50:3156–3168. https://doi.org/10.1016/j.neuropsychologia.2012.07.022
190. Ballinger EC, Ananth M, Talmage DA, Role LW (2016) Basal forebrain cholinergic circuits and signaling in cognition and cognitive decline. Neuron 91:1199–1218. https://doi.org/10.1016/j.neuron.2016.09.006
191. Bonini M, Fioretti D, Sargentini V, Del Giacco S, Rinaldi M, Tranquilli C, Bonini S (2013) Increased nerve growth factor serum levels in top athletes. Clin J Sport Med 23:228–231. https://doi.org/10.1097/JSM.0b013e31827ee6d5
192. Bansi J, Bloch W, Gamper U, Kesselring J (2013) Training in MS: influence of two different endurance training protocols (aquatic versus overland) on cytokine and neurotrophin concentrations during three week randomized controlled trial. Mult Scler 19:613–621. https://doi.org/10.1177/1352458512458605
193. Molteni R, Ying Z, Gomez-Pinilla F (2002) Differential effects of acute and chronic exercise on plasticity-related genes in the rat hippocampus revealed by microarray. Eur J Neurosci 16:1107–1116. https://doi.org/10.1046/j.1460-9568.2002.02158.x
194. Neeper SA, Gomez-Pinilla F, Choi J, Cotman CW (1996) Physical activity increases mRNA for brain-derived neurotrophic factor and nerve growth factor in rat brain. Brain Res 726:49–56
195. Hall JM, Savage LM (2016) Exercise leads to the re-emergence of the cholinergic/nestin neuronal phenotype within the medial septum/diagonal band and subsequent rescue of both hippocampal ACh efflux and spatial behavior. Exp Neurol 278:62–75. https://doi.org/10.1016/j.expneurol.2016.01.018
196. Shaywitz AJ, Greenberg ME (1999) CREB: a stimulus-induced transcription factor activated by a diverse array of extracellular signals. Annu Rev Biochem 68:821–861. https://doi.org/10.1146/annurev.biochem.68.1.821
197. Segi-Nishida E, Warner-Schmidt JL, Duman RS (2008) Electroconvulsive seizure and VEGF increase the proliferation of neural stem-like cells in rat hippocampus. Proc Natl Acad Sci USA 105:11352–11357. https://doi.org/10.1073/pnas.0710858105
198. Gora-Kupilas K, Josko J (2005) The neuroprotective function of vascular endothelial growth factor (VEGF). Folia Neuropathol 43:31–39
199. Ballard HJ (2017) Exercise makes your brain bigger: skeletal muscle VEGF and hippocampal neurogenesis. J Physiol 595:5721–5722. https://doi.org/10.1113/JP274658
200. Sun Y, Jin K, Xie L, Childs J, Mao XO, Logvinova A, Greenberg DA (2003) VEGF-induced neuroprotection, neurogenesis, and angiogenesis after focal cerebral ischemia. J Clin Invest 111:1843–1851. https://doi.org/10.1172/JCI17977
201. Nowacka MM, Obuchowicz E (2012) Vascular endothelial growth factor (VEGF) and its role in the central nervous system: a new element in the neurotrophic hypothesis of antidepressant drug action. Neuropeptides 46:1–10. https://doi.org/10.1016/j.npep.2011.05.005
202. Cao L, Jiao X, Zuzga DS, Liu Y, Fong DM, Young D, During MJ (2004) VEGF links hippocampal activity with neurogenesis, learning and memory. Nat Genet 36:827–835. https://doi.org/10.1038/ng1395
203. Jin K, Zhu Y, Sun Y, Mao XO, Xie L, Greenberg DA (2002) Vascular endothelial growth factor (VEGF) stimulates neurogenesis in vitro and in vivo. Proc Natl Acad Sci USA 99:11946–11950. https://doi.org/10.1073/pnas.182296499

204. Kiuchi T, Lee H, Mikami T (2012) Regular exercise cures depression-like behavior via VEGF-Flk-1 signaling in chronically stressed mice. Neuroscience 207:208–217. https://doi.org/10.1016/j.neuroscience.2012.01.023
205. Zarezadehmehrizi A, Hong J, Lee J, Rajabi H, Gharakhanlu R, Naghdi N, Azimi M, Park Y (2021) Exercise training ameliorates cognitive dysfunction in amyloid beta-injected rat model: possible mechanisms of angiostatin/VEGF signaling. Metab Brain Dis 36:2263–2271. https://doi.org/10.1007/s11011-021-00751-2
206. Stevenson ME, Behnke VK, Swain RA (2018) Exercise pattern and distance differentially affect hippocampal and cerebellar expression of FLK-1 and FLT-1 receptors in astrocytes and blood vessels. Behav Brain Res 337:8–16. https://doi.org/10.1016/j.bbr.2017.09.037
207. Petrova TV, Makinen T, Alitalo K (1999) Signaling via vascular endothelial growth factor receptors. Exp Cell Res 253:117–130. https://doi.org/10.1006/excr.1999.4707
208. Palmer TD, Willhoite AR, Gage FH (2000) Vascular niche for adult hippocampal neurogenesis. J Comp Neurol 425:479–494. https://doi.org/10.1002/1096-9861(20001002)425:4<479::aid-cne2>3.0.co;2-3
209. Fabel K, Fabel K, Tam B, Kaufer D, Baiker A, Simmons N, Kuo CJ, Palmer TD (2003) VEGF is necessary for exercise-induced adult hippocampal neurogenesis. Eur J Neurosci 18:2803–2812. https://doi.org/10.1111/j.1460-9568.2003.03041.x
210. Rich B, Scadeng M, Yamaguchi M, Wagner PD, Breen EC (2017) Skeletal myofiber vascular endothelial growth factor is required for the exercise training-induced increase in dentate gyrus neuronal precursor cells. J Physiol 595:5931–5943. https://doi.org/10.1113/JP273994
211. Heine VM, Zareno J, Maslam S, Joels M, Lucassen PJ (2005) Chronic stress in the adult dentate gyrus reduces cell proliferation near the vasculature and VEGF and Flk-1 protein expression. Eur J Neurosci 21:1304–1314. https://doi.org/10.1111/j.1460-9568.2005.03951.x
212. Warner-Schmidt JL, Duman RS (2007) VEGF is an essential mediator of the neurogenic and behavioral actions of antidepressants. Proc Natl Acad Sci USA 104:4647–4652. https://doi.org/10.1073/pnas.0610282104
213. Raff H, Sharma ST, Nieman LK (2014) Physiological basis for the etiology, diagnosis, and treatment of adrenal disorders: cushing's syndrome, adrenal insufficiency, and congenital adrenal hyperplasia. Compr Physiol 4:739–769. https://doi.org/10.1002/cphy.c130035
214. Raff H, Skelton MM, Cowley AW Jr (1989) Feedback control of vasopressin and corticotrophin secretion in conscious dogs: effect of hypertonic saline. J Endocrinol 122:41–48. https://doi.org/10.1677/joe.0.1220041
215. Usa K, Singh RJ, Netzel BC, Liu Y, Raff H, Liang M (2007) Renal interstitial corticosterone and 11-dehydrocorticosterone in conscious rats. Am J Physiol Renal Physiol 293:F186–F192. https://doi.org/10.1152/ajprenal.00484.2006
216. Yu T, Xu H, Wang W, Li S, Chen Z, Deng H (2015) Determination of endogenous corticosterone in rodent's blood, brain and hair with LC-APCI-MS/MS. J Chromatogr B Analyt Technol Biomed Life Sci 1002:267–276. https://doi.org/10.1016/j.jchromb.2015.08.035
217. Brummelte S, Galea LA (2010) Chronic high corticosterone reduces neurogenesis in the dentate gyrus of adult male and female rats. Neuroscience 168:680–690. https://doi.org/10.1016/j.neuroscience.2010.04.023
218. Wang TF, Tsai SF, Zhao ZW, Shih MM, Wang CY, Yang TT, Kuo YM (2021) Exercise-induced increases of corticosterone contribute to exercise-enhanced adult hippocampal neurogenesis in mice. Chin J Physiol 64:186–193. https://doi.org/10.4103/cjp.cjp_39_21
219. Gould E, Cameron HA, Daniels DC, Woolley CS, McEwen BS (1992) Adrenal hormones suppress cell division in the adult rat dentate gyrus. J Neurosci 12:3642–3650. https://doi.org/10.1523/JNEUROSCI.12-09-03642.1992
220. Chang YT, Chen YC, Wu CW, Yu L, Chen HI, Jen CJ, Kuo YM (2008) Glucocorticoid signaling and exercise-induced downregulation of the mineralocorticoid receptor in the induction of adult mouse dentate neurogenesis by treadmill running. Psychoneuroendocrinology 33:1173–1182. https://doi.org/10.1016/j.psyneuen.2008.05.014

221. Deng W, Aimone JB, Gage FH (2010) New neurons and new memories: how does adult hippocampal neurogenesis affect learning and memory? Nat Rev Neurosci 11:339–350. https://doi.org/10.1038/nrn2822
222. Radahmadi M, Alaei H, Sharifi MR, Hosseini N (2015) Effect of forced exercise and exercise withdrawal on memory, serum and hippocampal corticosterone levels in rats. Exp Brain Res 233:2789–2799. https://doi.org/10.1007/s00221-015-4349-y
223. Diamond DM, Bennett MC, Fleshner M, Rose GM (1992) Inverted-U relationship between the level of peripheral corticosterone and the magnitude of hippocampal primed burst potentiation. Hippocampus 2:421–430. https://doi.org/10.1002/hipo.450020409
224. Robison LS, Alessi L, Thanos PK (2018) Chronic forced exercise inhibits stress-induced reinstatement of cocaine conditioned place preference. Behav Brain Res 353:176–184. https://doi.org/10.1016/j.bbr.2018.07.009
225. Pal'chikova NA, Kuzminova OI, Selyatitskaya VG (2018) stress response to physical exercise in rats with alimentary obesity. Bull Exp Biol Med 164:587–590. https://doi.org/10.1007/s10517-018-4037-6

Potential Role of Extracellular Vesicles in Mediating Effects of Exercise on Brain Function

Hyo Youl Moon and Henriette van Praag

Abstract Exercise has a remarkable capacity to improve brain function by fostering neuronal plasticity, which enables us to better cope with various psychological and cognitive challenges. Numerous studies have demonstrated the neuroprotective effects of exercise. However, the underlying molecular mechanisms of the neuroprotective effects of exercise are not yet fully understood. In particular, the role of exercise-induced secretion of peripheral factors into circulation that influence the brain is understudied. Recent research has shown that extracellular vesicles (EVs), including microvesicles (MVs) and exosomes, are secreted during exercise. The discovery that EVs can mediate intracellular communication by delivering cargo signifies a promising area of research to understand the impact of exercise on the brain. In the present review, we provide an overview of recent advancements in understanding the regulatory mechanisms of EV biogenesis and discuss how EV molecular composition is influenced by exercise. Additionally, we highlight the potential role of EVs as exercise-specific mediators and as a promising therapeutic tool for neurodegenerative diseases, such as Alzheimer's disease.

Keywords Extracellular vesicles · Exercise · Neurogenesis · miRNAs · AD

H. Y. Moon (✉)
Department of Education, Seoul National University, Seoul, South Korea

Learning Sciences Research Institute, Seoul National University, Seoul, South Korea

Institute on Aging, Seoul National University, Seoul, South Korea
e-mail: skyman19@snu.ac.kr

H. van Praag
Stiles Nicholson Brain Institute and Charles E. Schmidt College of Medicine, Florida Atlantic University, Jupiter, FL, USA

H. Soya (ed.), *Exercise Brain Stimulation for Cognitive Function and Mental Health*, Advances in Neurobiology 44, https://doi.org/10.1007/978-981-95-0066-6_8

1 Introduction

Neuroplasticity is fundamental to cognitive flexibility, memory formation, and behavioral modulation [1, 2]. It plays a crucial role in shaping cortical maps, synaptic strength, and neural circuitry. A thorough understanding of neuroplasticity has implications for therapeutic strategies in neurorehabilitation, cognitive enhancement, and alleviating the effects of neurological disorders or injuries by harnessing the inherent malleability of the brain [3]. This is illustrated by lifestyle effects on the brain. Sedentary behavior causes various chronic diseases, including neurodegeneration [4, 5]. In particular, aging and age-related neurodegenerative disorders impact the hippocampus, leading to progressive atrophy of hippocampal volume and memory deficits [6, 7]. On the other hand, it is increasingly evident that active lifestyle and participating in exercise improves brain function [7, 8], potentially by mitigating these processes [9, 10]. Indeed, exercise-induced adaptations influence not only tissues responsible for energy utilization and oxygen transport—such as the heart, blood vessels, muscles, fat, and liver—but also impact the brain [11, 12]. Indeed, in humans exercise increases the grey matter volume in the hippocampus [7] and entorhinal cortex [10, 13]. Research in rodents indicates that enhanced cognition as a result of physical activity can be attributed to improved brain plasticity [7, 8]. Exercise leads to structural, neurochemical, mitochondrial, and vascular changes in particular in the hippocampus, a brain area that is essential for learning and memory [7, 8, 14].

Unique to the dentate gyrus (DG) of the hippocampus is a process called "adult neurogenesis." This subfield can generate new neurons in the adult brain [15]. Residual adult neural stem cells (aNSCs) in the subgranular zone of the DG generate proliferating neuronal progenitor cells that differentiate into mature granule cells over several weeks [16–18]. These new neurons undergo functional integration into existing intra-hippocampal and (sub) cortical neuronal networks 107 and are considered to play a functional role in spatial memory and pattern separation [19, 20]. The regulation of proliferation, survival, and integration of new neurons can be influenced by intrinsic and extrinsic factors, most strongly by exercise. Voluntary wheel running promotes adult hippocampal neurogenesis and optimizes the circuitry of adult-born neurons in rodents, in association with improved spatial memory, pattern separation, short- and long-term synaptic plasticity [7, 21–23]. The exercise-induced alterations in dentate gyrus (DG) synaptic plasticity are also directly influenced by glutamate or γ-aminobutyric acid (GABA), which play a role in the induction of long-term potentiation (LTP), a cellular model of learning and memory [11, 24, 25]. Furthermore, exercise can affect cognition and mood by quantitative and qualitative alterations in neurotransmitter receptors, including but not limited to glutamate N-methyl-D-aspartate (NMDA) receptors [26] and serotonin 5-HT3 receptors [27, 28]. Exercise-induced neurotrophins also play a significant role in brain plasticity. In rodents, running increases brain-derived neurotrophic factor (BDNF) expression levels in the hippocampus and cortex [29]. This increase is associated with improvements in hippocampal neurogenesis and spatial memory

[30]. Notably, the neurogenic response to running is abolished when the BDNF receptor, TrkB, in neural progenitor cells is ablated [31]. Macrophage migration inhibitory factor (MIF), a cytokine, undergoes alterations in the hippocampus due to exercise. Behavioral tests have demonstrated that MIF mediates the antidepressant effects of exercise [32].

While research into the changes within the brain following exercise have yielded important insights, the potential systemic triggers have only recently begun to be studied. Earlier work showed that circulating insulin-like growth factor (IGF) mediates effects of exercise on neurogenesis and that these are blocked upon administration of IGF-1 antiserum [33]. More recently, parabiosis studies have revealed that factors in the blood of young animals can enhance neurogenesis in their older counterparts [34, 35]. Recent research has identified additional blood-borne factors that may mediate neurogenic and pro-cognitive effects of physical activity on the brain [36–40]. Moon et al. [41] showed that the conditioned media (CM) collected from L6 rat skeletal muscle cells that were treated with a compound that enhances energy metabolism, AMP-kinase (AMPK) agonist 5-aminoimidazole-4-carboxamide-1-β-D-ribofuranoside (AICAR) increases the number of hippocampal aNSCs that differentiate into neurons and glia (Fig. 1). Such experiments provide strong evidence for inter-organ crosstalk and formed the basis for myokine Cathepsin B (Ctsb) identification [42]. Indeed, exercise facilitates the exchange of vital substrates for brain health through communication among peripheral organs, including muscle, liver, and adipose tissue. The discovery of exerkines such as irisin, Cathepsin B (Ctsb), Apelin, and other humoral factors are likely to become instrumental in designing improved therapies for neurodegenerative diseases [42–46].

There is increasing evidence for an important role for myokines in understanding the effects of exercise on the brain. Overexpression of PGC-1α in muscle leads to increased production of fibronectin type III domain-containing 5 (FNDC5), which undergoes cleavage to form irisin—a myokine implicated in oxygen consumption, browning of adipose tissue, and thermogenesis [45]. Exercise can increase the FNDC5 mRNA levels in skeletal muscle and this may contribute to the elevation of brain-derived neurotrophic factor (BDNF) levels in brain [47, 48]. In addition, central irisin administration has been shown to reverse deficits in pattern separation behavior in FDNC5 knockout mice [46]. Furthermore, irisin confers cognitive benefits in mouse models of Alzheimer's disease [46, 49]. Ctsb, a lysosomal thiol proteinase, is involved in numerous physiological and pathological processes. Various types of exercise lead to an elevation of plasma Ctsb across mammalian species [42, 50–53]. The administration of Ctsb in adult hippocampal progenitor cell cultures boosted the expression of neurotrophin BDNF and the marker for neuronal differentiation, doublecortin (DCX). On the other hand, Ctsb knockout mice exhibited compromised spatial memory function and changes in dentate granule cell physiology [42]. Several subsequent studies validate that increased levels of circulating Cathepsin B (CTSB) were specifically linked with enhanced cognitive control and information processing speed [51, 53, 54]. Another myokine, Apelin, was found to be stimulated by exercise, showing increased expression levels after an 8-week endurance training program involving 11 obese non-diabetic male subjects [55].

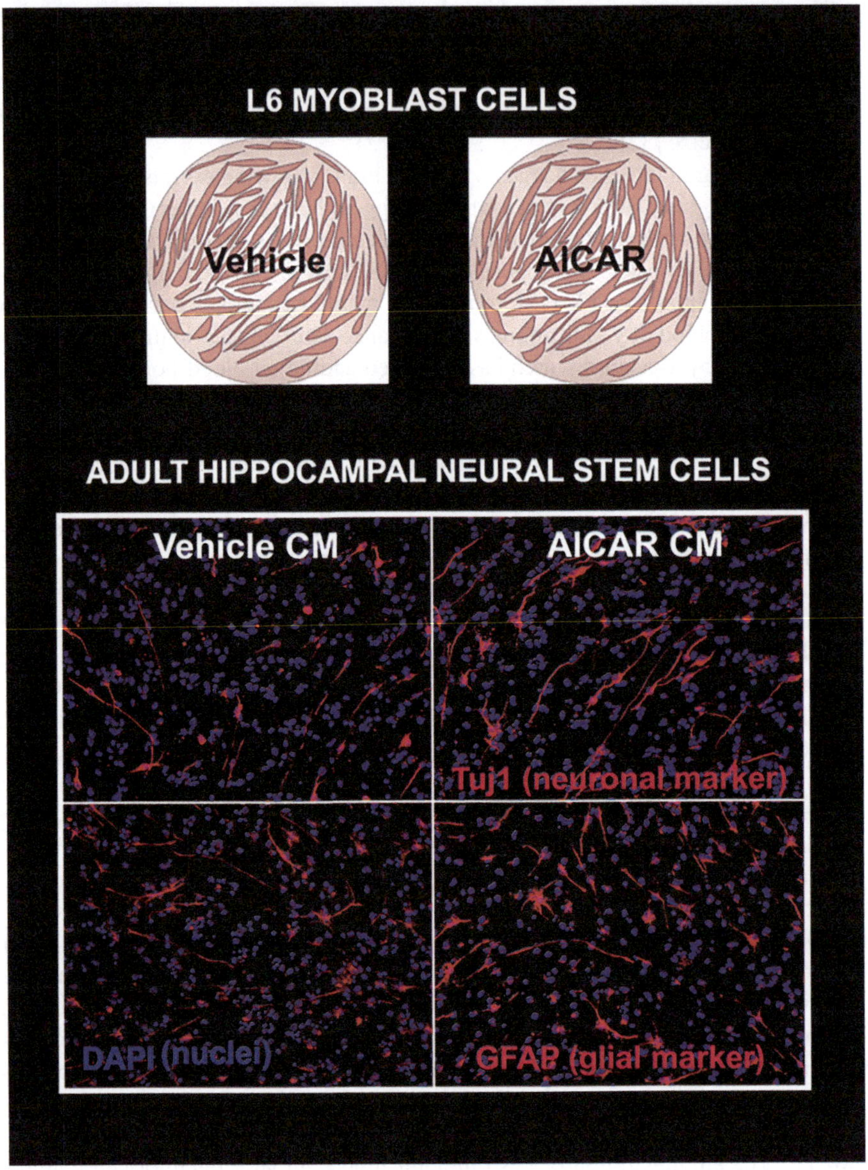

Fig. 1 AICAR conditioned media (CM) derived from L6 muscle cell cultures increases neural differentiation. Specifically, L6 rat muscle cell cultures were treated with vehicle or AICAR. AICAR CM significantly increased the proportion of aNSCs differentiating into Tuj1-positive neurons as compared to vehicle CM treatment conditions. AICAR CM also enhanced gliogenesis as evidenced by an increase in GFAP-positive astrocytes as compared to vehicle CM treatment. Representative photomicrographs of vehicle-CM- and AICAR-CM-treated aNPCs expressing neuronal marker Tuj1 (red), and of cells expressing astroglial marker GFAP (red). Nuclei were stained with 4′,6-diamidino-2-phenylindole (DAPI), blue. (Modified from Moon et al. [41])

Apelin has been demonstrated to play a protective and regulatory role in neuronal cells [44]. Additional studies have also indicated that Apelin can mitigate neural-cell death and alleviate neural dysfunction in the brain [56]. Myokines and myometabolites such as lactate [57], as well as ketone body beta-hydroxybutyrate [58, 59], enhance hippocampal BDNF levels and cognitive function. Nevertheless, the impact of these proteins has been demonstrated in exaggerated scenarios that do not reflect normal human physiology, such as studies using knockout mice or transgenic mice, or high-dosage injections. In addition, secreted proteins can also be easily degraded by various proteases during circulation in the blood. It is essential to note that no single protein, whether increased or decreased as a result of exercise, can singularly mediate the diverse effects of exercise on the body and brain.

2 Extracellular Vesicles and Neuronal Health

The complete understanding of the molecular mechanisms through which physical activity confers protective effects against neurological dysfunction remains elusive. Nevertheless, there is growing support for the idea that inter-tissue communication facilitated by extracellular vesicles (EVs) plays a significant role [60, 61]. Operating as a signaling modulators, EVs are produced and released by various tissues exerting autocrine, paracrine, and endocrine effects [60, 61]. They facilitate communication between muscles and various organs, fostering neurogenesis and vascularization in the brain [62], and promoting the metabolic adaptations in various tissues [63]. In the following sections, we will delve into studies exploring the impact of extracellular vesicles during exercise on neuronal health.

2.1 *Understanding of Extracellular Vesicles (EVs)*

Extracellular vesicles were initially considered cellular waste products resulting from cell damage or byproducts of cellular homeostasis [64]. Next, EVs were known to be mediators of skin diseases and respiratory diseases that are emitted from bacteria and bugs [64]. Recent research has revealed that these EVs are functional carriers that deliver complex cargo such as proteins, lipids, and miRNAs, and that these substances can be delivered to target cells that EVs encounter while traveling through the blood [65, 66]. Indeed, it has garnered interest as a means of intercellular communication capable of reprogramming recipient cells situated at a considerable distance from the point of EVs release. Exosomes, originating from endosomal pathways, typically range from 40 to 150 nm and carry a diverse cargo of proteins, lipids, and nucleic acids. Microvesicles, larger in size (to 1 μm), are shed directly from the cell membrane and contain membrane components along with genetic material. Recently, there has been a suggestion to categorize EVs into small (<200 nm) and large EVs [67]. EVs can be released by nearly any cell, albeit that the types and conditions of cells releasing exosomes can exhibit significant variability [65–67]. In general, exosomes have membrane proteins with various functions, such as tetraspanins (CD9, CD63, CD81, and CD82), which participate in cell penetration, invasion, and fusion processes. Moreover, the phospholipid bilayer structure of exosomes plays a role in prolonging the activation time of diverse substances encapsulated within them. A multitude of components has been discerned within diverse EVs types, encompassing approximately 4400 proteins, 194 lipids, 1639 mRNAs, and 764 miRNAs [68]. The components found in EVs, spanning areas such as metabolism, immune response, and angiogenesis, underscore the intricate nature of EVs and their potential for functional diversity [43]. This diversity hints at their capacity to mediate the effects of exercise. Notably, within the proteins encapsulated in EVs, a subset known as exerkines has been observed to increase during exercise [43]. Consequently, a growing community of researchers is actively engaged in uncovering the specific functions of EVs released during exercise.

Comprehending the details of loading, processes, and mechanisms related to exosome cargo sorting is indispensable for gaining insights into the physiological and pathological functions of the extracellular vesicles [69]. EVs are derived from the internal budding of endosomal multivesicular bodies (MVB). A core component of EVs biogenesis involves the ESCRT complexes. The ESCRT, a family of proteins, collaborates in successive complexes at the membrane of MVBs, facilitating cargo sorting and the generation of intraluminal vesicles (ILVs). Some MVBs formed are transported by associated RAB proteins, leading to fusion with the membrane. Upon fusion, MVBs can either undergo degradation through fusion with lysosomes or release ILVs into the extracellular space by fusing with the plasma membrane [69]. Once formed, EVs pinch off from the plasma membrane, preparing them for secretion through various modes. The occurrence of a membrane fusion event necessitates the involvement of SNAREs and is subject to regulation by tetraspanin complexes such as CD9 and CD81. Upon release, EVs undergo fusion with the plasma membranes of recipient cells [70]. Alternatively, surface proteins on EVs can interact with cell surface receptors on recipient cells, initiating intracellular signaling. Various uptake mechanisms have been suggested for exosomes, including endocytosis, membrane fusion, and interactions mediated by receptor-ligand interactions [70].

2.2 EVs and Exercise

Engaging in regular exercise leads to enduring adaptations that offer substantial health advantages and serve as a protective measure against various chronic diseases [71]. Aerobic exercise increases the heart rate to supply the oxygen the body needs [72]. Consistent aerobic exercise leads to adaptations of the cardiovascular and respiratory systems appropriate to the activity. Likewise, resistance exercise induces adaptation of the neuromuscular and musculoskeletal systems corresponding to the intensity. The mechanisms elucidating the intricate and multi-faceted systemic benefits of exercise are increasingly centered on the cross-talk between organs. Myokines originating from muscle are commonly investigated for their capacity to operate in a paracrine, autocrine, and/or endocrine fashion, as well as their involvement in signal transduction during exercise [73, 74]. To date, over 300 myokines have been identified, including a substantial number of inflammatory cytokines and proteins that are recognized for their involvement in signal transduction during exercise [75, 76] and several that are important for brain function. Exercise-induced responses involve the release of extracellular vesicles, predominantly exosomes and microvesicles. According to public databases that analyzed proteomes, including Vesiclepedia (www.microvesicles.org), EVpedia (www.evpedia.info), and ExoCarta (www.exocarta.org), EVs contain multiple myokines such as CTSB, IL-10, Apelin, brain-derived neurotrophic factor (BDNF), and so on. These discoveries provide support for the notion that EVs can mediate the effects of the multisystemic benefits of exercise.

Several research groups confirmed the changes in size distribution of plasma EVs with running in rodent models [77]. EVs also appeared different depending on the period and type of exercise [78]. Moon et al. [79] found the average diameter increased after 2 weeks of treadmill running [79], whereas Yoon et al. [80] noted no change in size alteration after 4 weeks of voluntary of exercise in C57BL/6 J mice [80]. In human study, Park et al. observed that urinary EVs were found to be associated with commonly used exercise fatigue or intensity markers such as rating of perceived exertion (RPE), heart rate, systolic blood pressure, and blood lactate level. Twenty meter shuttle run test elevated CD9-, CD63-, and CD81, positive markers of EVs, immediately after exercise. And about an hour after exercise, the amount of EVs markers returned to almost normal levels in urine [81]. Additional research has affirmed similar trends in the concentrations of EVs in the blood. Following a 1-hr bout of cycling exercise also affects the size of EVs particles with changes of over 300 proteins [82]. It has also been reported that the amount and content of these EVs change depending on physiological conditions. Elevated circulating levels of EVs are observed in both mice and humans with obesity [83]. Furthermore, EVs released by fat tissue macrophages in obese mice decrease insulin sensitivity in the liver and skeletal muscle by containing miR-155, which regulates PPARγ expression [84].

Indeed, research indicates that the attributes of EVs can differ based on physiological and environmental factors, such as metabolic diseases, gender, and altitude. Over a 12-week period of high-intensity interval training (HIIT), insulin sensitivity increases in males, both with and without type 2 diabetes (T2D). However, individuals with insulin resistance, specifically those with T2D and non-diabetic insulin-resistant individuals (IR NDM), exhibited noticeable changes of EVs in the blood [85]. Geoffrey Warnier and his colleagues discovered that a 6-week interval training regimen generally elevated the baseline levels of circulating EVs at sea level. However, this effect was not observed in hypoxic conditions, such as altitudes of 2000, 3000, and 4000 meters [86]. Gender is a factor that seems to play a role in the alterations of EVs induced by exercise [87]. In a study conducted by William R

Conkright et al., resistance exercise led to an elevation in the concentration of CD63+ EVs in both men and women. However, the increase in EV size was observed specifically in men [87].

2.3 Potential Role of Exercise-Induced EVs in Neuronal Health

In addition to proteins and peptides, altered levels of miRNA species both in tissues and circulation have been shown to play an important role in mediating neurodegenerative diseases [80, 88]. miRNAs, short for microRNAs, represent a category of small non-coding RNAs within EVs. These molecules, approximately 18–22 nucleotides in length, exert control over protein expression by binding to mRNA [89]. The distinct characteristics of miRNA expression not only signify the onset of neurodegenerative diseases in their early stages but also provide insights into the dynamic progression of diseases in later stages [88]. The mechanism through which miRNA communicates extracellularly was unclear. EVs are one of the confirmed route that these miRNA travel within various bodily fluids [90]. EVs carrying protein and miRNA cargo can traverse the blood-brain barrier (BBB) through both passive diffusion and active transport mechanisms like transcytosis [91]. Once inside the brain, miRNAs contribute to communication between various cell types, including neurons, astrocytes, and microglia [92]. Their regulatory roles encompass the modulation of neuronal inflammation and the promotion of neuronal growth following brain injury [92].

The involvement of EVs in neurodegenerative diseases like Alzheimer's disease (AD) or Parkinson's disease (PD) can be inferred by analyzing changes in the miRNAs, proteins, and peptides present in EVs. AD is a prevalent disorder affecting the elderly, encompassing over 60% of dementia cases [93]. AD is associated with a range of memory and cognitive impairments that evolve throughout the disease progression. EVs have the capacity to transport amyloid precursor protein (APP), C-terminal fragment of APP (CTFs-APP), proteases, Aβ oligomers, and tau protein. The accumulation of these elements is implicated in the development and progression of AD. The role that exercise-induced EVs might play in conferring benefits to patients with AD remains unknown. miR-21 has gained significant attention in the growing body of research investigating its mechanisms of involvement in central nervous system (CNS) disorders [94]. miR-21 attenuated apoptosis triggered by amyloid-β (Aβ) via modulating PDCD4/PI3K/AKT/GSK-3β pathway [94]. Ana Gámez-Valero and her colleagues found that hsa-miR-21-5p was significantly down-regulated in AD patients compared to dementia with Lewy bodies or controls [95]. Participating in exercise training has been shown to lead to an increase in circulating miR-21 level [96]. If the enhanced levels of miR-21 post-exercise are efficiently transmitted to brain through EVs, it holds the potential to alleviate AD symptoms by mitigating neuronal apoptosis induced by Aβ. Similarly, miR-342-3P, miR-338-3p, and miR-223 were also decreased in the blood of AD patients [97–99]. The levels of these miRNAs were increased in blood levels after exercise (Fig. 2), a pattern distinct from what is observed in Alzheimer's disease [100–102].

There are numerous advantages associated with delivering proteins or peptides through EVs. Apelin is broadly expressed in the central nervous system, yet its circulating levels rise with exercise [103]. Given its neuroprotective role through coupling to the G protein–coupled receptor (APJ), the transportation of Apelin via EVs could be beneficial in mitigating neurodegenerative diseases such as AD [104]. Exercise-induced CTSB also promotes neurogenesis through direct signal transduction [42]. Additionally, as a lysosomal protease, CTSB has been demonstrated to be neuroprotective in patients with AD by reducing levels of Aβ [105, 106]. Indeed, as previously mentioned, the presence of Cathepsin B (CTSB) in EVs suggests that it circulates in the blood and has been shown to cross the blood-brain barrier (BBB) [42].

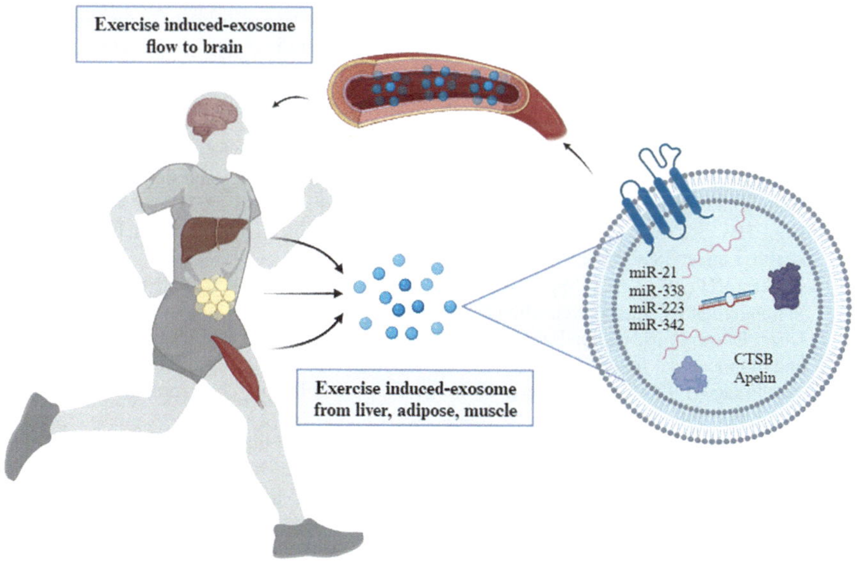

Fig. 2 Diagrammatic depiction of inter-organ communication facilitated through exercise-triggered EVs in neurodegenerative diseases. Engaging in physical activity has the potential to enhance the release of extracellular vesicles (EVs) into the bloodstream, leading to modifications in the levels of proteins and microRNAs (miRNAs) within extracellular vesicles targeting neurodegenerative diseases

3 Conclusion

In the absence of safe and effective drugs for neurodegenerative diseases, exercise stands out as a viable alternative for both prevention and treatment. Understanding the mechanisms that underlie the beneficial effects of exercise on neurodegenerative conditions is crucial and urgent. EVs offer advantageous properties for evaluating the impact of exercise. EVs, in particular, exhibit beneficial characteristics for gauging the effectiveness of exercise, such as stability in the bloodstream, the ability to target specific cells, and the capacity to traverse the blood-brain barrier (BBB). The identification of miRNAs and proteins within EVs post-exercise that contribute to enhancing neurodegenerative diseases further supports this notion. Despite many ongoing studies, a comprehensive understanding of how exercise influences the physiology of EVs is hindered by methodological limitations such as challenges in isolation and purification, along with the absence of high-resolution visualization techniques. Nevertheless, the exclusive information and distinct bioactive properties of transmitters and recipients make the study of EVs in exercise physiology worthwhile, given their potential effectiveness in preventing various brain diseases.

References

1. Wittenberg GF (2009) Neural plasticity and treatment across the lifespan for motor deficits in cerebral palsy. Dev Med Child Neurol 51(Suppl 4):130–133
2. Power JD, Schlaggar BL (2017) Neural plasticity across the lifespan. Wiley Interdiscip Rev Dev Biol 6(1)
3. Crosson B et al (2017) Advances in neurocognitive rehabilitation research from 1992 to 2017: the ascension of neural plasticity. Neuropsychology 31(8):900–920
4. Santiago JA, Potashkin JA (2023) Physical activity and lifestyle modifications in the treatment of neurodegenerative diseases. Front Aging Neurosci 15:1185671
5. Kohl HW 3rd et al (2012) The pandemic of physical inactivity: global action for public health. Lancet 380(9838):294–305
6. Hackert VH et al (2002) Hippocampal head size associated with verbal memory performance in nondemented elderly. NeuroImage 17(3):1365–1372
7. Duzel E, van Praag H, Sendtner M (2016) Can physical exercise in old age improve memory and hippocampal function? Brain 139:662–673
8. van Praag H (2009) Exercise and the brain: something to chew on. Trends Neurosci 32(5):283–290
9. Erickson KI et al (2009) Aerobic fitness is associated with hippocampal volume in elderly humans. Hippocampus 19(10):1030–1039
10. Erickson KI, Leckie RL, Weinstein AM (2014) Physical activity, fitness, and gray matter volume. Neurobiol Aging 35(Suppl 2):S20–S28
11. van Praag H et al (1999) Running enhances neurogenesis, learning, and long-term potentiation in mice. Proc Natl Acad Sci USA 96(23):13427–13431
12. Voss MW et al (2013) Bridging animal and human models of exercise-induced brain plasticity. Trends Cogn Sci 17(10):525–544
13. Rosario MA et al (2023) Cardiorespiratory fitness is associated with cortical thickness of medial temporal brain areas associated with spatial cognition in young but not older adults. Eur J Neurosci 59(1):82-100

14. Dietrich MO, Andrews ZB, Horvath TL (2008) Exercise-induced synaptogenesis in the hippocampus is dependent on UCP2-regulated mitochondrial adaptation. J Neurosci 28(42):10766–10771
15. Altman J, Das GD (1965) Autoradiographic and histological evidence of postnatal hippocampal neurogenesis in rats. J Comp Neurol 124(3):319–335
16. Kronenberg G et al (2003) Subpopulations of proliferating cells of the adult hippocampus respond differently to physiologic neurogenic stimuli. J Comp Neurol 467(4):455–463
17. Kempermann G et al (2004) Milestones of neuronal development in the adult hippocampus. Trends Neurosci 27(8):447–452
18. Bonaguidi MA et al (2011) In vivo clonal analysis reveals self-renewing and multipotent adult neural stem cell characteristics. Cell 145(7):1142–1155
19. Creer DJ et al (2010) Running enhances spatial pattern separation in mice. Proc Natl Acad Sci USA 107(5):2367–2372
20. Folsz O, Trouche S, Croset V (2023) Adult-born neurons add flexibility to hippocampal memories. Front Neurosci 17:1128623
21. Zhao C, Deng W, Gage FH (2008) Mechanisms and functional implications of adult neurogenesis. Cell 132(4):645–660
22. Vivar C et al (2023) Running throughout middle-age keeps old adult-born neurons wired. eNeuro 10(5):ENEURO.0084-23.2023
23. van Praag H (2008) Neurogenesis and exercise: past and future directions. Neuromolecular Med 10(2):128–140
24. Liu HL et al (2011) Treadmill exercise prevents decline in spatial learning and memory in APP/PS1 transgenic mice through improvement of hippocampal long-term potentiation. Behav Brain Res 218(2):308–314
25. Bliss T, Collingridge GL (2019) Persistent memories of long-term potentiation and the N-methyl-d-aspartate receptor. Brain Neurosci Adv 3:2398212819848213
26. Park JK, Lee SJ, Kim TW (2014) Treadmill exercise enhances NMDA receptor expression in schizophrenia mice. J Exerc Rehabil 10(1):15–21
27. Kondo M et al (2015) The 5-HT3 receptor is essential for exercise-induced hippocampal neurogenesis and antidepressant effects. Mol Psychiatry 20(11):1428–1437
28. Vivar C, van Praag H (2017) Running changes the brain: the long and the short of it. Physiology (Bethesda) 32(6):410–424
29. Neeper SA et al (1995) Exercise and brain Neurotrophins. Nature 373(6510):109–109
30. Marlatt MW et al (2012) Running throughout middle-age improves memory function, hippocampal neurogenesis, and BDNF levels in female C57BL/6J mice. Dev Neurobiol 72(6):943–952
31. Li Y et al (2008) TrkB regulates hippocampal neurogenesis and governs sensitivity to antidepressive treatment. Neuron 59(3):399–412
32. Moon HY et al (2012) Macrophage migration inhibitory factor mediates the antidepressant actions of voluntary exercise. Proc Natl Acad Sci USA 109(32):13094–13099
33. Trejo JL, Carro E, Torres-Aleman I (2001) Circulating insulin-like growth factor I mediates exercise-induced increases in the number of new neurons in the adult hippocampus. J Neurosci 21(5):1628–1634
34. Villeda SA et al (2014) Young blood reverses age-related impairments in cognitive function and synaptic plasticity in mice. Nat Med 20(6):659–663
35. Katsimpardi L et al (2014) Vascular and neurogenic rejuvenation of the aging mouse brain by young systemic factors. Science 344(6184):630–634
36. De Miguel Z et al (2021) Exercise plasma boosts memory and dampens brain inflammation via clusterin. Nature 600(7889):494–499
37. Horowitz AM et al (2020) Blood factors transfer beneficial effects of exercise on neurogenesis and cognition to the aged brain. Science 369(6500):167–173
38. Leiter O et al (2023) Platelet-derived exerkine CXCL4/platelet factor 4 rejuvenates hippocampal neurogenesis and restores cognitive function in aged mice. Nat Commun 14(1):4375

39. Leiter O et al (2019) Exercise-induced activated platelets increase adult hippocampal precursor proliferation and promote neuronal differentiation. Stem Cell Reports 12(4):667–679
40. Schroer AB et al (2023) Platelet factors attenuate inflammation and rescue cognition in ageing. Nature 620(7976):1071–1079
41. Moon HY et al (2019) Conditioned media from AICAR-treated skeletal muscle cells increases neuronal differentiation of adult neural progenitor cells. Neuropharmacology 145(Pt A):123–130
42. Moon HY et al (2016) Running-induced systemic Cathepsin B secretion is associated with memory function. Cell Metab 24(2):332–340
43. Chow LS et al (2022) Exerkines in health, resilience and disease. Nat Rev Endocrinol 18(5):273–289
44. Cheng B et al (2012) Neuroprotection of apelin and its signaling pathway. Peptides 37(1):171–173
45. Bostrom P et al (2012) A PGC1-alpha-dependent myokine that drives brown-fat-like development of white fat and thermogenesis. Nature 481(7382):463–468
46. Islam MR et al (2021) Author correction: exercise hormone irisin is a critical regulator of cognitive function. Nat Metab 3(10):1432
47. Wrann CD et al (2013) Exercise induces hippocampal BDNF through a PGC-1alpha/FNDC5 pathway. Cell Metab 18(5):649–659
48. Jedrychowski MP et al (2015) Detection and quantitation of circulating human Irisin by tandem mass spectrometry. Cell Metab 22(4):734–740
49. Lourenco MV et al (2019) Exercise-linked FNDC5/irisin rescues synaptic plasticity and memory defects in Alzheimer's models. Nat Med 25(1):165–175
50. Mazo CE et al (2022) High intensity acute aerobic exercise elicits alterations in circulating and skeletal muscle tissue expression of neuroprotective exerkines. Brain Plast 8(1):5–18
51. Gokce E et al (2021) Comparison of the effects of open- and closed-skill exercise on cognition and peripheral proteins: a cross-sectional study. PLoS One 16(6):e0251907
52. Mucher P et al (2021) Basal myokine levels are associated with quality of life and depressed mood in older adults. Psychophysiology 58(5):e13799
53. Gaitan JM et al (2021) Effects of aerobic exercise training on systemic biomarkers and cognition in late middle-aged adults at risk for Alzheimer's disease. Front Endocrinol (Lausanne) 12:660181
54. Kim J et al (2022) Cathepsin B and muscular strength are independently associated with cognitive control. Brain Plast 8(1):19–33
55. Besse-Patin A et al (2014) Effect of endurance training on skeletal muscle myokine expression in obese men: identification of apelin as a novel myokine. Int J Obes 38(5):707–713
56. Bao HJ et al (2015) Apelin-13 attenuates traumatic brain injury-induced damage by suppressing autophagy. Neurochem Res 40(1):89–97
57. Rai M, Demontis F (2022) Muscle-to-brain signaling via myokines and myometabolites. Brain Plast 8(1):43–63
58. Sleiman SF et al (2016) Exercise promotes the expression of brain derived neurotrophic factor (BDNF) through the action of the ketone body beta-hydroxybutyrate. Elife 5:e15092
59. Kwak SE et al (2021) Effects of exercise-induced beta-hydroxybutyrate on muscle function and cognitive function. Physiol Rep 9(3):e14497
60. Schnatz A et al (2021) Extracellular vesicles in neural cell interaction and CNS homeostasis. FASEB Bioadv 3(8):577–592
61. Thompson AG et al (2016) Extracellular vesicles in neurodegenerative disease - pathogenesis to biomarkers. Nat Rev Neurol 12(6):346–357
62. Zhang N et al (2021) Role of exosomes in brain diseases. Front Cell Neurosci 15:743353
63. Yang E et al (2020) Exosome-mediated metabolic reprogramming: the emerging role in tumor microenvironment remodeling and its influence on cancer progression. Signal Transduct Target Ther 5(1):242

64. Johnstone RM et al (1987) Vesicle formation during reticulocyte maturation. association of plasma membrane activities with released vesicles (exosomes). J Biol Chem 262(19):9412–9420
65. Thery C, Zitvogel L, Amigorena S (2002) Exosomes: composition, biogenesis and function. Nat Rev Immunol 2(8):569–579
66. Garcia-Martin R et al (2022) MicroRNA sequence codes for small extracellular vesicle release and cellular retention. Nature 601(7893):446–451
67. Thery C et al (2018) Minimal information for studies of extracellular vesicles 2018 (MISEV2018): a position statement of the International Society for Extracellular Vesicles and update of the MISEV2014 guidelines. J Extracell Vesicles 7(1):1535750
68. Kim DK et al (2013) EVpedia: an integrated database of high-throughput data for systemic analyses of extracellular vesicles. J Extracell Vesicles 2. https://doi.org/10.3402/jev.v2i0.20384
69. Wei H et al (2021) Regulation of exosome production and cargo sorting. Int J Biol Sci 17(1):163–177
70. Colombo M, Raposo G, Thery C (2014) Biogenesis, secretion, and intercellular interactions of exosomes and other extracellular vesicles. Annu Rev Cell Dev Biol 30:255–289
71. Korpelainen R et al (2016) Exercise capacity and mortality – a follow-up study of 3033 subjects referred to clinical exercise testing. Ann Med 48(5):359–366
72. Pedersen BK, Hoffman-Goetz L (2000) Exercise and the immune system: regulation, integration, and adaptation. Physiol Rev 80(3):1055–1081
73. Pedersen BK, Febbraio MA (2008) Muscle as an endocrine organ: focus on muscle-derived interleukin-6. Physiol Rev 88(4):1379–1406
74. Whitham M (2017) Skeletal muscle as an endocrine organ. Clin Endocrinol 86:5–6
75. Hartwig S et al (2014) Secretome profiling of primary human skeletal muscle cells. Biochim Biophys Acta 1844(5):1011–1017
76. Raschke S et al (2013) Identification and validation of novel contraction-regulated myokines released from primary human skeletal muscle cells. PLoS One 8(4):e62008
77. Lai ZJ et al (2023) Exosomes as a delivery tool of exercise-induced beneficial factors for the prevention and treatment of cardiovascular disease: a systematic review and meta-analysis. Front Physiol 14
78. Fischetti F et al (2023) The role of exercise parameters on small extracellular vesicles and microRNAs cargo in preventing neurodegenerative diseases. Front Physiol 14:1241010
79. Moon HY et al (2020) Neural maturation enhanced by exercise-induced extracellular derivatives. Sci Rep 10(1):3893
80. Yoon KJ et al (2021) Effects of voluntary running wheel exercise-induced extracellular vesicles on anxiety. Front Mol Neurosci 14:665800
81. Park S, Moon HY (2022) Urinary extracellular vesicle as a potential biomarker of exercise-induced fatigue in young adult males. Eur J Appl Physiol 122(10):2175–2188
82. Whitham M et al (2018) Extracellular vesicles provide a means for tissue crosstalk during exercise. Cell Metab 27(1):237–251 e4
83. Eguchi A et al (2016) Circulating adipocyte-derived extracellular vesicles are novel markers of metabolic stress. J Mol Med (Berl) 94(11):1241–1253
84. Ying W et al (2017) Adipose tissue macrophage-derived exosomal miRNAs can modulate and insulin sensitivity. Cell 171(2):372
85. Apostolopoulou M et al (2021) Metabolic responsiveness to training depends on insulin sensitivity and protein content of exosomes in insulin-resistant males. Sci Adv 7(41):eabi9551
86. Warnier G et al (2023) Effects of a 6-wk Sprint interval training protocol at different altitudes on circulating extracellular vesicles. Med Sci Sports Exerc 55(1):46–54
87. Conkright WR et al (2022) Resistance exercise differentially alters extracellular vesicle size and subpopulation characteristics in healthy men and women: an observational cohort study. Physiol Genomics 54(9):350–359

88. Li S, Lei Z, Sun T (2023) The role of microRNAs in neurodegenerative diseases: a review. Cell Biol Toxicol 39(1):53–83
89. Friedman RC et al (2009) Most mammalian mRNAs are conserved targets of microRNAs. Genome Res 19(1):92–105
90. Valadi H et al (2007) Exosome-mediated transfer of mRNAs and microRNAs is a novel mechanism of genetic exchange between cells. Nat Cell Biol 9(6):654–U72
91. Mori MA et al (2019) Extracellular miRNAs: from biomarkers to mediators of physiology and disease. Cell Metab 30(4):656–673
92. Zolboot N et al (2021) MicroRNAs instruct and maintain cell type diversity in the nervous system. Front Mol Neurosci 14:646072
93. Soliman HM et al (2021) Exosomes in Alzheimer's disease: from being pathological players to potential diagnostics and therapeutics. Int J Mol Sci 22(19):10794
94. Bai X, Bian Z (2022) MicroRNA-21 is a versatile regulator and potential treatment target in central nervous system disorders. Front Mol Neurosci 15:842288
95. Gamez-Valero A et al (2019) Exploratory study on microRNA profiles from plasma-derived extracellular vesicles in Alzheimer's disease and dementia with Lewy bodies. Transl Neurodegener 8:31
96. Valenti MT et al (2019) Physical exercise modulates miR-21-5p, miR-129-5p, miR-378-5p, and miR-188-5p expression in progenitor cells promoting Osteogenesis. Cells 8(7):742
97. Cheng L et al (2015) Prognostic serum miRNA biomarkers associated with Alzheimer's disease shows concordance with neuropsychological and neuroimaging assessment. Mol Psychiatry 20(10):1188–1196
98. Lugli G et al (2015) Plasma Exosomal miRNAs in persons with and without Alzheimer disease: altered expression and prospects for biomarkers. PLoS One 10(10):e0139233
99. Wei H et al (2018) Serum Exosomal miR-223 serves as a potential diagnostic and prognostic biomarker for dementia. Neuroscience 379:167–176
100. Mach N et al (2016) Integrated mRNA and miRNA expression profiling in blood reveals candidate biomarkers associated with endurance exercise in the horse. Sci Rep 6:22932
101. Nielsen S et al (2014) The miRNA plasma signature in response to acute aerobic exercise and endurance training. PLoS One 9(2):e87308
102. Zhou Q et al (2020) Circulating microRNAs in response to exercise training in healthy adults. Front Genet 11:256
103. Vinel C et al (2018) The exerkine apelin reverses age-associated sarcopenia. Nat Med 24(9):1360–1371
104. Tian Y et al (2020) The protective effects and mechanisms of Apelin/APJ system on ischemic stroke: a promising therapeutic target. Front Neurol 11:75
105. Mueller-Steiner S et al (2006) Antiamyloidogenic and neuroprotective functions of Cathepsin B: implications for Alzheimer's disease. Neuron 51(6):703–714
106. Cermak S et al (2016) Loss of Cathepsin B and L leads to lysosomal dysfunction, NPC-like cholesterol sequestration and accumulation of the key Alzheimer's proteins. PLoS One 11(11):e0167428
107. Vivar C, Potter MC, Choi J, Lee JY, Stringer TP, Callaway EM, Gage FH, Suh H, van Praag H (2012) Monosynaptic inputs to new neurons in the dentate gyrus. Nat Commun 3:1107

Acute Exercise Effects on Executive Function: Exploring the Relationship and Moderating Factors Through the 3W+1H Framework

Feng-Tzu Chen, Chen-Sin Hung, Nai-Chi Chen, and Yu-Kai Chang

Abstract This chapter delves into the impact of acute exercise on executive function—a key component of cognitive functions. Despite a robust body of evidence showcasing the substantial benefits of chronic exercise on executive function, a notable gap exists in our understanding of its acute effects. The chapter unfolds in four key segments. Firstly, it provides a succinct definition of executive function. Subsequently, it synthesizes findings from previous systematic reviews and meta-analyses, elucidating the overall impact of acute exercise on executive function. Despite occasional discrepancies in individual studies, a consistent positive association emerges. The third segment employs the 3W+1H framework to explore moderators shaping this relationship, scrutinizing the "Who, What, When, and How" factors. Through this lens, the chapter aims to uncover nuanced conditions under which acute exercise optimally enhances executive function. Lastly, the chapter outlines future research directions, emphasizing the necessity for targeted investigations to refine our understanding of the intricate interplay between acute exercise and executive function. This inquiry contributes to the ongoing discourse on the benefits of exercise for executive function, offering insights with potential applications in both research and practical contexts to promote cognitive well-being.

F.-T. Chen
Department of Kinesiology, National Tsing Hua University, Hsinchu, Taiwan

C.-S. Hung · N.-C. Chen
Department of Physical Education and Sports Sciences, National Taiwan Normal University, Taipei, Taiwan

Y.-K. Chang (✉)
Department of Physical Education and Sports Sciences, National Taiwan Normal University, Taipei, Taiwan

Social Emotional Education and Development Center, National Taiwan Normal University, Taipei, Taiwan

Institute of Sport Science and Innovations, Lithuanian Sports University, Kaunas, Lithuania

H. Soya (ed.), *Exercise Brain Stimulation for Cognitive Function and Mental Health*, Advances in Neurobiology 44, https://doi.org/10.1007/978-981-95-0066-6_9

Keywords Physical activity · Single bout of exercise · Cognitive function · Working memory

1 Introduction

Over the past four decades, a growing body of research has identified acute exercise (a single bout of exercise) as a treatment strategy for improving cognitive functions [1]. Although individual empirical studies have produced inconsistent findings, several reviews have consistently reported a positive effect of acute exercise on cognitive performance [2–4]. Indeed, the 2018 edition of the *Physical Activity Guidelines for Americans* (PAGA 18) marks the first time the topic of a single session of physical activity for cognitive function and brain health has been included [5]. However, the PAGA guideline for optimizing the effects of acute exercise is still in the developmental stage and to facilitate optimal acute exercise prescriptions, a commentary by Chang et al. [6] introduced the 3W+1H framework (i.e., what, who, when, and how, Fig. 1) to further understand the exercise-cognition relationship.

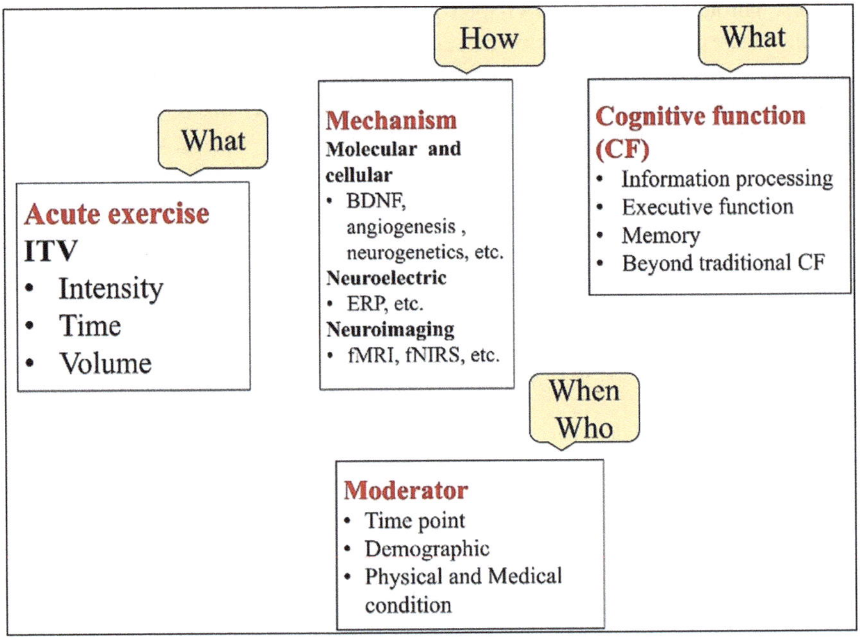

Fig. 1 The details of the 3W+1H framework. (Adapted from Chang et al. [6]). ITV intensity, time, and volume, ERP event-related potential, fMRI functional magnetic resonance imaging, fNIRS functional near-infrared spectroscopy

Colcombe and Kramer [7] suggested a substantial benefit in executive function (a component of cognitive function) with a high effect size (ES = 0.68) through chronic (long-term) exercise. This has sparked significant interest in executive function as a cognitive outcome in the context of acute exercise and cognitive functions. Despite our laboratory's recent updates on the chronic effects of exercise on executive function [8], relatively little is known about its acute effects on executive function. In this chapter, we introduce four main topics: (1) a brief overview of the definition of executive function; (2) the overall effects of acute exercise on executive function through previous systematic reviews and meta-analyses; (3) a discussion on the moderators influencing the positive relationship between acute exercise and executive function based on the 3W+1H framework; and (4) future directions for research on acute exercise and executive function.

2 What Is Executive Function?

Executive function, also known as executive control, is recognized as a meta- or higher-level cognitive function that regulates several fundamental cognitive processes [9]. Researchers have defined executive function as mental capacities dealing with novelty, planning, acting on appropriate strategies for conducting performance or activities, volition, purposeful behavior, and effective performance [10, 11]. Because executive function is an umbrella term rather than a unitary one, it can be divided into core and higher-level executive functions [12, 13].

In executive function, three components for core executive function (a-c) and one component for higher-level executive function (d) are included: (a) *Inhibitory control*: This component inhibits irrelevant information or stimuli and focuses on the aim objective [12]. (b) *Working memory* (also known as updating): This represents the ability to store or manipulate task-related information to regulate thoughts and behaviors within a short period [14]. It is associated with comprehension, learning, reasoning, and problem-solving in daily life. (c) *Switching* (also known as cognitive flexibility): This represents the ability to switch attention or strategies between mental sets according to task demands [15]. (d) *Planning and problem-solving*: This component commonly is regarded as metacognition [16] and they are considered to have the ability to model and predict actions before achieving goals that require skill or strategy [17]. In summary, core and high-level executive functions are frequently found to be negatively affected with increasing age but appear to maintain tractability.

To apply executive function to the issue of exercise and cognitive function, a commentary by Chang and Etnier [18] identified the most frequently used measures of executive function based on neuropsychological reviews and clinical investigations. Specifically, the Wisconsin card sorting test (WCST) assessed working memory for executive function, the Stroop test measured inhibition, the trail making test (TMT) evaluated shifting, verbal fluency gauged information retrieval from memory, and the Tower of Hanoi (TOH) and Tower of London (TOL) task assessed

Table 1 Classification of executive function assessments

Executive function			
Inhibition	Working memory	Switching	Planning
Stroop task The California older adult Stroop test—interference Auditory Stroop test Hayling sentence completion test Eriksen flanker task—incongruent Go/no-go test Stop signal task The random number generation task	N-back Spatial working memory Wisconsin card sorting task Digit span task Wechsler adult intelligence scale letter number sequencing Verbal fluency—FAS form Rey–Osterrieth complex figure test List sorting task Auditory verbal learning task COWAT—verbal fluency D-KEFS confirmed correct and free card sorting—shifting Running memory task—total recall Spatial running span task The list learning delayed recall test	Task switching Trail making test Shifting-naming Color trail test Spatial switching Picture switching Arrow switching Attentive matrices test D-KEFS verbal fluency The digit-letter task The plus-minus task Dimension-switching task	Tower of London task Six elements test Greenwich test Multiple errands test Hotel test Naturalistic action test RIPA problem solving Delay discounting task The Iowa gambling task Cambridge gambling task Raven's progressive matrices

Note: *COWAT* Control Oral Word Association Test, *D-KEFS* Delis-Kaplan Executive Function System, *RIPA* Ross Information Processing Assessment

planning and problem solving. Since then, these tasks have been highlighted for examining executive function in sport and exercise psychology and further applied to acute exercise and executive function research. For a more in-depth exploration in exercise and cognition studies, a systematic review and meta-analysis from our laboratory by Chen et al. [8] listed cognitive tasks for four executive function components (see Table 1).

3 Acute Exercise and Executive Function: Evidence from Meta-Analytic Review

In the context of meta-analytic reviews, the effects of acute exercise on cognitive functions have undergone scrutiny in multiple studies [2, 3, 19–21]. Initially, Etnier et al. [20] reported that acute exercise has a significant positive but small effect on cognitive performance (ES = 0.16). Sibley, Etnier [21] focused specifically on the child population, revealing a significant small effect (ES = 0.37) that demonstrates how acute exercise facilitates cognitive functions in children. Following this,

Lambourne et al. [3] concentrated on acute exercise studies ($n = 40$) and categorized them based on the timing of cognitive function task measurements, distinguishing between during exercise and after exercise. They found a detrimental effect on cognitive functions during exercise ($d = -0.14$) but improved cognitive functions in the post-exercise period ($d = 0.20$). This meta-analysis suggests that the timing of assessing cognitive tasks might influence the relationship between acute exercise and cognitive functions. In our laboratory's findings, Chang et al. [2], multiple moderators were considered to understand the effects of acute exercise on cognitive functions. They specifically noted a positive effect on executive function at three different examination times: during exercise, immediately following exercise, and after a delay longer than 1 minute following exercise. Most recently, Ishihara et al. [19] concentrated on acute aerobic exercise and its impact on executive function. Their results showed a beneficial effect of acute aerobic exercise on response time (RT) and difference score but not on accuracy rate in executive function performance.

Despite the existing evidence on the effects of acute exercise on executive function, these reviews highlight the importance of moderators in elucidating potential mechanisms and offering guidance for exercise recommendations to maximize cognitive benefits. Although two systematic reviews from our laboratory have extensively discussed and updated current knowledge regarding acute exercise on executive function [22, 23], questions about moderators remain largely unanswered. To fully comprehend the relationship between acute exercise and executive function, it is crucial to explore whether moderators could significantly contribute to establishing effective strategies for prescribing exercise. This exploration aims to enhance our understanding and facilitate the development of precise exercise prescriptions for acute exercise to optimize improvements in executive function.

4 The Moderators to Influence the Relationship Between Acute Exercise and Executive Function

The 2018 iteration of the Physical Activity Guidelines for Americans (PAGA 18) represents a noteworthy development, as it marks the inaugural inclusion of a discourse on the impact of a singular session of physical activity on cognitive function and brain health [5]. Nevertheless, the PAGA guideline pertaining to the optimization of the effects of acute exercise remains in the developmental phase. To advance precision in the prescription of optimal acute exercise regimens, Chang et al. [6] presented a commentary introducing the 3W+1H framework, addressing the considerations of what, who, when, and how (Fig. 1). This framework is posited to ensure, on empirical grounds, that acute exercise prescriptions are meticulously tailored to augment executive function.

4.1 What

The FITT-VP principle, guided by the American College of Sports Medicine [24], encompasses frequency, intensity, time, type, volume, and progression, providing essential considerations for tailoring optimal acute exercise prescriptions. In studies on acute exercise and executive function, three dimensions (intensity, time, and volume) within the "what" category are primarily associated with acute exercise investigations. Additionally, previous studies explore the question of "what cognitive function category" is most beneficial from acute exercise. In the following discussion, we will delve into the experimental methods employed in past studies and present their results.

Exercise intensity is frequently considered a moderator that influences the effect of acute exercise on executive function. In the early stages of research on this topic, Kamijo et al. [25] investigated three exercise intensities (high, medium, low) of acute aerobic exercise on the inhibition aspect of executive function. Their results revealed an inverted U-shaped behavioral performance, suggesting that participants engaged in medium-intensity exercise (rating of perceived exertion, RPE = 12–14) demonstrated better performance than those in high- or low-intensity exercises. Subsequently, Chang, Etnier [18] focused on resistance exercise and further examined the dose relationship between acute exercise and executive function. Their findings indicated that participants in the moderate-intensity group (70% of 10RM) exhibited better Stroop test performance than the other three groups (40% of 10RM, 100% of 10RM, and reading control). In a similar experimental design, Chang et al. [26] applied the dose-response effect of acute exercise on the planning aspect of executive function in middle-aged adults. Their results also supported the inverted U-shaped relationship between acute exercise and executive function. Specifically, moderate-intensity resistance exercise was recommended for improving executive function in this population.

Exercise time (duration) is another moderator influencing the effect of acute exercise on executive function. Evidence provided by Chang et al. [27] indicates that acute exercise for 30 minutes, including warm-up and cool-down, has a more significant impact compared to shorter (20 minutes) and longer durations (45 minutes). This suggests that the duration of acute exercise engagement is a crucial factor when prescribing exercise to improve executive function. Furthermore, in a study targeting late middle-aged adults (55–65 years old), Chen et al. (2018) found that acute exercise for 30 and 45 minutes resulted in superior switching performance in executive function. Collectively, these two studies from our laboratory suggest that moderate acute exercise (30–45 minutes) effectively enhances executive function in both younger and late middle-aged adults.

Despite the limited number of studies examining exercise volume, this variable appears to be a significant factor in acute exercise studies. For instance, Geng et al. [28] suggested that acute exercise, irrespective of exercise volume (considering variations in intensity and duration), enhances move- and time-related performance in Tower of London (TOL) tasks. Similarly, Chang et al. [29] indicated that acute exercise, regardless of exercise volume, improves the planning aspect of executive function (TOL task) in older adults with APOE-ε4 carriers, who have a higher risk of Alzheimer's disease. While this study offers a new perspective on the moderating effect of exercise volume on executive function, additional research is needed to establish a clearer understanding of the relationship between acute exercise and executive function.

The "what" aspect also encompasses the question of which cognitive function categories benefit most from acute exercise. Previous meta-analyses have suggested that acute exercise appears to be selectively sensitive to specific cognitive functions, including information processing, memory, and executive function [2]. Specifically, the results from the meta-analysis indicated that acute exercise has a positive impact on executive function across all age groups. Given that executive function is considered a higher-level cognitive function with consistent positive effects from acute exercise, it has frequently been employed as the primary cognitive function assessment in subsequent studies [22, 23]. However, a study from our laboratory found no significant improvements in the executive function of working memory (using the WSCT) after acute exercise when compared to a reading control group [30]. Therefore, we suggest that the beneficial effect of acute exercise on executive function might be selective rather than general.

In brief, recent research has shown that acute exercise is effective in enhancing executive function. Studies have also delved into the characteristics of acute exercise, such as intensity, time, and volume, to unravel the intricacies of the relationship between acute exercise and executive function. The positive effects observed in executive function following acute exercise engagements have led to the common application of the cognitive category (i.e., executive function) in studies investigating the effects of acute exercise on cognitive function.

4.2 When and Who

When considering the "when" aspect, the time point for assessing executive function becomes a central topic in investigating the effects of acute exercise. Specifically, three different time points (during exercise, immediately following exercise, and after a delay of longer than 1 minute) are extensively discussed to comprehend the relationship between acute exercise and executive function [2]. During exercise, a study from our laboratory by Wang et al. [30] indicated that the executive function of working memory (WSCT) was impaired when participants engaged in acute exercise at high intensity (80% heart rate reserve). This finding suggested that acute exercise with high intensity might lead to transient hypofrontality, wherein both individual executive body movement and cognitive function assessment concurrently declined during exercise. In the timeline following exercise or after exercise for longer than 1 minute, a study by our laboratory further suggested that acute exercise benefits executive function immediately after exercise, and this positive effect can be sustained up to 80 minutes after acute exercise [31].

Regarding "who," a few studies have explored whether participants' characteristics influence the benefits of acute exercise, with demographic background being one dimension to understand the moderating relationship between acute exercise and executive function. Specifically, fitness level (high fit, low fit) has been considered a moderator in examining the effect of acute exercise on executive function. In a study targeting older adults, Chu et al. [32] from our laboratory indicated that acute exercise facilitated Stroop task performance in both high- and low-fit individuals, representing a general improvement in the inhibition aspect of executive function. Conversely, Li et al. [33] suggested that fitness level moderates the effect of acute exercise on executive function, implying that participants in the high-fitness group exhibited a higher accuracy rate than those in the low-fitness group. While fitness level appears to moderate the effects on executive function, the outcomes may vary when targeting different age groups, such as older adults and younger adults. We advocate for additional studies to delve further into this issue, fostering a better understanding of the moderating role (e.g., fitness level) in the relationship between acute exercise and executive function.

Another potential moderator under the category of "who" that influences the effect of acute exercise on executive function is the individual's medical condition. Specifically, the beneficial effects of acute exercise appear to be particularly sensitive for those diagnosed with worse mental/cognitive or physical conditions, such as mild cognitive impairment (MCI), preterm birth, or diabetes. A recent study by Devenney et al. [34] proposed that participants with MCI exhibited more significant changes in executive function, as assessed by the Stroop task, after acute exercise compared to normative older adults. Furthermore, two recent studies from our laboratory demonstrated that acute exercise can enhance executive function (inhibition, planning) in children with preterm birth [35, 36]. Similarly, Vincent, Hall [37] investigated the effect of acute exercise on executive function among adults with diabetes, finding improvements in executive function, particularly in inhibition (Stroop Task), but only for female participants.

4.3 How

Previous evidence supporting a positive relationship between acute exercise and executive function is grounded in the understanding that exercise's physiological responses can have a beneficial impact on executive functioning. Numerous physiological responses, including increased levels of brain-derived neurotrophic factor (BDNF) [38, 39], alterations in angiogenesis or neurogenetics [40], and the activation of brain functioning [41, 42], have been implicated in the literature concerning acute exercise and its effects on executive function.

In animal studies, brain-derived neurotrophic factor (BDNF) has been implicated in neural development and functioning, playing roles in neurogenesis, dendritic growth, and the long-term potentiation of neurons [43, 44]. Higher levels of BDNF have been associated with more significant cognitive functions in human studies

[45, 46]. A meta-analysis in human studies reported a moderate effect size (ES = 0.46) for increases in BDNF following acute exercise [47]. Previously, two studies have examined the relationship between BDNF and executive function after acute exercise. From our laboratory, Chang et al. [48] first investigated whether acute exercise influences BDNF levels and executive function in young adults. The study found that, after acute exercise, participants exhibited improved Stroop task performance, although the level of BDNF did not show significant changes. In contrast, Devenney et al. [34] demonstrated that acute high-intensity exercise increased BDNF levels in a population with mild cognitive impairment (MCI), but they did not observe benefits in executive function performance. As this research direction continues to develop, we suggest more studies are needed to fully understand the role of BDNF in the relationship between acute exercise and executive function.

The underlying mechanism of acute exercise on executive function may involve the activation of brain functioning. In the neuroelectric approach, event-related potential (ERP) has been considered a valuable strategy for evaluating changes in brain activation during the performance of executive function tasks. Previous evidence from our laboratory, as reported by Chu et al. [49], demonstrated that participants exhibited a larger P3 amplitude and shorter latency after acute exercise. These findings suggest that acute exercise effectively increases attentional resources and enhances information processing speed efficiently. In the field of neuroimaging, functional magnetic resonance imaging (fMRI) is the primary approach for investigating the effects of exercise on executive function. However, to date, only a few studies have examined the impact of acute exercise on both executive function performance and brain functioning assessed by fMRI [41, 42]. These studies predominantly indicate that acute exercise facilitates activation in the prefrontal cortex region, which is linked to the performance of executive function.

5 Considerations for Future Directions

While we have outlined evidence supporting the effectiveness of acute exercise in enhancing executive function, it is crucial to acknowledge several unanswered or unexplored questions. These include inquiries into exercise modalities, lifestyle and genetic factors, neuroimaging approaches, and underlying mechanisms. Addressing these aspects can contribute to a more comprehensive understanding of the intricate relationship between acute exercise and executive function.

5.1 Exercise Modality

When considering exercise characteristics, high-intensity interval exercise (HIIE) or high-intensity interval training (HIIT) is characterized by brief, intermittent bursts of intensive aerobic exercise interspersed with brief periods of low-intensity exercise or recovery [50]. This modality has recently been employed to explore the relationship between acute exercise and executive function. For instance, Tsai et al. [51] compared the effects of acute HIIE and moderate-intensity continuous exercise (MICE) on executive function in healthy late middle-aged and older adults. Their findings revealed that both HIIE and MICE could benefit executive function, suggesting that the positive effects on executive function from HIIE are comparable to those from MICE. A current systematic review from our lab also indicated the positive effects of acute HIIT on multiple executive functions, including inhibition, working memory, and shifting [22]. However, HIIE or HIIT is still a relatively new exercise modality in the context of acute exercise and executive function, warranting further research to deepen our understanding of this relationship.

Another approach to exercise modality is acute exercise with blood flow restriction (BFR). This exercise generally involves low load/intensity in combination with partial restriction of blood flow (e.g., arm, leg). Previous studies investigating the effect of acute resistance exercise with BFR on executive function have shown mixed results. For example, Sardeli et al. [52] found no benefits in inhibiting executive function through the Stroop task. Currently, Sugimoto et al. [53] explored the effect of acute exercise with BFR on executive function and found an improvement in interference control after walking with BFR. A recent review by Yamada et al. [54] introduces BFR into research related to acute exercise and executive function. In the review, several possible mechanisms (e.g., lactate, mRNA expression, norepinephrine, cortical excitability) are proposed to explain how exercise with BFR could enhance executive function.

5.2 *Lifestyle and Genetic Factor*

Lifestyle and genetic factors play a significant role in shaping the relationship between acute exercise and executive function. In examining lifestyle factors, previous studies have commonly explored whether the level of physical activity expenditure moderates the effects of acute exercise on executive function. Two specific lifestyle interventions, such as reducing sedentary behavior time [55] and improving sleep quality ([56, 57]), are currently being applied in studies investigating the relationship between acute exercise and executive function. Moreover, genetic factors have been considered in this context, with a focus on the apolipoprotein E (APOE ε4) allele—a genetic factor associated with the development of Alzheimer's disease. The impact of APOE ε4 on the effects of acute exercise on executive function has been explored in recent research [29]. As we move forward, it is essential to further explore the role of lifestyle behaviors and genetic risk factors as moderators to gain a more comprehensive understanding of the complex relationship between exercise and cognition.

5.3 *Neuroimaging Approach*

Functional near-infrared spectroscopy (fNIRS), a neuroimaging and non-invasive approach for assessing cerebral hemodynamic responses during cognitive tasks, is currently gaining traction in research on the acute effects of exercise on executive function. For instance, Ji et al. [58] delved into the effects of acute exercise on executive function and concurrent fNIRS changes, revealing that acute exercise not only enhances Stroop task performance but also increases cerebral oxygenation. Similarly, Suwabe et al. [59] extended this exploration by investigating the effects of moderate-intensity acute exercise with music on mood regulation and executive function, utilizing fNIRS. Their findings suggested a positive mood and greater activation of the prefrontal cortex after acute exercise with music, indicating that the affective response to exercise influences the rapid enhancement of executive function. Despite these promising findings, the use of the fNIRS approach in exploring acute exercise and executive function is still relatively limited, with only a few studies to date. Consequently, additional future research is needed to establish and further understand the relationship, as well as to uncover potential underlying mechanisms.

5.4 *Neurobiological Mechanism*

To delve deeper into the neurobiological mechanisms, a recent systematic review by Zou et al. [60] recommends the incorporation of eye-based measurements (e.g., eye movement measures such as saccades, pupillary measures like pupil dilation, and vascular measures such as retinal vessel diameter) in studies investigating acute exercise and executive function. This direction is proposed as a fruitful avenue for future research to gain a more comprehensive understanding of the underlying mechanisms. In previous studies, two investigations demonstrated increased tonic pupil dilation during executive function tasks after acute exercise in younger adults [61, 62]. However, beyond pupil dilation, there is a paucity of studies exploring other eye-based measurements, and none have investigated their effects in different populations (e.g., children, older adults). Consequently, we advocate for more future research utilizing eye-based measurements to further illuminate the neurobiological mechanisms involved in the relationship between acute exercise and executive function.

6 Conclusion

In this chapter, we have provided a comprehensive overview, beginning with the definition of executive function and supported it with evidence from meta-analytic articles affirming the positive impact of acute exercise on executive function.

Introducing the "3W+1H framework," we have outlined potential moderators that influence the relationship between acute exercise and executive function. However, despite the progress made, certain aspects of the acute exercise and executive function relationship remain unanswered and unexplored. To address these gaps, we propose future perspectives for follow-up studies, including exercise modalities, lifestyle and genetic factors, neuroimaging approaches, and underlying mechanisms. Combining the existing knowledge and research evidence, we conclude that acute exercise is an effective means of improving executive function.

Acknowledgments This work was financially supported by the National Science and Technology Council, Taiwan (MOST 110-2410-H-003-142-MY3; 111-2918-I-003-001, 111-2410-H-039-012, 112-2410-H-007-108-, 113-2628-H-007-005-MY3) and by the National Taiwan Normal University (NTNU) within the framework of the Higher Education Sprout Project (HESP) by the Ministry of Education (MOE) in Taiwan, as well as "Institute for Research Excellence in Learning Sciences" and "Social Emotional Education and Development Center" of the Featured Areas Research Center Program within the framework of the HESP of the MOE in Taiwan by the National Taiwan Normal University (NTNU) within the framework of the Higher Education Sprout Project (HESP) by the Ministry of Education (MOE) in Taiwan, by Social Emotional Education and Development Center and Institute for Research Excellence in Learning Sciences of the Featured Areas Research Center Program within the framework of the HESP of the MOE in Taiwan, as well as Initiative for Excellence in Universities program (No. S-A-UEI-23-2, SMSM) from the Ministry of Education, Science, and Sport of the Republic of Lithuania.

References

1. Tomporowski PD, Ellis NR (1986) Effects of exercise on cognitive processes: a review. Psychol Bull 99(3):338
2. Chang YK, Labban JD, Gapin JI, Etnier JL (2012) The effects of acute exercise on cognitive performance: a meta-analysis. Brain Res 1453:87–101. https://doi.org/10.1016/j.brainres.2012.02.068
3. Lambourne K, Audiffren M, Tomporowski PD (2010) Effects of acute exercise on sensory and executive processing tasks. Med Sci Sports Exerc 42(7):1396–1402. https://doi.org/10.1249/MSS.0b013e3181cbee11
4. Angevaren M, Aufdemkampe G, Verhaar HJ, Aleman A, Vanhees L (2008) Physical activity and enhanced fitness to improve cognitive function in older people without known cognitive impairment. Cochrane Database Syst Rev (3):Cd005381. https://doi.org/10.1002/14651858.CD005381.pub3
5. Piercy KL, Troiano RP, Ballard RM, Carlson SA, Fulton JE, Galuska DA et al (2018) The physical activity guidelines for Americans. JAMA 320(19):2020–2028
6. Chang YK, Erickson KI, Stamatakis E, Hung TM (2019) How the 2018 US physical activity guidelines are a call to promote and better understand acute physical activity for cognitive function gains. Sports Med 49:1625-1627. https://doi.org/10.1007/s40279-019-01190-x
7. Colcombe S, Kramer AF (2003) Fitness effects on the cognitive function of older adults: a meta-analytic study. Psychol Sci 14(2):125–130. https://doi.org/10.1111/1467-9280.t01-1-01430
8. Chen FT, Etnier JL, Chan KH, Chiu PK, Hung TM, Chang YK (2020) Effects of exercise training interventions on executive function in older adults: a systematic review and meta-analysis. Sports Med (Auckland, NZ). 50(8):1451–1467. https://doi.org/10.1007/s40279-020-01292-x
9. Alvarez JA, Emory E (2006) Executive function and the frontal lobes: a meta-analytic review. Neuropsychol Rev 16(1):17–42. https://doi.org/10.1007/s11065-006-9002-x

10. Salthouse T (2008) Exercise and mental resources: methodological problems. Exerc Mediat Eff Cogn 2:111–116
11. Rabbitt P (2004) Introduction: Methodologies and models in the study of executive function. In: Methodology of frontal and executive function. Routledge, London, pp 9–45
12. Miyake A, Friedman NP, Emerson MJ, Witzki AH, Howerter A, Wager TD (2000) The unity and diversity of executive functions and their contributions to complex "frontal lobe" tasks: a latent variable analysis. Cogn Psychol 41(1):49–100. https://doi.org/10.1006/cogp.1999.0734
13. Romine CB, Reynolds CR (2005) A model of the development of frontal lobe functioning: findings from a meta-analysis. Appl Neuropsychol 12(4):190–201. https://doi.org/10.1207/s15324826an1204_2
14. Baddeley A (2012) Working memory: theories, models, and controversies. Annu Rev Psychol 63(1):1–29. https://doi.org/10.1146/annurev-psych-120710-100422
15. Diamond A (2013) Executive functions. Annu Rev Psychol 64(1):135–168. https://doi.org/10.1146/annurev-psych-113011-143750
16. Tomporowski, P. D., McCullick, B., Pendleton, D. M., & Pesce, C. (2015). Exercise and children's cognition: The role of exercise characteristics and a place for metacognition. J Sport Health Sci 4(1):47–55.
17. Unterrainer JM, Owen AM (2006) Planning and problem solving: from neuropsychology to functional neuroimaging. J Physiol Paris 99(4–6):308–317. https://doi.org/10.1016/j.jphysparis.2006.03.014
18. Chang YK, Etnier JL (2009) Exploring the dose-response relationship between resistance exercise intensity and cognitive function. J Sport Exerc Psychol 31(5):640–656. https://doi.org/10.1123/jsep.31.5.640
19. Ishihara T, Drollette ES, Ludyga S, Hillman CH, Kamijo K (2021) The effects of acute aerobic exercise on executive function: a systematic review and meta-analysis of individual participant data. Neurosci Biobehav Rev 128:258–269. https://doi.org/10.1016/j.neubiorev.2021.06.026
20. Etnier JL, Salazar W, Landers DM, Petruzzello SJ, Han M, Nowell P (1997) The influence of physical fitness and exercise upon cognitive functioning: a meta-analysis. J Sport Exerc Psychol 19:249–277
21. Sibley BA, Etnier JL (2003) The relationship between physical activity and cognition in children: a meta-analysis. Pediatr Exerc Sci 15(3):243–256
22. Ai JY, Chen FT, Hsieh SS, Kao SC, Chen AG, Hung TM, Chang YK (2021) The effect of acute high-intensity interval training on executive function: a systematic review. Int J Environ Res Public Health 18(7). https://doi.org/10.3390/ijerph18073593
23. Huang TY, Chen FT, Li RH, Hillman CH, Cline TL, Chu CH et al (2022) Effects of acute resistance exercise on executive function: a systematic review of the moderating role of intensity and executive function domain. Sports Med Open 8(1):141. https://doi.org/10.1186/s40798-022-00527-7
24. ACSM (2013) ACSM's guidelines for exercise testing and prescription. Lippincott Williams & Wilkins
25. Kamijo K, Nishihira Y, Hatta A, Kaneda T, Wasaka T, Kida T, Kuroiwa K (2004) Differential influences of exercise intensity on information processing in the central nervous system. Eur J Appl Physiol 92(3):305–311. https://doi.org/10.1007/s00421-004-1097-2
26. Chang YK, Chu IH, Chen FT, Wang CC (2011) Dose-response effect of acute resistance exercise on Tower of London in middle-aged adults. J Sport Exerc Psychol 33(6):866–883. https://doi.org/10.1123/jsep.33.6.866
27. Chang YK, Chen FT, Kuan G, Wei GX, Chu CH, Yan J et al (2019) Effects of acute exercise duration on the inhibition aspect of executive function in late middle-aged adults. Front Aging Neurosci 11:227. https://doi.org/10.3389/fnagi.2019.00227
28. Geng X, Ai JY, Wang CC, Chen FT, Chang YK (2023) Effects of volume-matched acute exercise on metacognition in late middle-aged adults: the roles of exercise intensity and duration. Prog Brain Res 283:1–19

29. Chang YK, Karageorghis CI, Wang CC, Li RH, Chen FT, Fang RY, Hung TM (2022) Effects of exercise intensity and duration at a predetermined exercise volume on executive function among Apolipoprotein E (APOE)-ε4 carriers. Curr Psychol 42(25):1–12. https://doi.org/10.1007/s12144-022-03257-x
30. Wang CC, Chu CH, Chu IH, Chan KH, Chang YK (2013) Executive function during acute exercise: the role of exercise intensity. J Sport Exerc Psychol 35(4):358–367. https://doi.org/10.1123/jsep.35.4.358
31. Hung T-M, Tsai C-L, Chen F-T, Wang C-C, Chang Y-K (2013) The immediate and sustained effects of acute exercise on planning aspect of executive function. Psychol Sport Exerc 14(5):728–736
32. Chu CH, Chen AG, Hung TM, Wang CC, Chang YK (2015) Exercise and fitness modulate cognitive function in older adults. Psychol Aging 30(4):842–848. https://doi.org/10.1037/pag0000047
33. Li L, Zhang S, Cui J, Chen LZ, Wang X, Fan M, Wei GX (2019) Fitness-dependent effect of acute aerobic exercise on executive function. Front Physiol 10:902. https://doi.org/10.3389/fphys.2019.00902
34. Devenney KE, Guinan EM, Kelly ÁM, Mota BC, Walsh C, Olde Rikkert M et al (2019) Acute high-intensity aerobic exercise affects brain-derived neurotrophic factor in mild cognitive impairment: a randomised controlled study. BMJ Open Sport Exerc Med 5(1):e000499. https://doi.org/10.1136/bmjsem-2018-000499
35. Chen FT, Feng SH, Nien JT, Cheng YT, Chen YC, Chang YK (2023) Effects of acute moderate-intensity exercise on executive function in children with preterm birth: a randomized crossover study. Early Hum Dev 183:105795. https://doi.org/10.1016/j.earlhumdev.2023.105795
36. Ren F-F, Feng S-H, Li R-H, Chu C-H, Chang Y-K, Chen F-T (2023) The effects of acute aerobic and resistance exercise on the planning aspect of executive functions in children with preterm birth. Int J Sport Exerc Psychol:1–14
37. Vincent CM, Hall PA (2017) Cognitive effects of a 30-min aerobic exercise bout on adults with overweight/obesity and type 2 diabetes. Obes Sci Pract 3(3):289–297. https://doi.org/10.1002/osp4.112
38. Ferris LT, Williams JS, Shen CL (2007) The effect of acute exercise on serum brain-derived neurotrophic factor levels and cognitive function. Med Sci Sports Exerc 39(4):728–734. https://doi.org/10.1249/mss.0b013e31802f04c7
39. Winter B, Breitenstein C, Mooren FC, Voelker K, Fobker M, Lechtermann A et al (2007) High impact running improves learning. Neurobiol Learn Mem 87(4):597–609. https://doi.org/10.1016/j.nlm.2006.11.003
40. Chmura J, Nazar K, Kaciuba-Uściłko H (1994) Choice reaction time during graded exercise in relation to blood lactate and plasma catecholamine thresholds. Int J Sports Med 15(4):172–176. https://doi.org/10.1055/s-2007-1021042
41. Voss MW, Weng TB, Narayana-Kumanan K, Cole RC, Wharff C, Reist L et al (2020) Acute exercise effects predict training change in cognition and connectivity. Med Sci Sports Exerc 52(1):131–140. https://doi.org/10.1249/mss.0000000000002115
42. Weng TB, Pierce GL, Darling WG, Falk D, Magnotta VA, Voss MW (2017) The acute effects of aerobic exercise on the functional connectivity of human brain networks. Brain Plast 2(2):171–190. https://doi.org/10.3233/bpl-160039
43. Gorski JA, Zeiler SR, Tamowski S, Jones KR (2003) Brain-derived neurotrophic factor is required for the maintenance of cortical dendrites. J Neurosci 23(17):6856–6865. https://doi.org/10.1523/jneurosci.23-17-06856.2003
44. Lu B, Pang PT, Woo NH (2005) The yin and yang of neurotrophin action. Nat Rev Neurosci 6(8):603–614. https://doi.org/10.1038/nrn1726
45. Erickson KI, Miller DL, Roecklein KA (2012) The aging hippocampus: interactions between exercise, depression, and BDNF. Neuroscientist 18(1):82–97

46. Erickson KI, Prakash RS, Voss MW, Chaddock L, Hu L, Morris KS et al (2009) Aerobic fitness is associated with hippocampal volume in elderly humans. Hippocampus 19(10):1030–1039. https://doi.org/10.1002/hipo.20547
47. Szuhany KL, Bugatti M, Otto MW (2015) A meta-analytic review of the effects of exercise on brain-derived neurotrophic factor. J Psychiatr Res 60:56–64. https://doi.org/10.1016/j.jpsychires.2014.10.003
48. Chang YK, Alderman BL, Chu CH, Wang CC, Song TF, Chen FT (2017) Acute exercise has a general facilitative effect on cognitive function: a combined ERP temporal dynamics and BDNF study. Psychophysiology 54(2):289–300. https://doi.org/10.1111/psyp.12784
49. Chu CH, Kramer AF, Song TF, Wu CH, Hung TM, Chang YK (2017) Acute exercise and neurocognitive development in preadolescents and young adults: an ERP study. Neural Plast 2017:2631909. https://doi.org/10.1155/2017/2631909
50. Laursen PB, Jenkins DG (2002) The scientific basis for high-intensity interval training: optimising training programmes and maximising performance in highly trained endurance athletes. Sports Med (Auckland, NZ) 32(1):53–73. https://doi.org/10.2165/00007256-200232010-00003
51. Tsai CL, Chang YC, Pan CY, Wang TC, Ukropec J, Ukropcová B (2021) Acute effects of different exercise intensities on executive function and oculomotor performance in middle-aged and older adults: moderate-intensity continuous exercise vs high-intensity interval exercise. Front Aging Neurosci 13:743479. https://doi.org/10.3389/fnagi.2021.743479
52. Sardeli AV, Ferreira MLV, Santos LC, Rodrigues MS, Damasceno A, Cavaglieri CR, Chacon-Mikahil MPT (2018) Low-load resistance exercise improves cognitive function in older adults. Rev Bras Med Esporte 24:125–129
53. Sugimoto T, Suga T, Tomoo K, Dora K, Mok E, Tsukamoto H et al (2021) Blood flow restriction improves executive function after walking. Med Sci Sports Exerc 53(1):131–138
54. Yamada Y, Frith EM, Wong V, Spitz RW, Bell ZW, Chatakondi RN et al (2021) Acute exercise and cognition: a review with testable questions for future research into cognitive enhancement with blood flow restriction. Med Hypotheses 151:110586. https://doi.org/10.1016/j.mehy.2021.110586
55. Sun B et al (2019) Physical activity and sedentary time in relation to semen quality in healthy men screened as potential sperm donors. Hum Reprod 34(12):2330–2339.
56. Sewell KR et al (2023) The interaction between physical activity and sleep on cognitive function and brain beta-amyloid in older adults. Behav Brain Res 437:114108
57. Won J et al (2019) Caudate Volume Mediates the Interaction between Total Sleep Time and Executive Function after Acute Exercise in Healthy Older Adults. Brain Plast 5(1):69–82
58. Ji Z, Feng T, Mei L, Li A, Zhang C (2019) Influence of acute combined physical and cognitive exercise on cognitive function: an NIRS study. PeerJ 7:e7418. https://doi.org/10.7717/peerj.7418
59. Suwabe K, Hyodo K, Fukuie T, Ochi G, Inagaki K, Sakairi Y, Soya H (2021) Positive mood while exercising influences beneficial effects of exercise with music on prefrontal executive function: a functional NIRS study. Neuroscience 454:61–71. https://doi.org/10.1016/j.neuroscience.2020.06.007
60. Zou L, Herold F, Ludyga S, Kamijo K, Müller NG, Pontifex MB et al (2023) Look into my eyes: what can eye-based measures tell us about the relationship between physical activity and cognitive performance? J Sport Health Sci. https://doi.org/10.1016/j.jshs.2023.04.003
61. Ayala N, Heath M (2021) Pupillometry reveals the role of arousal in a postexercise benefit to executive function. Brain Sci 11(8). https://doi.org/10.3390/brainsci11081048
62. Shigeta TT, Morris TP, Henry DH, Kucyi A, Bex P, Kramer AF, Hillman CH (2021) Acute exercise effects on inhibitory control and the pupillary response in young adults. Int J Psychophysiol 170:218–228. https://doi.org/10.1016/j.ijpsycho.2021.08.006

Involvement of Dopamine in Cognitive Improvement by Aerobic Exercise

Soichi Ando, Toshihiko Fujimoto, Mizuki Sudo, and Manabu Tashiro

Abstract It has been suggested that acute physical exercise at low to moderate intensity improves cognitive function, as shown by improvements in cognitive performance. Decades of research have explored or discussed physiological mechanisms underlying cognitive improvements induced by acute exercise. However, the precise physiological mechanisms responsible for improvements in cognitive function remain to be elucidated. There is a large body of evidence to suggest that cognitive function is linked with dopamine (DA) in the brain. Rodent studies have shown that acute exercise releases neurotransmitters in the brain. Recent studies using positron emission tomography (PET) have also suggested that acute exercise released endogenous DA in humans. Furthermore, it appears that endogenous DA release is linked with improvements in cognitive function induced by acute exercise. Therefore, in this chapter, we focus on DA and discuss it as a promising candidate to account for exercise-cognition interaction, particularly improvement in cognitive function induced by acute exercise. We propose that further studies using PET would be helpful to progress our understanding of improvements in cognitive function induced by acute exercise.

S. Ando (✉)
Graduate School of Informatics and Engineering, The University of Electro-Communications, Chofu, Tokyo, Japan
e-mail: soichi.ando@uec.ac.jp

T. Fujimoto
Institute of Excellence in Higher Education, Tohoku University, Sendai, Miyagi, Japan
e-mail: toshihiko.fujimoto.d3@tohoku.ac.jp

M. Sudo
Shinjyuku Satellite Office, Physical Fitness Research Institute, Meiji Yasuda Life Foundation of Health and Welfare, Shinjuku, Tokyo, Japan

M. Tashiro
Research Center for Accelerator and Radioisotope of Science, Sendai, Miyagi, Japan
e-mail: manabu.tashiro.a2@tohoku.ac.jp

H. Soya (ed.), *Exercise Brain Stimulation for Cognitive Function and Mental Health*, Advances in Neurobiology 44, https://doi.org/10.1007/978-981-95-0066-6_10

Keywords Acute exercise · Cognition · Dopamine · Positron emission tomography

1 Introduction

Cognitive function refers processes such as memory, attention, language, problem solving, and planning [1]. Cognitive function is classified into central executive and non-executive domains. Central executive function is a higher-order cognitive ability that controls purposeful and goal-directed behavior [2], which is a sophisticated function specific to humans. The concept of executive function has been established by many researchers. For example, Miyake and colleagues suggested three separable but partially correlated components: shifting, updating, and inhibition [3]. Diamond regarded inhibition (response inhibition and interference control), working memory, and cognitive flexibility as core executive functions [4]. Conversely, non-executive function is basic and includes processing speed, memory, and sustained attention [5]. Both executive and non-executive functions are thought to play an important role in our daily life and physical activity (e.g., sporting activities). Hence, considerable attention has been devoted to examining the association between exercise and cognitive function.

It has been suggested that regular habitual exercise has beneficial effects on a broad array of cognitive functions [6–8]. A large body of evidence also indicates that acute physical exercise at low to moderate intensity improves cognitive function [9–12], as shown by improvements in cognitive performance. Hence, despite some arguments [13], there is a consensus that acute exercise has the potential to improve cognitive function. A recent review reported that improvements in cognitive function following moderate-intensity exercise are frequently accompanied by the changes in brain activation assessed by electroencephalogram [14]. Thus, it is reasonable to consider that acute exercise alters brain activity and that altered brain activity is associated with cognitive function.

Decades of research have explored or discussed the physiological mechanisms underlying cognitive improvements induced by acute exercise. Several potential candidates have been proposed and discussed by many researchers. These include alterations in neuromodulation by neurotransmitters and neurotrophic factors, functional connectivity, cerebral blood flow, cerebral oxygenation, cerebral metabolism, and various psychological factors [9, 12, 15–27]. Nevertheless, the precise physiological mechanisms responsible for improvements in cognitive function induced by acute exercise remain to be elucidated (Fig. 1). Therefore, a central question is still unclear: how does a single bout of exercise improve cognitive function?

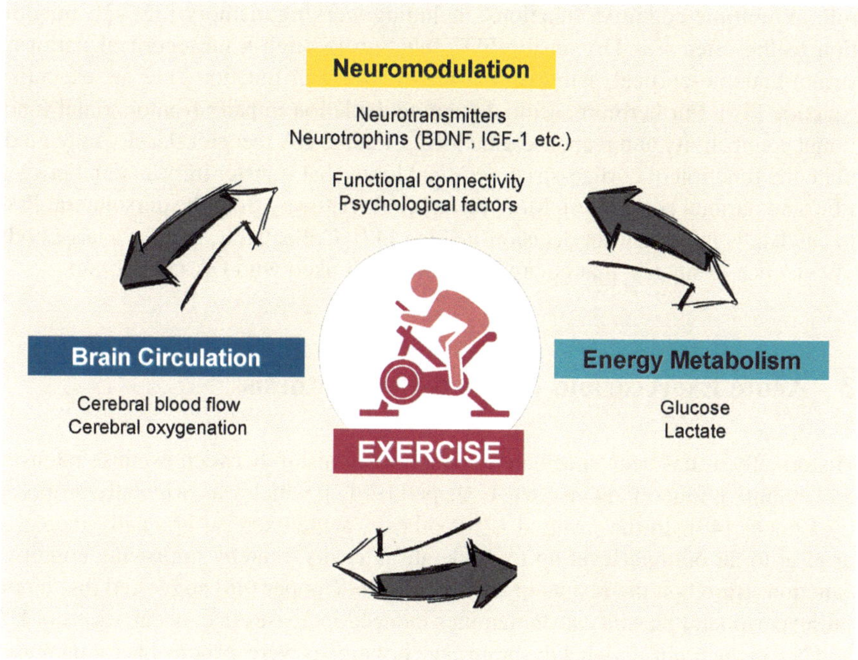

Fig. 1 Potential physiological mechanisms underlying improvements in cognitive function induced by acute exercise

2 Dopamine (DA) and Cognitive function

Catecholamines act both as neurotransmitters and hormones vital to the maintenance of homeostasis through the autonomic nervous system [28]. The main types of catecholamines are dopamine (DA), noradrenaline (NA), and adrenaline. DA, which is also known as 3,4-dihydroxytyramine, is a neurotransmitter produced by dopaminergic neurons in the brain. DA is synthesized by the tyrosine hydroxylase enzyme, which, by adding a hydroxyl group, transforms tyrosine into L-DOPA, which in turn is decarboxylated into DA [29]. DA receptors are widely distributed within the brain. DA binds to five receptor subtypes, pharmacologically grouped as D1- (D1 and D5) or D2-like (D2Short, D2Long, D3, and D4) [30]. In humans, DA densely innervates the striatum and, to a lesser degree, limbic and cortical areas via separate pathways. Dopaminergic projections from the substantia nigra (SN) reach the associative (largely the caudate) and sensorimotor (largely the putamen) part of the striatum via the nigrostriatal pathway. The ventral tegmental area (VTA) provides dopaminergic input to the ventral striatum (largely the nucleus accumbens) and the limbic system via the mesolimbic pathway and to the lateral prefrontal cortex (PFC) and other neocortical areas via the mesocortical pathway [31].

Many researchers have been suggested that DA is essential for cognitive control of the PFC [32–34]. Indeed, there is abundant evidence that the DA plays a central role in multiple cognitive functions, including working memory [35–43]. In addition to the effects of DA on the PFC function through a mesocortical pathway, striato-thalamo-cortical pathway modulates the PFC function such as executive function [44]. Furthermore, acute dopamine depletion impairs fronto-striatal functional connectivity and response time [45]. Caudate DA receptor density may modulate the function of cortico-striato-cortical loops that transfer information between PFC and parietal cortex [46]. More recently, the pathway from the dorsolateral PFC to caudate is important for decision-making [47]. Collectively, there is a large body of evidence to suggest that cognitive function is linked with DA in the brain.

3 Acute Exercise and Inverted-U Hypothesis

Historically, it has been speculated that the relationship between exercise intensity and cognitive function is inverted-U shaped [9–12], which was originally proposed by Cooper [48]. In the inverted-U hypothesis, acute exercise gradually increases arousal to an optimal level up to moderate intensity, thereby improving cognitive function. Based on the results of animal studies, Cooper [48] suggested that circulating peripheral plasma catecholamines induced increases in concentrations of DA and NA in the brain, and that the brain catecholamines were responsible for increases in arousal by activating the reticular formation. McMorris extended this view in his review articles [12, 49]. He recently suggested an interoception model. In this model, during moderate intensity exercise, the afferent interoceptive system and the proprioceptive somatosensory system, through interactions between the anterior insula, anterior cingulate, orbitofrontal and lateral prefrontal cortices, increase tonic release of catecholamines [12]. As a result, phasic release of DA and NA improves cognitive function.

Interestingly, it has been thought that there is an inverted-U relationship between the DA system and cognitive function with insufficient or excessive DA level impairing cognitive function [41, 50]. Primate studies also suggest that either too much or too little DA is reported to weaken the signal-to-noise ratio [51]. Furthermore, neurocomputational models predict that DA regulates the signal-to-noise ratio of neural information processing [52], and that DA facilitates the responsivity of neural networks [53]. Therefore, although the inverted-U hypothesis is not always the case [12], it is tempting to suggest that DA also contributes to improvements in cognitive function in response to acute exercise. In the following sections, we discuss this notion elaborately, together with available literature. We focused on DA and discuss as a promising candidate to account for exercise-cognition interaction, particularly improvements in cognitive function induced by acute exercise (see below).

4 Acute Exercise and DA System

Rodent studies have shown that acute exercise releases neurotransmitters in the brain [54–57]. In humans, despite lack of empirical evidence, acute exercise is thought to influence brain circuits involving a number of neurotransmitters including DA [9, 12]. Nevertheless, it is less clear how acute exercise alters central neurotransmitter release in the human brain. This is probably due to technical and methodological challenges.

Positron emission tomography (PET) is a noninvasive imaging technique that produces high-resolution images of the physiological function of human tissues, including measurements of blood perfusion and metabolic processes, as well as quantification of neurotransmitter/hormonal receptor densities [58]. [^{11}C]raclopride is a selective antagonist of DA D2 receptor and it is a widely used tracer to assess striatal D2 dopamine receptor binding characteristics (Fig. 2). Since [^{11}C]raclopride competes with endogenous DA release in the brain (Fig. 3), reduction in binding potential means endogenous DA release [59]. With this established technique, it is feasible to evaluate DA release in response to acute exercise.

Sacheli and colleagues measured endogenous DA release in response to acute stationary cycling at 60% VO_2 reserve for 30 min [60]. Participants were habitual exercisers and sedentary subjects with Parkinson's disease, The participants

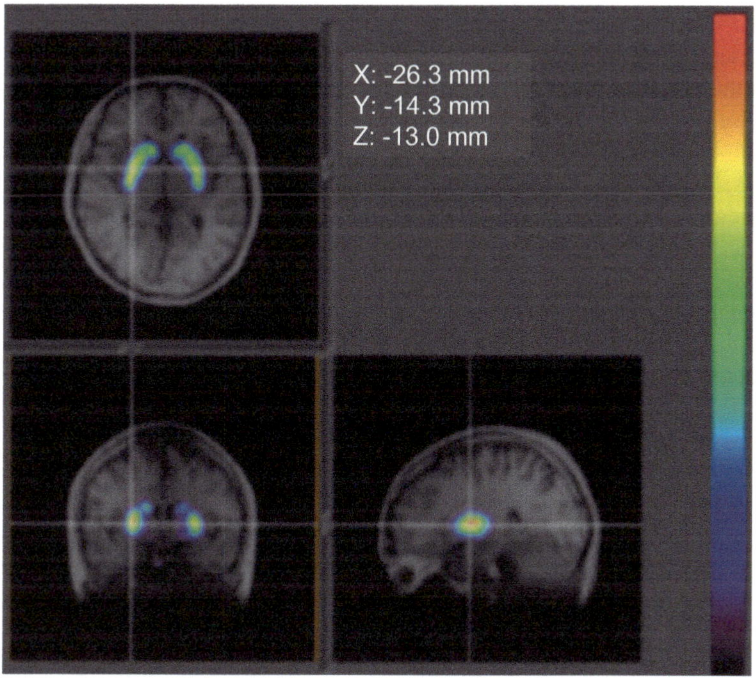

Fig. 2 An example of positron emission tomography image showing dopamine (DA) D2 receptor binding potential (BP_{ND}). Color bar represents intensity of BP_{ND}

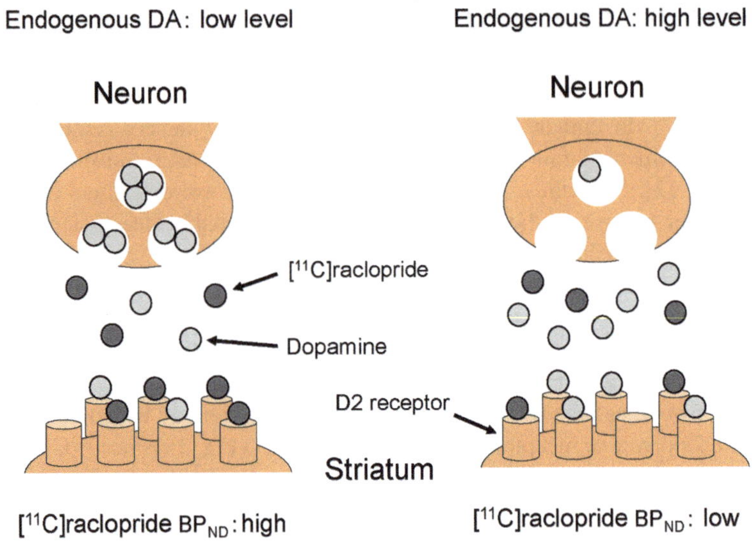

Fig. 3 Illustration of competition between dopamine (DA) D2 receptor antagonist [^{11}C]raclopride and endogenous DA. BP_{ND} binding potential

completed [^{11}C]raclopride PET scans before and after cycling to determine exercise-induced release of endogenous DA in the dorsal striatum. Intriguingly, this study indicated that acute exercise induces greater endogenous DA release in habitual exercisers with Parkinson's disease compared with sedentary counterparts.

In the previous study by Sacheli et al., PET measurement was performed after exercise. Notably, a few studies have demonstrated the feasibility of using PET during low- to moderate-intensity exercise [61, 62]. Therefore, using PET scan with [^{11}C]raclopride, our research group examined whether acute exercise increased endogenous DA release in the brain in healthy individuals (Fig. 4) [63]. In our study, PET scans were performed in the exercise and control (rest) conditions. In the exercise condition, the participants cycled in a supine position at ~35% VO_{2peak} for 40 min. [^{11}C]raclopride was injected intravenously 10 min after the start of exercise, and emission scan for the brain was performed 90 min after intravenous injection. We observed reductions in binding potential in the caudate and putamen, which indicates that acute exercise released endogenous DA [63].

Nevertheless, these results are not consistent with the previous findings by Wang and colleagues [64]. The authors have shown that acute exercise did not change endogenous DA release in healthy individuals. However, in the study by Wang and colleague, the PET scan was started after the cool down period (5 min) following exercise (~10–15 min after exercise). Rodent studies have previously indicated that DA release in response to acute exercise is temporary in the brain [54–56]. Furthermore, as Wang and colleagues also discussed, the absence of increase in endogenous DA release may be, at least partly, affected by the timing of PET scan

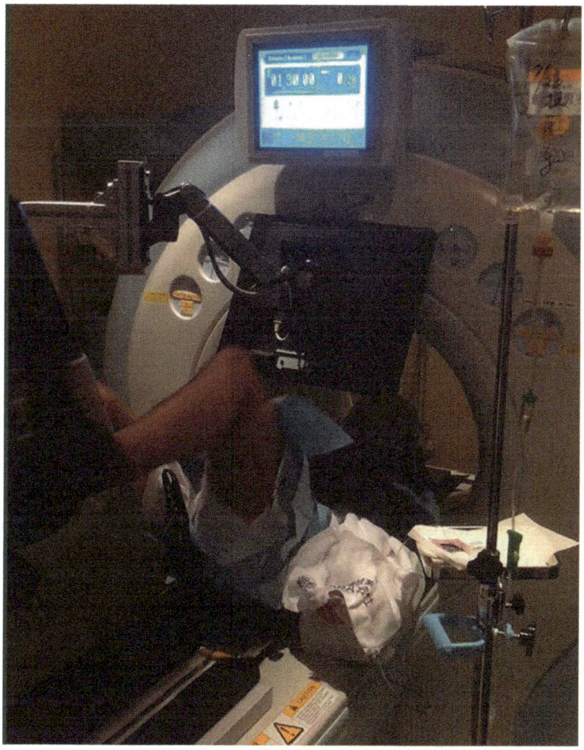

Fig. 4 Picture of our PET experiment

[64]. Therefore, it is necessary to accumulate empirical evidence to conclude that acute exercise releases endogenous DA.

5 Association Between Cognitive Function and Endogenous DA Release Induced by Acute Exercise

As has been noted, DA is one of the promising candidates to account for exercise-induced cognitive improvement. However, it has not been investigated whether DA is involved in cognitive improvement induced by acute exercise. Therefore, in our study [63], we tested if cognitive improvement is linked with endogenous DA release induced by acute exercise and observed that endogenous DA release induced by acute exercise is correlated with cognitive function. As far as we know, this is the first study showing that improvements in cognitive function are associated with DA release in response to acute exercise in humans.

Berse and colleagues [65] investigated whether single nucleotide polymorphisms of DA related genes would predict the effectiveness of exercise-cognition interaction in 297 adolescents [65]. The authors demonstrated that improved cognitive function following acute exercise was associated with single nucleotide polymorphisms of DA-related genes. Hence, the results of our PET study are compatible with the previous findings by Berse and colleagues showing that DA may be linked with cognitive function in response to acute exercise.

The main topic of this chapter is association between endogenous DA and cognitive improvement induced by acute exercise. However, it should be noted that previous studies have been suggested the linkage between physical activity and DA. For example, age-related DA loss is attenuated by higher physical activity [66]. DA receptor availability is associated with physical activity [67] or aerobic fitness level [68]. Furthermore, although a previous study only partially supported the DA-mediated improvements in cognitive function after exercise intervention [68], exercise intervention induces neuroplasticity in dopaminergic signaling in early stage Parkinson's disease patients [69] and methamphetamine users [70]. Collectively, further studies are required to understand the role of endogenous DA in cognitive function in response to acute/chronic exercise.

In this chapter, we focused on only endogenous DA release in response to acute exercise. However, contributions of NA are not ignored. NA is released from the locus coeruleus (LC), and NA are associated with behavioral states such as arousal, mood, and attention [71, 72]. Indeed, the DA and NA appear to act in a parallel and overlapping manner [73]. Interestingly, however, a recent imaging study indicated attenuated VTA activation and impairment in cognitive function (interference time in the Stroop task) in schizophrenia patients as compared with healthy controls [74]. Conversely, activations in the LC were comparable between the patients and controls. Thus, the results of this study may emphasize the crucial role of the DA in exercise-cognition interaction.

6 Summary and Future Perspective

In this chapter, we started by asking one simple question. How does a single bout of exercise improve cognitive function? To answer this question, we focused on DA and discussed how DA potentially mediates cognitive function. Specifically, we proposed a hypothesis that improvements in cognitive function is ascribed to increases in endogenous DA release in the brain. However, as it stands, empirical evidence is still lacking and this hypothesis warrants further investigation. As noted above, PET is the powerful imaging tools to investigate this research topic under physiological conditions in humans. Additional studies using PET is critically important to progress our understanding of improvements in cognitive function induced by a single bout of exercise. In particular, cognitive control processes that endogenous DA modulates are of great interest. Given that acute exercise improves performance in cognitively demanding as well as less demanding tasks [9–12], it is likely that DA affects multiple cognitive processes. Contributions of other neurotransmitters are also interesting topics for further studies.

References

1. Pessoa L (2008) On the relationship between emotion and cognition. Nat Rev Neurosci 9(2):148–158. https://doi.org/10.1038/nrn2317
2. Etnier JL, Chang YK (2009) The effect of physical activity on executive function: a brief commentary on definitions, measurement issues, and the current state of the literature. J Sport Exerc Psychol 31(4):469–483. https://doi.org/10.1123/jsep.31.4.469
3. Miyake A, Friedman NP, Emerson MJ, Witzki AH, Howerter A, Wager TD (2000) The unity and diversity of executive functions and their contributions to complex "frontal lobe" tasks: a latent variable analysis. Cogn Psychol 41(1):49–100. https://doi.org/10.1006/cogp.1999.0734
4. Diamond A (2013) Executive functions. Annu Rev Psychol 64:135–168. https://doi.org/10.1146/annurev-psych-113011-143750
5. Spann MN, Silberman A, Feldman J, Korzeniewski SJ, Turner JB, Whitaker AH (2020) Executive and non-executive functions in low birthweight/preterm adolescents with differing temporal patterns of inattention. PLoS One 15(4):e0231648. https://doi.org/10.1371/journal.pone.0231648
6. Erickson KI, Hillman C, Stillman CM, Ballard RM, Bloodgood B, Conroy DE et al (2019) Physical activity, cognition, and brain outcomes: a review of the 2018 physical activity guidelines. Med Sci Sports Exerc 51(6):1242–1251. https://doi.org/10.1249/MSS.0000000000001936
7. Kramer AF, Hahn S, Cohen NJ, Banich MT, McAuley E, Harrison CR et al (1999) Ageing, fitness and neurocognitive function. Nature 400(6743):418–419. https://doi.org/10.1038/22682
8. Ludyga S, Gerber M, Puhse U, Looser VN, Kamijo K (2020) Systematic review and meta-analysis investigating moderators of long-term effects of exercise on cognition in healthy individuals. Nat Hum Behav 4(6):603–612. https://doi.org/10.1038/s41562-020-0851-8
9. Ando S, Komiyama T, Sudo M, Higaki Y, Ishida K, Costello JT et al (2020) The interactive effects of acute exercise and hypoxia on cognitive performance: a narrative review. Scand J Med Sci Sports 30(3):384–398. https://doi.org/10.1111/sms.13573
10. Chang YK, Labban JD, Gapin JI, Etnier JL (2012) The effects of acute exercise on cognitive performance: a meta-analysis. Brain Res 1453:87–101. https://doi.org/10.1016/j.brainres.2012.02.068
11. Lambourne K, Tomporowski P (2010) The effect of exercise-induced arousal on cognitive task performance: a meta-regression analysis. Brain Res 1341:12–24. https://doi.org/10.1016/j.brainres.2010.03.091
12. McMorris T (2021) The acute exercise-cognition interaction: from the catecholamines hypothesis to an interoception model. Int J Psychophysiol 170:75–88. https://doi.org/10.1016/j.ijpsycho.2021.10.005
13. Ishihara T, Drollette ES, Ludyga S, Hillman CH, Kamijo K (2021) The effects of acute aerobic exercise on executive function: a systematic review and meta-analysis of individual participant data. Neurosci Biobehav Rev 128:258–269. https://doi.org/10.1016/j.neubiorev.2021.06.026
14. Kao SC, Cadenas-Sanchez C, Shigeta TT, Walk AM, Chang YK, Pontifex MB et al (2020) A systematic review of physical activity and cardiorespiratory fitness on P3b. Psychophysiology 57(7):e13425. https://doi.org/10.1111/psyp.13425
15. Ando S, Kokubu M, Yamada Y, Kimura M (2011) Does cerebral oxygenation affect cognitive function during exercise? Eur J Appl Physiol 111(9):1973–1982. https://doi.org/10.1007/s00421-011-1827-1
16. Basso JC, Suzuki WA (2017) The effects of acute exercise on mood, cognition, neurophysiology, and neurochemical pathways: a review. Brain Plast 2(2):127–152. https://doi.org/10.3233/BPL-160040
17. Byun K, Hyodo K, Suwabe K, Ochi G, Sakairi Y, Kato M et al (2014) Positive effect of acute mild exercise on executive function via arousal-related prefrontal activations: an fNIRS study. Neuroimage 98:336–345. https://doi.org/10.1016/j.neuroimage.2014.04.067

18. Cantelon JA, Giles GE (2021) A review of cognitive changes during acute aerobic exercise. Front Psychol 12:653158. https://doi.org/10.3389/fpsyg.2021.653158
19. Hashimoto T, Tsukamoto H, Takenaka S, Olesen ND, Petersen LG, Sorensen H et al (2018) Maintained exercise-enhanced brain executive function related to cerebral lactate metabolism in men. FASEB J 32(3):1417–1427. https://doi.org/10.1096/fj.201700381RR
20. Hwang J, Brothers RM, Castelli DM, Glowacki EM, Chen YT, Salinas MM et al (2016) Acute high-intensity exercise-induced cognitive enhancement and brain-derived neurotrophic factor in young, healthy adults. Neurosci Lett 630:247–253. https://doi.org/10.1016/j.neulet.2016.07.033
21. Kuwamizu R, Suwabe K, Damrongthai C, Fukuie T, Ochi G, Hyodo K et al (2021) Spontaneous eye blink rate connects missing link between aerobic fitness and cognition. Med Sci Sports Exerc 53(7):1425–1433. https://doi.org/10.1249/MSS.0000000000002590
22. Ogoh S, Tsukamoto H, Hirasawa A, Hasegawa H, Hirose N, Hashimoto T (2014) The effect of changes in cerebral blood flow on cognitive function during exercise. Physiol Rep 2(9). https://doi.org/10.14814/phy2.12163
23. Shirzad M, Tari B, Dalton C, Van Riesen J, Marsala MJ, Heath M (2022) Passive exercise increases cerebral blood flow velocity and supports a postexercise executive function benefit. Psychophysiology 59(12):e14132. https://doi.org/10.1111/psyp.14132
24. Sudo M, Costello JT, McMorris T, Ando S (2022) The effects of acute high-intensity aerobic exercise on cognitive performance: a structured narrative review. Front Behav Neurosci 16:957677. https://doi.org/10.3389/fnbeh.2022.957677
25. Suwabe K, Byun K, Hyodo K, Reagh ZM, Roberts JM, Matsushita A et al (2018) Rapid stimulation of human dentate gyrus function with acute mild exercise. Proc Natl Acad Sci USA 115(41):10487–10492. https://doi.org/10.1073/pnas.1805668115
26. Tari B, Vanhie JJ, Belfry GR, Shoemaker JK, Heath M (2020) Increased cerebral blood flow supports a single-bout postexercise benefit to executive function: evidence from hypercapnia. J Neurophysiol 124(3):930–940. https://doi.org/10.1152/jn.00240.2020
27. Winter B, Breitenstein C, Mooren FC, Voelker K, Fobker M, Lechtermann A et al (2007) High impact running improves learning. Neurobiol Learn Mem 87(4):597–609. https://doi.org/10.1016/j.nlm.2006.11.003
28. Paravati S, Rosani A, Warrington SJ (2023) Physiology, Catecholamines. StatPearls, Treasure Island
29. Speranza L, di Porzio U, Viggiano D, de Donato A, Volpicelli F (2021) Dopamine: the neuromodulator of long-term synaptic plasticity, reward and movement control. Cells 10(4). https://doi.org/10.3390/cells10040735
30. Martel JC, Gatti MAS (2020) Dopamine receptor subtypes, physiology and pharmacology: new ligands and concepts in schizophrenia. Front Pharmacol 11:1003. https://doi.org/10.3389/fphar.2020.01003
31. Rieckmann A, Karlsson S, Karlsson P, Brehmer Y, Fischer H, Farde L et al (2011) Dopamine D1 receptor associations within and between dopaminergic pathways in younger and elderly adults: links to cognitive performance. Cereb Cortex 21(9):2023–2032. https://doi.org/10.1093/cercor/bhq266
32. Floresco SB, Magyar O (2006) Mesocortical dopamine modulation of executive functions: beyond working memory. Psychopharmacology 188(4):567–585. https://doi.org/10.1007/s00213-006-0404-5
33. Ko JH, Strafella AP (2012) Dopaminergic neurotransmission in the human brain: new lessons from perturbation and imaging. Neuroscientist 18(2):149–168. https://doi.org/10.1177/1073858411401413
34. Ott T, Nieder A (2019) Dopamine and cognitive control in prefrontal cortex. Trends Cogn Sci 23(3):213–234. https://doi.org/10.1016/j.tics.2018.12.006
35. Backman L, Ginovart N, Dixon RA, Wahlin TB, Wahlin A, Halldin C et al (2000) Age-related cognitive deficits mediated by changes in the striatal dopamine system. Am J Psychiatry 157(4):635–637. https://doi.org/10.1176/ajp.157.4.635

36. Backman L, Nyberg L, Soveri A, Johansson J, Andersson M, Dahlin E et al (2011) Effects of working-memory training on striatal dopamine release. Science 333(6043):718. https://doi.org/10.1126/science.1204978
37. Karlsson S, Rieckmann A, Karlsson P, Farde L, Nyberg L, Backman L (2011) Relationship of dopamine D1 receptor binding in striatal and extrastriatal regions to cognitive functioning in healthy humans. Neuroimage 57(2):346–351. https://doi.org/10.1016/j.neuroimage.2011.04.047
38. McNab F, Varrone A, Farde L, Jucaite A, Bystritsky P, Forssberg H et al (2009) Changes in cortical dopamine D1 receptor binding associated with cognitive training. Science 323(5915):800–802. https://doi.org/10.1126/science.1166102
39. Nyberg L, Karalija N, Salami A, Andersson M, Wahlin A, Kaboovand N et al (2016) Dopamine D2 receptor availability is linked to hippocampal-caudate functional connectivity and episodic memory. Proc Natl Acad Sci USA 113(28):7918–7923. https://doi.org/10.1073/pnas.1606309113
40. Takahashi H, Yamada M, Suhara T (2012) Functional significance of central D1 receptors in cognition: beyond working memory. J Cereb Blood Flow Metab 32(7):1248–1258. https://doi.org/10.1038/jcbfm.2011.194
41. Takahashi H, Kato M, Takano H, Arakawa R, Okumura M, Otsuka T et al (2008) Differential contributions of prefrontal and hippocampal dopamine D(1) and D(2) receptors in human cognitive functions. J Neurosci 28(46):12032–12038. https://doi.org/10.1523/JNEUROSCI.3446-08.2008
42. Vijayraghavan S, Wang M, Birnbaum SG, Williams GV, Arnsten AF (2007) Inverted-U dopamine D1 receptor actions on prefrontal neurons engaged in working memory. Nat Neurosci 10(3):376–384. https://doi.org/10.1038/nn1846
43. Wang M, Vijayraghavan S, Goldman-Rakic PS (2004) Selective D2 receptor actions on the functional circuitry of working memory. Science 303(5659):853–856. https://doi.org/10.1126/science.1091162
44. Klostermann EC, Braskie MN, Landau SM, O'Neil JP, Jagust WJ (2012) Dopamine and frontostriatal networks in cognitive aging. Neurobiol Aging 33(3):623 e15–623 e24. https://doi.org/10.1016/j.neurobiolaging.2011.03.002
45. Nagano-Saito A, Leyton M, Monchi O, Goldberg YK, He Y, Dagher A (2008) Dopamine depletion impairs frontostriatal functional connectivity during a set-shifting task. J Neurosci 28(14):3697–3706. https://doi.org/10.1523/JNEUROSCI.3921-07.2008
46. Rieckmann A, Karlsson S, Fischer H, Backman L (2011) Caudate dopamine D1 receptor density is associated with individual differences in frontoparietal connectivity during working memory. J Neurosci 31(40):14284–14290. https://doi.org/10.1523/JNEUROSCI.3114-11.2011
47. Oyama K, Hori Y, Nagai Y, Miyakawa N, Mimura K, Hirabayashi T et al (2021) Chemogenetic dissection of the primate prefronto-subcortical pathways for working memory and decision-making. Sci Adv 7(26). https://doi.org/10.1126/sciadv.abg4246
48. Cooper CJ (1973) Anatomical and physiological mechanisms of arousal, with special reference to the effects of exercise. Ergonomics 16(5):601–609. https://doi.org/10.1080/00140137308924551
49. McMorris T (2016) Developing the catecholamines hypothesis for the acute exercise-cognition interaction in humans: Lessons from animal studies. Physiol Behav 165:291–299. https://doi.org/10.1016/j.physbeh.2016.08.011
50. Cools R, D'Esposito M (2011) Inverted-U-shaped dopamine actions on human working memory and cognitive control. Biol Psychiatry 69(12):e113–e125. https://doi.org/10.1016/j.biopsych.2011.03.028
51. Arnsten AF (2011) Catecholamine influences on dorsolateral prefrontal cortical networks. Biol Psychiatry 69(12):e89–e99. https://doi.org/10.1016/j.biopsych.2011.01.027
52. Li SC, Lindenberger U, Sikstrom S (2001) Aging cognition: from neuromodulation to representation. Trends Cogn Sci 5(11):479–486. https://doi.org/10.1016/s1364-6613(00)01769-1

53. MacDonald SW, Cervenka S, Farde L, Nyberg L, Backman L (2009) Extrastriatal dopamine D2 receptor binding modulates intraindividual variability in episodic recognition and executive functioning. Neuropsychologia 47(11):2299–2304. https://doi.org/10.1016/j.neuropsychologia.2009.01.016
54. Chen C, Nakagawa S, Kitaichi Y, An Y, Omiya Y, Song N et al (2016) The role of medial prefrontal corticosterone and dopamine in the antidepressant-like effect of exercise. Psychoneuroendocrinology 69:1–9. https://doi.org/10.1016/j.psyneuen.2016.03.008
55. Goekint M, Bos I, Heyman E, Meeusen R, Michotte Y, Sarre S. Acute running stimulates hippocampal dopaminergic neurotransmission in rats, but has no influence on brain-derived neurotrophic factor. J Appl Physiol (1985). 2012;112(4):535-541. doi: https://doi.org/10.1152/japplphysiol.00306.2011.
56. Hasegawa H, Takatsu S, Ishiwata T, Tanaka H, Sarre S, Meeusen R (2011) Continuous monitoring of hypothalamic neurotransmitters and thermoregulatory responses in exercising rats. J Neurosci Methods 202(2):119–123. https://doi.org/10.1016/j.jneumeth.2011.05.024
57. Meeusen R, Piacentini MF, De Meirleir K (2001) Brain microdialysis in exercise research. Sports Med 31(14):965–983. https://doi.org/10.2165/00007256-200131140-00002
58. Rudroff T, Kindred JH, Kalliokoski KK (2015) [18F]-FDG positron emission tomography--an established clinical tool opening a new window into exercise physiology. J Appl Physiol (1985) 118(10):1181–1190. https://doi.org/10.1152/japplphysiol.01070.2014
59. Laruelle M (2000) Imaging synaptic neurotransmission with in vivo binding competition techniques: a critical review. J Cereb Blood Flow Metab 20(3):423–451. https://doi.org/10.1097/00004647-200003000-00001
60. Sacheli MA, Murray DK, Vafai N, Cherkasova MV, Dinelle K, Shahinfard E et al (2018) Habitual exercisers versus sedentary subjects with Parkinson's disease: multimodal PET and fMRI study. Mov Disord 33(12):1945–1950. https://doi.org/10.1002/mds.27498
61. Hiura M, Nariai T, Ishii K, Sakata M, Oda K, Toyohara J et al (2014) Changes in cerebral blood flow during steady-state cycling exercise: a study using oxygen-15-labeled water with PET. J Cereb Blood Flow Metab 34(3):389–396. https://doi.org/10.1038/jcbfm.2013.220
62. Hiura M, Nariai T, Sakata M, Muta A, Ishibashi K, Wagatsuma K et al (2018) Response of cerebral blood flow and blood pressure to dynamic exercise: a study using PET. Int J Sports Med 39(3):181–188. https://doi.org/10.1055/s-0043-123647
63. Ando S, Fujimoto T, Sudo M, Watanuki S, Hiraoka K, Takeda K et al (2024) The neuromodulatory role of dopamine in improved reaction time by acute cardiovascular exercise. J Physiol 602(3):461–484. https://doi.org/10.1113/JP285173
64. Wang GJ, Volkow ND, Fowler JS, Franceschi D, Logan J, Pappas NR et al (2000) PET studies of the effects of aerobic exercise on human striatal dopamine release. J Nucl Med 41(8):1352–1356
65. Berse T, Rolfes K, Barenberg J, Dutke S, Kuhlenbaumer G, Volker K et al (2015) Acute physical exercise improves shifting in adolescents at school: evidence for a dopaminergic contribution. Front Behav Neurosci 9:196. https://doi.org/10.3389/fnbeh.2015.00196
66. Dang LC, Castrellon JJ, Perkins SF, Le NT, Cowan RL, Zald DH et al (2017) Reduced effects of age on dopamine D2 receptor levels in physically active adults. Neuroimage 148:123–129. https://doi.org/10.1016/j.neuroimage.2017.01.018
67. Kohncke Y, Papenberg G, Jonasson L, Karalija N, Wahlin A, Salami A et al (2018) Self-rated intensity of habitual physical activities is positively associated with dopamine D2/3 receptor availability and cognition. Neuroimage 181:605–616. https://doi.org/10.1016/j.neuroimage.2018.07.036
68. Jonasson LS, Nyberg L, Axelsson J, Kramer AF, Riklund K, Boraxbekk CJ (2019) Higher striatal D2-receptor availability in aerobically fit older adults but non-selective intervention effects after aerobic versus resistance training. Neuroimage 202:116044. https://doi.org/10.1016/j.neuroimage.2019.116044

69. Fisher BE, Li Q, Nacca A, Salem GJ, Song J, Yip J et al (2013) Treadmill exercise elevates striatal dopamine D2 receptor binding potential in patients with early Parkinson's disease. Neuroreport 24(10):509–514. https://doi.org/10.1097/WNR.0b013e328361dc13
70. Robertson CL, Ishibashi K, Chudzynski J, Mooney LJ, Rawson RA, Dolezal BA et al (2016) Effect of exercise training on striatal dopamine D2/D3 receptors in methamphetamine users during behavioral treatment. Neuropsychopharmacology 41(6):1629–1636. https://doi.org/10.1038/npp.2015.331
71. Arnsten AF, Pliszka SR (2011) Catecholamine influences on prefrontal cortical function: relevance to treatment of attention deficit/hyperactivity disorder and related disorders. Pharmacol Biochem Behav 99(2):211–216. https://doi.org/10.1016/j.pbb.2011.01.020
72. Schwarz LA, Luo L (2015) Organization of the locus coeruleus-norepinephrine system. Curr Biol 25(21):R1051–R10R6. https://doi.org/10.1016/j.cub.2015.09.039
73. Ranjbar-Slamloo Y, Fazlali Z (2019) Dopamine and noradrenaline in the brain; overlapping or dissociate functions? Front Mol Neurosci 12:334. https://doi.org/10.3389/fnmol.2019.00334
74. Kohler S, Wagner G, Bar KJ (2019) Activation of brainstem and midbrain nuclei during cognitive control in medicated patients with schizophrenia. Hum Brain Mapp 40(1):202–213. https://doi.org/10.1002/hbm.24365

Part II
Brain-Body Communication

The Dual Role of Lactate in the Exercising Brain: Fueling Energy and Signaling Fatigue

Takashi Matsui and Hideaki Soya

Abstract Historically regarded as a "fatigue substance" that impedes muscle contraction, lactate has undergone a paradigm shift in its scientific perception. Currently, it is acknowledged as an indispensable substrate for aerobic ATP synthesis, functioning as a pivotal intermediate in the glycolysis pathway. Within the brain context, lactate derived from glycogen in astrocytes is instrumental in augmenting exercise endurance and enhancing memory formation via the astrocyte-neuron lactate shuttle. This glia-neuron interaction mechanism underscores the expansive influence of lactate beyond mere muscular physiology. Contemporary investigations have pivoted toward elucidating a novel hypothesis: the role of lactate as a ligand for the G protein-coupled receptor (GPR) 81, posited to orchestrate a "fatigue signal" amid exercise capabilities. This hypothesized function in neural inhibition underscores the comprehensive impact of lactate on physiological responses to prolonged exercise. This evolving perspective of lactate not only accentuates its central role in metabolic processes but also paves the way for advanced exploration into the whole-body's sophisticated mechanisms for managing and adapting to physical exhaustion and central fatigue.

T. Matsui (✉)
Institute of Health and Sport Sciences, University of Tsukuba, Tsukuba, Ibaraki, Japan

Sport Neuroscience Division, Kokoro Division, Advanced Research Initiative for Human High Performance (ARIHHP), Institute of Health and Sport Sciences, University of Tsukuba, Tsukuba, Ibaraki, Japan
e-mail: matsui.takashi.ga@u.tsukuba.ac.jp

H. Soya
Institute of Health and Sport Sciences, University of Tsukuba, Tsukuba, Ibaraki, Japan

Sport Neuroscience Division, Kokoro Division, Advanced Research Initiative for Human High Performance (ARIHHP), Institute of Health and Sport Sciences, University of Tsukuba, Tsukuba, Ibaraki, Japan

Laboratory of Exercise Biochemistry and Neuroendocrinology, Institute of Health and Sport Sciences, University of Tsukuba, Tsukuba, Ibaraki, Japan

Center for Cybernics Research, University of Tsukuba, Tsukuba, Ibaraki, Japan

H. Soya (ed.), *Exercise Brain Stimulation for Cognitive Function and Mental Health*, Advances in Neurobiology 44, https://doi.org/10.1007/978-981-95-0066-6_11

Keywords Exercise · Fatigue · Glycogen · GPR81 · Lactate

1 Introduction

Lactate, historically known for its rapid accumulation in fatigued muscles and its role in inhibiting muscle contraction by lowering pH, has long been recognized as a fatigue substance. However, in recent years, lactate has gained renewed attention as an important energy source in skeletal muscles, cardiac muscles, and the brain [1]. As a primary intermediate metabolite in glycogen-based glucose metabolism, lactate is taken up by the mitochondria, aiding ATP synthesis. Lactate produced in active muscles during exercise is transported via blood flow to the muscles, heart, and brain, serving as an energy source that embodies the concept of an organ network.

Additionally, our research discovered that the brain itself, particularly astrocytes, stores glycogen and uses lactate as an energy source during exercise to enhance endurance [2, 3]. More recently, we have explored a novel hypothesis: lactate acts as a ligand for G protein-coupled receptor 81 (GPR81, also known as HCAR1), potentially forming a "fatigue signal" during exercise. This chapter introduces the latest trends in lactate research, including our own work on energy metabolism, suggesting multifaceted roles for lactate in the body, especially in the context of exercise-induced fatigue and neural regulation.

2 Lactate as Energy: Inter-Organ Cooperation via Blood Flow

Initially identified by Nobel Laureate A.V. Hill, in 1922, as a fatigue substance that accumulates in tired muscles, inhibiting contraction by reducing pH and lactate has since been reconceptualized in the realm of exercise physiology [4]. Presently, it is recognized as a crucial energy substrate within the glycolysis pathway, serving as an intermediary metabolite that facilitates substantial ATP synthesis within the mitochondria upon oxidation to pyruvate. This role is particularly pronounced during physical exertion in the skeletal and cardiac muscles. Lactate is abundantly generated in fast-twitch muscle fibers, rich in glycogen, and is then transported and oxidized in slow-twitch fibers, containing numerous mitochondria, a process described as the "fast-twitch–slow-twitch muscle fiber lactate shuttle" (Fig. 1a) [5]. The heart, which is also rich in mitochondria, utilizes blood-derived lactate for oxidative metabolism during increased physical activity (Fig. 1b) [6].

Concurrently, the brain, activated during exercise for various functionalities, such as cortical muscle contraction control, hippocampal spatial cognition, and hypothalamic energy metabolism, exhibits an increased demand for energy.

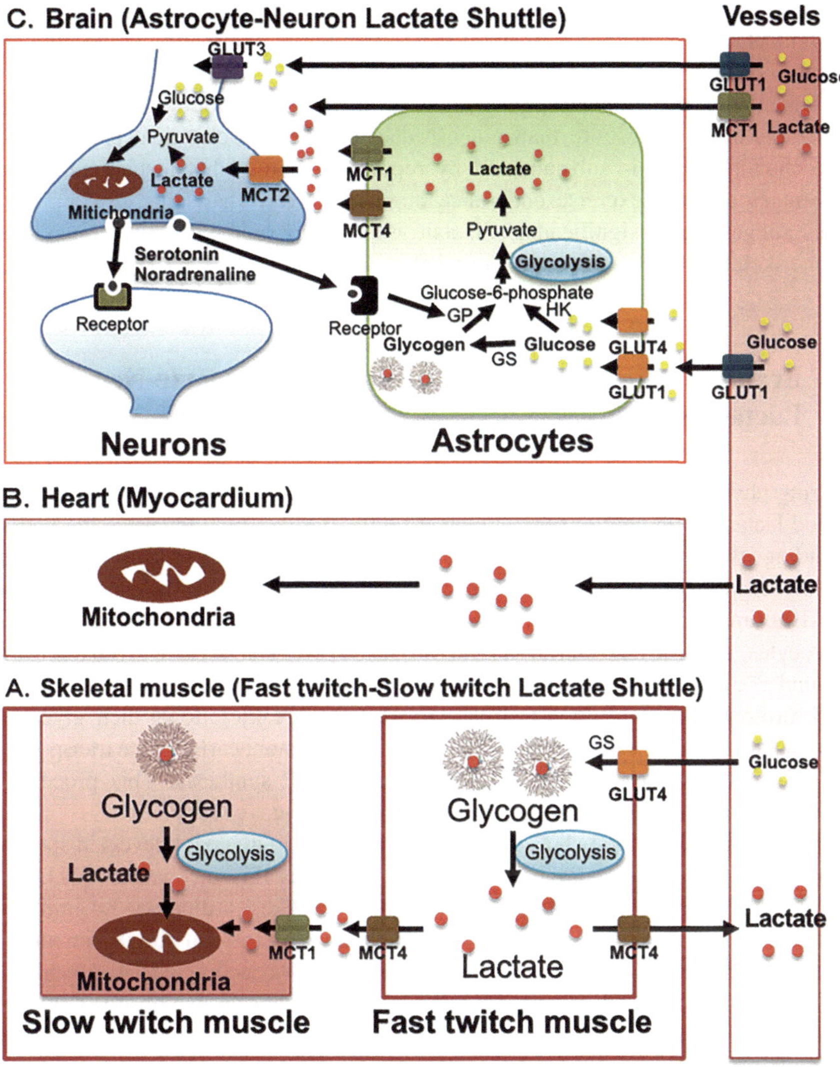

Fig. 1 The organ network of lactate as an energy source. (**a**) The fast-twitch–slow-twitch muscle fiber lactate shuttle in skeletal muscles during exercise, (**b**) lactate transport to cardiac muscle via blood flow, (**c**) lactate metabolism in the brain (astrocyte-neuron lactate shuttle). Key enzymes and transporters are labeled: GP (glycogen phosphorylase), GS (glycogen synthase), HK (hexokinase), MCT (monocarboxylate transporter)

Traditionally perceived as relying solely on glucose (and ketones during starvation), the brain, as proposed by intensive studies from the University of Copenhagen, likely assimilates and oxidizes lactate, particularly under high-intensity exercise conditions [7]. This is evidenced by increased lactate release from active muscles and elevated blood lactate levels. Interestingly, in non-exercise scenarios, such as whisker stimulation in rats, activated neurons in the cerebral cortex's sensory areas prefer lactate over glucose, suggesting that astrocytes predominantly uptake glucose to convert it into lactate for neuronal supply [8].

Given its production through glycolysis, lactate is considered a more readily utilizable energy substrate than glucose due to its closer chemical proximity to mitochondria. This is substantiated by reports indicating that lactate-containing beverages enhance exercise endurance more effectively than glucose-containing ones, suggesting the significance of lactate as an energy source in active organs such as the skeletal muscle, heart, and brain [9].

3 Brain Lactate Production System: The Astrocyte-Neuron Lactate Shuttle

During physical activity, the brain, which becomes more active, utilizes increased blood lactate as an energy source. Concurrently, our research advances the understanding of the significance of lactate produced from glycogen within the brain during exercise. The brain, composed of approximately 10 billion neurons and 50–100 times more glial cells, stores glycogen primarily in a type of glia called astrocytes, with the concentration ratio of muscle glycogen to brain glycogen being around 2–3:1 [3]. Astrocytes respond to neurotransmitters such as noradrenaline and serotonin, released upon neuron activation, by breaking down their glycogen into lactate. Lactate is then transported to neurons via monocarboxylate transporters (MCT), entering the mitochondria to facilitate ATP synthesis. This process is referred to as the "astrocyte-neuron lactate shuttle" (Fig. 1c) [10].

Empirical evidence, such as the observed decrease in glycogen levels in specific brain areas during memory tasks in chicks, supports this mechanism [11]. The administration of glycogen phosphorylase inhibitors (like 1,4-dideoxy-1,4-imino-d-arabinitol, DAB) or MCT2 inhibitors (such as α-cyano-4-hydroxy-cinnamate, 4-CIN), which block lactate uptake in neurons, leads to a decline in spatial cognitive functions governed by the hippocampus [12]. This effect was also observed in mice lacking brain-specific glycogen synthase, indicating the absence of brain glycogen [13]. These findings collectively underscore the crucial role of glycogen-derived lactate as an energy substrate for maintaining neural function.

4 Brain Lactate Production During Prolonged Exercise

The rapid metabolic turnover of brain lactate and glycogen, coupled with the challenges in their quantification, has long obscured the understanding of brain metabolism during exercise. However, the introduction of microwave irradiation in the early 2000s, a technique that rapidly elevates rat brain temperature to approximately

90 °C to deactivate metabolic enzymes, facilitated accurate measurements of brain lactate and glycogen levels. This method indicated a decrease in brain glycogen during sleep deprivation (a model for brain fatigue) and under hypoglycemic conditions.

Our research, utilizing microwave irradiation, investigated the impact of various durations of moderate-intensity exercise (30, 60, and 120 min at a speed of 20 m/min) on rat brain lactate and glycogen levels [3]. The results showed that brain lactate levels increased in tandem with blood lactate levels (Fig. 2a–c). Notably, while shorter exercise durations did not significantly reduce brain glycogen, a considerable decrease of approximately 60% in brain glycogen was observed after 120 min of exercise, coinciding with muscle glycogen depletion, hypoglycemia, and fatigue (Fig. 2d, e). Additionally, this decrease showed a negative correlation with increased serotonin metabolic turnover and brain lactate levels (Fig. 2f), suggesting that brain glycogen contributes to lactate production via serotoninergic neuron activation, especially toward the end of prolonged, fatigue-inducing exercise.

Given that depletion of brain glycogen can lead to neuronal death [14], the relatively lower reduction rate of brain glycogen compared to muscles and the sustained high levels of lactate might represent a biological defense mechanism against central fatigue through serotonin elevation. The observed increase in monocarboxylate transporters (MCT) in the cerebral cortex and hippocampus further supports this hypothesis. Therefore, we propose that lactate derived from brain glycogen plays a crucial role in sustaining endurance and neural survival during prolonged exercise.

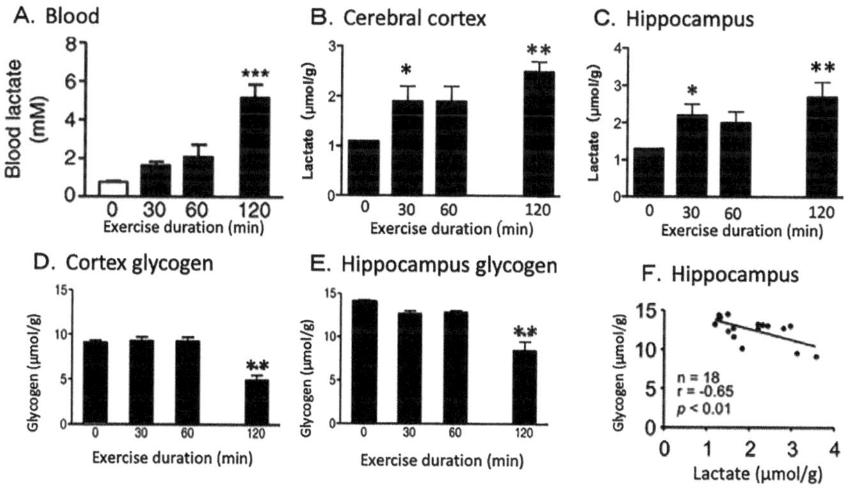

Fig. 2 Brain lactate production during prolonged exercise (Modified from Matsui et al. [3]). (**a**) Blood lactate levels, (**b**) lactate concentration in the cerebral cortex, (**c**) lactate concentration in the hippocampus, (**d**) glycogen concentration in the cerebral cortex, (**e**) glycogen concentration in the hippocampus, (**f**) correlation between lactate and glycogen concentrations in the hippocampus. *: $p < 0.05$, **: $p < 0.01$ vs. 0 (pre-exercise group)

5 Brain Lactate Is a Key to Exercise Endurance

To validate the aforementioned hypothesis, we conducted a study in which catheters were implanted into the lateral ventricles of rats. DAB (a glycogen breakdown inhibitor) was administered through these catheters, followed by treadmill exercise until the rats reached exhaustion (Fig. 3a). This approach revealed that DAB inhibited the decrease in brain glycogen typically caused by exhaustive exercise, especially outside the cerebral cortex, and notably suppressed the increase in brain lactate levels in the hippocampus (Fig. 3b, c). During exhaustive exercise, hippocampal ATP levels, which usually do not decrease, were reduced following DAB administration, and the duration of exercise leading to exhaustion was shortened by 20.6% (Fig. 3d, e). Furthermore, the administration of 4-CIN (an MCT2 inhibitor) to similar rats resulted in decreased hippocampal ATP levels at exhaustion and accelerated the onset of exhaustion by 34.8% (Fig. 3g, h). These findings support

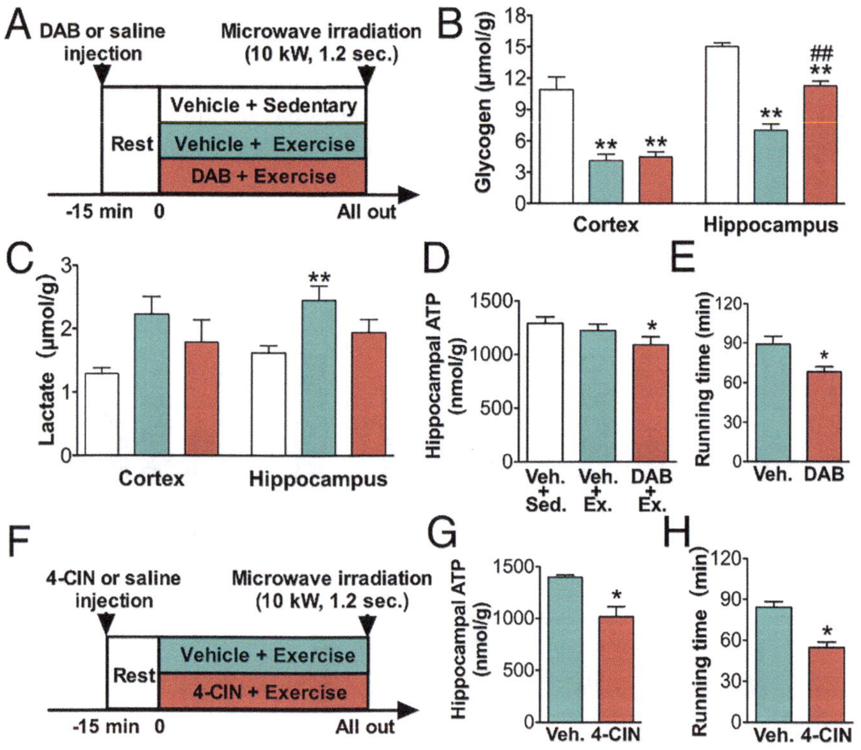

Fig. 3 Contribution of brain glycogen-derived lactate to exercise endurance (Modified from Matsui et al. [2]). (**a**) DAB administration experimental protocol, (**b**) brain glycogen concentration, (**c**) brain lactate concentration, (**d**) ATP concentration in the hippocampus, (**e**) running time to exhaustion. *: p 0.05, **: p 0.01 vs. solvent administration + rest. #: p 0.05, ##: p 0.01 vs. saline + exercise. (**f**) 4-CIN administration experimental protocol, (**g**) ATP concentration in the hippocampus, (**h**) running time to exhaustion. *: p 0.05 vs. solvent administration + exercise

our working hypothesis, suggesting that brain glycogen, through lactate production and supply, plays a vital role in maintaining brain energy metabolism, thereby contributing to exercise endurance.

Contrasting the metabolomic results of exhausted skeletal muscles and the brain, we observed that while lactate and ATP levels decreased in the muscles, the brain maintained ATP levels and exhibited increased lactate levels during fatigue [2]. Therefore, central fatigue during endurance exercise under physiological conditions may not arise from energy depletion like muscle fatigue, but could instead be formed through a fatigue signal that prevents brain overactivity.

6 Lactate as an Inhibitory Signal for Cellular Activity

Recent studies have revealed the expression of lactate receptors (HCAR1, also known as GPR81) in neurons and astrocytes of the cerebral cortex and hippocampus. These receptors have been found to decrease cAMP synthesis via adenylate cyclase [15]. Given that cAMP acts as a facilitator for noradrenaline release in neurons and glycogen breakdown in astrocytes, it is understood that brain lactate functions as a negative feedback mechanism, curbing overactivity in neurons and astrocytes. Therefore, during exercise, lactate derived from brain glycogen, while supplying fuel to the brain, acts as an inhibitory signal for cellular activity via HCAR1. This mechanism contributes to the formation of central fatigue, ultimately helping to maintain ATP concentration, thereby serving as a vital biological defense mechanism (Fig. 4).

7 Fatigue Induced by Brain Lactate Signaling

A pioneering study addressed this topic by examining the effects of 3-chloro-5-hydroxybenzoic acid (CHBA), a GPR81 agonist, administered into the brain ventricles. Researchers have reported a significant reduction in swimming endurance until fatigue following CHBA administration (Fig. 5). After 30 min of swimming with the same weight load, there were no significant changes in blood lactate levels or muscle glycogen between the CHBA and control groups, suggesting that the level of peripheral fatigue remained comparable. However, in the exercise cortex of the CHBA-administered group, extracellular γ-aminobutyric acid (GABA) levels were higher and extracellular glutamate concentrations and glutamate/GABA ratios were lower than those in the control group. These findings suggest that an increase in brain lactate levels through GPR81 activation could act as a signal exacerbating central fatigue during exercise.

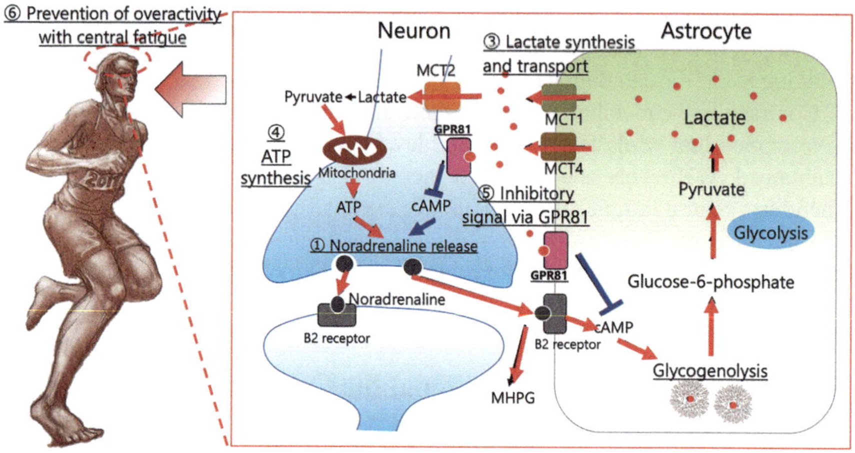

Fig. 4 The role of brain lactate receptor HCAR1 in forming inhibitory signals for cellular activity (conceptual diagram) *HCAR1 is also known as GPR81

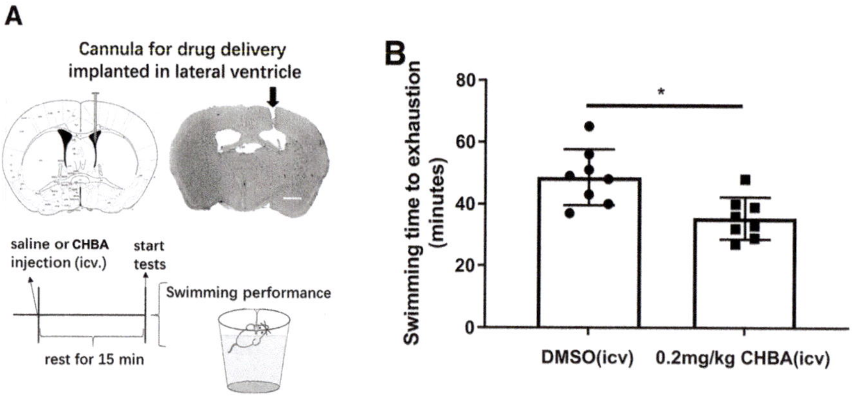

Fig. 5 Acceleration of fatigue onset during exercise by lactate receptor GPR81 (Modified from Li et al. [16]). (**a**) Experimental protocol, (**b**) swimming time to exhaustion

8 Conclusion

In the referenced study [16], while GPR81 agonists shortened exercise duration, the levels of blood lactate and muscle glycogen at the point of exhaustion remained similar to those in the control group. This observation, which was not explicitly noted by the authors, suggests a potential enhancement in carbohydrate metabolism (and a corresponding decrease in lipid metabolism) during exercise, leading to an accelerated depletion of carbohydrate energy.

To investigate this further, we administered DHBA (a GPR81 agonist) locally to the ventromedial hypothalamus (VMH), a central hub of energy metabolism in rats.

We then conducted a 30-minute endurance exercise in a treadmill metabolic chamber and measured systemic energy metabolism indicators: oxygen consumption (VO_2), carbon dioxide production (VCO_2), and respiratory exchange ratio (RER). The results showed that DHBA administration increased VCO_2 and RER during exercise. These findings suggest that lactate in the VMH, acting through GPR81, inhibits whole-body fatty acid oxidation during endurance exercise, thereby promoting fatigue [17]. This mechanism might represent a "selfish brain" strategy, inducing fatigue to protect the brain and nervous system from overactivity by halting exercise. Future efforts will focus on validating this hypothesis.

References

1. Brooks GA, Arevalo JA, Osmond AD, Leija RG, Curl CC, Tovar AP (2022) Lactate in contemporary biology: a phoenix risen. J Physiol 600(5):1229–1251
2. Matsui T, Omuro H, Liu YF, Soya M, Shima T, McEwen BS, Soya H (2017) Astrocytic glycogen-derived lactate fuels the brain during exhaustive exercise to maintain endurance capacity. Proc Natl Acad Sci USA 114(24):6358–6363
3. Matsui T, Soya S, Okamoto M, Ichitani Y, Kawanaka K, Soya H (2011) Brain glycogen decreases during prolonged exercise. J Physiol 589(Pt 13):3383–3393
4. Bassett DR Jr (2002) Scientific contributions of A. V. Hill: exercise physiology pioneer. J Appl Physiol (1985) 93(5):1567–1582
5. Brooks GA (1986) The lactate shuttle during exercise and recovery. Med Sci Sports Exerc 18(3):360–368
6. Brooks GA, Brown MA, Butz CE, Sicurello JP, Dubouchaud H (1999) Cardiac and skeletal muscle mitochondria have a monocarboxylate transporter MCT1. J Appl Physiol 87(5):1713–1718
7. Ide K, Schmalbruch IK, Quistorff B, Horn A, Secher NH (2000) Lactate, glucose and O2 uptake in human brain during recovery from maximal exercise. J Physiol 522 Pt 1(Pt 1):159–164
8. Chuquet J, Quilichini P, Nimchinsky EA, Buzsáki G (2010) Predominant enhancement of glucose uptake in astrocytes versus neurons during activation of the somatosensory cortex. J Neurosci 30(45):15298–15303
9. Azevedo JL, Tietz E, Two-Feathers T, Paull J, Chapman K (2007) Lactate, fructose and glucose oxidation profiles in sports drinks and the effect on exercise performance. PLoS One 2(9):e927
10. Pellerin L, Pellegri G, Bittar PG, Charnay Y, Bouras C, Martin JL, Stella N, Magistretti PJ (1998) Evidence supporting the existence of an activity-dependent astrocyte-neuron lactate shuttle. Dev Neurosci 20(4–5):291–299
11. O'Dowd BS, Gibbs ME, Ng KT, Hertz E, Hertz L (1994) Astrocytic glycogenolysis energizes memory processes in neonate chicks. Brain Res Dev Brain Res 78(1):137–141
12. Suzuki A, Stern SA, Bozdagi O, Huntley GW, Walker RH, Magistretti PJ, Alberini CM (2011) Astrocyte-neuron lactate transport is required for long-term memory formation. Cell 144(5):810–823
13. Duran J, Saez I, Gruart A, Guinovart JJ, Delgado-García JM (2013) Impairment in long-term memory formation and learning-dependent synaptic plasticity in mice lacking glycogen synthase in the brain. J Cereb Blood Flow Metab 33(4):550–556
14. Suh SW, Bergher JP, Anderson CM, Treadway JL, Fosgerau K, Swanson RA (2007) Astrocyte glycogen sustains neuronal activity during hypoglycemia: studies with the glycogen phosphorylase inhibitor CP-316,819 ([R-R*,S*]-5-chloro-N-[2-hydroxy-3-(methoxymethylamino)-3-oxo-1-(phenylmethyl)propyl]-1H-indole-2-carboxamide). J Pharmacol Exp Ther 321(1):45–50

15. Lauritzen KH, Morland C, Puchades M, Holm-Hansen S, Hagelin EM, Lauritzen F, Attramadal H, Storm-Mathisen J, Gjedde A, Bergersen LH (2014) Lactate receptor sites link neurotransmission, neurovascular coupling, and brain energy metabolism. Cereb Cortex 24(10):2784–2795
16. Li J, Xia Y, Xu H, Xiong R, Zhao Y, Li P, Yang T, Huang Q, Shan F (2022) Activation of brain lactate receptor GPR81 aggravates exercise-induced central fatigue. Am J Physiol Regul Integr Comp Physiol 323(5):R822–R831
17. Matsui T (2023) Lactate receptor GPR81 in the ventromedial hypothalamus suppresses whole body fat oxidation during endurance exercise: implications for brain lactate as a central fatigue signal. J Physiol Sci 73(Suppl 1):232

Interoceptive Signaling by Circulating Insulin Like Growth Factor I and Neuroprotection by Exercise

Jonathan Zegarra-Valdivia, Estrella Fernandez de Sevilla, Jaime Pignatelli, and Ignacio Torres Aleman

Abstract Physical activity has been proven to be beneficial for brain function. Due to a lack of appropriate therapies for the majority of brain diseases, exercise has become a favored alternative to prevent and even treat several of these pathologies. Thus, the mechanisms underlying the neuroprotective actions of exercise are under intense scrutiny. Furthermore, since many patients afflicted with different neurological conditions are not able to perform exercise, development of pharmacological mimics based on knowledge of underlying cellular and molecular mechanisms is of therapeutic interest (Narkar VA, Downes M, Yu RT, Embler E, Wang YX, Banayo E, Cell 134:405–415, 2008). As part of these mechanisms, we will examine the role of insulin-like growth factor I (IGF-I), a pleiotropic neuroprotective signal, and one of the established mediators of the beneficial actions of exercise in the brain. Exercise stimulates the entrance of circulating IGF-I into the brain where it mediates pro-neurogenic, pro-cognitive, and mood modulatory effects known to be associated to exercise. Through its potent cytoprotective actions (anti-apoptotic, anti-oxidant, anti-inflammatory), IGF-I participates in reparative and homeostatic processes associated to exercise. We postulate that circulating IGF-I, a regulator of

J. Zegarra-Valdivia
Achucarro Basque Center for Neuroscience, Leioa, Spain

Ciberned, Madrid, Spain

Universidad Señor de Sipán, Chiclayo, Perú

Global Brain Health Institute, University of California, San Francisco, CA, USA

E. F. de Sevilla · J. Pignatelli
Ciberned, Madrid, Spain

Cajal Institute, CSIC, Madrid, Spain

I. T. Aleman (✉)
Achucarro Basque Center for Neuroscience, Leioa, Spain

Ciberned, Madrid, Spain

Ikerbasque Foundation for Science, Bilbao, Spain
e-mail: ignacio.torres@achucarro.org

H. Soya (ed.), *Exercise Brain Stimulation for Cognitive Function and Mental Health*, Advances in Neurobiology 44, https://doi.org/10.1007/978-981-95-0066-6_12

muscle and bone mass, forms part of an interoceptive system within a humoral branch informing the brain of muscle/bone mass. In this way, IGF-I conveys interoceptive signaling to brain areas involved in orchestrating physical activity to adapt them to available vigor, i.e., muscle strength. Because exercise engages the activity of many brain areas, neuroprotection by exercise-elicited entrance of circulating IGF-I is brain-wide.

Keywords Exercise neuroprotection · IGF-I · Interoception · Physical vigor

1 Introduction

It is known from ancient times that physical exercise is widely beneficial and exerts powerful actions on mind and body [2]. In modern times, initially anecdotal evidence of positive actions of exercise on brain health has blossomed into a frenzy of studies analyzing its multiple actions on brain physiology, and, most significantly, on neuropathological conditions of all types [3]. Increasingly abundant evidence points to a therapeutic utility of exercise in conditions such as depression, Alzheimer's disease (AD), mild cognitive impairment, anxiety, and many others [4–6].

While cardiometabolic fitness, rather than direct actions on the brain, is a favored mechanism underlying neuroprotective actions of exercise [7], there are many exercise-associated mechanisms that directly impinge on brain structure and function. Most prominently, tissue maintenance, neurogenesis, angiogenesis, or synaptic plasticity [3, 8], all of which in all probability also contribute to neuroprotection by exercise. In this review, we will focus on the role of circulating IGF-I as a key mediator of several of the beneficial actions of exercise on the brain, although it is clear that many other signals from the periphery influence brain function during exercise [9]—the so-called "exerkines" [10]. The reason we focus on this particular humoral signal is explained in detail.

2 Exercise Stimulates Entrance of Circulating IGF-I into the Brain

Initial studies on the anatomical distribution of IGF-I and its receptor (IGF-IR) in the brain noticed a wide distribution of the latter, while its ligand shows a time-dependent pattern, with abundant, wide expression during development, and highly localized, very limited expression in the adult [11, 12]. The mismatch between ample brain IGF-IR distribution and local, low brain IGF-I mRNA expression was coupled with the observation that brain protein levels of IGF-I remain at relatively high levels throughout life [13], although a gradual decline, as seen in the rest of the body [14], takes place along aging [15–17].

This prompted us to search for mechanisms of entrance of IGF-I from the periphery into the brain as a source of this brain growth factor. While other humoral signals present in the circulation enter the brain through the blood-brain barrier (BBB), this specialized structure allows only highly restricted passage of molecules [18]; even water molecules require aquaporins, specific transport proteins in the neurovascular unit comprising the BBB [19] to get access to the brain. Although pioneer work indicated that different hormones, including the close relative insulin [20], traversed the BBB [21], it was considered that only a very selective group of proteins would cross this functional and anatomical barrier between the brain and the blood circulation.

Since we observed that systemic injection of IGF-I produced potent reparative effects on brain pathology [22], we hypothesized that circulating IGF-I could reach the brain. Considering that in response to exercise, skeletal muscle takes up circulating IGF-I, even though it also locally produces this growth factor [23], and that systemically administered IGF-I reaches ample areas in the brain [24], we postulated that exercise is a physiological stimulus of the entrance of circulating IGF-I in the brain. Indeed, after exercise, many brain areas show labelled IGF-I systemically injected [24]. Broadly speaking, the areas that show increased IGF-I after exercise were the same that took up systemic IGF-I when exogenously administered [24], pointing to an area-specific uptake probably reflecting particular routes of entrance into the brain. Hence, exercise stimulates uptake of circulating IGF-I not only by skeletal muscle, but also by the brain. We postulate that other target organs in the body take up circulating IGF-I in response to exercise, but this needs to be studied.

Although a detailed account of the effects of exercise on circulating IGF-I is beyond the scope of this review, a cautionary note on this respect is necessary. While there is abundant information on the effects of different types of exercise, i.e., aerobic vs. anaerobic; high-intensity vs. low-intensity, etc., on serum IGF-I levels, available data so far indicate either unchanged, decreased, or increased serum IGF-I after exercise. This points to a context-dependent effect of exercise on serum IGF-I levels which require accounting for multiple variables such as sex, age, fitness, feeding status, type of exercise, and probably many other aspects that remain to be characterized. Sufficient to say that at present, determining the levels of IGF-I after a given exercise protocol does not provide straightforward information on its possible actions on target organs such as the brain.

3 Overlapping Neuroprotective Actions of IGF-I and Exercise

During past decades, abundant evidence has established that IGF-I is a prototype neuroprotective signal [25, 26], and that circulating IGF-I is an important component of its neurotrophic activity [22, 27]. Similarly, exercise has proven widely neuroprotective in humans and experimental animals and is currently considered an important life-style factor for the promotion of brain health [28–32]. While factors other than IGF-I participate in neuroprotection by exercise [33–37], all the known neuroprotective actions of circulating IGF-I has also been associated to exercise. This include angiogenesis [38, 39], synaptogenesis [30, 40], lowering amyloidosis [28, 41, 42], anxiolysis [6, 43], amelioration of posttraumatic stress [44, 45], enhanced synaptic plasticity [46, 47], improved memory [27, 31], increased survival in inherited [48–50] and sporadic neurodegeneration [49, 51], and even greater hypothalamic insulin sensitivity [52, 53]. Moreover, in several of these activities, IGF-I has been shown to be directly involved. Thus, exercise-induced adult hippocampal neurogenesis [54] depends on circulating IGF-I [55], while anti-depressive and anxiolytic effects of exercise [56–58] also involve IGF-I [59], in both human and animal studies. Exercise-induced amelioration of spinal cord injury in conjunction with stem cell transplantation also requires IGF-I [60], where exercise is neuroprotective [49].

A thorough explanation of the involvement of IGF-I on exercise neuroprotection in conjunction with other factors will demand a detailed analysis of the pathways involved (Fig. 1). Thus, increased adult hippocampal neurogenesis after exercise not only requires IGF-I [55], but also VEGF [61] and serotonin [62]. Accordingly, future studies would require considering the combined actions of all known signals and whether they act through common or independent pathways. For example, in the case of the induction of BDNF by exercise, the use of an anti-IGF-I blocking antibody showed that this neurotrophic factor acts downstream of IGF-I, at least under specific neuroprotective protocols [59]. A similar scenario is seen in the induction of proteasome activity downstream of IGF-I, in relation to exercise-induced neurogenesis [63]. In this vein, whether exercise-induced angiogenesis involving VEGF production requires IGF-I [39], muscle-derived lactate [64], or both, would also require further evaluation.

A particularly relevant neuroprotective activity of exercise refers to brain deterioration in aging. Exercise protects and prevents against many of the age-associated neurological disturbances [8, 65–67], including cognitive deterioration [4, 68–70], a condition of high prevalence due to the increasing proportion of the elderly in the world population [71]. Although there is evidence that IGF-I also protects against age-associated brain deterioration [72] and improves cognition [27], it has been argued that the brain IGF-I receptor is detrimental for brain diseases usually associated to aging [73, 74]. Again, the numerous studies analyzing the association of serum IGF-I levels with cognition along the aging process have mostly reported a positive correlation [75–77], but also a negative [78], or even no correlation [79] has

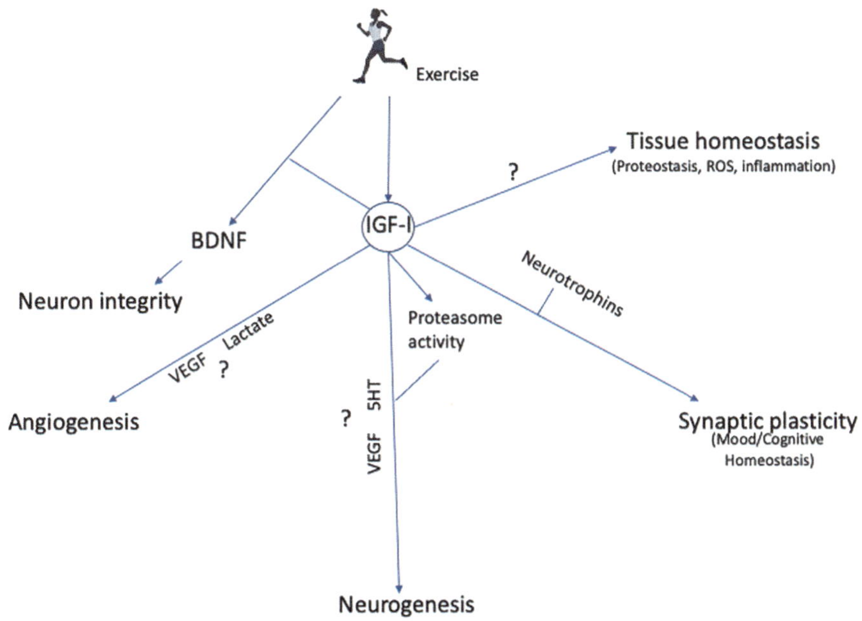

Fig. 1 Neuroprotective actions of exercise involve interactions between different mediators. A central role for IGF-I in several of the effects of exercise in the brain is illustrated for the sake of simplicity. Among neuroprotective actions of exercise, we emphasize promotion of changes in brain architecture (neuronal integrity, formation of new vessels and neurons), general brain tissue homeostasis (protein turnover, oxidative radicals scavenging and inflammation), and higher skills (mood and cognition) homeostasis. While several of the neuroprotective effects involving IGF-I are known to include other downstream mediators (such as BDNF in specific neuroprotective protocols or proteasome activity in neurogenesis), possible interactions, if any, with other mediators participating in exercise neuroprotection (VEGF, lactate, 5HT, neurotrophins) are not yet known. Assuming that exercise neuroprotection is mediated by the interaction of different mediators leading to effects that cannot be deduced from the analysis of each of them individually, once we describe all factors involved, we will require experiments combining all of them to fully understand physiological exercise neuroprotection

been described, which confirms that measuring serum IGF-I levels, which present wide individual variation within the population [80], does not provide sufficiently robust information of the biological role of this growth factor in the aging brain. Conversely, studies analyzing neuroprotection by exercise in elderly population where serum IGF-I levels are measured consistently show increased serum IGF-I after exercise [58, 81–83]. We consider that large-scale proteomic/transcriptomic analyses in cross-sectional studies of aging individuals will be needed to understand the context-dependent (age, sex, etc.) role of serum IGF-I in neuroprotection by exercise during aging.

4 IGF-I as an Interoceptive Signal of Physical Vigor

The brain receives external (sensory) and internal (interoceptive) information and processes it for behavioral output and maintenance of body homeostasis. Perception of internal sensations or interoception is considered an additional sense [84] where the vagus nerve, with over 80% of sensory neurons, plays an essential role [85], together with a wide variety of humoral signals. Among these, hormones constitute key informers to the brain of the activity of all organs in the body. This body-to-brain flow of information, which is bidirectional, is required for maintenance of body health [84] and internal states shaping behavior [86].

The fact that physical activity is important for the maintenance of proper brain function suggests that mental and physical fitness constitute a continuum, i.e., well-being, with shared processes, including a set of as yet undefined interoceptive signals that modulate brain centers involved in orchestrating mental and physical activity accordingly. We postulate that IGF-I is one such interoceptive "fitness signal." For instance, while circulating insulin signals the hypothalamus to provide information on adiposity status [87], among other actions, circulating IGF-I may provide information on muscle mass status [88] on hypothalamic nuclei (and elsewhere in the brain), as active muscles release IGF-I into the circulation [89]. Thus, circulating IGF-I acts as a peripheral signal to inform brain homeostatic centers involved in regulation of physical activity, muscle tone, energy balance, mood, and cognition to gauge physical performance and adapt to it accordingly. This information may be loosely categorized as a "physical fitness" cue which is known to impinge on homeostatic emotional tone or "wellness" [90] (Fig. 2). IGF-I should be considered a positive modulator of wellness as it is anxiolytic [43], promotes social activity [91], and improves cognition and sensory perception [47, 92]. Indeed, well-being is an internal brain state, and as such, will be modulated, as is known at least in relation to energy availability [93]. Interoceptive signals of physical status will also modulate affective/cognitive interactions leading to wellbeing constructs in a way resembling how interoceptive signals modulate internal states such as hunger [94]. Naturally, not only insulin and IGF-I, but other humoral signals form part of this fitness information. For instance, serum leptin, that informs the brain of nutritional status, interferes with the rewarding effects of exercise [95].

Under pathological conditions, modulation by internal/external inputs of brain activity may be distorted, including this proposed "fitness" loop. More dramatically, absence of regular physical activity will drastically diminish IGF-I input to the brain. In turn, non-exercise-dependent interoceptive signals impacting on brain centers involved in integrating appropriate responses to physical activity will be unbalanced and, in all probability, will disturb normal brain function. The growing sedentary world population [96] may help explain the unprecedented growth in the incidence of mental illnesses in recent times [97, 98]. Thus, it is not that exercise promotes brain health; it is that its absence is deleterious to brain health, making it an essential component of a healthy lifestyle.

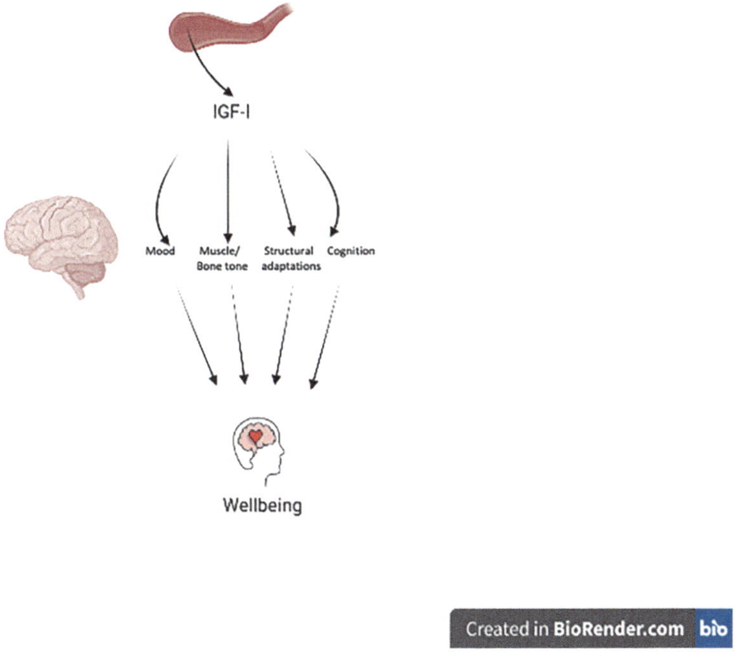

Fig. 2 Interoceptive signaling by IGF-I and wellbeing. Circulating IGF-I is captured by the brain through a dedicated process dubbed "neurotrophic coupling" at the blood-brain barrier in response to brain activity [100]. In the brain, IGF-I modulates mood acting on different areas within the limbic system (hippocampus, hypothalamus, cortex, amygdala), informs brain centers involved in the control of physical activity (hypothalamus, motor cortex, brain stem), stimulates neurogenesis, angiogenesis, and neuronal arborization, and modulates cognition (cortex, hippocampus, subcortical regions)

5 Conclusions and Perspectives

Analysis of central actions of circulating IGF-I eventually led to its adscription as an important mediator of exercise neuroprotection. Since all neuroprotective actions of IGF-I overlap with those produced by exercise, this growth factor should be routinely accounted for when exploring underlying mechanisms of exercise neuroprotection. We also need to keep in mind that the simultaneous actions of many mediators of exercise effects in the brain need to be considered, and, significantly, their interactions will result in step-wise/multilayered, complex processes not yet fully understood.

A step forward in our conceptual understanding of the role of IGF-I in exercise neuroprotection may be provided by explaining exercise neuroprotection as part of an interoceptive mechanism linking the periphery with brain states. Specifically, neuromodulators such as IGF-I provides information of physical vigor (muscle and bone mass) to distributed brain areas orchestrating wellbeing states. As a bona fide

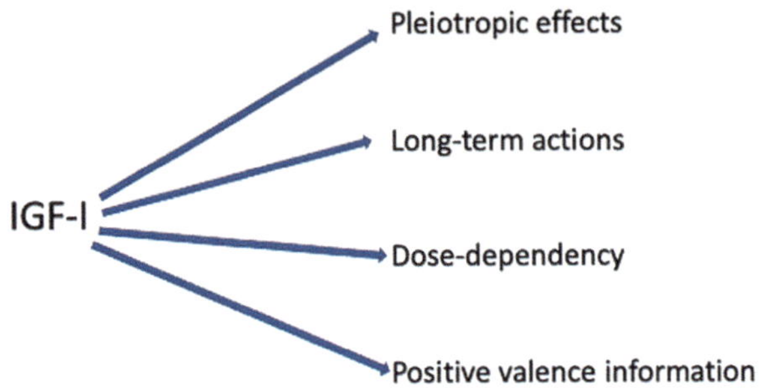

Fig. 3 Analyzing the role of IGF-I from a systems neurobiology perspective. IGF-I actions are pleiotropic, long-term, dose-dependent and related to positive valence stimuli. A major characteristic of IGF-I actions in brain is its pleiotropy. IGF-I modulates all major brain region, cell types, and contexts. Long-term actions of IGF-I have been documented, with prolonged effects provided by in vivo exposure to the growth factor [27]. Dose dependency is also routinely observed when using IGF-I in neuroprotective schemes, such as recovery of neurogenesis in IGF-I deficient mice [55]. That IGF-I provides a positive valence is inferred from its pro-cognitive [27] and anxiolytic actions [43], enhancement of social intercourse [91], or promotion of playful behavior [101]

neuromodulator of brain states, IGF-I actions are pleiotropic [25], long-term [27], dose-dependent [99], and of positive valence [43], which fits reasonably well with the definition of brain states by Flavell et al. [86]. Future studies on the role of IGF-I as a mediator of exercise neuroprotection will require a systems neurobiology approach analyzing its actions, brain- and cell-wide, within very specific contexts (Fig. 3).

Acknowledgements Work reviewed in this chapter was funded by CIBERNED (Instituto de Salud Carlos III, Spain), forms part of the project PID2019-104376RB-I00 funded by MCIN/AEI/10.13039/501100011033, and a grant by Comunidad de Madrid through the European Social Fund (ESF)-financed program NEUROMETAB-CM (B2017/BMD-3700).

References

1. Narkar VA, Downes M, Yu RT, Embler E, Wang YX, Banayo E et al (2008) AMPK and PPAR[delta] agonists are exercise Mimetics. Cell 134(3):405–415
2. Booth FW, Chakravarthy MV, Spangenburg EE (2002) Exercise and gene expression: physiological regulation of the human genome through physical activity. J Physiol 543(Pt 2):399–411
3. Huuha AM, Norevik CS, Moreira JBN, Kobro-Flatmoen A, Scrimgeour N, Kivipelto M et al (2022) Can exercise training teach us how to treat Alzheimer's disease? Ageing Res Rev 75:101559
4. Baker LD, Frank LL, Foster-Schubert K, Green PS, Wilkinson CW, McTiernan A et al (2010) Aerobic exercise improves cognition for older adults with glucose intolerance: a risk factor for Alzheimer's disease. J Alzheimer's Dis 22(2):569–579
5. Bartoletti A, Medini P, Berardi N, Maffei L (2004) Environmental enrichment prevents effects of dark-rearing in the rat visual cortex. Nat Neurosci 7(3):215–216

6. Carek PJ, Laibstain SE, Carek SM (2011) Exercise for the treatment of depression and anxiety. Int J Psychiatry Med 41(1):15–28
7. Tari AR, Nauman J, Zisko N, Skjellegrind HK, Bosnes I, Bergh S et al (2019) Temporal changes in cardiorespiratory fitness and risk of dementia incidence and mortality: a population-based prospective cohort study. Lancet Public Health 4(11):e565–ee74
8. Colcombe SJ, Erickson KI, Raz N, Webb AG, Cohen NJ, McAuley E et al (2003) Aerobic fitness reduces brain tissue loss in aging humans. J Gerontol A Biol Sci Med Sci 58(2):176–180
9. Tari AR, Norevik CS, Scrimgeour NR, Kobro-Flatmoen A, Storm-Mathisen J, Bergersen LH et al (2019) Are the neuroprotective effects of exercise training systemically mediated? Prog Cardiovasc Dis 62(2):94–101
10. Chow LS, Gerszten RE, Taylor JM, Pedersen BK, van Praag H, Trappe S et al (2022) Exerkines in health, resilience and disease. Nat Rev Endocrinol 18(5):273–289
11. Bondy CA, Lee WH (1993) Patterns of insulin-like growth factor and IGF receptor gene expression in the brain. Functional implications. Ann N Y Acad Sci 692:33–43
12. Bondy C, Werner H, Roberts CT Jr, LeRoith D (1992) Cellular pattern of type-I insulin-like growth factor receptor gene expression during maturation of the rat brain: comparison with insulin-like growth factors I and II. Neuroscience 46(4):909–923
13. Pons S, Rejas MT, Torres-Aleman I (1991) Ontogeny of insulin-like growth factor I, its receptor, and its binding proteins in the rat hypothalamus. Brain Res Dev Brain Res 62(2):169–175
14. Xu X, Bennett SA, Ingram RL, Sonntag WE (1995) Decreases in growth hormone receptor signal transduction contribute to the decline in insulin-like growth factor I gene expression with age. Endocrinology 136(10):4551–4557
15. Piriz J, Muller A, Trejo JL, Torres-Aleman I (2011) IGF-I and the aging mammalian brain. Exp Gerontol 46(2–3):96–99
16. Shetty AK, Hattiangady B, Shetty GA (2005) Stem/progenitor cell proliferation factors FGF-2, IGF-1, and VEGF exhibit early decline during the course of aging in the hippocampus: role of astrocytes. Glia 51(3):173–186
17. Sonntag WE, Lynch CD, Bennett SA, Khan AS, Thornton PL, Cooney PT et al (1999) Alterations in insulin-like growth factor-1 gene and protein expression and type 1 insulin-like growth factor receptors in the brains of ageing rats. Neuroscience 88(1):269–279
18. Zlokovic BV (2008) The blood-brain barrier in health and chronic neurodegenerative disorders. Neuron 57(2):178–201
19. Nielsen S, Nagelhus EA, Amiry-Moghaddam M, Bourque C, Agre P, Ottersen OP (1997) Specialized membrane domains for water transport in glial cells: high-resolution immunogold cytochemistry of aquaporin-4 in rat brain. J Neurosci 17(1):171–180
20. Reinhardt RR, Bondy CA (1994) Insulin-like growth factors cross the blood-brain barrier. Endocrinology 135(5):1753–1761
21. Banks WA (2019) The blood–brain barrier as an endocrine tissue. Nat Rev Endocrinol 15(8):444–455
22. Fernandez AM, de la Vega AG, Torres-Aleman I (1998) Insulin-like growth factor I restores motor coordination in a rat model of cerebellar ataxia. Proc Natl Acad Sci USA 95(3):1253–1258
23. Eliakim A, Moromisato M, Moromisato D, Brasel JA, Roberts C, Cooper DM (1997) Increase in muscle IGF-I protein but not IGF-I mRNA after 5 days of endurance training in young rats. Am J Phys 273(4 Pt 2):R1557–R1R61
24. Carro E, Nunez A, Busiguina S, Torres-Aleman I (2000) Circulating insulin-like growth factor I mediates effects of exercise on the brain. J Neurosci 20(8):2926–2933
25. Fernandez AM, Torres-Aleman I (2012) The many faces of insulin-like peptide signalling in the brain. Nat Rev Neurosci 13(4):225–239
26. Dudek H, Datta SR, Franke TF, Birnbaum MJ, Yao R, Cooper GM et al (1997) Regulation of neuronal survival by the serine-threonine protein kinase Akt. Science 275(5300):661–665

27. Trejo JL, Piriz J, Llorens-Martin MV, Fernandez AM, Bolos M, LeRoith D et al (2007) Central actions of liver-derived insulin-like growth factor I underlying its pro-cognitive effects. Mol Psychiatry 12(12):1118–1128
28. Brown BM, Peiffer JJ, Taddei K, Lui JK, Laws SM, Gupta VB et al (2013) Physical activity and amyloid-beta plasma and brain levels: results from the Australian imaging, biomarkers and lifestyle study of ageing. Mol Psychiatry 18(8):875–881
29. Cotman CW, Berchtold NC (2002) Exercise: a behavioral intervention to enhance brain health and plasticity. Trends Neurosci 25(6):295–301
30. Dietrich MO, Andrews ZB, Horvath TL (2008) Exercise-induced synaptogenesis in the hippocampus is dependent on UCP2-regulated mitochondrial adaptation. J Neurosci 28(42):10766–10771
31. Erickson KI, Voss MW, Prakash RS, Basak C, Szabo A, Chaddock L et al (2011) Exercise training increases size of hippocampus and improves memory. Proc Natl Acad Sci USA 108(7):3017–3022
32. Hillman CH, Erickson KI, Kramer AF (2008) Be smart, exercise your heart: exercise effects on brain and cognition. Nat Rev Neurosci 9(1):58–65
33. Llorens-Martin M, Torres-Aleman I, Trejo JL (2010) Exercise modulates insulin-like growth factor 1-dependent and -independent effects on adult hippocampal neurogenesis and behaviour. Mol Cell Neurosci 44(2):109–117
34. Horowitz AM, Fan X, Bieri G, Smith LK, Sanchez-Diaz CI, Schroer AB et al (2020) Blood factors transfer beneficial effects of exercise on neurogenesis and cognition to the aged brain. Science 369(6500):167–173
35. Moon Hyo Y, Becke A, Berron D, Becker B, Sah N, Benoni G et al (2016) Running-induced systemic Cathepsin B secretion is associated with memory function. Cell Metab 24(2):332–340
36. Neeper SA, Gomez-Pinilla F, Choi J, Cotman C (1995) Exercise and brain neurotrophins. Nature 373(6510):109
37. Cervenka I, Agudelo LZ, Ruas JL (2017) Kynurenines: Tryptophan's metabolites in exercise, inflammation, and mental health. Science 357(6349):eaaf9794
38. Isaacs KR, Anderson BJ, Alcantara AA, Black JE, Greenough WT (1992) Exercise and the brain: angiogenesis in the adult rat cerebellum after vigorous physical activity and motor skill learning. J Cereb Blood Flow Metab 12(1):110–119
39. Lopez-Lopez C, LeRoith D, Torres-Aleman I (2004) Insulin-like growth factor I is required for vessel remodeling in the adult brain. Proc Natl Acad Sci USA 101(26):9833–9838
40. O'Kusky JR, Ye P, D'Ercole AJ (2000) Insulin-like growth factor-I promotes neurogenesis and synaptogenesis in the hippocampal dentate gyrus during postnatal development. J Neurosci 20(22):8435–8442
41. Adlard PA, Perreau VM, Pop V, Cotman CW (2005) Voluntary exercise decreases amyloid load in a transgenic model of Alzheimer's disease. J Neurosci 25(17):4217–4221
42. Carro E, Trejo JL, Gomez-Isla T, LeRoith D, Torres-Aleman I (2002) Serum insulin-like growth factor I regulates brain amyloid-beta levels. Nat Med 8(12):1390–1397
43. Santi A, Bot M, Aleman A, Penninx BWJH, Aleman IT (2018) Circulating insulin-like growth factor I modulates mood and is a biomarker of vulnerability to stress: from mouse to man. Transl Psychiatry 8(1):142
44. Fetzner MG, Asmundson GJ (2014) Aerobic exercise reduces symptoms of posttraumatic stress disorder: a randomized controlled trial. Cogn Behav Ther 44:1–13
45. Fernandez de Sevilla ME, Pignatelli J, Zegarra-Valdivia JA, Mendez P, Nunez A, Torres AI (2022) Insulin-like growth factor I mitigates post-traumatic stress by inhibiting AMP-kinase in orexin neurons. Mol Psychiatry 27(4):2182–2196
46. Kalogeraki E, Greifzu F, Haack F, L+Âwel S. (2014) Voluntary physical exercise promotes ocular dominance plasticity in adult mouse primary visual cortex. J Neurosci 34(46):15476–15481

47. Noriega-Prieto JA, Maglio LE, Zegarra-Valdivia JA, Pignatelli J, Fernandez AM, Martinez-Rachadell L et al (2021) Astrocytic IGF-IRs induce adenosine-mediated inhibitory downregulation and improve sensory discrimination. J Neurosci 41(22):4768–4781
48. Fryer JD, Yu P, Kang H, Mandel-Brehm C, Carter AN, Crespo-Barreto J et al (2011) Exercise and genetic rescue of SCA1 via the transcriptional repressor Capicua. Science 334(6056):690–693
49. Carro E, Trejo JL, Busiguina S, Torres-Aleman I (2001) Circulating insulin-like growth factor I mediates the protective effects of physical exercise against brain insults of different etiology and anatomy. J Neurosci 21(15):5678–5684
50. Ilg W, Synofzik M, Brotz D, Burkard S, Giese MA, Schols L (2009) Intensive coordinative training improves motor performance in degenerative cerebellar disease. Neurology 73(22):1823–1830
51. Huttenrauch M, Brauss A, Kurdakova A, Borgers H, Klinker F, Liebetanz D et al (2016) Physical activity delays hippocampal neurodegeneration and rescues memory deficits in an Alzheimer disease mouse model. Transl Psychiatry 6:e800
52. Ropelle ER, Flores MB, Cintra DE, Rocha GZ, Pauli JR, Morari J et al (2010) IL-6 and IL-10 anti-inflammatory activity links exercise to hypothalamic insulin and leptin sensitivity through IKKβ and ER stress inhibition. PLoS Biol 8(8):e1000465
53. Huffman DM, Farias QG, Mao K, Zhang X, Wan J, Apontes P et al (2015) Central insulin-like growth factor-1 (IGF-1) restores whole-body insulin action in a model of age-related insulin resistance and IGF-1 decline. Aging Cell 15(1):181–186
54. van Praag H, Shubert T, Zhao C, Gage FH (2005) Exercise enhances learning and hippocampal neurogenesis in aged mice. J Neurosci 25(38):8680–8685
55. Trejo JL, Carro E, Torres-Aleman I (2001) Circulating insulin-like growth factor I mediates exercise-induced increases in the number of new neurons in the adult hippocampus. J Neurosci 21(5):1628–1634
56. Adamson BC, Ensari I, Motl RW (2015) Effect of exercise on depressive symptoms in adults with neurologic disorders: a systematic review and meta-analysis. Arch Phys Med Rehabil 96(7):1329–1338
57. Duman CH, Schlesinger L, Terwilliger R, Russell DS, Newton SS, Duman RS (2009) Peripheral insulin-like growth factor-I produces antidepressant-like behavior and contributes to the effect of exercise. Behav Brain Res 198(2):366–371
58. Cassilhas RC, Antunes HK, Tufik S, de Mello MT (2010) Mood, anxiety, and serum IGF-1 in elderly men given 24 weeks of high resistance exercise. Percept Mot Skills 110(1):265–276
59. Chen MJ, Russo-Neustadt AA (2007) Running exercise- and antidepressant-induced increases in growth and survival-associated signaling molecules are IGF-dependent. Growth Factors 25(2):118–131
60. Hwang DH, Shin HY, Kwon MJ, Choi JY, Ryu BY, Kim BG (2014) Survival of neural stem cell grafts in the Lesioned spinal cord is enhanced by a combination of treadmill locomotor training via insulin-like growth Factor-1 signaling. J Neurosci 34(38):12788–12800
61. Fabel K, Fabel K, Tam B, Kaufer D, Baiker A, Simmons N et al (2003) VEGF is necessary for exercise-induced adult hippocampal neurogenesis. Eur J Neurosci 18(10):2803–2812
62. Klempin F, Beis D, Mosienko V, Kempermann G, Bader M, Alenina N (2013) Serotonin is required for exercise-induced adult hippocampal neurogenesis. J Neurosci 33(19):8270–8275
63. Niu X, Zhao Y, Yang N, Zhao X, Zhang W, Bai X et al (2020) Proteasome activation by insulin-like growth factor-1/nuclear factor erythroid 2-related factor 2 signaling promotes exercise-induced neurogenesis. Stem cells (Dayton, Ohio) 38(2):246–260
64. Morland C, Andersson KA, Haugen OP, Hadzic A, Kleppa L, Gille A et al (2017) Exercise induces cerebral VEGF and angiogenesis via the lactate receptor HCAR1. Nat Commun 8:15557
65. Han C, Ding D, Lopez M-C, Manohar S, Zhang Y, Kim M-J et al (2016) Effects of long-term exercise on age-related hearing loss in mice. J Neurosci 36(44):11308–11319

66. Endres M, Gertz K, Lindauer U, Katchanov J, Schultze J, Schrock H et al (2003) Mechanisms of stroke protection by physical activity. Ann Neurol 54(5):582–590
67. Hyodo K, Dan I, Suwabe K, Kyutoku Y, Yamada Y, Akahori M et al (2012) Acute moderate exercise enhances compensatory brain activation in older adults. Neurobiol Aging 33(11):2621–2632
68. Barrientos RM, Frank MG, Crysdale NY, Chapman TR, Ahrendsen JT, Day HEW et al (2011) Little exercise, big effects: reversing aging and infection-induced memory deficits, and underlying processes. J Neurosci 31(32):11578–11586
69. Byun K, Hyodo K, Suwabe K, Ochi G, Sakairi Y, Kato M et al (2014) Positive effect of acute mild exercise on executive function via arousal-related prefrontal activations: an fNIRS study. NeuroImage 98:336–345
70. Cassilhas RC, Viana VA, Grassmann V, Santos RT, Santos RF, Tufik S et al (2007) The impact of resistance exercise on the cognitive function of the elderly. Med Sci Sports Exerc 39(8):1401–1407
71. Christensen K, Doblhammer G, Rau R, Vaupel JW (2009) Ageing populations: the challenges ahead. Lancet 374(9696):1196–1208
72. Salzmann A, James SN, Williams DM, Richards M, Cadar D, Schott JM et al (2021) Investigating the relationship between IGF-I, IGF-II, and IGFBP-3 concentrations and later-life cognition and brain volume. J Clin Endocrinol Metab 106(6):1617–1629
73. Cohen E, Paulsson JF, Blinder P, Burstyn-Cohen T, Du D, Estepa G et al (2009) Reduced IGF-1 signaling delays age-associated proteotoxicity in mice. Cell 139(6):1157–1169
74. Gontier G, George C, Chaker Z, Holzenberger M, Aid S (2015) Blocking IGF signaling in adult neurons alleviates Alzheimer's disease pathology through amyloid-beta clearance. J Neurosci 35(33):11500–11513
75. Arai Y, Hirose N, Yamamura K, Shimizu K, Takayama M, Ebihara Y et al (2001) Serum insulin-like growth factor-1 in centenarians: implications of IGF-1 as a rapid turnover protein. J Gerontol A Biol Sci Med Sci 56(2):M79–M82
76. Bellar D, Glickman EL, Juvancic-Heltzel J, Gunstad J (2011) Serum IGF-1 is associated with working memory, executive function and selective attention in a sample of healthy, fit older adults. Neuroscience 178:133–137
77. Doi T, Shimada H, Makizako H, Tsutsumimoto K, Hotta R, Nakakubo S et al (2014) Association of insulin-like growth factor-1 with mild cognitive impairment and slow gait speed. Neurobiol Aging 36:942–947
78. Tumati S, Burger H, Martens S, van der Schouw YT, Aleman A (2016) Association between cognition and serum insulin-like growth Factor-1 in Middle-Aged & Older men: an 8 year follow-up study. PLoS One 11(4):e0154450
79. Frater J, Lie D, Bartlett P, McGrath JJ (2018) Insulin-like Growth Factor 1 (IGF-1) as a marker of cognitive decline in normal ageing: a review. Ageing Res Rev 42:14–27
80. Elmlinger MW, Kühnel W, Weber MM, Ranke MB (2004) Reference ranges for two automated chemiluminescent assays for serum insulin-like growth factor I (IGF-I) and IGF-binding protein 3 (IGFBP-3). Clin Chem Lab Med 42(6):654–664
81. Tsai CL, Ukropec J, Ukropcova B, Pai MC (2018) An acute bout of aerobic or strength exercise specifically modifies circulating exerkine levels and neurocognitive functions in elderly individuals with mild cognitive impairment. Neuroimage Clin 17:272–284
82. Tsai CL, Wang CH, Pan CY, Chen FC (2015) The effects of long-term resistance exercise on the relationship between neurocognitive performance and GH, IGF-1, and homocysteine levels in the elderly. Front Behav Neurosci 9:23
83. Arazi H, Babaei P, Moghimi M, Asadi A (2021) Acute effects of strength and endurance exercise on serum BDNF and IGF-1 levels in older men. BMC Geriatr 21(1):50
84. Chen WG, Schloesser D, Arensdorf AM, Simmons JM, Cui C, Valentino R et al (2021) The emerging science of Interoception: sensing, integrating, interpreting, and regulating signals within the self. Trends Neurosci 44(1):3–16
85. Prescott SL, Liberles SD (2022) Internal senses of the vagus nerve. Neuron 110(4):579–599

86. Flavell SW, Gogolla N, Lovett-Barron M, Zelikowsky M (2022) The emergence and influence of internal states. Neuron 110(16):2545–2570
87. Woods SC, D'Alessio DA (2008) Central control of body weight and appetite. J Clin Endocrinol Metab 93(11 Suppl 1):S37–S50
88. Nindl BC, Alemany JA, Kellogg MD, Rood J, Allison SA, Young AJ et al (2007) Utility of circulating IGF-I as a biomarker for assessing body composition changes in men during periods of high physical activity superimposed upon energy and sleep restriction. J Appl Physiol (1985) 103(1):340–346
89. Berg U, Bang P (2004) Exercise and circulating insulin-like growth factor I. Horm Res 62(Suppl 1):50–58
90. Kringelbach ML, Berridge KC (2009) Towards a functional neuroanatomy of pleasure and happiness. Trends Cogn Sci 13(11):479–487
91. Zegarra-Valdivia JA, Santi A, Fernandez de Sevilla ME, Nunez A, Torres AI (2019) Serum insulin-like growth factor I deficiency associates to Alzheimer's disease co-morbidities. J Alzheimer's Dis 69(4):979–987
92. Garcia-Magro N, Zegarra-Valdivia JA, Troyas-Martinez S, Torres-Aleman I, Nunez A (2022) Response facilitation induced by insulin-like growth factor-I in the primary somatosensory cortex of mice was reduced in aging. Cells 11(4):717
93. Lockie SH, Andrews ZB (2013) The hormonal signature of energy deficit: increasing the value of food reward. Mol Metab 2(4):329–336
94. Padilla SL, Qiu J, Soden ME, Sanz E, Nestor CC, Barker FD et al (2016) Agouti-related peptide neural circuits mediate adaptive behaviors in the starved state. Nat Neurosci 19(5):734–741
95. Fernandes MF, Matthys D, Hryhorczuk C, Sharma S, Mogra S, Alquier T et al (2015) Leptin suppresses the rewarding effects of running via STAT3 signaling in dopamine neurons. Cell Metab 22(4):741–749
96. Hallal PC, Andersen LB, Bull FC, Guthold R, Haskell W, Ekelund U (2012) Global physical activity levels: surveillance progress, pitfalls, and prospects. Lancet 380(9838):247–257
97. Hoare E, Milton K, Foster C, Allender S (2016) The associations between sedentary behaviour and mental health among adolescents: a systematic review. Int J Behav Nutr Phys Act 13(1):108
98. Teychenne M, Costigan SA, Parker K (2015) The association between sedentary behaviour and risk of anxiety: a systematic review. BMC Public Health 15:513
99. Noriega-Prieto JA, Maglio LE, Dávila JC, Gutiérrez A, Alemán IT, de Sevilla DF (2024) Bidirectional modulation of synaptic transmission by insulin-like growth factor I. Front Cell Neurosci 18:1390663. bioRxiv. 2023:2023.12.29.573627
100. Nishijima T, Piriz J, Duflot S, Fernandez AM, Gaitan G, Gomez-Pinedo U et al (2010) Neuronal activity drives localized blood-brain-barrier transport of serum insulin-like growth factor-I into the CNS. Neuron 67(5):834–846
101. Burgdorf J, Kroes RA, Beinfeld MC, Panksepp J, Moskal JR (2010) Uncovering the molecular basis of positive affect using rough-and-tumble play in rats: a role for insulin-like growth factor I. Neuroscience 126:769–777

Exploring the Impact of High-Intensity Interval Training on Cognitive Functions—Muscle and Brain Interaction

Sylwester Kujach, Robert Antoni Olek, and Radosław Laskowski

Abstract Recent research highlights the relationship between physical activity and cognitive functions. It has been shown that aerobic and resistance exercises, including a wide range of intensity and duration, can evoke a positive impact on cognitive functions and mental health in various age groups. Also, high-intensity interval training (HIT) has been recognized as an exercise modality inducing desired adaptive changes at the level of physical performance (muscle) as well as cognitive functions (brain). Previous research has also shown HIT to be an effective strategy due to its minimal time commitment and significant multiple health benefits. The mechanism behind the cognitive function facilitation as a result of acute and chronic HIT may involve the induction of neurotransmitters, as well as the synthesis of neuroprotective factors and increased activation of brain areas critical for cognitive functioning. Moreover, HIT also causes robustly increased lactate production, recently identified as the "first myokine" modulating cerebral metabolism. Additionally, HIT may initially disrupt the redox balance where the moderate formation of reactive oxygen species (ROS) may act as a signaling mechanism, also improving cognitive functions. Although research supports the potential of HIT to improve cognitive function, challenges remain due to differences in exercise structure, duration, and

S. Kujach (✉)
Department of Physiology, Faculty of Physical Education, Gdansk University of Physical Education and Sport, Gdańsk, Poland

Department of Physiology, Faculty of Medicine, Medical University of Gdańsk, Gdańsk, Poland
e-mail: sylwester.kujach@awf.gda.pl

R. A. Olek
Department of Athletics, Strength and Conditioning, Poznań University of Physical Education, Poznań, Poland

R. Laskowski
Department of Physiology, Faculty of Physical Education, Gdansk University of Physical Education and Sport, Gdańsk, Poland

H. Soya (ed.), *Exercise Brain Stimulation for Cognitive Function and Mental Health*, Advances in Neurobiology 44, https://doi.org/10.1007/978-981-95-0066-6_13

intensity of HIT protocols as well as cognitive domains and cognitive testing timing that make it difficult to draw firm conclusions.

To summarize, despite many variables that may influence differences in adaptive changes, existing research highlights the potential health benefits of HIT, also suggesting its effectiveness in enhancing human cognitive functions.

Keywords Cognitive function · High-intensity interval exercise · Neuroplasticity · Neurogenesis

1 Introduction

Physical activity is widely recognized as an effective and inexpensive strategy for preventing and treating various diseases [1]. Epidemiological studies have demonstrated an inverse association between peak oxygen uptake (VO_{2peak}), an indicator of cardio-respiratory fitness, and conditions such as cardiovascular and coronary heart disease, cancer, and overall mortality [1, 2]. It has also been indicated that reduced physical capacity in the elderly is linked to sarcopenia and a decline in physical performance, which restricts the diffusion of oxygen. This leads to increased metabolic and oxidative stresses in skeletal muscles, resulting in muscle fiber loss and worsened exercise performance [4]. Given that aerobic capacity and muscle fitness are crucial markers of health [5], strategies aimed at preventing their decline could mitigate muscle loss and associated functional decline, particularly among older adults. Nevertheless, regular physical activity also provides a wide range of physical health benefits for young individuals, including improvements in body composition, physical capacity, and various health-related factors like blood pressure, insulin resistance, and lipid profile [6].

Furthermore, a growing body of research suggests a positive link between physical activity and cognitive function across the lifespan [6, 7]. Studies have indicated that acute, regular moderate-intensity aerobic exercise as well as resistance exercise have a beneficial impact on cognition and other mental health outcomes such as depression, anxiety, and/or executive functions across different age groups [9–14]. These scientific findings are particularly relevant in light of the significant health threats posed by cognitive decline and dementia, which can affect various aspects of human thinking, decision-making, language use, and memory [15]. It is estimated that the incidence of dementia will nearly triple by 2050 due to the increasing older adult population, and by 2060, healthcare spending on dementia is expected to surpass that of all other conditions [16, 17]. As current curative treatments do not appear to be fully effective or readily available, the main focus should be on prevention, with an emphasis on promoting physical activity as modifiable risk factor.

Despite understanding the positive health impact of physical activity, most adults fail to meet the minimum physical activity guidelines due to a lack of time or, paradoxically, health status [18, 19]. Therefore, there is still a need for strategies that can achieve the desired changes while requiring less time commitment.

High-intensity interval training (HIT) is a popular training strategy among coaches, physiotherapists, and exercisers as it can produce positive health effects [3,20, 21]. Considering the significantly lower training volume (especially in comparison to traditional endurance training), HIT seems to be an excellent method for inducing rapid adaptive changes with less time commitment [22].

HIT involves high-intensity work periods separated by passive rest or low-intensity exercises [23]. The forgotten principle of interval training is that subsequent exercises should be performed without fully eliminating fatigue from the previous effort. The modifiable components of interval training include the number and type of exercise bouts, type of recovery periods, exercise intensity, exercise and recovery time, and duration of the intervention [24]. This multitude of elements allows the HIT to be adjusted to the participant's abilities and needs, but it also generates difficulties in assessing the effectiveness of HIT. Over the past decade, we have observed a significant rise in HIT-related publications [25]. Study shows various protocols for interval training, with HIT and sprint interval training (SIT) being the most commonly described. HIT involves "near maximal" exercises at an intensity that evokes 80–95% of maximal heart rate (HRmax), while SIT includes exercises performed at intensities eliciting maximal oxygen uptake or peak ($VO_{2max}/_{peak}$) and "all-out" or "supramaximal" efforts. It is worth mentioning that the definition of high-intensity exercise (HIE) varies among studies. HIT has been shown to induce adaptive changes similar to those seen with traditional high-volume endurance training and thus is often considered as a viable alternative for promoting pro-health effects [22, 26–29] (Fig. 1). Even short-term interventions, such as six sessions of 4–6 30-second all-out repetitions over 2 weeks, can increase maximal activity of mitochondrial enzymes [30, 31]. HIT has also been shown to increase muscle protein content related to the transport and oxidation of glucose and fatty acids, while decreasing the supply of energy for muscle work in a non-oxidative manner [32]. Observed post-HIT adaptations have shown rapid changes in muscle remodeling by recruiting type II muscle fibers, where these fibers demonstrate higher recruit thresholds and are entirely active during HIT. Also, the molecular changes are mainly related to the activation of several kinases and phosphatases involved in signal transduction, including the AMP-activated protein kinase, the calcium/calmodulin-dependent protein kinase, and the mitogen-activated protein kinase cascade. These signaling pathways play important role in the induction of co-activators involved in mitochondrial biogenesis and metabolism, including the expression of peroxisome proliferator-activated receptor co-activator-1 (PGC-1a).

However, the safety of HIT is still a concern when it comes to its broader use in clinical settings. This is particularly true in terms of its impact on vascular health and the rate of injuries. There is a perception that HIE could increase the risks of cardiovascular issues, especially among sedentary individuals. There are also concerns about the potential effects of HIT on blood pressure and vasoconstriction, which could raise the risk of cerebral injury, stroke, or blood-brain barrier disruption [25]. Furthermore, some studies have reported that HIT can be uncomfortable and cause displeasure resulting in lower enjoyment [33, 34] and reduced likelihood of consistent participation [35, 36]. However, these negative effects and feelings seem to be primarily associated with more intense modalities, e.g., Sprint Interval Trainning (SIT) performed by inactive individuals. They can potentially be minimized by modifying or adapting the interval exercise protocols [37, 38]. Nevertheless, it is worth mentioning that HIT-based interventions may increase engagement in a regular exercise programs because they are perceived as more enjoyable compared to moderate-intensity endurance training [39] (Fig. 1).

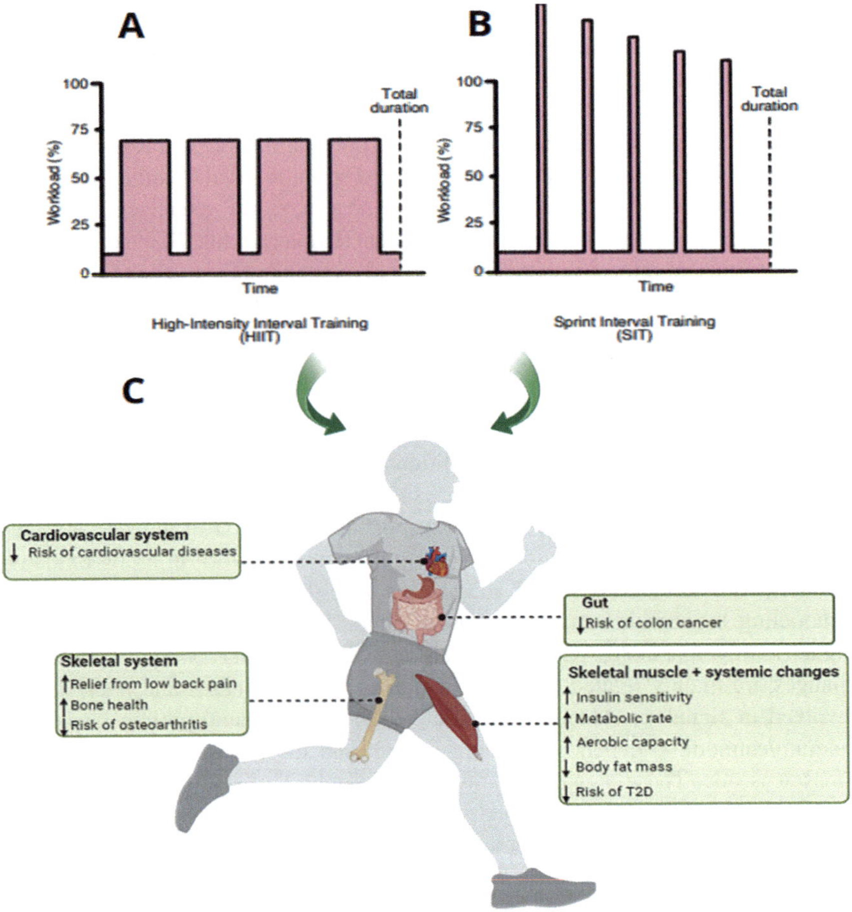

Fig. 1 HIT and SIT modalities and pro-health benefits

2 High-Intensity Interval Training and Brain Functions

Physical exercise has a positive impact on improving the functional and structural changes that contribute to cognitive function improvement [9, 10, 12, 14]. Generally, exercise increases arousal by promoting production and release of neurotransmitters, hormones, or trophic factors [37, 38, 40, 41]. This increased arousal can consistently enhance cerebral blood flow, which is dependent on the intensity of the exercise [37, 42]. The enriched blood circulation, containing oxygen, glucose, and peripherally synthesized molecules, affects the metabolism of brain tissue, including regions critical for cognitive functioning [37, 38, 43, 44]. Long-term physical activity promotes synaptic plasticity and the formation of new neurons, known as neurogenesis [45, 46]. As a result, cognitive function is improved [44]. These processes are regulated by various mechanisms affected by both acute and chronic exercise.

Furthermore, the contractions of skeletal muscles during exercise lead to the release of proteins called exerkines (exercise: cyto-kine, myo-kine) [38, 44, 47]. These proteins can act in an autocrine, paracrine, and endocrine manner. Among the many molecules released during exercise, some have pro-inflammatory potential, while others have anti-inflammatory effects. These opposing properties also have an impact on synaptic plasticity and neurogenesis. Additionally, several neurochemicals, including brain-derived neurotrophic factor (BDNF), insulin-like growth factor-1 (IGF-1), and vascular endothelial growth factor (VEGF), play key roles in modulating the effects of exercise on the brain and cognition [48].

An acute exercise can increase the levels of BDNF, IGF-1, and VEGF [38, 49, 50], which can contribute to improvements in mood, tissue protection, and cognitive function [51–53]. However, these molecules also play important role in angiogenesis, synaptogenesis, and neurogenesis following chronic exercise [54, 55]. The type, duration, and intensity of exercise can also influence the synthesis and release of these molecules. While the functional and structural adaptations are generally associated with physical activity, they may also be observed in response to HIT.

2.1 Arousal-Induced Neuromodulatory Response Effect of HIT

It has been demonstrated that exercise can increase arousal levels depending on the intensity of exercise [41, 56]. The exercise-induced arousal has an impact on both physiological and psychological states of being awake, which is thought to involve the activation of the reticular-activating system (RAS), controlling ascending projections to the prefrontal cortex [57–59]. It has been shown that RAS as a neuronal network connected with several nuclei in the brain steam can be modulated through HIT [37]. Based on animal studies, it is speculated that acute HIT has the potential to stimulate monoamine systems and increase the release of neurotransmitters such as dopamine (DA) and noradrenaline (NA) into brain regions involved in cognitive functioning [41, 60]. Additionally, the release of NA by HIT may enhance neural activation in the prefrontal cortex, as the noradrenergic system is located in the locus coeruleus (LC). In vivo microdialysis studies suggest that acute exercise stimulates NA synthesis and release, activating brain regions including the dorsolateral prefrontal cortex (DLPFC), anterior cingulate cortex (ACC), and hippocampus, all of which play a role in cognitive processing [61, 62].

Additionally, the dopaminergic system originates in the ventral tegmental area (VTA) and has also neural ascending projections to several brain regions such as the prefrontal cortex, amygdala, and hippocampus. DA has a modulatory effect on executive cognitive information processing, decision-making, learning, and motor control [62]. Interestingly, DA and NA work together to improve the signal-to-noise ratio (see [41]).

It is important to acknowledge that HIE has also been linked to a robust increase in arousal, which may potentially result in a decline in cognitive performance. Prolonged exposure to HIE has been shown to cause impairments in cognitive control and have neurotoxic effects on the human brain [34].

To summarize, HIT can increase the synthesis and release of neurotransmitters such as NA and DA, leading to cognitive improvement. However, excessive neurotransmitter synthesis may also lead to cognitive impairment.

2.2 *HIT and Hemodynamic Response*

Despite the growing body of research on hemodynamic changes triggered by intense exercise, there are still few direct conclusions derived from experiments involving HIT protocols. Preliminary evidence, limited to the systemic circulation, suggests that repeated exposure to mechanical forces linked with acute exertional hyperemia stimulates complex alterations in pattern of pressure, strain, and shear stress that could induce functional and structural adaptation of the vascular wall through endothelial cell mechanotransduction. Consequently, this adaptation could enhance processes such as neurogenesis and neuroplasticity, potentially facilitating cognitive performance [25, 63, 64]. Shear stress refers to the frictional force that is exerted on the walls of blood vessels as a result of blood flow. This force can also stimulate the release of signaling molecules, such as nitric oxide (NO), which promote vasodilation and improve neuronal function [63, 65, 66].

Also, exercise increases blood flow to the brain, and the extent of this increase is related to the intensity of exercise. Cerebral blood flow (CBF) is regulated through complex interactions involving neural activity, metabolism, sympathetic nervous system activity, blood pressure, and cardiac output [67, 68]. However, during HIE, CBF is decreased, likely due to cerebral vasoconstriction induced by hypocapnia resulting from hyperventilation [67]. It is important to note that the decrease in CBF during intense exercise may vary across different areas of the brain [63, 69]. Although there are reports suggesting that reduced CBF during intense exercise may directly impact cognitive functions, research findings are not conclusive [70, 71]. Considering CBF, it is important to differentiate between HIT and high-intensity continuous exercise. The intermittent nature of HIT could significantly alter hemodynamic responses [25]. Recently, prof. Soya's group tested the hypothesis that a single session of HIT improves executive information processing, specifically selective attention and inhibitory control, in sedentary young adults. Using functional near-infrared spectroscopy (fNIRS), enhanced executive performance by increasing cortical activation in the lateral prefrontal cortex after a single session of HIT has been demonstrated.

2.3 Neurobiological Response to HIT

Multiple studies have indicated that exercise can enhance human cognition by boosting the synthesis and release of neuroprotective factors, which are crucial components modulating neural plasticity and brain energy metabolism [40]. Brain-derived neurotrophic factor (BDNF), insulin-like growth factor-1 (IGF-1), and vascular endothelial growth factor (VEGF) are recognized as main proteins mediating the effects of exercise on the brain and cognitive function [48]. Physical activity, even in a single session, can elevate BDNF levels, contributing to the alleviation of mood disorders, promoting tissue protection and regeneration, and ultimately enhancing cognitive performance in humans [51, 52]. Recent studies have highlighted that SIT potentially triggers muscle remodeling, which might consequently promote BDNF production [43, 72, 73]. Both HIE and strength training stimulate the secretion of IGF-1 [74, 75]. Additionally, IGF-1 is crucial for converting pro-BDNF into BDNF within the central nervous system, capable of crossing the blood-brain barrier, influencing neurogenesis, and synaptic plasticity [76, 77]. Furthermore, IGF-1 mediates exercise-induced angiogenesis, which augments the synthesis of BDNF and VEGF [76, 78]. Circulating VEGF not only promotes neurogenesis and synaptic plasticity but also stimulates neural stem cell proliferation and differentiation. Moreover, it enhances the production of VEGF, BDNF, and IGF-1 in central endothelial cells and astrocytes [78, 79].

It has been revealed that exercise activates PGC-1a in skeletal muscle cells, which could trigger the expression of FNDC5 gene, a positive regulator of BDNF levels in the brain, particularly in the hippocampus [73]. Notably, PGC1-a elevation has been observed following HIT [72]. Additionally, Ferris et al. (2007) have noted that vigorous exercise results in higher increases of peripheral BDNF concentration compared to low-intensity exercise [49]. It has also been shown that HIT-induced skeletal muscle contractions might activate this biochemical pathway, leading to elevated BDNF levels in the brain [80]. SaucedoMarquez et al. (2015) indicated that the SIT protocol was more effective at raising BDNF levels compared to traditional, moderate-intensity continuous exercise [80]. Furthermore, studies have found that BDNF could influence presynaptic neurotransmitter release and initiate excitatory postsynaptic currents via TrkB receptors, directly affecting neuronal depolarization

[81, 82]. Additionally, numerous animal and human studies have indicated significant positive associations between serum BDNF and cortical BDNF, along with heightened BDNF levels and enhanced cognitive performance post-acute exercise [38, 50, 83, 84]. A recent meta-analysis indicated single bout of HIT as the most effective training method for increasing peripheral BDNF concentration in adults. Additionally, chronic HIT intervention may also effectively increase BDNF levels [85]. These findings suggest that interval exercise might stimulate BDNF, leading to cognitive enhancement and changes in neural activation in the prefrontal cortex [37, 86].

While some studies have indicated a rise in serum IGF-1 following HIE [87], others have reported no significant increase in peripheral IGF-1 concentration after exercise [88, 89]. We observed an increase in peripheral IGF-1 concentration following acute SIT [38]. IGF-1 is primarily produced in the liver but is also secreted in the brain, skeletal muscles, and bones [90]. Both peripheral and cerebral sources contribute to post-exercise IGF-1 concentration in some extent [90]. Moreover, acute exercise triggers the expression and release of liver-derived IGF-1, leading to increased brain uptake [91]. Similarly to IGF-1, we also observed a significant increase in VEGF concentrations. The effects of acute exercise on peripheral VEGF have generated conflicting results, with some researchers reporting an increase following acute exercise [92, 93], while others found no significant differences in VEGF concentration after exercise [94]. It is worth noting that VEGF levels in the hippocampus decline with age, and raising in post-exercise VEGF levels could potentially be a therapeutic approach to enhance brain function [95]. Circulating IGF-1 and VEGF in the periphery can influence neurogenesis and synaptic plasticity by stimulating the proliferation and differentiation of neural stem cells [79, 96].

The responses of BDNF, IGF-1, and VEGF seem to be closely linked to exercise intensity, and variations in study designs regarding exercise type, intensity, duration, and timing of post-exercise blood sampling could help clarify some of the conflicting findings [50, 93, 97].

2.3.1 Lactate Involvement

Another crucial molecular mechanism that could explain the cognitive advantages of HIT is the heightened production of metabolic factors like lactate. These metabolites have demonstrated neuroprotective effects and play a role in boosting energy production in the brain.

Recently, SIT has been linked to elevated glucose metabolism and lactate (LA) production, both crucial energy sources for the human brain [38, 98–100]. Among various HIT protocols, SIT most significantly increases blood LA levels [39, 101]. Research by Glenn et al. (2015) highlighted that peripherally produced LA serves as an energy source for the brain, especially after traumatic brain injury [102]. While the brain typically relies on glucose at rest, during HIE, there is a decrease in cerebral glucose consumption, accompanied by an increase in blood LA, which affects cerebral uptake [103]. Neuronal activation demands more energy [104], and the elevated peripheral LA concentration resulting from HIE promotes the supply of LA as an energy substrate to meet neuronal energy requirements [105, 106].

The increased blood LA resulting from HIT or SIT might play a pivotal role in enhancing cognitive function, along with the aforementioned elevation in neuroprotective protein levels. Recent research by Hashimoto et al. (2018) demonstrated that after SIE, both arterial LA and brain LA uptake escalate, suggesting that systemic LA influences brain LA uptake and subsequently affects post-exercise executive function facilitation [100, 107]. Furthermore, our study identified a positive correlation between blood LA and neuro-supporting protein concentrations. Ferris et al. (2007) also observed a correlation between increased blood LA levels after exercise and elevated peripheral BDNF levels [49]. Additionally, LA has been identified as a significant factor in stimulating BDNF synthesis [108]. El Hayek et al. (2019) revealed that LA influences neuron redox status by altering the NADH/NAD+ ratio, activating SIRT1, and ultimately engaging the hippocampal PGC1-a/FNDC5 pathway, leading to BDNF expression [73, 108, 109]. Moreover, LA released from contracting muscles contributes to cerebral angiogenesis by activating the LA hydroxycarboxylic acid receptor 1 (HCAR1), a crucial regulator of VEGF [110]. Furthermore, LA induces IGF-1 mRNA expression by stimulating the somatotropic axis [111]. Given the substantial increase in blood LA following SIE in our study, it is plausible that the enhancement in cognitive functions could be attributed to accelerated cerebral LA metabolism and the induction of neuroprotective proteins [38].

Different HIT protocols have varying effects on blood LA levels [101, 112]. While SIE is characterized by high LA production, longer protocols might rely more on aerobic metabolism, resulting in lower LA production [21]. Hence, different interval training protocols could modulate central nervous system activation and subsequently influence cognitive performance differently.

2.3.2 Reactive Oxygen Species (ROS)

Exercise disrupts redox homeostasis through the formation of reactive oxygen species (ROS) [113]. Initially, it was suggested that ROS production induced by exercise causes damage to the structural tissues [114]. Recent evidence has shown that at physiological concentrations, ROS act as crucial signaling molecules that mediate adaptation [115]. It has been suggested that enhanced production of ROS during exercise can induce BDNF synthesis in the brain [116]. Increased mitochondrial activity in neurons during exercise leads to excessive ROS production. As a result, ROS enhance the activity of CRE-binding protein, activate CREB and BDNF transcription [116]. Although CBF decreases during HIE [67], which reduces oxygen supply, the transient nature of interval exercise can accelerate ROS formation while blood flow is restored [117]. In addition, exercise also increases Ca^{2+} levels in neurons. This ion, in combination with calpain and xanthine oxidase, induces higher ROS production in the brain [118].

In fact, a single sprint anaerobic exercise lasting 30 s has induced oxidative stress and changes in plasma non-enzymatic antioxidant status [119, 120]. Moreover, 6 weeks of HIT in animal model elevated hydrogen peroxide (H_2O_2) in the brain, which correlated with elevated BDNF and glial cell line-derived neurotrophic factor (GDNF) levels [117]. The excess generation of ROS during exercise can be partially regulated by PGC-1α, which activates the expression of proteins involved in the regulation of oxidative stress. HIT enhances the antioxidant defenses in brain tissues [121–124]. Interestingly, in physiological state, these changes are not associated with cognitive modifications [122, 125]. By contrast, in pathological conditions, HIT alleviates cognitive decline parallel to alterations in brain oxidant/antioxidant status [123, 124, 126]. In a transgenic mice model of Alzheimer's disease (AD), 12 weeks of HIT elevated hippocampal activities of superoxide dismutase (SOD) and catalase (CAT), while ROS and H_2O_2 levels decreased [123]. Moreover, HIT improved learning and memory functions [123]. Similarly, in a rat AD model, amyloid-beta injection impaired learning and memory performances, reduced exploratory/locomotor activities, and enhanced anxiety-like behavior. These behavioral deficits accompanied hippocampal oxidative stress and neuronal loss. HIT continued for eight consecutive weeks improved anxiety-like behavior, attenuated total oxidant status, and prevented the neuronal loss in the hippocampal areas [124]. Additionally, in female rats with post-traumatic stress disorder, memory decline was protected by HIT via decreasing oxidative stress, anxiety levels and by improving antioxidant capacity as a protective system for neuronal damage [126].

The brain is susceptible to oxidative stress because it harnesses chemically diverse reactive species to perform heterogeneous signaling functions [127]. Oxidative stress has been linked to age-related cognitive decline and neurodegenerative diseases. However, it is crucial to note that exercise-induced ROS generation rarely reaches levels detrimental to health [128]. What is more, exercise-induced increases in production of ROS may protect against ROS-associated diseases. Many studies have demonstrated that regular exercise up-regulates the antioxidant system and is able to stimulate the oxidative damage repair system [128] (Fig. 2). Overall, extensive data support the health-promoting effects of regular exercise, regardless of exercise intensity (Fig. 2).

3 HIT and Cognition: Acute and Chronic Effects of HIT on Cognitive Functions

Acute high-intensity aerobic exercise affects the brain's metabolic, circulatory, and neurohormonal systems [41, 60, 67, 129].

The Inverted-U theory predicts that HIE increases arousal above the optimal level leading to a transient cognitive performance impairment. However, the evidence from current research does not firmly support this theory [130–133].

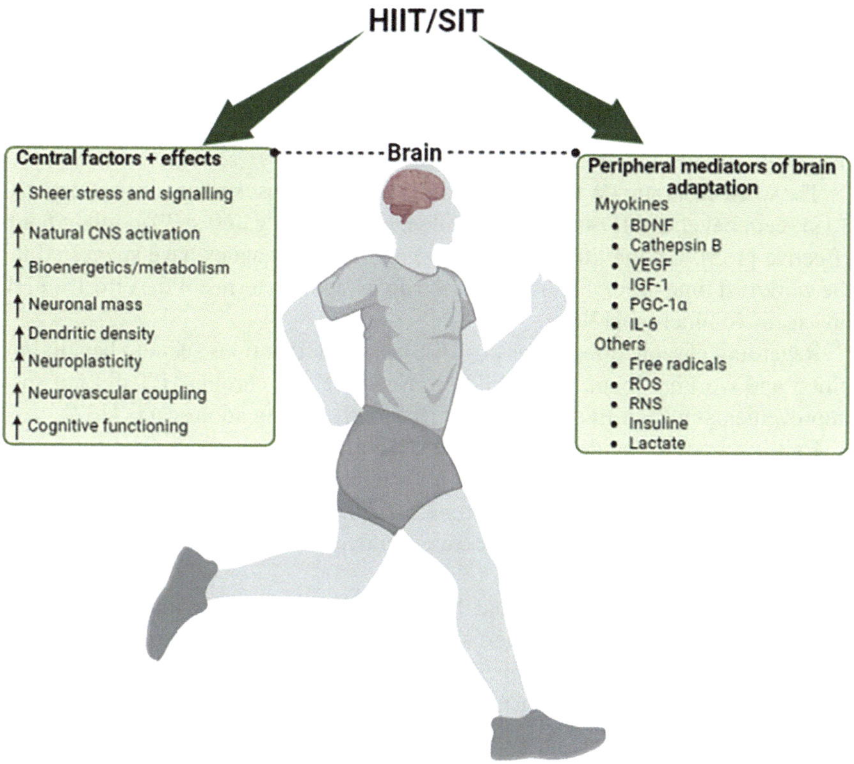

Fig. 2 Neurobiological response to HIT

Additionally, the hypofrontality hypothesis speculates that extensive activation of motor and sensory systems during HIE could attenuate higher-order cognitive functions of the prefrontal cortex (PFC) as the brain has limited metabolic resources [134]. Also, in the context of HIE and cognitive performance interactions, the interoceptive model has been introduced [131]. This model comprises motivation, perceived effort costs, and perceived availability of resources, jointly with regional activations and neurotransmitter releases in the brain.

Despite these theoretical frameworks, the precise neurobiological and psychological mechanisms linking acute HIE to cognitive performance remain unclear and poorly understood. It is worth noting, however, that those hypotheses were proposed based mostly on high-intensity continuous exercise, which are different from intermittent HIT structures, suggesting that HIT might differently affect cognitive functions. Interestingly, recent studies have demonstrated that acute HIT tends to have a positive effect on cognitive functions regardless of its intensity and or modalities, which is not entirely consistent with previously mentioned hypotheses.

3.1 Children and Adolescent

Research shows acute HIT can positively impact inhibition in children and adolescents, sustaining benefits for up to an hour post-exercise, but its effect on cognitive flexibility remains uncertain.

The work-rest ratio (WRR) of HIT could influence this: smaller ratios (e.g., 30 s: 30 s) seem beneficial for inhibition, while larger ones (e.g., 60 s: 30 s) might not be effective [135]. On the other hand, Jing-Yi Ai and colleagues have suggested that the work/rest time ratio of HIT is not a sensitive factor in terms of the effects of HIT on executive function [136].

Regarding chronic effects, various HIT-based interventions seem to benefit inhibition and working memory in children. Nevertheless, chronic HIT did not show improvements in cognitive flexibility performance among adolescents [137].

Existing body of evidence evaluating HIT effect on cognitive function consistently supports the positive impacts of both acute and chronic interventions among children and adolescents [8, 14]. Moreover, findings from the present study suggest that HIT has positive effects on a core executive function such as working memory [138].

Notably, it has been mentioned by Hsieh and colleagues that the similarity between the natural activity patterns of youths and the HIT structure, associated with its time-efficient nature, suggests the feasibility of integrating HIT-based physical activity into school routines [14]. This approach could potentially enhance brain health, presenting a viable option for school settings [14].

3.2 Adults

We conclude from two recent reviews that both acute as well as chronic HIT interventions may be beneficial in improving cognitive performance in adults [14, 71, 130].

The authors of these reports point out the challenges associated with drawing clear conclusions because the analyzed experimental studies are characterized by significant discrepancies, i.e., the structure of exercise, its duration, and intensity.

Attention is also paid to different effects resulting from sexual dimorphism, age, and health status of the participants, the cognitive domain, and finally, the timing in which the cognitive tests were conducted.

However, despite the above-mentioned challenges, current research indicates the benefits of HIT/SIT on cognitive functions among adults. In the review by Hsieh et al., focusing on cognitive domains such as inhibition, memory, and cognitive flexibility, positive effects were observed in over 50% of the analyzed interventions [14]. It should be mentioned that they included both single and chronic HIT interventions. However, in a very recent review that assessed a wide range of cognitive domains including attention, inhibition, memory, information processing speed, reaction time, cognitive flexibility, intelligence, and learning, improvements were observed in over 30% of the analyzed HIT/SIT interventions. The authors indicate that more pronounced effects have been observed in acute than chronic HIT/SIT programs [139].

4 Conclusion

Research strongly supports the positive impact of various forms of exercise on cognitive function and mental health. High-intensity interval training (HIT) specifically shows promise for quick and effective cognitive enhancement by promoting neurotransmitter release and increasing neuro-supportive molecules such as BDNF. This type of training also triggers the production of metabolites that may modulate brain metabolism.

Although research supports the potential of HIT to improve cognitive function, challenges remain due to differences in exercise structure, duration, and intensity of HIT protocols as well as cognitive domains and cognitive testing timing that make it difficult to draw firm conclusions.

Therefore, further research should focus on chronic interventions targeting older adults, assessing various cognitive domains (working memory, cognitive flexibility). This necessitates stronger research designs, exploration of HIT dosage/modality effects, consideration of individual differences, and understanding of neurobiological mechanisms linking HIT to cognition.

References

1. Miller KR, McClave SA, Jampolis MB, Hurt RT, Krueger K, Landes S, Collier B (2016) The health benefits of exercise and physical activity. Curr Nutr Rep 5:204–212
2. Knowles A-M, Herbert P, Easton C, Sculthorpe N, Grace FM (2015) Impact of low-volume, high-intensity interval training on maximal aerobic capacity, health-related quality of life and motivation to exercise in ageing men. Age (Dordr) 37:25
3. Wu Z-J, Wang Z-Y, Gao H-E, Zhou X-F, Li F-H (2021) Impact of high-intensity interval training on cardiorespiratory fitness, body composition, physical fitness, and metabolic parameters in older adults: a meta-analysis of randomized controlled trials. Exp Gerontol 150:111345
4. Kaminsky LA, Arena R, Beckie TM et al (2013) The importance of cardiorespiratory fitness in the United States: the need for a national registry: a policy statement from the American Heart Association. Circulation 127:652–662
5. Lee D, Artero EG, Sui X, Blair SN (2010) Mortality trends in the general population: the importance of cardiorespiratory fitness. J Psychopharmacol 24:27–35
6. Janssen I, Le Blanc AG (2010) Systematic review of the health benefits of physical activity and fitness in school-aged children and youth. Int J Behav Nutr Phys Act 7:40
7. Hillman CH, Erickson KI, Kramer AF (2008) Be smart, exercise your heart: exercise effects on brain and cognition. Nat Rev Neurosci 9:58–65

8. Kao S-C, Wang C-H, Hillman CH (2020) Acute effects of aerobic exercise on response variability and neuroelectric indices during a serial n-back task. Brain Cogn 138:105508
9. Chang YK, Labban JD, Gapin JI, Etnier JL (2012) The effects of acute exercise on cognitive performance: a meta-analysis. Brain Res 1453:87–101
10. Guiney H, Machado L (2013) Benefits of regular aerobic exercise for executive functioning in healthy populations. Psychon Bull Rev 20:73–86
11. Li JW, O'Connor H, O'Dwyer N, Orr R (2017) The effect of acute and chronic exercise on cognitive function and academic performance in adolescents: a systematic review. J Sci Med Sport 20:841–848
12. McMorris T, Tomporowski P, Audiffren M (2009) Exercise and cognitive function. Wiley, Chichester
13. Wilke J (2020) Functional high-intensity exercise is more effective in acutely increasing working memory than aerobic walking: an exploratory randomized, controlled trial. Sci Rep 10:12335–12335
14. Hsieh S-S, Chueh T-Y, Huang C-J, Kao S-C, Hillman CH, Chang Y-K, Hung T-M (2021) Systematic review of the acute and chronic effects of high-intensity interval training on executive function across the lifespan. J Sports Sci 39:10–22
15. Bishop NA, Lu T, Yankner BA (2010) Neural mechanisms of ageing and cognitive decline. Nature 464:529–535
16. Prince M, Ali G-C, Guerchet M, Prina AM, Albanese E, Wu Y-T (2016) Recent global trends in the prevalence and incidence of dementia, and survival with dementia. Alz Res Therapy 8:23
17. Wimo A, Seeher K, Cataldi R et al (2023) The worldwide costs of dementia in 2019. Alzheimers Dement 19:2865–2873
18. Godin GR, Desharnais P, Valois P, Lepage J, Jobin BR (1994) Differences in perceived barriers to exercise between high and low intenders: observations among different populations. Am J Health Promot 8:279–285
19. Stutts WC (2002) Physical activity determinants in adults. Perceived benefits, barriers, and self efficacy. AAOHN J 50:499–507
20. Gibala MJ, Little JP (2010) Just HIT it! A time-efficient exercise strategy to improve muscle insulin sensitivity. J Physiol 588:3341–3342
21. Astorino TA, Sheard AC (2019) Does sex mediate the affective response to high intensity interval exercise? Physiol Behav 204:27–32
22. Gibala MJ (2007) High-intensity interval training: a time-efficient strategy for health promotion? Curr Sports Med Rep 6:211–213
23. Eddolls WTB, McNarry MA, Stratton G, Winn CON, Mackintosh KA (2017) High-intensity interval training interventions in children and adolescents: a systematic review. Sports Med 47:2363–2374
24. Billat LV (2001) Interval training for performance: a scientific and empirical practice. Special recommendations for middle- and long-distance running. Part I: aerobic interval training. Sports Med 31:13–31
25. Calverley TA, Ogoh S, Marley CJ et al (2020) HIITing the brain with exercise: mechanisms, consequences and practical recommendations. J Physiol 598:2513–2530
26. Nilsson BB, Hellesnes B, Westheim A, Risberg MA (2008) Group-based aerobic interval training in patients with chronic heart failure: Norwegian Ullevaal Model. Phys Ther 88:523–535
27. Tjonna AE, Lee SJ, Rognmo O et al (2008) Aerobic interval training versus continuous moderate exercise as a treatment for the metabolic syndrome: a pilot study. Circulation 118:346–354
28. Warburton DE, McKenzie DC, Haykowsky MJ, Taylor A, Shoemaker P, Ignaszewski AP, Chan SY (2005) Effectiveness of high-intensity interval training for the rehabilitation of patients with coronary artery disease. Am J Cardiol 95:1080–1084

29. Wisloff U, Stoylen A, Loennechen JP et al (2007) Superior cardiovascular effect of aerobic interval training versus moderate continuous training in heart failure patients: a randomized study. Circulation 115:3086–3094
30. Burgomaster KA, Hughes SC, Heigenhauser GJ, Bradwell SN, Gibala MJ (2005) Six sessions of sprint interval training increases muscle oxidative potential and cycle endurance capacity in humans. J Appl Physiol 98:1985–1990
31. Gibala MJ, Little JP, van Essen M, Wilkin GP, Burgomaster KA, Safdar A, Raha S, Tarnopolsky MA (2006) Short-term sprint interval versus traditional endurance training: similar initial adaptations in human skeletal muscle and exercise performance. J Physiol 575:901–911
32. Burgomaster KA, Howarth KR, Phillips SM, Rakobowchuk M, Macdonald MJ, McGee SL, Gibala MJ (2008) Similar metabolic adaptations during exercise after low volume sprint interval and traditional endurance training in humans. J Physiol 586:151–160
33. Decker ES, Ekkekakis P (2017) More efficient, perhaps, but at what price? Pleasure and enjoyment responses to high-intensity interval exercise in low-active women with obesity. Psychol Sport Exerc 28:1–10
34. Farias-Junior LF, Browne RAV, Freire YA, Oliveira-Dantas FF, Lemos TMAM, Galvão-Coelho NL, Hardcastle SJ, Okano AH, Aoki MS, Costa EC (2019) Psychological responses, muscle damage, inflammation, and delayed onset muscle soreness to high-intensity interval and moderate-intensity continuous exercise in overweight men. Physiol Behav 199:200–209
35. Hardcastle SJ, Ray H, Beale L, Hagger MS (2014) Why sprint interval training is inappropriate for a largely sedentary population. Front Psychol 5:1505
36. Biddle SJH, Batterham AM (2015) High-intensity interval exercise training for public health: a big HIT or shall we HIT it on the head? Int J Behav Nutr Phys Act 12:95
37. Kujach S, Byun K, Hyodo K, Suwabe K, Fukuie T, Laskowski R, Dan I, Soya H (2017) A transferable high-intensity intermittent exercise improves executive performance in association with dorsolateral prefrontal activation in young adults. NeuroImage 169:117–125
38. Kujach S, Olek RA, Byun K, Suwabe K, Sitek EJ, Ziemann E, Laskowski R, Soya H (2019) Acute Sprint interval exercise increases both cognitive functions and peripheral neurotrophic factors in humans: the possible involvement of lactate. Front Neurosci 13:1455–1455
39. Olney N, Wertz T, LaPorta Z, Mora A, Serbas J, Astorino TA (2018) Comparison of acute physiological and psychological responses between moderate-intensity continuous exercise and three regimes of High-intensity Interval Training. J Strength Cond Res 32:2130–2138
40. Gomez-Pinilla F, Hillman C (2013) The influence of exercise on cognitive abilities. Compr Physiol 3:403–428
41. McMorris T (2016) Developing the catecholamines hypothesis for the acute exercise-cognition interaction in humans: lessons from animal studies. Physiol Behav 165:291–299
42. Kashihara K, Maruyama T, Murota M, Nakahara Y (2009) Positive effects of acute and moderate physical exercise on cognitive function. J Physiol Anthropol 28:155–164
43. Phillips C, Baktir MA, Srivatsan M, Salehi A (2014) Neuroprotective effects of physical activity on the brain: a closer look at trophic factor signaling. Front Cell Neurosci 8:170–170
44. Kujach S, Chroboczek M, Jaworska J, Sawicka A, Smaruj M, Winklewski P, Laskowski R (2022) Judo training program improves brain and muscle function and elevates the peripheral BDNF concentration among the elderly. Sci Rep 12:13900
45. van Praag H (2009) Exercise and the brain: something to chew on. Trends Neurosci 32:283–290
46. Erickson KI, Voss MW, Prakash RS et al (2011) Exercise training increases size of hippocampus and improves memory. Proc Natl Acad Sci USA 108:3017–3022
47. Kim S, Choi J-Y, Moon S, Park D-H, Kwak H-B, Kang J-H (2019) Roles of myokines in exercise-induced improvement of neuropsychiatric function. Pflugers Arch 471:491–505
48. Cotman CW, Berchtold NC, Christie LA (2007) Exercise builds brain health: key roles of growth factor cascades and inflammation. Trends Neurosci 30:464–472

49. Ferris LT, Williams JS, Shen CL (2007) The effect of acute exercise on serum brain-derived neurotrophic factor levels and cognitive function. Med Sci Sports Exerc 39:728–734
50. Griffin EW, Mullally S, Foley C, Warmington SA, O'Mara SM, Kelly AM (2011) Aerobic exercise improves hippocampal function and increases BDNF in the serum of young adult males. Physiol Behav 104:934–941
51. Hansen CJ, Stevens LC, Coast JR (2001) Exercise duration and mood state: how much is enough to feel better? Health Psychol 20:267–275
52. Rojas Vega S, Abel T, Lindschulten R, Hollmann W, Bloch W, Struder HK (2008) Impact of exercise on neuroplasticity-related proteins in spinal cord injured humans. Neuroscience 153:1064–1070
53. Zoladz JA, Pilc A (2010) The effect of physical activity on the brain derived neurotrophic factor: from animal to human studies. J Physiol Pharmacol 61:533–541
54. Cotman CW, Berchtold NC (2002) Exercise: a behavioral intervention to enhance brain health and plasticity. Trends Neurosci 25:295–301
55. Voss MW, Erickson KI, Prakash RS et al (2013) Neurobiological markers of exercise-related brain plasticity in older adults. Brain Behav Immun 28:90–99
56. Yerkes RM, Dodson JD (1908) The relation of strength of stimulus to rapidity of habit formation. J Comp Neurol Psychol 18:459–482
57. Audiffren M, Tomporowski PD, Zagrodnik J (2008) Acute aerobic exercise and information processing: energizing motor processes during a choice reaction time task. Acta Psychol 129:410–419
58. Sanders AF (1983) Towards a model of stress and human performance. Acta Psychol 53:61–97
59. Garcia-Rill E (2015) Waking and the reticular activating system in health and disease. Academic, London
60. Meeusen R, Piacentini MF, De Meirleir K (2001) Brain microdialysis in exercise research. Sports Med 31:965–983
61. Aston-Jones G, Cohen JD (2005) An integrative theory of locus coeruleus-norepinephrine function: adaptive gain and optimal performance. Annu Rev Neurosci 28:403–450
62. Arnsten AFT (2011) Catecholamine influences on dorsolateral prefrontal cortical networks. Biol Psychiatry 69:e89–e99
63. Lucas SJ, Cotter JD, Brassard P, Bailey DM (2015) High-intensity interval exercise and cerebrovascular health: curiosity, cause, and consequence. J Cereb Blood Flow Metab 35:902–911
64. Green DJ, Hopman MTE, Padilla J, Laughlin MH, Thijssen DHJ (2017) Vascular adaptation to exercise in humans: role of hemodynamic stimuli. Physiol Rev 97:495–528
65. Tinken TM, Thijssen DHJ, Hopkins N, Black MA, Dawson EA, Minson CT, Newcomer SC, Laughlin MH, Cable NT, Green DJ (2009) Impact of shear rate modulation on vascular function in humans. Hypertension 54:278–285
66. Bolduc V, Thorin-Trescases N, Thorin E (2013) Endothelium-dependent control of cerebrovascular functions through age: exercise for healthy cerebrovascular aging. Am J Phys Heart Circ Phys 305:H620–H633
67. Ogoh S, Ainslie PN (2009) Cerebral blood flow during exercise: mechanisms of regulation. J Appl Physiol (1985) 107:1370–1380
68. Smith KJ, Ainslie PN (2017) Regulation of cerebral blood flow and metabolism during exercise. Exp Physiol 102:1356–1371
69. Sato K, Ogoh S, Hirasawa A, Oue A, Sadamoto T (2011) The distribution of blood flow in the carotid and vertebral arteries during dynamic exercise in humans. J Physiol 589:2847–2856
70. Komiyama T, Tanoue Y, Sudo M, Costello JT, Uehara Y, Higaki Y, Ando S (2020) Cognitive impairment during high-intensity exercise: influence of cerebral blood flow. Med Sci Sports Exerc 52:561–568
71. Sudo M, Costello JT, McMorris T, Ando S (2022) The effects of acute high-intensity aerobic exercise on cognitive performance: a structured narrative review. Front Behav Neurosci 16:957677

72. Little JP, Safdar A, Bishop D, Tarnopolsky MA, Gibala MJ (2011) An acute bout of high-intensity interval training increases the nuclear abundance of PGC-1alpha and activates mitochondrial biogenesis in human skeletal muscle. Am J Physiol Regul Integr Comp Physiol 300:R1303–R1310
73. Wrann CD, White JP, Salogiannnis J, Laznik-Bogoslavski D, Wu J, Ma D, Lin JD, Greenberg ME, Spiegelman BM (2013) Exercise induces hippocampal BDNF through a PGC-1alpha/FNDC5 pathway. Cell Metab 18:649–659
74. Cassilhas RC, Viana VA, Grassmann V, Santos RT, Santos RF, Tufik S, Mello MT (2007) The impact of resistance exercise on the cognitive function of the elderly. Med Sci Sports Exerc 39:1401–1407
75. Bermon S, Ferrari P, Bernard P, Altare S, Dolisi C (1999) Responses of total and free insulin-like growth factor-I and insulin-like growth factor binding protein-3 after resistance exercise and training in elderly subjects. Acta Physiol Scand 165:51–56
76. Ding Q, Vaynman S, Akhavan M, Ying Z, Gomez-Pinilla F (2006) Insulin-like growth factor I interfaces with brain-derived neurotrophic factor-mediated synaptic plasticity to modulate aspects of exercise-induced cognitive function. Neuroscience 140:823–833
77. Nishijima T, Piriz J, Duflot S et al (2010) Neuronal activity drives localized blood-brain-barrier transport of serum insulin-like growth factor-I into the CNS. Neuron 67:834–846
78. Lopez-Lopez C, LeRoith D, Torres-Aleman I (2004) Insulin-like growth factor I is required for vessel remodeling in the adult brain. Proc Natl Acad Sci USA 101:9833–9838
79. Ruiz de Almodovar C, Lambrechts D, Mazzone M, Carmeliet P (2009) Role and therapeutic potential of VEGF in the nervous system. Physiol Rev 89:607–648
80. Saucedo Marquez CM, Vanaudenaerde B, Troosters T, Wenderoth N (2015) High-intensity interval training evokes larger serum BDNF levels compared with intense continuous exercise. J Appl Physiol (1985) 119:1363–1373
81. Kafitz KW, Rose CR, Thoenen H, Konnerth A (1999) Neurotrophin-evoked rapid excitation through TrkB receptors. Nature 401:918–921
82. Jovanovic JN, Czernik AJ, Fienberg AA, Greengard P, Sihra TS (2000) Synapsins as mediators of BDNF-enhanced neurotransmitter release. Nat Neurosci 3:323–329
83. Karege F, Schwald M, Cisse M (2002) Postnatal developmental profile of brain-derived neurotrophic factor in rat brain and platelets. Neurosci Lett 328:261–264
84. Winter B, Breitenstein C, Mooren FC et al (2007) High impact running improves learning. Neurobiol Learn Mem 87:597–609
85. Rodríguez-Gutiérrez E, Torres-Costoso A, Saz-Lara A, Bizzozero-Peroni B, Guzmán-Pavón MJ, Sánchez-López M, Martínez-Vizcaíno V (2023) Effectiveness of high-intensity interval training on peripheral brain-derived neurotrophic factor in adults: a systematic review and network meta-analysis. Scand Med Sci Sports 34(1):e14496
86. Yanagisawa H, Dan I, Tsuzuki D, Kato M, Okamoto M, Kyutoku Y, Soya H (2010) Acute moderate exercise elicits increased dorsolateral prefrontal activation and improves cognitive performance with Stroop test. NeuroImage 50:1702–1710
87. Kraemer RR, Durand RJ, Acevedo EO, Johnson LG, Kraemer GR, Hebert EP, Castracane VD (2004) Rigorous running increases growth hormone and insulin-like growth factor-I without altering ghrelin. Exp Biol Med (Maywood) 229:240–246
88. Jahreis G, Hesse V, Schmidt HE, Scheibe J (1989) Effect of endurance exercise on somatomedin-C/insulin-like growth factor I concentration in male and female runners. Exp Clin Endocrinol 94:89–96
89. Schiffer T, Schulte S, Hollmann W, Bloch W, Struder HK (2009) Effects of strength and endurance training on brain-derived neurotrophic factor and insulin-like growth factor 1 in humans. Horm Metab Res 41:250–254
90. Dall R, Lange KH, Kjaer M, Jorgensen JO, Christiansen JS, Orskov H, Flyvbjerg A (2001) No evidence of insulin-like growth factor-binding protein 3 proteolysis during a maximal exercise test in elite athletes. J Clin Endocrinol Metab 86:669–674

91. Carro E, Nunez A, Busiguina S, Torres-Aleman I (2000) Circulating insulin-like growth factor I mediates effects of exercise on the brain. J Neurosci 20:2926–2933
92. Kraus RM, Stallings HW, Yeager RC, Gavin TP (2004) Circulating plasma VEGF response to exercise in sedentary and endurance-trained men. J Appl Physiol 96:1445–1450
93. Wahl P, Zinner C, Achtzehn S, Behringer M, Bloch W, Mester J (2011) Effects of acid-base balance and high or low intensity exercise on VEGF and bFGF. Eur J Appl Physiol 111:1405–1413
94. Landers-Ramos RQ, Jenkins NT, Spangenburg EE, Hagberg JM, Prior SJ (2014) Circulating angiogenic and inflammatory cytokine responses to acute aerobic exercise in trained and sedentary young men. Eur J Appl Physiol 114:1377–1384
95. Shetty AK, Hattiangady B, Shetty GA (2005) Stem/progenitor cell proliferation factors FGF-2, IGF-1, and VEGF exhibit early decline during the course of aging in the hippocampus: role of astrocytes. Glia 51:173–186
96. Zacchigna S, Lambrechts D, Carmeliet P (2008) Neurovascular signalling defects in neurodegeneration. Nat Rev Neurosci 9:169–181
97. Skriver K, Roig M, Lundbye-Jensen J, Pingel J, Helge JW, Kiens B, Nielsen JB (2014) Acute exercise improves motor memory: exploring potential biomarkers. Neurobiol Learn Mem 116:46–58
98. van Hall G, Stromstad M, Rasmussen P, Jans O, Zaar M, Gam C, Quistorff B, Secher NH, Nielsen HB (2009) Blood lactate is an important energy source for the human brain. J Cereb Blood Flow Metab 29:1121–1129
99. Tsukamoto H, Suga T, Takenaka S, Tanaka D, Takeuchi T, Hamaoka T, Isaka T, Hashimoto T (2016) Greater impact of acute high-intensity interval exercise on post-exercise executive function compared to moderate-intensity continuous exercise. Physiol Behav 155:224–230
100. Hashimoto T, Tsukamoto H, Takenaka S, Olesen ND, Petersen LG, Sorensen H, Nielsen HB, Secher NH, Ogoh S (2018) Maintained exercise-enhanced brain executive function related to cerebral lactate metabolism in men. FASEB J 32:1417–1427
101. Wood KM, Olive B, LaValle K, Thompson H, Greer K, Astorino TA (2016) Dissimilar physiological and perceptual responses between Sprint interval training and high-intensity interval training. J Strength Cond Res 30:244–250
102. Glenn TC, Martin NA, Horning MA, McArthur DL, Hovda DA, Vespa P, Brooks GA (2015) Lactate: brain fuel in human traumatic brain injury: a comparison with normal healthy control subjects. J Neurotrauma 32:820–832
103. Kemppainen J, Aalto S, Fujimoto T, Kalliokoski KK, Langsjo J, Oikonen V, Rinne J, Nuutila P, Knuuti J (2005) High intensity exercise decreases global brain glucose uptake in humans. J Physiol 568:323–332
104. Dalsgaard MK, Ide K, Cai Y, Quistorff B, Secher NH (2002) The intent to exercise influences the cerebral O(2)/carbohydrate uptake ratio in humans. J Physiol 540(2):681–689
105. Dienel GA (2017) The metabolic trinity, glucose-glycogen-lactate, links astrocytes and neurons in brain energetics, signaling, memory, and gene expression. Neurosci Lett 637:18–25
106. Barros LF (2013) Metabolic signaling by lactate in the brain. Trends Neurosci 36:396–404
107. Weston M, Taylor KL, Batterham AM, Hopkins WG (2014) Effects of low-volume High-intensity Interval Training (HIT) on fitness in adults: a meta-analysis of controlled and non-controlled trials. Sports Med 44:1005–1017
108. El Hayek L, Khalifeh M, Zibara V et al (2019) Lactate mediates the effects of exercise on learning and memory through SIRT1-dependent activation of hippocampal Brain-Derived Neurotrophic Factor (BDNF). J Neurosci 39:2369–2382
109. Koltai E, Szabo Z, Atalay M, Boldogh I, Naito H, Goto S, Nyakas C, Radak Z (2010) Exercise alters SIRT1, SIRT6, NAD and NAMPT levels in skeletal muscle of aged rats. Mech Ageing Dev 131:21–28
110. Morland C, Andersson KA, Haugen ØP et al (2017) Exercise induces cerebral VEGF and angiogenesis via the lactate receptor HCAR1. Nat Commun 8:15557

111. Salgueiro RB, Peliciari-Garcia RA, do Carmo Buonfiglio D, Peroni CN, Nunes MT (2014) Lactate activates the somatotropic axis in rats. Growth Hormon IGF Res 24:268–270
112. Warr-di Piero D, Valverde-Esteve T, Redondo-Castán JC, Pablos-Abella C, Sánchez-Alarcos Díaz-Pintado JV (2018) Effects of work-interval duration and sport specificity on blood lactate concentration, heart rate and perceptual responses during high intensity interval training. PLoS One 13:e0200690
113. Jackson MJ (2005) Reactive oxygen species and redox-regulation of skeletal muscle adaptations to exercise. Philos Trans R Soc Lond Ser B Biol Sci 360:2285–2291
114. Davies KJ, Quintanilha AT, Brooks GA, Packer L (1982) Free radicals and tissue damage produced by exercise. Biochem Biophys Res Commun 107:1198–1205
115. Radak Z, Pan L, Zhou L, Mozaffaritabar S, Gu Y, Pinho RA, Zheng X, Ba X, Boldogh I (2024) Epigenetic and "redoxogenetic" adaptation to physical exercise. Free Radic Biol Med 210:65–74
116. Radak Z, Suzuki K, Higuchi M, Balogh L, Boldogh I, Koltai E (2016) Physical exercise, reactive oxygen species and neuroprotection. Free Radic Biol Med 98:187–196
117. Afzalpour ME, Chadorneshin HT, Foadoddini M, Eivari HA (2015) Comparing interval and continuous exercise training regimens on neurotrophic factors in rat brain. Physiol Behav 147:78–83
118. Kahlert S, Zündorf G, Reiser G (2005) Glutamate-mediated influx of extracellular Ca2+ is coupled with reactive oxygen species generation in cultured hippocampal neurons but not in astrocytes. J Neurosci Res 79:262–271
119. Groussard C, Machefer G, Rannou F, Faure H, Zouhal H, Sergent O, Chevanne M, Cillard J, Gratas-Delamarche A (2003) Physical fitness and plasma non-enzymatic antioxidant status at rest and after a Wingate test. Can J Appl Physiol 28:79–92
120. Groussard C, Rannou-Bekono F, Machefer G, Chevanne M, Vincent S, Sergent O, Cillard J, Gratas-Delamarche A (2003) Changes in blood lipid peroxidation markers and antioxidants after a single sprint anaerobic exercise. Eur J Appl Physiol 89:14–20
121. Freitas DA, Rocha-Vieira E, Soares BA et al (2018) High intensity interval training modulates hippocampal oxidative stress, BDNF and inflammatory mediators in rats. Physiol Behav 184:6–11
122. Feter N, Spanevello RM, Soares MSP et al (2019) How does physical activity and different models of exercise training affect oxidative parameters and memory? Physiol Behav 201:42–52
123. Li B, Liang F, Ding X, Yan Q, Zhao Y, Zhang X, Bai Y, Huang T, Xu B (2019) Interval and continuous exercise overcome memory deficits related to β-Amyloid accumulation through modulating mitochondrial dynamics. Behav Brain Res 376:112171
124. Gholipour P, Komaki A, Parsa H, Ramezani M (2022) Therapeutic effects of high-intensity interval training exercise alone and its combination with Ecdysterone against amyloid Beta-induced rat model of Alzheimer's disease: a behavioral, biochemical, and histological study. Neurochem Res 47:2090–2108
125. Freitas DA, Rocha-Vieira E, De Sousa RAL et al (2019) High-intensity interval training improves cerebellar antioxidant capacity without affecting cognitive functions in rats. Behav Brain Res 376:112181
126. Koyuncuoğlu T, Sevim H, Çetrez N, Meral Z, Gönenç B, Kuntsal Dertsiz E, Akakın D, Yüksel M, Kasımay Çakır Ö (2021) High intensity interval training protects from post traumatic stress disorder induced cognitive impairment. Behav Brain Res 397:112923
127. Cobley JN, Fiorello ML, Bailey DM (2018) 13 reasons why the brain is susceptible to oxidative stress. Redox Biol 15:490–503
128. Radak Z, Chung HY, Goto S (2005) Exercise and hormesis: oxidative stress-related adaptation for successful aging. Biogerontology 6:71–75
129. Ide K, Secher NH (2000) Cerebral blood flow and metabolism during exercise. Prog Neurobiol 61:397–414

130. Moreau D, Chou E (2019) The acute effect of high-intensity exercise on executive function: a meta-analysis. Perspect Psychol Sci 14:734–764
131. McMorris T (2021) The acute exercise-cognition interaction: from the catecholamines hypothesis to an interoception model. Int J Psychophysiol 170:75–88
132. Cantelon JA, Giles GE (2021) A review of cognitive changes during acute aerobic exercise. Front Psychol 12:653158
133. Zheng K, Zou L, Wei G, Huang T (2021) Concurrent performance of executive function during acute bouts of exercise in adults: a systematic review. Brain Sci 11:1364
134. Dietrich A, Audiffren M (2011) The Reticular-Activating Hypofrontality (RAH) model of acute exercise. Neurosci Biobehav Rev 35:1305–1325
135. Ludyga S, Pühse U, Lucchi S, Marti J, Gerber M (2019) Immediate and sustained effects of intermittent exercise on inhibitory control and task-related heart rate variability in adolescents. J Sci Med Sport 22:96–100
136. Ai J-Y, Chen F-T, Hsieh S-S, Kao S-C, Chen A-G, Hung T-M, Chang Y-K (2021) The effect of acute high-intensity interval training on executive function: a systematic review. Int J Environ Res Public Health 18:3593
137. Costigan SA, Eather N, Plotnikoff RC, Taaffe DR, Lubans DR (2015) High-intensity interval training for improving health-related fitness in adolescents: a systematic review and meta-analysis. Br J Sports Med 49:1253–1261
138. Tottori N, Morita N, Ueta K, Fujita S (2019) Effects of high intensity interval training on executive function in children aged 8-12 years. Int J Environ Res Public Health 16:4127
139. Gilson ND, Andersson D, Papinczak ZE, Rutherford Z, John J, Coombes JS, Brown WJ (2023) High intensity and sprint interval training, and work-related cognitive function in adults: a systematic review. Scand J Med Sci Sports 33:814–833

Part III
Advanced Research for Elucidating the Effect of Exercise

Assessing Exercise and Fitness-Related Brain Health Using Multi-modal Magnetic Resonance Imaging Outcomes

J. Carson Smith, Daniel D. Callow, Gabriel S. Pena, and Jeremy J. Purcell

Abstract Several magnetic resonance imaging (MRI)-based modalities have provided insight into how acute and chronic exercise and physical activity impact brain health. Structural MRI-based measures provide information regarding brain volume and cortical thickness, diffusion-weighted imaging measures provide indices of white matter and gray microstructure, and cerebrovascular effects that may be assessed through measures of regional cerebral blood flow, and brain activation and neural networks can be indexed using task-based and connectivity-based functional MRI, respectively. In this chapter, a series of studies are described in which these multi-modal MRI outcomes, along with indices of neurocognitive function, have been documented after a brief walking exercise intervention in healthy older adults and those diagnosed with mild cognitive impairment. Finally, we provide some additional insight into the neurophysiological mechanisms that may be foundational for these effects, but are yet not measurable in humans. Multi-modal neuroimaging is a non-invasive method in humans to assess the potential mechanisms whereby acute and chronic exercise may exert benefits to brain function and neural networks related to cognition that may protect older adults from age-related cognitive decline and dementia.

Keywords Alzheimer's Disease · Cognition · Exercise · Magnetic resonance imaging · Mild cognitive impairment · Physical activity

---

J. C. Smith (✉) · D. D. Callow · G. S. Pena · J. J. Purcell
University of Maryland, College Park, MD, USA
e-mail: carson@umd.edu

H. Soya (ed.), *Exercise Brain Stimulation for Cognitive Function and Mental Health*, Advances in Neurobiology 44, https://doi.org/10.1007/978-981-95-0066-6_14

1 Introduction

A firm understanding of the effects of exercise on human brain function has been elusive due to both the inability to use invasive measurement techniques and the pleiotropic nature of exercise. Magnetic resonance imaging (MRI) has provided useful tools that have both allowed scientists to measure the internal structure and function of the brain, and also enabled a variety of modalities and capabilities through the manipulation of the magnetic properties of tissues that permit documentation of the varied and nuanced ways that exercise impacts the brain (see Fig. 1). Here, we provide an overview of several MRI-based modalities that have provided insight into how acute and chronic exercise and physical activity impact structural brain volume and cortical thickness; diffusion-weighted imaging to assess white matter and gray microstructure; cerebrovascular effects that may influence regional cerebral blood flow; and brain activation using task-based and connectivity-based functional MRI. We describe a series of studies in which these multi-modal MRI outcomes, along with indices of neurocognitive function, have been documented after a brief walking exercise intervention in healthy older adults and those diagnosed with mild cognitive impairment. Finally, we provide some additional insight into the neurophysiological mechanisms that may be foundational for these effects, but are yet not measurable in humans.

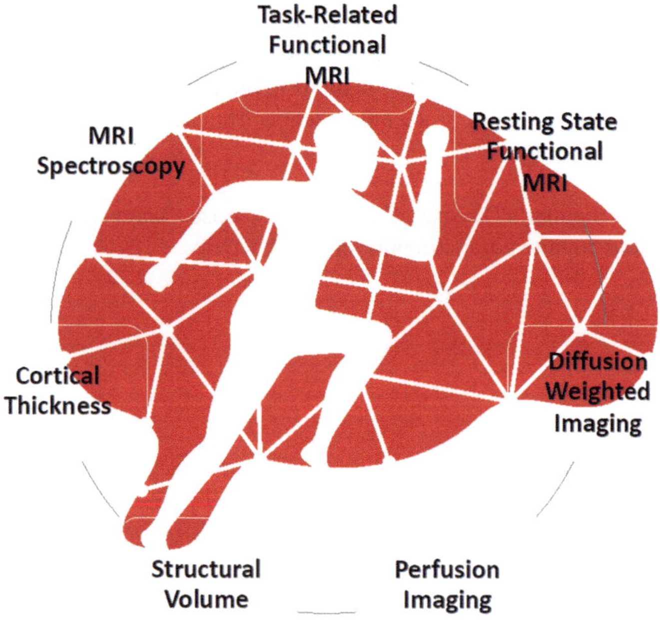

Fig. 1 MRI provides multiple modalities that may provide insight into the effects of acute and chronic exercise on brain structure and function

2 Structural Volumetric Imaging

Magnetic resonance imaging (MRI)-based volumetry is commonly used to assess brain volume changes in aging, particularly that of gray matter. The resolution of these structural images has improved over time, and submillimeter scans are now commonly acquired for research purposes. Additionally, several well-established processing software can take a single T1 or T2 scan and provide researchers with information about the size of 100's of subcortical and cortical structures. Allowing for a thorough characterization of brain volume trajectories in healthy and pathological aging. Notably, the prefrontal cortex and the hippocampus are specific gray matter brain regions that consistently exhibit atrophy of approximately 1–2% a year in older age, with more rapid declines experienced after the onset of cognitive impairment. Indeed, executive function and memory are two cognitive processes that consistently decline with age and are tightly linked to the prefrontal cortex and hippocampus volume. Gray matter volume has, thus, become a valid biomarker for healthy cognitive aging and cognitive decline across the lifespan. Given the extensive characterization of brain volumetry throughout the aging process, understanding how exercise may influence age-related changes in brain volume is critical to understanding the relationship between exercise and healthy brain aging.

2.1 *Physical Activity, Fitness, and Volumetry*

Physical inactivity (less than 150 min of moderate to vigorous activity) is associated with smaller gray, but not white matter volume in older adults [1, 2]. However, this relationship appears most specific to preserving hippocampal and prefrontal cortex volume against age-related decline, with reports of benefits for other brain regions being far less consistent [3]. For example, a recent 12-year longitudinal study found that greater physical activity was associated with a larger hippocampal volume. Notably, hippocampal preservation for older adults was also largest for those who were APOE-e4 positive (a genetic risk factor for Alzheimer's disease), as shown in Fig. 2 [4]. Finally, a recent study found that objectively measured moderate-intensity physical activity is associated with larger bilateral dorsolateral prefrontal cortex volume in older adults [5], while a longitudinal study also reported that physical activity levels in older adults predicts frontal lobe volume at a 9-year follow-up [6].

Cardiorespiratory fitness may be even more closely related to gray matter volume in aging than physical activity levels. Again, this benefit appears particularly distinct to brain regions influencing executive function and memory, such as the hippocampus and prefrontal cortex [3]. Erickson et al. (2009), found that hippocampal volume mediated the relationship between cardiorespiratory fitness and spatial memory in healthy older adults [7]. While Weinstein et al. (2012) found that prefrontal cortex volume mediated the relationship between aerobic fitness and executive function processes [8]. Within these and other cross-sectional studies, the

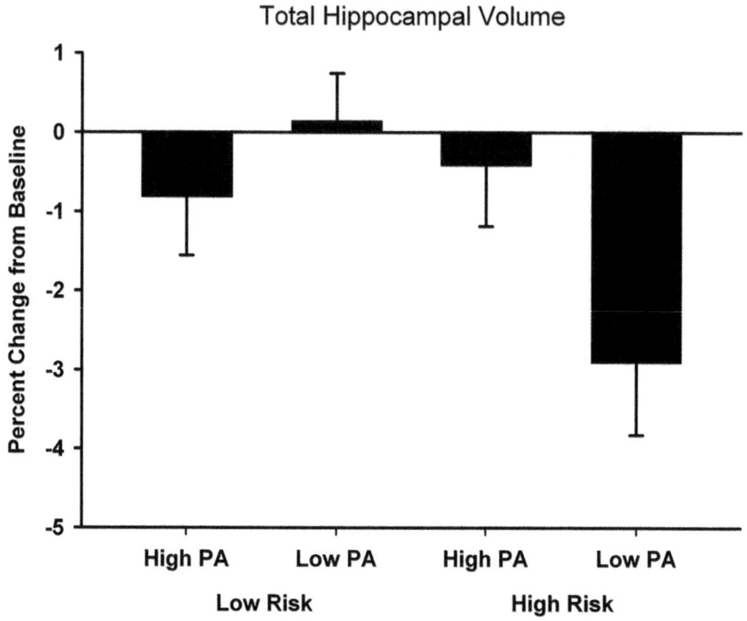

Fig. 2 Percent change from baseline in total hippocampal volume in groups of healthy older adults that varied in self-reported PA (high PA, low PA) and genetics risk for Alzheimer's disease through carriage of the APOE-e4 allele (low risk, high risk). Error bars represent SEM. From [4]

relationship between fitness, gray matter volume, and cognition was independent of age, although evidence is limited to older adults [9]. These results suggest that maintaining adequate levels of physical activity and cardiorespiratory fitness may be essential for preserving brain structures critical for age-susceptible higher-order cognitive processes.

2.2 *Exercise Training and Volumetry*

Two groundbreaking studies provided the first evidence that exercise could alter gray matter volume in older adults. The first was a 6-month aerobic exercise training program for healthy older adults which reported increases in frontal lobe volume compared to a stretching and toning control group [10]. Furthermore, Erickson et al. (2011) conducted a randomized controlled trial in which 120 healthy older adults underwent 1 year of aerobic exercise training or a stretching control group [11] and found that the aerobic training group saw an increase in anterior hippocampal volume of 2% and that improvements in fitness were related to increases in volume. However, a recent meta-analysis of such exercise training RCTs on the hippocampus found only a small retention of left hippocampal volume compared to control group conditions [12]. Furthermore, a meta-analysis of 14 exercise RCTs looking at the effect of exercise on gray matter volume in older adults found that exercise was not linked to significant changes in gray matter volume compared to those performing control conditions [13]. These results suggest that the benefits of exercise and improved cardiorespiratory fitness may be specific to certain age susceptible brain regions, such as the hippocampus.

2.3 *Volumetry, Cognition, and Future Directions*

Unfortunately, while fitness and physical activity-based preservation of the hippocampus and prefrontal cortex are associated with better executive functioning and memory performance independent of age, most exercise interventions in older adults have failed to establish a link between training-based changes in gray matter volume with changes in cognitive performance. This lack of evidence could be attributed to several factors, including differentiation, employing more biophysically specific multimodal neuroimaging approaches and measures in tandem with volumetric MRI [14], and high-resolution structural imaging approaches, particularly of the hippocampus [15, 16]. Finally, future studies need to determine the cognitive domains and processes exercise-related preservation of gray matter tissue specifically mediates.

3 White Matter Diffusion-Weighted Imaging

Diffusion-weighted imaging (DWI) is a commonly acquired MRI technique often used to measure the strength of the brain's structural network connections. DWI captures the movement of water molecules within brain tissue and provides a glimpse into the integrity and microstructural properties of the underlying neural environment. Water generally diffuses more rapidly within and along cellular structures, making DWI a popular approach for evaluating the integrity of the brain's white matter tracts. The long directional nature of axon fiber bundles that compose these tracts, often insulated by myelin, leads water to diffuse more readily along the primary direction of a healthy tract than perpendicular to it. However, myelin loss or loss in the number or density of axons within a tract leads to fewer neural barriers to the diffusion of water and, thus, greater overall and perpendicular diffusion to the fiber tract, which is synonymous with poor tract integrity [17].

These large white matter tracts facilitate the efficient movement of information between gray matter brain regions, similar to how traffic flows between cities along a highway system. Construction work, accidents, or road wear and tear can slow or impede the efficient flow of traffic. Similarly, aging results in degenerating white matter fiber tracts via the "last-in-first-out" hypothesis, which argues that the last brain network connections to develop are often the first to deteriorate [18]. Consistent with this hypothesis, aging exhibits a pattern of deterioration in which the posterior white matter tracts exhibit a decline in diffusion-based integrity measures before the anterior tracts. Notably, the disconnected hypothesis argues that this loss of white matter axonal integrity and density plays an essential role in age-related cognitive decline [19]. Thus, understanding how physical activity, fitness, and exercise training relate to white matter fiber tract integrity with DWI can provide nuanced insight into how physical activity and exercise impact the structural connectivity of the aging brain.

3.1 White Matter DWI and Physical Activity and Fitness

The benefits of physical activity and fitness for healthy brain aging may extend to white matter. Cross-sectional studies consistently report that subjective and objective measures of PA and cardiorespiratory fitness are positively associated with structural network connectivity and white matter tract integrity across the lifespan [20]. However, the most robust evidence for benefits of higher physical activity and fitness is in older adults. Numerous studies report a positive relationship between older adults' fitness and physical activity with the integrity of the primary association and projection fiber tracts along the posterior to anterior gradient, where there is often the most significant degree of deterioration in aging [21]. In fact, higher-fit older adults exhibit white matter tract integrity much closer to fit younger adults than their low-fit older counterparts [22], suggesting that maintaining physical activity and fitness levels may preserve the brain's white matter tracts and connections against age-related deterioration.

3.2 White Matter DWI and Exercise Interventions

While cross-sectional studies may indicate that exercise can reverse age-related white matter integrity deterioration, the evidence from exercise interventions is far less conclusive. Randomized controlled trials conducted in older adults have predominantly failed to find significant differences in white matter tract integrity between intervention and control groups [20]. The only intervention to show significant improvements in white matter integrity compared to control conditions was a dance intervention, which had unique physical, cognitive, and social components to all other interventions [23]. Unfortunately, it is therefore challenging to determine whether the improvements were specific to the exercise alone. Nevertheless, individuals in many of these exercise interventions who exhibit improvements in cardiorespiratory fitness often exhibited concurrent improvements in white matter integrity. For example, a 5-year exercise intervention in a subsample of the 100-generation study tested for changes in white matter diffusion among older adults undergoing a high-intensity interval training or moderate continuous intensity training intervention vs. a third group that followed the national physical activity guidelines of at least 30 minutes of physical activity a day [24]. Importantly, this study did not have a proper non-exercise control group, potentially contributing to the lack of an effect of the intervention group on changes in white matter tract integrity. However, cardiorespiratory fitness, exercise intensity, and exercise duration were independently related to positive white matter tract integrity at the 1,3, and 5-year follow-ups. Thus, current exercise training intervention studies suggest that individual differences contributing to someone being an exercise responder vs. non-responder may play an essential role in unraveling how exercise preserves the integrity of the aging brain's structural network connections.

3.3 White Matter DWI, Cognition, and Future Directions

As previously described, age-related white matter deterioration generally follows a posterior-to-anterior gradient and is associated with age-related declines in both processing speed and executive function [25]. However, white matter tract integrity can mediate age-related differences in a broad spectrum of cognitive processes and predict cognitive status, cognitive decline, and the incidence of mild cognitive impairment [26]. While scarce, several cross-sectional studies have tested for a mediating effect of white matter integrity on the relationship between physical activity and fitness with a broad range of cognitive domains. These studies report that fitness mediates the relationship between white matter tract integrity and global, fluid, and crystallized cognition, as well as reaction time and spatial memory [27]. However, exercise intervention studies have failed to report similar mediating effects of white matter integrity on cognitive performance.

4 Gray Matter DWI

While DWI is commonly used to study the integrity of white matter tracts, advances in imaging protocols and techniques allow researchers to employ DWI to probe the microstructural integrity of underlying gray matter tissue. Brain structure size or thickness often characterizes gray matter tissue health measures and age-related changes. However, gray matter DWI provides additional information about the composition and integrity of these cortical and subcortical brain structures. Gray matter usually consists of more neuronal cell bodies, synapses, dendrites, and glial bodies than axons. Thus, the microstructural environment is generally less directionally cohesive, and diffusion measures often represent different physiological processes than within white matter. Higher diffusion within subcortical regions like the hippocampus is associated with aging and may represent a decline in barriers to the diffusion of water via lower synaptic, glial, and dendritic density. Higher cortical and subcortical diffusion also predicts cognitive decline, impairment, and neurodegenerative disease, potentially due to reduced neural density. However, while cross-sectional differences in mean diffusion indicate reduced tissue density and health, short-term changes in gray matter diffusion (on the order of minutes or hours) are sensitive to different neuroplastic mechanisms, such as the upregulation of neurotrophic factors and inflammatory mediators. Therefore, it is essential to consider the context within which the measure is acquired when interpreting gray matter diffusion results.

New multi-shell imaging and multi-compartment diffusion modeling approaches now allow researchers to surpass simple voxel summary metrics and decompose the diffusion signal into more biophysically specific components. While previous modeling approaches might produce a single summary of the overall diffusion within a white or gray matter voxel, new approaches, such as neurite orientation dispersion density imaging, can separate the diffusion signal into free water, intracellular, and extracellular diffusion components. These more neurophysiologically specific compartments are essential for gray matter diffusion imaging as more microstructural factors may contribute to the diffusion signals in gray matter than white matter. Gray matter microstructural alterations appear to precede age-related volumetric changes and can predict cognitive decline and changes in cognitive status independent of gray matter volume loss. Interestingly, a large cross-sectional study in twins found that while volumetric differences were more attributable to an individual's genetics, gray matter microstructure differences were more closely linked to genetics and environmental factors [28]. Thus, gray matter diffusion imaging provides more specific and sometimes more sensitive information about gray matter tissue health and can provide additional mechanistic insight into how exercise preserves the aging brain.

4.1 Physical Activity, Fitness, and Gray Matter DWI

Subjective measures of physical activity are related to lower diffusion in the medial temporal lobe and cingulate cortex of older adults between the ages of 70 and 79 years, while higher cardiorespiratory fitness is associated with lower diffusion within the hippocampus and entorhinal cortex of individuals over 80 years [29]. Meanwhile, a recent longitudinal study in cognitively healthy older adults found that higher objective measures of moderate to vigorous physical activity were associated with less microstructural deterioration within the entorhinal cortex and parahippocampal gyrus over a 4-year period [30]. All three studies suggest that physical activity and fitness may help preserve the microstructural integrity of age-susceptible gray matter structures for healthy older adults, particularly the medial temporal lobe. Finally, a recent study in younger adults utilized a multi-shell diffusion imaging and modeling approach and reported that higher fitness was related to elevated extracellular diffusion across cortical and subcortical gray matter structures. Notably, higher extracellular diffusion mediated the positive relationship between fitness and crystallized cognition and has previously been linked to higher intelligence in younger adults [27]. Interestingly, in younger adults, greater extra-cellular diffusion may indicate reduced dendritic and synaptic density, with these neurophysiological differences representing the development of more efficient brain networks that facilitate better cognitive performance [31]. Therefore, age is an essential factor in future studies.

4.2 Acute and Chronic Exercise Interventions and Gray Matter DWI

The implementation of gray matter DWI in exercise interventions is limited. A randomized controlled trial in healthy older adults found that fitness improvements following a 6-month aerobic training program were associated with reduced hippocampal gray matter diffusion [32]. Improved hippocampal tissue density mediates training-based improvements in hippocampal volume, suggesting that fitness-related changes in hippocampal volume may occur through initial tissue density changes. Similarly, 9 weeks of freewheel running in rats showed a relationship between the extent of exercise performed and a more dense and complex hippocampal microstructural environment [33]. Additionally, a recent exercise intervention looking at older adults with and without mild cognitive impairment found that 12 weeks of aerobic exercise training increased gray matter diffusion within the cerebellum and insular cortex, which was associated with improved verbal memory and fluency performance. This increase in gray matter diffusion indicated decreased synaptic density within these cortical regions but was associated with behavioral improvements [14]. The authors suggest that this may have been due to exercise-induced improvements in network efficiency, given a study in the same cohort found reduced task-based fMRI activity in similar cortical structures [34]. Notably, this and previously discussed training studies did not employ a proper non-exercise control condition, and therefore there is a need for additional randomized controlled trials that include adequate control conditions to determine the effects of exercise on gray matter microstructure in older adults.

Gray matter DWI also provides a novel and exciting opportunity to explore the effects of an acute bout of exercise on neurophysiological processes within the aging brain. Previous animal and human studies have found that improvements when learning a short bout of spatial navigation is associated with hippocampal mean diffusion differences before and after completing the task. Even more exciting is that in two paired animal studies, hippocampal diffusion changes were associated with morphological changes in astrocytes and the expression of neurotrophic factors such as synaptophysin, glial fibrillary acidic protein, and brain-derived neurotrophic factors [35]. Central measures of neurotrophic factors are only possible in animals; gray matter DWI may allow us to index such neuroplastic factors resulting from a single exercise session in the human brain. Two recent studies looked to address this question by looking at the effects of acute moderate-intensity aerobic exercise on hippocampal microstructural diffusion. The first study used a coarser single shell diffusion imaging approach and found that exercise was associated with elevated mean diffusion and lower water dispersion within the hippocampal gray matter of their older adults sample [36]. A second follow-up study then used a high-resolution multi-shell DWI approach to examine the effects of a similar acute exercise session on hippocampal subfield microstructure (Callow et al. in review). They found elevated extracellular diffusion in the hippocampus approximately 30 min following exercise cessation, but also that this elevated extracellular diffusion was specific to the DG/CA3 subfield. These acute exercise-induced increases in extracellular diffusion are consistent with greater glial activity, neuroinflammatory mediators, and expression of neurotrophic factors. All of which increase in the hippocampus and DG/CA3 following higher intensity wheel running in rats. Importantly, in these animal studies, this acute increase in hippocampal neuroinflammation is thought to coincide with the upregulation of neurotrophic cascades and lead to long-term neurogenic effects that can promote optimal brain health [37]. Future studies that link these short-term acute effects to longer-term training adaptations are needed to fully characterize the neurophysiological effects of exercise on the aging brain.

5 Cerebral Blood Flow by Arterial Spin Labeling MRI

Although acute exercise and exercise training have been shown to benefit cognitive performance [38, 39], the physiological mechanisms underpinning these relationships in humans remain poorly understood. One hypothesis is that aerobic exercise likely improves cerebrovascular health by enhancing endothelial function and vascular conductivity [40, 41]. For example, regular exercise in animals has been shown to result in greater microvascular density, and in humans altered patterns of cerebral perfusion are accompanied by brain volume changes [42]. While several studies have examined the effects of exercise training on the cerebrovascular system, fewer studies have analyzed cognition and brain perfusion after exercise.

Arterial spin labeling (ASL) is a specialized MRI sequencing technique that non-invasively uses arterial blood water as endogenous imaging tracer [43, 44]. During the ASL sequence, radiofrequency (RF) pulsations invert blood magnetization as it flows through the imaging plane [45]. A short delay period follows inversion (post-label delay), providing time for the labeled blood to perfuse the brain [46]. After this period, images are collected that contain labeled blood and signal from static tissue water [47]. Additionally, a series of control (non-labeled) images are collected that do not contain magnetic inversion [48]. The amount of arterial blood delivered to the tissue through perfusion (also known as cerebral perfusion or cerebral blood flow; CBF) is estimated based on the signal difference between the control and labeled images [49]. The duration of the label—magnetic tracer—can last between 1300 and 1750 ms, depending on the longitudinal relaxation time of blood, which is similar to the arterial transport time (ATT), or the duration of time it takes for blood to flow from the labeling plane to brain tissue [50]. Variability in ATT, both within and between individuals, influences ASL implementation as long delays can diminish signal-to-noise ratio (SNR), and short delays can prevent complete delivery of the labeled blood to brain tissue [43].

Pulsed and continuous ASL sequences are the two primary labeling methods used in human research [43, 45, 51]. While both are common, they differ in both labeling duration and spatial extent. Continuous ASL (CASL) involves a labeling period lasting between 1000 and 3000 ms. During the labeling period, blood is inverted by continuous radiofrequency energy as it flows through a single labeling plane. Pseudo-continuous ASL (PCASL) involves a similar labeling plane, in which blood is inverted by multiple RF pulses (e.g., often a thousand or more pulses per second) and imparts superior labeling efficiency [43, 52].

One of the first studies to utilize ASL after exercise found that acute low-moderate intensity cycling exercise led to elevated CBF both at rest and during simple motor activity [53]. A 30-minute bout of acute low-moderate intensity bicycle exercise (Borg's rating scale of perceived exertion; RPE = 13) was performed in five young (age ~25) healthy adults. Regional CBF was assessed before exercise and again immediately afterward, during rest and a simple motor task (finger tapping). Results demonstrated a 20% increase in motor cortex CBF at rest, and during the finger-tapping task. These findings suggest that low-moderate intensity aerobic exercise leads to elevated CBF at both rest and during subsequent simple motor activity; however, the sample was small and the analysis was constrained to the motor cortex.

In opposition to previous results, there is also evidence that a single session of exercise may decrease rCBF. MacIntosh and colleagues explored the effects of acute moderate intensity bicycle exercise in healthy young (age 20–35) adults using pseudo-continuous ASL to assess rCBF before and after exercise (10-minutes and 40-minutes post) [41]. Task-based fMRI was also used to assess functional brain activation during the Go/No Go task to measure executive function. Results revealed, compared to baseline, that gray matter rCBF at 10-minutes post-exercise was 11% lower and then returned to baseline by 40-minutes post-exercise. On the other hand, white matter rCBF was significantly greater during both post-exercise assessments. Additionally, region of interest analyses demonstrated decreased perfusion in the hippocampus and insular cortex. Functional brain activation during the executive function task was similar but was significantly lower after exercise in the left parietal operculum. These findings highlight the complexity of changes in perfusion after exercise, which may be regionally heterogeneous and task dependent.

Acute exercise-induced changes in rCBF also may be intensity dependent. A group of researchers studied the effects of acute bicycle exercise (20-minutes), at two different intensities (40–50% and 60–70% of HRR) on rCBF in 13 older stroke survivors (age ~64) [54]. Using a within-subjects repeated-measures design, low and moderate intensity exercise were performed on separate days, and participants underwent perfusion-weighted MRI before and immediately after exercise. In general, rCBF decreased after low intensity exercise and was elevated after moderate intensity exercise, although these changes were not linked to any change in cognitive function.

Short-term exercise interventions in older adults can improve cerebrovascular function and lead to improved cognition. In a randomized controlled trial examining the effects of aerobic exercise training on 37 healthy older adults (age 57–75) [55], half of the volunteers performed exercise three times per week for 3 months, while the other half served as a waitlist control. Fitness, rCBF, and cognition were assessed three times throughout the trial. After 3 months, it was found that those who exercise had increased rCBF in the anterior cingulate cortex and also improved performance in immediate and delayed memory recall. The effects of longer-term participation in exercise has also been examined. Using a unique sample of highly trained older athletes, researchers performed a cross-sectional analysis examining the effects of life-long aerobic exercise on rCBF and cerebrovascular reactivity (CVR) [56]. Ten master athletes (age ~75), ten sedentary older adults (age ~75), and nine younger adults participated in the study. Results demonstrated that life-long exercise preserved rCBF in the posterior cingulate and precuneus, which are regions of the DMN and known to be susceptible to the AD pathology. However, CVR was decreased in the master athletes compared to their sedentary counterparts. This is an important study as it demonstrates the protective effects of a healthy and aerobically fit lifestyle on cerebrovascular health. However, it remains poorly understood if these effects translate to cognitive performance. Nevertheless, it seems that consistent exercise training throughout life protects the older brain from decreases in rCBF such that absolute rCBF remains comparable to younger adults.

In summary, exercise boosts improve cognitive performance but the physiological underpinnings of this relationship remain unclear. Resting CBF has a propensity to decrease with age, but higher levels of aerobic fitness and exercise training ameliorate this effect. The effects of acute aerobic exercise in older adults have been understudied and appear to be more ambiguous. In animals, exercise-induced increases in blood flow and volume have been associated with structural changes in the hippocampus, motor cortex, and cerebellum [57]; and in monkeys, physical inactivity facilitates the dissipation of these improvements. While exercise may boost cognition, the pattern of changes in CBF that underlie this relationship are likely complex and require further research.

6 Task-Based fMRI

Task-based functional magnetic resonance imaging (fMRI) relies on the blood oxygenation-dependent (BOLD) imaging [58] which is used as a proxy measure for neural activity when performing a task [59]. Given the consistent reports of cognitive improvements in executive function, memory, and mood due to exercise [60–62], it is not surprising these tasks have been used to study the neural changes associated with exercise [63, 64].

Executive function improvements due to exercise are the most robust effects [60] and have included selective attention and working memory fMRI tasks [64]. There have been at least a couple different selective attention tasks used including the Flanker [65] and Go/No-Go task [66]. These tasks involve selectively attending to a task element in favor of a prepotent response and typically involve comparing fMRI activity for incongruent versus congruent stimuli.

The most prevalently used executive function fMRI task in the exercise literature is the Flanker task. For both acute and long term exercise training, much of these exercise-related differences were in brain areas typically observed in flanker task (i.e., inferior frontal gyrus, parietal, insula, anterior cingulate) [67, 68]. This is the case for a study on a single session of 30 min of cycling in older adults [69], and in another study involving a 12 month resistance training intervention [70]. However, one longer-term study involving a 6-month aerobic exercise intervention in sedentary adults did not observe any exercise-induced Flanker task differences but did observe a correlation of fMRI activation differences with the fitness gains in frontal cortices [71]. Overall, this work suggests that exercise can lead to an enhancement of regions associated with executive function as measured with the flanker task, and that some of these effects may be dependent on the magnitude of fitness gains.

Another common selective attention task is the stop signal Go/No-Go task. Despite some inconsistencies of the direction of effects, broadly speaking, the differences due to exercise were observed within areas typically reported for the Go/No-Go task such as the inferior frontal gyrus, inferior parietal, superior frontal, middle frontal, and middle occipital gyrus [72]. For instance, whereas 30 min of moderate intensity cycling in young adults has been associated with increased superior frontal, precentral, and inferior frontal gyrus activation in one study [73]; in another, it was associated with decreased activation only in the parietal operculum [41]. One study went a step further and compared differences in Go/No-Go activation for high versus low intensity exercise, and reported a high-intensity reduction in activation in the superior frontal gyrus and insula [73]. This suggests that there may be multiple different ways in which exercise may lead to beneficial selective attention processing, and that some areas may be uniquely sensitive to the intensity of exercise engagement.

Working memory is another common executive function task [74] used in fMRI studies of exercise. In healthy young adults, 30 min of moderate intensity cycling was associated with 2-back working memory activation increases in the right middle frontal gyrus, right lingual gyrus, and left fusiform gyrus [75], yet lower activation in the anterior cingulate gyrus, paracentral lobule, and inferior frontal gyrus [75, 76]. In a longer term, 12-week low intensity exercise intervention in older adults, there were marked increases in the bilateral lateral prefrontal cortex activation during an 1-back task [77]. It is worth noting that most of these reported regions are typically associated with n-back working memory fMRI task-related activation [78]. This suggests a complex relationship between the directional effects of exercise on the regions associated with working memory.

Long-term memory, including semantic and episodic memory, has been commonly studied with regards to the benefits of exercise. For instance, a single session of 30 min of cycling is associated with increases in activation for a semantic memory task (i.e., memory for famous written names) in inferior temporal gyrus, middle temporal gyrus, fusiform gyrus, and hippocampus [79]. Similarly, in young adults after a 6-week aerobic exercise intervention, there was also an increase in hippocampal activation during an fMRI paired-association recollection task [80]. Interestingly, a 12-week aerobic exercise intervention in older adults observed exercise-induced differences in the inferior parietal lobule and precuneus during a semantic memory task [34]. Importantly, both of these regions are known to be part of the episodic memory network [81]. More nuanced memory studies that involve a mnemonic similarity task (MST) reported that a single 10 min session of light intensity cycling can lead to increases in memory pattern separation both behaviorally and neurally [82]. These findings are supported by extensive literature showing that exercise can induce an upregulation of the memory system in humans and non-humans [83, 84].

The least studied are task-based fMRI studies of exercise, mood, and reward regulation. Previous work has shown that 30 minutes of treadmill running lead to lower activation during a reward processing monetary incentive delay (MID) task in the dorsal striatum [85] and was interpreted as facilitating a lower sensitivity to money-based rewards. A more recent study found that 20 minutes of cycling lead to an increase in activation in the bilateral medial superior frontal gyrus and inferior parietal lobule regions (Yash et al. 2023) both of which have been associated with the cognitive control of emotional process [86].

Regarding more advanced theories and methods, one prominent neurobiology of aging theory is that cognitive and neural decline is because multiple neural systems become de-differentiated with age [87–91]. One method used to quantify this de-differentiation involves demonstration of the relative lessening of distinctness of spatial patterns of fMRI signal responses for different stimuli [87, 92]. After 6 months of aerobic exercise training, participants demonstrated greater visual system neural distinctness, supporting the idea that physical activity may protect against age-dependent de-differentiation across multiple neural systems [93].

Besides focusing on specific task, another approach is to aggregate findings across multiple studies to identify consistent exercise effects that are not necessarily task specific. One such meta-analysis aggregated all exercise intervention fMRI task-based studies and identified regions with consistent effects across different tasks, samples, and intervention durations (ranging from acute to multiple weeks) [64]. Notably, the two most consistent exercise-induced differences were in the left intraparietal sulcus region and the right precuneus. First, the left intraparietal sulcus region was prominent across all of the exercise intervention studies, and even when only focused on studies with participants over 35 years of age or on long-term interventions (>12 weeks). This suggests that the left intraparietal sulcus effect may be largely driven by older samples and longer interventions and is an area often associated with dorsal attentional network processing [94]. The right precuneus effect was observed in studies of interventions less than 12 weeks, in studies with participants less than 35 years old, and in studies focused only on physically inactive subjects. This region has been associated with a variety of processes that integrate internal and externally driven input [95] and has been associated with a default mental state not associated with a particular task [96, 97]. This leads to an interpretation that exercise interventions at different ages and durations impact networks differently, i.e., younger and shorter interventions impact the default mode processing, and older and longer interventions impact the attentional networks. This leads to important questions pertaining to how exercise specifically impacts networks processing rather than local BOLD changes.

7 Connectivity-Based fMRI

It is becoming increasingly important to go beyond looking at local changes in the brain to examine the effects of exercise on brain connectivity and networks. One basic form of this approach examines the correlations in the BOLD fMRI time series between non-adjacent voxels in the brain in order to characterize different networks such as the default mode, dorsal attention, sensory, and motor networks [98]. This type of connectivity can either be task-based [99] or it can involve what is referred to as resting state fMRI. Resting state fMRI involves examining the spontaneous fluctuations in the BOLD response and quantifying the connectivity between brain areas in the absence of any task or stimulus besides a visual fixation cue [100].

The findings from studies examining the effects of exercise on fMRI network connectivity fMRI are mixed [20]. One of the more prominent networks associated with the effects of exercise is the default mode network (DMN) including the bilateral posterior cingulate cortex, precuneus, inferior parietal lobe, medial prefrontal cortex, middle frontal gyrus, and middle temporal cortex [96], and the frontal executive network (FE), including the anterior prefrontal cortex, insula, and frontal operculum [101]. Whereas the FE network has been involved in higher-level cognitive functions such as attentional control and selective attention [101], the DMN has been closely related to the episodic memory network, which also includes medial temporal lobe structures along with the aforementioned DMN network regions [102].

It is interesting that one intervention study in older adults examining the effects of 1 year of aerobic (walking) exercise intervention on fMRI network connectivity reported increases in connectivity of both the DMN and FE networks [103]. Furthermore, the magnitude of changes in functional connectivity in the FE was specifically associated with the magnitude of improvement in executive function task performance [103]. In support of this interventional finding is a cross-sectional study reporting that older adults who were more physically active had a stronger DMN connectivity [104, 105]. Looking more specifically at memory network connectivity, after just 10 min of cycling in young adults, there was an increase in task-based connectivity between the hippocampus and the parahippocampal cortex, fusiform gyrus, and inferior parietal lobule [82]. All of these regions have been associated with a brain-wide episodic memory network and overlap with the aforementioned DMN [106]. This work suggests that both single and multiple sessions of exercise can lead to enhancement of functional networks associated with episodic memory.

It is worth noting, though, that at least a couple long-term interventions studies in older adults did not report direct significant differences in typical connectivity measures due to exercise; instead, they report that brain network changes are contingent on the changes in fitness level [107, 108]. For instance, one study involving a 6-month aerobic exercise intervention found no significant difference in the connectivity patterns due to intervention, yet did find that aerobic capacity (VO_2-peak) was positively related to hippocampal network connectivity, and negatively related to somatosensory network connectivity [109]. In another study that involved a 12-week aerobic exercise intervention, there also was no significant difference in connectivity patterns, however, they did observe that the magnitude of connectivity effects prior to the intervention predicted long-term intervention connectivity and working memory improvements [107]. Overall, although the effect of exercise on connectivity was not clear in some studies, there is support for the idea that the magnitude of the benefits of exercise can be an important for indexing and predicting the magnitude of positive effects of exercise.

Finally, another approach to examining connectivity patterns is to quantify how segregated or modular each of the networks (e.g., default mode, dorsal attention, sensorimotor, etc.) is from each other. This approach quantifies network modularity during fMRI resting state. One study observed that after 6 months of either aerobic exercise intervention or even a stretching and balancing intervention, there was higher brain network modularity and that this was related to improvements in executive function as measured by reasoning, task-switching, and working memory task performance [108].

In summary, there are notable effects of exercise on brain network connectivity, particularly in functional networks related to executive function or memory processing. Still, some of these effects require nuanced approaches to detect and, in turn, provide intriguing insights into how exercise affects functional brain network processing.

8 Multi-modal Imaging Outcomes After Exercise in Mild Cognitive Impairment

Mild cognitive impairment (MCI) due to Alzheimer's disease (AD) is associated with early memory loss, neuropathology, inefficient or ineffective neural processing, and increased risk for conversion to Alzheimer's dementia. Unfortunately, treatments aimed at improving clinical symptoms or biomarkers related to neural function have not shown much promise. Physical exercise is often recommended for people diagnosed with MCI, primarily because of its widely reported cognitive benefits in healthy older adults. However, before exercise can be widely prescribed in MCI, it is important to document whether or not exercise alters brain structure and function. In a series of published papers among the same cohort of participants, we examined the effects of exercise training on multi-modal magnetic resonance imaging outcomes. Seventeen MCI participants and 18 cognitively intact controls, similar in sex, age, education, genetic risk, and medication use, volunteered for a 12-week exercise intervention consisting of supervised treadmill walking at a moderate intensity. Both MCI and control participants significantly increased their cardiorespiratory fitness by approximately 10% on a treadmill exercise test. Before and after the exercise intervention, participants completed a comprehensive neuropsychological battery and an MRI scan. We found consistent evidence that exercise training impacts both cognition and induces neuroplasticity among individuals diagnosed with MCI.

Participants completed a fMRI famous name discrimination task, which engages the hippocampal memory system (aka, default mode network (DMN)) [34]. Performance was consistently high (by design) and did not change after 12 weeks of treadmill walking. However, 11 brain regions activated during the semantic memory task showed a significant decrease in activation intensity following the intervention that was similar between the MCI and healthy groups, including the posterior cingulate cortex, which is considered the hub of the DMN and a location where amyloid plaque accumulates. Using the bilateral posterior cingulate/precuneus as a seed region for a resting state functional connectivity analysis, both groups showed increased connectivity with the left postcentral gyrus. Moreover, the MCI group showed increased connectivity in 10 additional regions spanning frontal, parietal, temporal and insular lobes, and the cerebellum [110]. In a subsequent analysis focused on hippocampal connectivity, MCI participants evidenced significantly less posterior hippocampal connectivity prior to the intervention, and both groups showed increased hippocampal connectivity after 12 weeks of walking. Greater changes in anterior and posterior hippocampal connectivity with right posterior cingulate cortex were associated with improved verbal recognition memory performance in the MCI participants [111].

Utilizing perfusion imaging, a technique termed arterial spin labeling, we also found that MCI participants, but not healthy controls, showed a decrease in CBF in the left insula, which showed hyperperfusion relative to healthy older adults at baseline [112]. Additionally, exercise improved episodic memory (measured by the Rey Auditory Verbal Learning Test) and verbal fluency (measured by the Controlled Oral Word Association Test) in both groups. In the MCI participants, reduced CBF in the anterior cingulate cortex and left insula was associated with improved verbal fluency performance. We further investigated associations between changes in network connectivity with more detailed metrics of phonemic verbal fluency, including the frequency of word use, number of syllables in each word, and switches to different categories of words. We found significant associations between exercise-induced increases in cerebellum connectivity with the bilateral precuneus and greater word frequency, increases in the number of category switches, and a reduction in the number of syllables of the words generated, but these effects were observed in the healthy controls and not in the MCI [113].

Looking at MRI measures of brain structure, we found that improvement in cardiorespiratory fitness (VO_{2peak}) due to the intervention was positively correlated with cortical thickness change in the bilateral insula, precentral gyri, precuneus, posterior cingulate, and inferior and superior frontal cortices [114]. Moreover, MCI participants exhibited stronger positive correlations compared to healthy elders in the left insula and superior temporal gyrus. The cortical gray matter layer has been shown to decline with age, and the decline in cortical thickness is greater in MCI and AD patients. Exercise training and improved fitness may help boost gray matter thickness, although the mechanisms for these changes have not been established. Diffusion-weighted MR imaging is a method that provides additional insight into gray matter microstructure and involves tracking the directionality and rate of water diffusion across gray matter regions. After the 12-week walking intervention, both MCI and healthy control participants showed improved performance on Trial 1 of an auditory verbal learning task, and the Controlled Oral Word Association Test (COWAT), a measure of verbal fluency. An analysis of cortical gray matter diffusivity showed individuals diagnosed with MCI, compared to healthy controls, had a greater increase in mean diffusivity (MD) in the left insular cortex, which was positively associated with improvement in verbal fluency performance [14]. Additionally, both groups showed an increase in left insular and left and right cerebellar MD. Increases in left insular diffusivity were similarly found to be positively associated with improvements in COWAT performance in both groups, while increases in cerebellar MD were related to gains in episodic memory performance. Overall, these results suggest that walking exercise may impact neural networks that mediate verbal fluency performance in older adults by remodeling of the microstructural properties of gray matter in the cerebral cortex.

Ultimately complex cognition is governed by the interactions between large-scale brain networks. Three core brain networks are salience network, the hippocampal memory network also known as the DMN, and the fronto-parietal network, which coordinate to engage in a range of cognitive functions. We investigated the effects of the exercise intervention on within- and between-network functional connectivity of the DMN, frontoparietal network (FPN), and salience network (SAL) in our sample of healthy older adults and those diagnosed with MCI, and further examined the associations between changes in network connectivity and changes in cognitive performance. We observed increased connectivity within the DMN and the salience network, and increased connectivity between all three networks after exercise training [115]. Interestingly, increased connectivity within the salience network and increased connectivity between the salience network and FPN were associated with enhanced story recall performance in both groups (Fig. 3). Collectively, these studies provide evidence that large scale networks are modified after exercise training in older adults, even those with mild memory loss.

These findings suggest exercise may improve neural efficiency during memory retrieval in MCI and cognitively intact older adults and may lead to improvement in cognitive function. Clinical trials are needed to determine if exercise is effective in delaying conversion to AD.

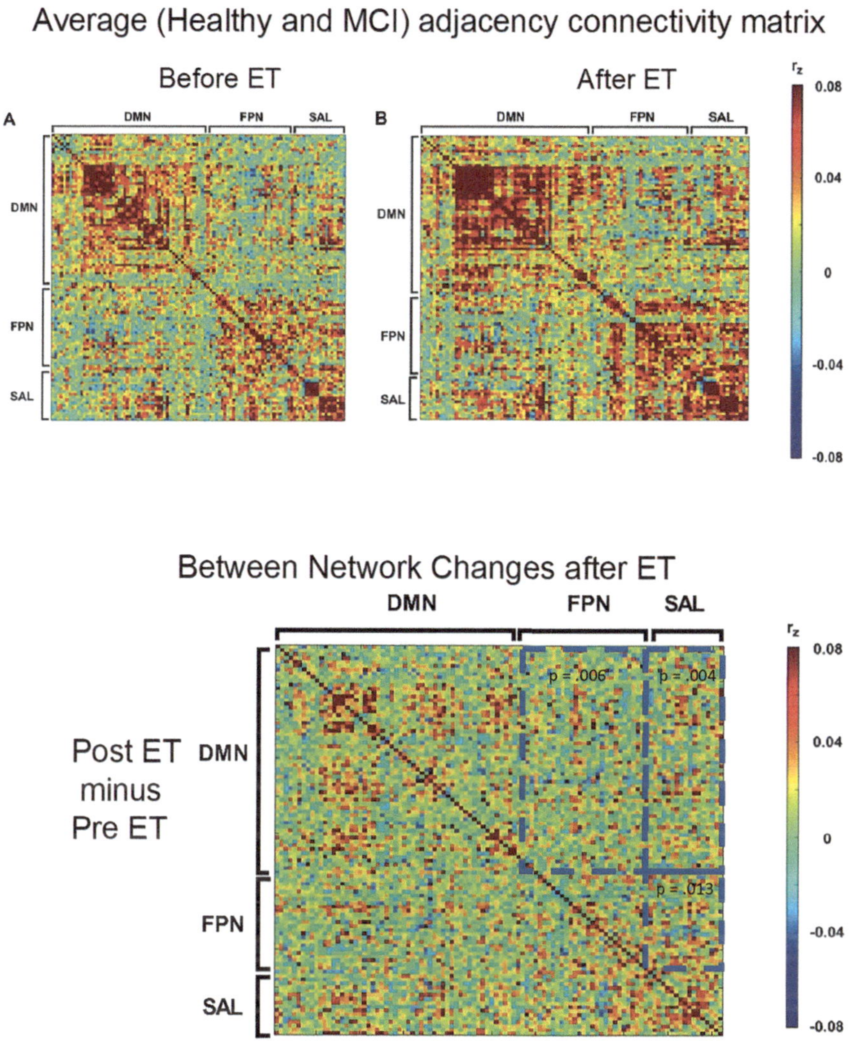

Fig. 3 Group-averaged connectivity adjacency matrices before ET (top left), after ET (top right), and the change from pre- to post-ET (bottom). The shaded areas (bottom) highlight significantly increased connectivity between default mode network (DMN), fronto-parietal network (FPN), and salience network (SAL) in both healthy older adults and older adults diagnosed with MCI. Changes between the SAL and FPN were related to improved episodic memory performance after ET in both groups. (Adapted from Won et al. [115])

9 Potential Mechanisms from Animal Models

Elucidating the molecular mechanisms as to how chronic exercise promotes brain health in humans remains challenging to address, and research efforts often rely on circulating peripheral biomarkers that may not parallel the cerebral environment [116]. Nevertheless, animal models of biological and pathological aging suggest exercise is, in part, neuroprotective through the upregulation of potent neurotrophins, growth factors, and myokines that, over time, promote the proliferation and differentiation of quiescent adult neural stem cells in a process termed adult neurogenesis [117, 118]. Of note, animal studies have shown that in the mammalian brain, adult neurogenesis takes place in the subventricular zone of the lateral ventricles, and the sub granular zone of the dentate gyrus in the hippocampus. Importantly, because the hippocampus is a sub-cortical structure that is particularly susceptible to neurodegeneration in pathologies such as Alzheimer's Disease [119], adult neurogenesis is regarded as a promising way to treat neurodegenerative pathologies in humans.

Extensively studied, neurotrophins are growth factors that are widely expressed throughout the central nervous system and that play fundamental roles in neural plasticity as well as the development and maintenance of neurons [120]. Though there are multiple neurotrophins, brain derived neurotrophic factor (BDNF) continues to be heavily studied in the field of exercise neuroscience given that it is a positive regulator of adult neurogenesis and highly inducible by aerobic exercise [121]. In this context, aerobic exercise can upregulate mRNA expression of synaptic proteins such as synapsin I [122] and synaptotagmin [123], as well as of BDNF's high affinity receptor TrkB [122]. Moreover, exercise can also upregulate signal transduction pathways such as the MAPK ERK 1/2 pathway and transcription factors such as cyclic AMP response element (CREB) that are closely associated with neuroplasticity. It is important to highlight here that these findings are not exclusive to acute exercise, and animal models consistently show chronic aerobic exercise upregulates the expression of BDNF [124], promotes hippocampal neurogenesis, and improves performance in behavioral assays related to memory and learning [125]. Beyond BDNF, exercise can facilitate neuroprotection through insulin growth factor 1 (IGF-1) and vascular endothelial growth factor (VEGF) [126]. IGF-1 is a growth factor that is essential for cellular growth, and IGF-1 signaling pathways have been implicated in neurogenesis and cognitive health. In aging rats, IGF-1 upregulation is associated with higher synaptic plasticity and improved performance in spatial memory tasks [127]. As it relates to physical activity, exercise increases the secretion of IGF-1 [128] and promotes the activation of cellular signaling pathways such as MAPK that ultimately facilitate long-term potentiation and neuroplasticity [126]. Further, IGF-1 may also interact with BDNF, as peripheral IGF-1 blocking can reduce spatial recall, decrease phosphorylation of MAPK, and attenuate synaptic plasticity [129]. Lastly, VEGF is an angiogenic growth factor that, when upregulated, can promote adult neurogenesis, presumably via increased angiogenesis within the neurogenic niche [130]. Early studies show exercise-related improvements in cognitive tasks and neurogenesis are related to the VEGF2/Filk-1 signaling pathway [131], which is known to upregulate the expression of cell survival mediators such as PI3K/AKT1 signaling.

Increased interest in skeletal muscle as a secretory organ has led to the identification of exercise-related myokines that facilitate the neuroprotective benefits of physical activity and exercise [117]. Myokines are muscle-derived factors that can play autocrine, paracrine, and endocrine roles to regulate physiological processes including metabolism and brain health. Among these, irisin and cathepsin-B (CTSB) are promising myokines, given their responsiveness to exercise and their ability to cross the blood-brain barrier and affect brain function and neurogenesis [117]. Encoded by the *FNDC5* gene, irisin is a myokine that can promote neuronal proliferation and differentiation [132] as well as engage neuroprotective signaling pathways such as PI3K/AKT and ERK 1/2 [133]. Moreover, in cortical neurons, irisin overexpression is accompanied by BDNF overexpression and peripheral administration of irisin can lead to hippocampal neurogenesis through BDNF [134]. Importantly, while there is evidence to suggest irisin can be upregulated following exercise, the degree to which irisin is involved in a endocrine skeletal muscle-brain crosstalk is currently under debate and investigation. Meanwhile, CTSB is a family of lysosomal cysteine proteases that recent studies may function as an exercise-inducible myokine mediating the neuroprotective effects of exercise [135]. Indeed, in vitro studies show AMPK-agonist AICAR leads to CTSB secretion from L6 muscle cells, which has been replicated in gastrocnemius muscle from mice that underwent 30 days of voluntary aerobic exercise [136]. Additionally, aerobic exercise can also increase CTSB gene expression in the hippocampus and may be involved in hippocampal adult neurogenesis via upregulation of *doublecortin* and *BDNF* gene expression. Moreover, because exercise can induce hypoxia within the brain, it may further lead to CTSB upregulation and facilitate neuroprotection via the clearance of cytotoxic debris such as amyloid peptides [137]. However, it is worth highlighting here that CTSB is a novel candidate myokine that needs to be further studied as there is limited evidence supporting the reported possible neuroprotective benefits. Intriguingly, though not fully understood, there is some evidence to suggest skeletal muscle-derived VEGF may play an important role in exercise-related hippocampal neurogenesis [138], and thus, VEGF may also function as a myokine following physical activity.

References

1. Batouli SAH, Saba V (2017) At least eighty percent of brain grey matter is modifiable by physical activity: a review study. Behav Brain Res 332:204–217
2. Rovio S, Spulber G, Nieminen LJ, Niskanen E, Winblad B, Tuomilehto J, Nissinen A, Soininen H, Kivipelto M (2010) The effect of midlife physical activity on structural brain changes in the elderly. Neurobiol Aging 31:1927–1936
3. Erickson KI, Leckie RL, Weinstein AM (2014) Physical activity, fitness, and gray matter volume. Neurobiol Aging 35:S20–S28

4. Smith JC, Nielson KA, Woodard JL et al (2014) Physical activity reduces hippocampal atrophy in elders at genetic risk for Alzheimer's disease. Front Aging Neurosci 6:61
5. Northey JM, Rattray B, Pumpa KL, Pryor DJ, Fraser MA, Shaw ME, Anstey KJ, Cherbuin N (2020) Objectively measured physical activity is associated with dorsolateral prefrontal cortex volume in older adults. NeuroImage 221:117150
6. Erickson KI, Raji CA, Lopez OL, Becker JT, Rosano C, Newman AB, Gach HM, Thompson PM, Ho AJ, Kuller LH (2010) Physical activity predicts gray matter volume in late adulthood: the cardiovascular health study. Neurology 75:1415–1422
7. Erickson KI, Prakash RS, Voss MW, Chaddock L, Hu L, Morris KS, White SM, Wójcicki TR, McAuley E, Kramer AF (2009) Aerobic fitness is associated with hippocampal volume in elderly humans. Hippocampus 19:1030–1039
8. Weinstein AM, Voss MW, Prakash RS et al (2012) The association between aerobic fitness and executive function is mediated by prefrontal cortex volume. Brain Behav Immun 26:811–819
9. Stillman CM, Esteban-Cornejo I, Brown B, Bender CM, Erickson KI (2020) Effects of exercise on brain and cognition across age groups and health states. Trends Neurosci 43:533–543
10. Colcombe SJ, Erickson KI, Scalf PE, Kim JS, Prakash R, McAuley E, Elavsky S, Marquez DX, Hu L, Kramer AF (2006) Aerobic exercise training increases brain volume in aging humans. J Gerontol A Biol Sci Med Sci 61:1166–1170
11. Erickson KI, Voss MW, Prakash RS et al (2011) Exercise training increases size of hippocampus and improves memory. Proc Natl Acad Sci USA 108:3017–3022
12. Firth J, Stubbs B, Vancampfort D, Schuch F, Lagopoulos J, Rosenbaum S, Ward PB (2018) Effect of aerobic exercise on hippocampal volume in humans: a systematic review and meta-analysis. NeuroImage 166:230–238
13. Gogniat MA, Robinson TL, Miller LS (2021) Exercise interventions do not impact brain volume change in older adults: a systematic review and meta-analysis. Neurobiol Aging 101:230–246
14. Callow DD, Won J, Pena GS, Jordan LS, Arnold-Nedimala NA, Kommula Y, Nielson KA, Smith JC (2021) Exercise training-related changes in cortical gray matter diffusivity and cognitive function in mild cognitive impairment and healthy older adults. Front Aging Neurosci 13:645258
15. Frodl T, Strehl K, Carballedo A, Tozzi L, Doyle M, Amico F, Gormley J, Lavelle G, O'Keane V (2020) Aerobic exercise increases hippocampal subfield volumes in younger adults and prevents volume decline in the elderly. Brain Imaging Behav 14:1577–1587
16. Duzel E, van Praag H, Sendtner M (2016) Can physical exercise in old age improve memory and hippocampal function? Brain 139:662–673
17. Walhovd KB, Johansen-Berg H, Káradóttir RT (2014) Unraveling the secrets of white matter--bridging the gap between cellular, animal and human imaging studies. Neuroscience 276:2–13
18. Davis SW, Dennis NA, Buchler NG, White LE, Madden DJ, Cabeza R (2009) Assessing the effects of age on long white matter tracts using diffusion tensor tractography. NeuroImage 46:530–541
19. Coelho A, Fernandes HM, Magalhães R et al (2021) Signatures of white-matter microstructure degradation during aging and its association with cognitive status. Sci Rep 11:4517
20. Won J, Callow DD, Pena GS, Gogniat MA, Kommula Y, Arnold-Nedimala NA, Jordan LS, Smith JC (2021) Evidence for exercise-related plasticity in functional and structural neural network connectivity. Neurosci Biobehav Rev 131:923
21. Oberlin LE, Verstynen TD, Burzynska AZ et al (2016) White matter microstructure mediates the relationship between cardiorespiratory fitness and spatial working memory in older adults. NeuroImage 131:91–101
22. Hayes SM, Salat DH, Forman DE, Sperling RA, Verfaellie M (2015) Cardiorespiratory fitness is associated with white matter integrity in aging. Ann Clin Transl Neurol 2:688–698

23. Burzynska AZ, Jiao Y, Knecht AM, Fanning J, Awick EA, Chen T, Gothe N, Voss MW, McAuley E, Kramer AF (2017) White matter integrity declined over 6-months, but dance intervention improved integrity of the fornix of older adults. Front Aging Neurosci 9:59
24. Pani J, Eikenes L, Reitlo LS, Stensvold D, Wisløff U, Håberg AK (2022) Effects of a 5-year exercise intervention on white matter microstructural organization in older adults. A generation 100 substudy. Front Aging Neurosci 14:859383
25. Madden DJ, Bennett IJ, Song AW (2009) Cerebral white matter integrity and cognitive aging: contributions from diffusion tensor imaging. Neuropsychol Rev 19:415–435
26. Power MC, Su D, Wu A et al (2019) Association of white matter microstructural integrity with cognition and dementia. Neurobiol Aging 83:63–72
27. Callow DD, Purcell JJ, Won J, Smith JC (2022) Neurite dispersion and density mediates the relationship between cardiorespiratory fitness and cognition in healthy younger adults. Neuropsychologia 169:108207
28. Baxi M, Di Biase MA, Lyall AE, Cetin-Karayumak S, Seitz J, Ning L, Makris N, Rosene D, Kubicki M, Rathi Y (2020) Quantifying genetic and environmental influence on gray matter microstructure using diffusion MRI. Cereb Cortex 30:6191–6205
29. Tian Q, Erickson KI, Simonsick EM et al (2014) Physical activity predicts microstructural integrity in memory-related networks in very old adults. J Gerontol A Biol Sci Med Sci 69:1284–1290
30. Tian Q, Schrack JA, Landman BA, Resnick SM, Ferrucci L (2022) Longitudinal associations of absolute versus relative moderate-to-vigorous physical activity with brain microstructural decline in aging. Neurobiol Aging 116:25–31
31. Genç E, Fraenz C, Schlüter C, Friedrich P, Hossiep R, Voelkle MC, Ling JM, Güntürkün O, Jung RE (2018) Diffusion markers of dendritic density and arborization in gray matter predict differences in intelligence. Nat Commun 9:1905
32. Kleemeyer MM, Kühn S, Prindle J, Bodammer NC, Brechtel L, Garthe A, Kempermann G, Schaefer S, Lindenberger U (2016) Changes in fitness are associated with changes in hippocampal microstructure and hippocampal volume among older adults. NeuroImage 131:155–161
33. Islam MR, Luo R, Valaris S et al (2020) Diffusion tensor-MRI detects exercise-induced neuroplasticity in the hippocampal microstructure in mice. Brain Plast 5:147–159
34. Smith JC, Nielson KA, Antuono P, Lyons J-A, Hanson RJ, Butts AM, Hantke NC, Verber MD (2013) Semantic memory functional MRI and cognitive function after exercise intervention in mild cognitive impairment. J Alzheimer's Dis 37:197–215
35. Sagi Y, Tavor I, Hofstetter S, Tzur-Moryosef S, Blumenfeld-Katzir T, Assaf Y (2012) Learning in the fast lane: new insights into neuroplasticity. Neuron 73:1195–1203
36. Callow DD, Won J, Alfini AJ, Purcell JJ, Weiss LR, Zhan W, Smith JC (2021) Microstructural plasticity in the hippocampus of healthy older adults after acute exercise. Med Sci Sports Exerc 53:1928–1936
37. Packer N, Hoffman-Goetz L (2015) Acute exercise increases hippocampal TNF-α, Caspase-3 and Caspase-7 expression in healthy young and older mice. J Sports Med Phys Fitness 55:368–376
38. Colcombe S, Kramer AF (2003) Fitness effects on the cognitive function of older adults: a meta-analytic study. Psychol Sci 14:125–130
39. Tomporowski PD (2003) Effects of acute bouts of exercise on cognition. Acta Psychol 112:297–324
40. Leung FP, Yung LM, Laher I, Yao X, Chen ZY, Huang Y (2008) Exercise, vascular wall and cardiovascular diseases: an update (Part 1). Sports Med 38:1009–1024
41. MacIntosh BJ, Crane DE, Sage MD, Rajab AS, Donahue MJ, McIlroy WE, Middleton LE (2014) Impact of a single bout of aerobic exercise on regional brain perfusion and activation responses in healthy young adults. PLoS One 9(1):e85163
42. Zatorre RJ, Fields RD, Johansen-Berg H (2012) Plasticity in gray and white: neuroimaging changes in brain structure during learning. Nat Neurosci 15:528–536

43. Alsop DC, Detre JA, Golay X et al (2015) Recommended implementation of arterial spin-labeled perfusion MRI for clinical applications: a consensus of the ISMRM perfusion study group and the European consortium for ASL in dementia. Magn Reson Med 73:102–116
44. Williams DS, Detre JA, Leigh JS, Koretsky AP (1992) Magnetic resonance imaging of perfusion using spin inversion of arterial water. Proc Natl Acad Sci USA 89:212–216
45. Detre JA, Leigh JS, Williams DS, Koretsky AP (1992) Perfusion imaging. Magn Reson Med 23:37–45
46. Alsop DC, Dai W, Grossman M, Detre JA (2010) Arterial spin labeling blood flow MRI: its role in the early characterization of Alzheimer's disease. J Alzheimer's Dis 20:871–880
47. Buxton RB (2005) Quantifying CBF with arterial spin labeling. J Magn Reson Imaging 22:723–726
48. Aslan S, Xu F, Wang PL, Uh J, Yezhuvath US, van Osch M, Lu H (2010) Estimation of labeling efficiency in pseudocontinuous arterial spin labeling. Magn Reson Med 63:765–771
49. Jezzard P, Chappell MA, Okell TW (2018) Arterial spin labeling for the measurement of cerebral perfusion and angiography. J Cereb Blood Flow Metab 38:603–626
50. Lu H, Clingman C, Golay X, van Zijl PCM (2004) Determining the longitudinal relaxation time (T1) of blood at 3.0 Tesla. Magn Reson Med 52:679–682
51. Kwong KK, Chesler DA, Weisskoff RM, Donahue KM, Davis TL, Ostergaard L, Campbell TA, Rosen BR (1995) MR perfusion studies with T1-weighted echo planar imaging. Magn Reson Med 34:878–887
52. Dai W, Garcia D, de Bazelaire C, Alsop DC (2008) Continuous flow-driven inversion for arterial spin labeling using pulsed radio frequency and gradient fields. Magn Reson Med 60:1488–1497
53. Smith JC, Paulson ES, Cook DB, Verber MD, Tian Q (2010) Detecting changes in human cerebral blood flow after acute exercise using arterial spin labeling: implications for fMRI. J Neurosci Methods 191:258–262
54. Robertson AD, Crane DE, Rajab AS, Swardfager W, Marzolini S, Shirzadi Z, Middleton LE, MacIntosh BJ (2015) Exercise intensity modulates the change in cerebral blood flow following aerobic exercise in chronic stroke. Exp Brain Res 233:2467–2475
55. Chapman SB, Aslan S, Spence JS, DeFina LF, Keebler MW, Didehbani N, Lu H (2013) Shorter term aerobic exercise improves brain, cognition, and cardiovascular fitness in aging. Front Aging Neurosci 5:1–9
56. Thomas BP, Yezhuvath US, Tseng BY, Liu P, Levine BD, Zhang R, Lu H (2013) Life-long aerobic exercise preserved baseline cerebral blood flow but reduced vascular reactivity to CO2. J Magn Reson Imaging 38:1177–1183
57. Swain RA, Harris AB, Wiener EC, Dutka MV, Morris HD, Theien BE, Konda S, Engberg K, Lauterbur PC, Greenough WT (2003) Prolonged exercise induces angiogenesis and increases cerebral blood volume in primary motor cortex of the rat. Neuroscience 117:1037–1046
58. Ogawa S, Lee TM, Nayak AS, Glynn P (1990) Oxygenation-sensitive contrast in magnetic resonance image of rodent brain at high magnetic fields. Magn Reson Med 14:68–78
59. Logothetis NK, Pauls J, Augath M, Trinath T, Oeltermann A (2001) Neurophysiological investigation of the basis of the fMRI signal. Nature 412:150–157
60. Sanders LMJ, Hortobágyi T, la Bastide-van GS, van der Zee EA, van Heuvelen MJG (2019) Dose-response relationship between exercise and cognitive function in older adults with and without cognitive impairment: a systematic review and meta-analysis. PLoS One 14:e0210036
61. Voss MW, Nagamatsu LS, Liu-Ambrose T, Kramer AF (2011) Exercise, brain, and cognition across the life span. J Appl Physiol 111:1505–1513
62. Basso JC, Suzuki WA (2017) The effects of acute exercise on mood, cognition, neurophysiology, and neurochemical pathways: a review. Brain Plast 2:127–152
63. Herold F, Aye N, Lehmann N, Taubert M, Müller NG (2020) The contribution of functional magnetic resonance imaging to the understanding of the effects of acute physical exercise on cognition. Brain Sci 10:175

64. Yu Q, Herold F, Becker B, Klugah-Brown B, Zhang Y, Perrey S, Veronese N, Müller NG, Kramer AF, Zou L (2021) Cognitive benefits of exercise interventions: an fMRI activation likelihood estimation meta-analysis. Brain Struct Funct 226:601–619
65. Eriksen BA, Eriksen CW (1974) Effects of noise letters upon the identification of a target letter in a nonsearch task. Percept Psychophys 16:143–149
66. Verbruggen F, Logan GD (2008) Response inhibition in the stop-signal paradigm. Trends Cogn Sci 12:418–424
67. Siemann J, Herrmann M, Galashan D (2018) The effect of feature-based attention on flanker interference processing: an fMRI-constrained source analysis. Sci Rep 8:1580
68. Siemann J, Herrmann M, Galashan D (2016) fMRI-constrained source analysis reveals early top-down modulations of interference processing using a flanker task. NeuroImage 136:45–56
69. Won J, Alfini AJ, Weiss LR, Callow DD, Smith JC (2019) Brain activation during executive control after acute exercise in older adults. Int J Psychophysiol 146:240–248
70. Liu-Ambrose T, Nagamatsu LS, Voss MW, Khan KM, Handy TC (2012) Resistance training and functional plasticity of the aging brain: a 12-month randomized controlled trial. Neurobiol Aging 33:1690–1698
71. Pensel MC, Daamen M, Scheef L, Knigge HU, Rojas Vega S, Martin JA, Schild HH, Strüder HK, Boecker H (2018) Executive control processes are associated with individual fitness outcomes following regular exercise training: blood lactate profile curves and neuroimaging findings. Sci Rep 8(1):4893
72. Simmonds DJ, Pekar JJ, Mostofsky SH (2008) Meta-analysis of Go/No-go tasks demonstrating that fMRI activation associated with response inhibition is task-dependent. Neuropsychologia 46:224–232
73. Mehren A, Diaz Luque C, Brandes M, Lam AP, Thiel CM, Philipsen A, Özyurt J (2019) Intensity-dependent effects of acute exercise on executive function. Neural Plast 2019:8608317
74. Diamond A (2013) Executive functions. Annu Rev Psychol 64:135–168
75. Li L, Men W-W, Chang Y-K, Fan M-X, Ji L, Wei G-X (2014) Acute aerobic exercise increases cortical activity during working memory: a functional MRI study in female college students. PLoS One 9:e99222
76. Li L, Zhang S, Cui J, Chen L-Z, Wang X, Fan M, Wei G-X (2019) Fitness-dependent effect of acute aerobic exercise on executive function. Front Physiol 10:902
77. Nishiguchi S, Yamada M, Tanigawa T et al (2015) A 12-week physical and cognitive exercise program can improve cognitive function and neural efficiency in community-dwelling older adults: a randomized controlled trial. J Am Geriatr Soc 63:1355–1363
78. Owen AM, McMillan KM, Laird AR, Bullmore E (2005) N-back working memory paradigm: a meta-analysis of normative functional neuroimaging studies. Hum Brain Mapp 25:46–59
79. Won J, Alfini AJ, Weiss LR, Michelson CS, Callow DD, Ranadive SM, Gentili RJ, Smith JC (2020) Semantic memory activation after acute exercise in healthy older adults. J Int Neuropsychol Soc 25(6):557–568
80. Wagner G, Herbsleb M, de la Cruz F, Schumann A, Köhler S, Puta C, Gabriel HW, Reichenbach JR, Bär K-J (2017) Changes in fMRI activation in anterior hippocampus and motor cortex during memory retrieval after an intense exercise intervention. Biol Psychol 124:65–78
81. Jeong W, Chung CK, Kim JS (2015) Episodic memory in aspects of large-scale brain networks. Front Hum Neurosci 9:454
82. Suwabe K, Byun K, Hyodo K et al (2018) Rapid stimulation of human dentate gyrus function with acute mild exercise. Proc Natl Acad Sci 115:10487–10492
83. Voss MW, Soto C, Yoo S, Sodoma M, Vivar C, van Praag H (2019) Exercise and hippocampal memory systems. Trends Cogn Sci 23:318–333
84. Jennen L, Mazereel V, Lecei A, Samaey C, Vancampfort D, van Winkel R (2022) Exercise to spot the differences: a framework for the effect of exercise on hippocampal pattern separation in humans. Rev Neurosci 33:555–582

85. Bothe N, Zschucke E, Dimeo F, Heinz A, Wüstenberg T, Ströhle A (2013) Acute exercise influences reward processing in highly trained and untrained men. Med Sci Sports Exerc 45:583–591
86. Ochsner KN, Silvers JA, Buhle JT (2012) Functional imaging studies of emotion regulation: a synthetic review and evolving model of the cognitive control of emotion. Ann N Y Acad Sci 1251:E1–E24
87. Carp J, Park J, Polk TA, Park DC (2011) Age differences in neural distinctiveness revealed by multi-voxel pattern analysis. NeuroImage 56:736–743
88. Dennis NA, Cabeza R (2011) Age-related dedifferentiation of learning systems: an fMRI study of implicit and explicit learning. Neurobiol Aging 32:2318.e17–2318.e30
89. Koen JD, Srokova S, Rugg MD (2020) Age-related neural dedifferentiation and cognition. Curr Opin Behav Sci 32:7–14
90. Koen JD, Rugg MD (2019) Neural dedifferentiation in the aging brain. Trends Cogn Sci 23:547–559
91. Park DC, Polk TA, Park R, Minear M, Savage A, Smith MR (2004) Aging reduces neural specialization in ventral visual cortex. Proc Natl Acad Sci USA 101:13091–13095
92. Srokova S, Hill PF, Koen JD, King DR, Rugg MD (2020) Neural differentiation is moderated by age in scene-selective, but not face-selective, cortical regions. eNeuro 7:ENEURO.0142-20.2020
93. Kleemeyer MM, Polk TA, Schaefer S, Bodammer NC, Brechtel L, Lindenberger U (2017) Exercise-induced fitness changes correlate with changes in neural specificity in older adults. Front Hum Neurosci 11:123
94. Vossel S, Geng JJ, Fink GR (2014) Dorsal and ventral attention systems. Neuroscientist 20:150–159
95. Cavanna AE, Trimble MR (2006) The precuneus: a review of its functional anatomy and behavioural correlates. Brain 129:564–583
96. Smallwood J, Bernhardt BC, Leech R, Bzdok D, Jefferies E, Margulies DS (2021) The default mode network in cognition: a topographical perspective. Nat Rev Neurosci 22:503–513
97. Raichle ME (2015) The brain's default mode network. Annu Rev Neurosci 38:433–447
98. Van Dijk KRA, Hedden T, Venkataraman A, Evans KC, Lazar SW, Buckner RL (2010) Intrinsic functional connectivity as a tool for human connectomics: theory, properties, and optimization. J Neurophysiol 103:297–321
99. Friston KJ (2011) Functional and effective connectivity: a review. Brain Connect 1:13–36
100. Fox MD, Raichle ME (2007) Spontaneous fluctuations in brain activity observed with functional magnetic resonance imaging. Nat Rev Neurosci 8:700–711
101. Dosenbach NUF, Visscher KM, Palmer ED, Miezin FM, Wenger KK, Kang HC, Burgund ED, Grimes AL, Schlaggar BL, Petersen SE (2006) A core system for the implementation of task sets. Neuron 50:799–812
102. Sestieri C, Corbetta M, Romani GL, Shulman GL (2011) Episodic memory retrieval, parietal cortex, and the default mode network: functional and topographic analyses. J Neurosci 31:4407–4420
103. Voss MW, Prakash RS, Erickson KI et al (2010) Plasticity of brain networks in a randomized intervention trial of exercise training in older adults. Front Aging Neurosci 2:32
104. Boraxbekk CJ, Salami A, Wåhlin A, Nyberg L (2016) Physical activity over a decade modifies age-related decline in perfusion, gray matter volume, and functional connectivity of the posterior default-mode network-a multimodal approach. NeuroImage 131:133–141
105. Voss MW, Erickson KI, Prakash RS et al (2010) Functional connectivity: a source of variance in the association between cardiorespiratory fitness and cognition? Neuropsychologia 48:1394–1406
106. Cooper RA, Ritchey M (2019) Cortico-hippocampal network connections support the multidimensional quality of episodic memory. Elife 8:e45591
107. Voss MW, Weng TB, Narayana-Kumanan K et al (2020) Acute exercise effects predict training change in cognition and connectivity. Med Sci Sports Exerc 52:131–140

108. Baniqued PL, Gallen CL, Voss MW et al (2017) Brain network modularity predicts exercise-related executive function gains in older adults. Front Aging Neurosci 9:426
109. Flodin P, Jonasson LS, Riklund K, Nyberg L, Boraxbekk CJ (2017) Does aerobic exercise influence intrinsic brain activity? An aerobic exercise intervention among healthy old adults. Front Aging Neurosci 9:267
110. Chirles TJ, Reiter K, Weiss LR, Alfini AJ, Nielson KA, Smith JC (2017) Exercise training and functional connectivity changes in mild cognitive impairment and healthy elders. J Alzheimer's Dis 57:845–856
111. Won J, Callow DD, Pena SG, Jordan LS, Arnold-Nedimala NA, Nielson KA, Smith JC (2021) Hippocampal functional connectivity and memory performance after exercise intervention in older adults with mild cognitive impairment. J Alzheimer's Dis 82:1015
112. Alfini AJ, Weiss LR, Nielson KA, Verber MD, Smith JC (2019) Resting cerebral blood flow after exercise training in mild cognitive impairment. J Alzheimer's Dis 67:671–684
113. Won J, Faroqi-Shah Y, Callow DD, Williams A, Awoyemi A, Nielson KA, Smith JC (2021) Association between greater cerebellar network connectivity and improved phonemic fluency performance after exercise training in older adults. Cerebellum 20:542
114. Reiter K, Nielson KA, Smith TJ, Weiss LR, Alfini AJ, Smith JC (2015) Improved cardiorespiratory fitness is associated with increased cortical thickness in mild cognitive impairment. J Int Neuropsychol Soc 21:757–767
115. Won J, Nielson KA, Smith JC (2023) Large-scale network connectivity and cognitive function changes after exercise training in older adults with intact cognition and mild cognitive impairment. J Alzheimers Dis Rep 7:399–413
116. Calverley TA, Ogoh S, Marley CJ et al (2020) HIITing the brain with exercise: mechanisms, consequences and practical recommendations. J Physiol 598:2513–2530
117. Severinsen MCK, Pedersen BK (2020) Muscle-organ crosstalk: the emerging roles of myokines. Endocr Rev 41:594–609
118. Zhao C, Deng W, Gage FH (2008) Mechanisms and functional implications of adult neurogenesis. Cell 132:645–660
119. Henneman WJP, Sluimer JD, Barnes J, van der Flier WM, Sluimer IC, Fox NC, Scheltens P, Vrenken H, Barkhof F (2009) Hippocampal atrophy rates in Alzheimer disease: added value over whole brain volume measures. Neurology 72:999–1007
120. Noble EE, Billington CJ, Kotz CM, Wang C (2011) The lighter side of BDNF. Am J Physiol Regul Integr Comp Physiol 300:R1053–R1069
121. Szuhany KL, Bugatti M, Otto MW (2015) A meta-analytic review of the effects of exercise on brain-derived neurotrophic factor. J Psychiatr Res 60:56–64
122. Gómez-Pinilla F, Ying Z, Roy RR, Molteni R, Edgerton VR (2002) Voluntary exercise induces a BDNF-mediated mechanism that promotes neuroplasticity. J Neurophysiol 88:2187–2195
123. Molteni R, Ying Z, Gómez-Pinilla F (2002) Differential effects of acute and chronic exercise on plasticity-related genes in the rat hippocampus revealed by microarray. Eur J Neurosci 16:1107–1116
124. Loprinzi PD (2019) Does brain-derived neurotrophic factor mediate the effects of exercise on memory? Phys Sportsmed 47:395–405
125. Marlatt MW, Potter MC, Lucassen PJ, van Praag H (2012) Running throughout middle-age improves memory function, hippocampal neurogenesis, and BDNF levels in female C57BL/6J mice. Dev Neurobiol 72:943–952
126. Cotman CW, Berchtold NC, Christie LA (2007) Exercise builds brain health: key roles of growth factor cascades and inflammation. Trends Neurosci 30:464–472
127. Pardo J, Abba MC, Lacunza E, Ogundele OM, Paiva I, Morel GR, Outeiro TF, Goya RG (2018) IGF-I gene therapy in aging rats modulates hippocampal genes relevant to memory function. J Gerontol A Biol Sci Med Sci 73:459–467
128. Ratey JJ, Loehr JE (2011) The positive impact of physical activity on cognition during adulthood: a review of underlying mechanisms, evidence and recommendations. Rev Neurosci 22:171–185

129. Ding Q, Vaynman S, Akhavan M, Ying Z, Gomez-Pinilla F (2006) Insulin-like growth factor I interfaces with brain-derived neurotrophic factor-mediated synaptic plasticity to modulate aspects of exercise-induced cognitive function. Neuroscience 140:823–833
130. Fabel K, Fabel K, Tam B, Kaufer D, Baiker A, Simmons N, Kuo CJ, Palmer TD (2003) VEGF is necessary for exercise-induced adult hippocampal neurogenesis. Eur J Neurosci 18:2803–2812
131. Cao L, Jiao X, Zuzga DS, Liu Y, Fong DM, Young D, During MJ (2004) VEGF links hippocampal activity with neurogenesis, learning and memory. Nat Genet 36:827–835
132. Moon H-S, Dincer F, Mantzoros CS (2013) Pharmacological concentrations of irisin increase cell proliferation without influencing markers of neurite outgrowth and synaptogenesis in mouse H19-7 hippocampal cell lines. Metabolism 62:1131–1136
133. Chen W, Wang L, You W, Shan T (2021) Myokines mediate the cross talk between skeletal muscle and other organs. J Cell Physiol 236:2393–2412
134. Wrann CD, White JP, Salogiannnis J, Laznik-Bogoslavski D, Wu J, Ma D, Lin JD, Greenberg ME, Spiegelman BM (2013) Exercise induces hippocampal BDNF through a PGC-1α/FNDC5 pathway. Cell Metab 18:649–659
135. Kim S, Choi J-Y, Moon S, Park D-H, Kwak H-B, Kang J-H (2019) Roles of myokines in exercise-induced improvement of neuropsychiatric function. Pflugers Arch 471:491–505
136. Moon HY, Becke A, Berron D et al (2016) Running-induced systemic cathepsin B secretion is associated with memory function. Cell Metab 24:332–340
137. Mueller-Steiner S, Zhou Y, Arai H et al (2006) Antiamyloidogenic and neuroprotective functions of cathepsin B: implications for Alzheimer's disease. Neuron 51:703–714
138. Vonderwalde I, Kovacs-Litman A (2018) Aerobic exercise promotes hippocampal neurogenesis through skeletal myofiber-derived vascular endothelial growth factor. J Physiol 596:761–763

The Effects of Exercise on Hippocampus-Based Cognitive Dysfunction in Type 2 Diabetes Mellitus: A Possible Involvement of Hippocampal Monocarboxylate Transporter 2

Takeru Shima and Hideaki Soya

Abstract Cognitive dysfunction in type 2 diabetes mellitus (T2DM) poses a significant peril not only to compromised human well-being but also to the onset of dementia, Alzheimer's disease, and depression. Given the present research findings utilizing animal models, various potential biochemical mechanisms have been reported for hippocampus-based cognitive dysfunction in T2DM. This chapter focuses on the relationship between cognitive dysfunction in T2DM and dysregulation of the hippocampal astrocyte-neuron lactate shuttle, with specific emphasis on monocarboxylate transporter 2 (MCT2). Furthermore, it provides a summary of the evidence suggesting the potentiality of exercise as a viable therapeutic intervention, encompassing not solely glycemic control but also the amelioration of cognitive dysfunction in T2DM.

Keyword Type 2 diabetes mellitus · Hippocampus · Exercise · Monocarboxylate transporter 2

T. Shima (✉)
Department of Health and Physical Education, Cooperative Faculty of Education, Gunma University, Maebashi, Gunma, Japan
e-mail: ta-shima@gunma-u.ac.jp

H. Soya
Laboratory of Exercise Biochemistry and Neuroendocrinology, Institute of Health and Sport Sciences, University of Tsukuba, Tsukuba, Ibaraki, Japan

Sport Neuroscience Division, Kokoro Division, Advanced Research Initiative for Human High Performance (ARIHHP), Institute of Health and Sport Sciences, University of Tsukuba, Tsukuba, Ibaraki, Japan

Center for Cybernics Research, University of Tsukuba, Tsukuba, Ibaraki, Japan
e-mail: soya.hideaki.gt@u.tsukuba.ac.jp

H. Soya (ed.), *Exercise Brain Stimulation for Cognitive Function and Mental Health*, Advances in Neurobiology 44, https://doi.org/10.1007/978-981-95-0066-6_15

1 Introduction

The International Diabetes Federation has reported that the global prevalence of diabetes mellitus in 2021 reached 537 million adults, with a projected increase to 783 million by 2045. Among the various types of diabetes mellitus, Type 2 diabetes mellitus (T2DM) is the predominant form, accounting for over 90% of all cases. T2DM gives rise to several complications in peripheral organs, including the kidney and retina. In addition to affecting peripheral organs, T2DM exerts detrimental effects on the brain, leading to dysfunction, specifically manifested as hippocampus-based cognitive dysfunction [1–3]. Notably, patients diagnosed with T2DM exhibit more pronounced deficits in hippocampal learning and memory function compared to individuals with Type 1 diabetes mellitus [3]. Such cognitive dysfunction associated with T2DM represents a significant disability, adversely impacting both the well-being of affected individuals and their susceptibility to debilitating conditions such as dementia, Alzheimer's disease, and depression [2, 4, 5]. Therefore, the development of an efficacious therapeutic approach targeting the hippocampus in T2DM is imperative to ensure optimal human well-being.

Since exercise has been shown to ameliorate dysfunction in both peripheral organs and the brain [6, 7], it is widely employed as a therapeutic strategy for individuals with T2DM [8, 9]. Notably, it has been suggested that exercise can effectively promote overall health in obesity, a primary contributor to the development of T2DM, whereas the use of diabetic medications may not yield comparable results [10]. Therefore, the inclusion of exercise becomes indispensable for ensuring the well-being of individuals with T2DM. This chapter presents a concise overview of the evidence linking hippocampal impairments and cognitive dysfunction in T2DM, along with the effects of exercise on mitigating cognitive dysfunction dependent on the hippocampus.

2 Impairment of Hippocampus with T2DM

The hippocampus is susceptible to biological and pathophysiological changes [11, 12], including those associated with T2DM. In terms of morphological alterations, a significant decrease in hippocampal volumes has been observed in relation to T2DM [13]. Specifically, reductions in volumes have been noted in the bilateral molecular layer, dentate gyrus (DG), cornu ammonis 4, and fimbria [14]. These morphological changes are closely related to cognitive dysfunction in T2DM. The reduction in hippocampal volumes is believed to be influenced by various molecular signaling pathways, such as decreased neurogenesis and angiogenesis, as well as increased inflammation and oxidative stress. Neurogenesis, a process commonly observed in the hippocampal DG [15, 16], has been shown to be decreased in the hippocampal DG of animal models with T2DM [17–19]. Similarly, hippocampal angiogenesis, a critical process for the formation of new microvessels, is downregulated in T2DM [20]. Animal models of T2DM demonstrate an increase in the expression of inflammation and oxidative stress factors in the hippocampus [21, 22]. These disturbances in the hippocampus contribute to the reduction in hippocampal volumes observed in T2DM. Although there are multiple pathological factors underlying these neuronal damages, insulin resistance is considered a common pathological factor in the neuronal alterations. Insulin resistance, a characteristic symptom in the peripheral organs of individuals with T2DM [23, 24], can also be observed in the hippocampus. Previous studies have suggested a potential association between hippocampal insulin resistance and cognitive dysfunction in T2DM [25–27]. Moreover, central insulin resistance may be a shared pathology between T2DM and Alzheimer's disease [28, 29]. However, some studies have indicated that hippocampal insulin signaling remains unaltered not only in early-stage T2DM mice but also in older T2DM mice [30, 31]. Thus, factors other than insulin are likely to regulate hippocampal alterations associated with T2DM. Identifying the key targets for developing therapeutic strategies targeting the hippocampus in T2DM is importance.

3 Hippocampal Lactate Shuttle in T2DM

There is a possibility that lactate plays a role in regulating hippocampal neurogenesis, angiogenesis, and inflammation. Lactate is supplied to neurons through the blood circulation and astrocytes via glycolysis and/or glycogenolysis [32–35]. Astrocytes store glycogen, which serves as a source of lactate. Lactate generated in astrocytes is released into the extracellular fluid through MCT1 and MCT4. MCT1 is also expressed in endothelial cells to facilitate the entry of lactate into the extracellular fluid. Subsequently, neurons take up lactate from the extracellular fluid through MCT2 [35]. Lactate serves as an energy substrate for neuronal activity and acts as a neuromodulator, enhancing neuronal plasticity, including neurogenesis [32, 36–39]. Compared to glucose, lactate is believed to provide support for high-demand neural activity [40]. Previous studies have reported that intracerebroventricular or hippocampal injection of lactate enhances hippocampal plasticity in healthy rodents [38, 41, 42]. Even intraperitoneal injection of lactate promotes hippocampal plasticity and cognitive function in healthy rodents [39, 43–46]. However, hippocampal lactate levels in T2DM mice have been found to be higher than those in control mice [47], indicating that simply increasing hippocampal lactate levels may not be sufficient to enhance cognitive function in T2DM. Therefore, the role of lactate transporters, specifically the MCTs, would be crucial. Pharmacological inhibition of lactate transport through MCTs in the hippocampus leads to cognitive dysfunction in healthy animals [37, 38, 48–54]. In our research, we have observed lower expression of MCT2 in the hippocampus of T2DM animals compared to control animals [55–57]. On the other hand, hippocampal expressions of MCT1 and MCT4 remain unchanged in T2DM (Fig. 1; Shima et al. [55]), suggesting that the reduced utilization of lactate via MCT2 is likely a crucial factor in hippocampal

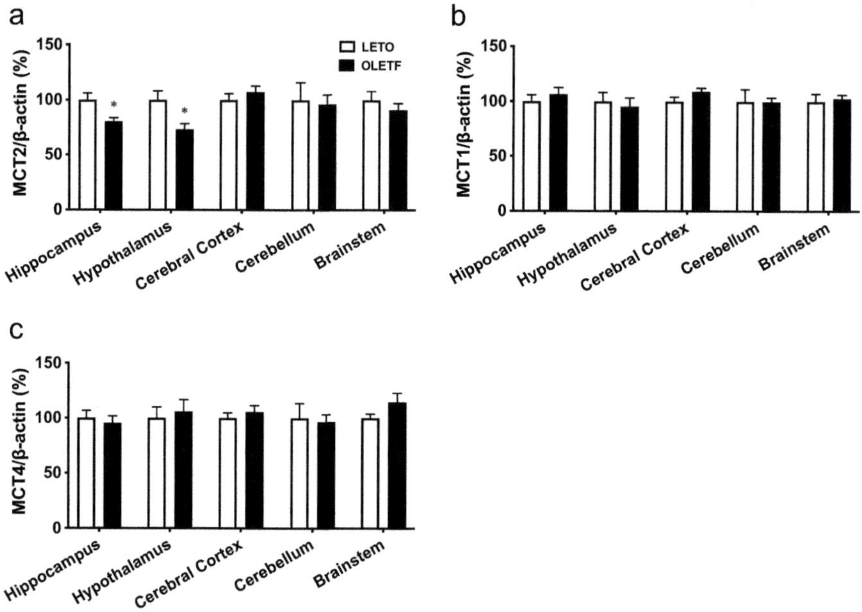

Fig. 1 Impaired expression of hippocampal MCTs in a rat model of T2DM. Specifically, the protein expression of MCT2 in the hippocampus and hypothalamus is shown to be downregulated in OLETF rats compared to LETO rats, despite MCT1 and MCT4 expressions remaining unchanged in the presence of T2DM. (**a**) MCT2 expressions, (**b**) MCT1 expressions, and (**c**) MCT4 expressions in the hippocampus, hypothalamus, cerebral cortex, cerebellum, and brainstem. (Referenced by Shima et al. [55])

complications associated with cognitive dysfunction in T2DM. Similarly, rodent models of Alzheimer's disease also exhibit decreased levels of MCT2 in the hippocampus [58, 59]. The downregulation of lactate transport through the MCT2 transporters into neurons could be a potential underlying mechanism linking T2DM and Alzheimer's disease.

Lactate also works as a ligand for its receptor, G-protein–coupled receptor 81 (GPR81; also known as Hydroxycarboxylic Acid Receptor 1 [HCAR1]) [60–62]. Activation of the lactate receptor has been shown to enhance hippocampal neuroplasticity and function in healthy rodents [63–66]. Inflammation has been found to inhibit the expression of lactate transporters [67, 68]; however, the changes in lactate receptor expression in the brain of individuals with T2DM are still unclear. Therefore, further research is needed to explore this aspect.

Lactate also plays a role in lactylation. Histones and proteins can undergo lactate-induced lactylation, which in turn contributes to the regulation of gene transcription and metabolism [69, 70]. Recently, there have been reports suggesting that lactate-induced lactylation may be involved in the regulation of insulin resistance in skeletal muscle [71], potentially linking it to the complications associated with

T2DM. However, the impact of lactate-induced lactylation in the hippocampus of individuals with T2DM remains unclear. Similar to the lactate receptor, further research is needed to investigate the effects of lactylation in the T2DM brain.

4 Regulation of MCT2 Expression in the Hippocampus

Previous studies have identified several molecules that can modulate the expression of MCTs. In vitro and in vivo studies have shown that brain-derived neurotrophic factor (BDNF) can increase neuronal MCT2 expression [72, 73]. Noradrenaline, insulin, and insulin-like growth factor-1 (IGF-1) have also been found to enhance MCT2 levels in neurons, although these effects have only been demonstrated in vitro experiments [74–77]. Moreover, treatment with fibroblast growth factor 21 (FGF21) has been shown to restore hippocampal MCT2 expression in mice with type 1 diabetes [78]. The enhancement of MCT2 expression by these molecules is mediated through the activation of the PI3K/Akt pathway. Glutamate has also been found to enhance MCT2 expression [79]; in contrast, long-term administration of N-methyl-D-aspartate (NMDA) receptor antagonist downregulates MCT1 and MCT4 levels but upregulates MCT2 in the hippocampus of mice [80]. Recently, the influence of microRNAs (small non-coding RNAs) has also been investigated. It has been reported that the expression of MCT1 in the mouse brain is downregulated by rno-miR-124-3p, while it is upregulated by miR-219 [81, 82]. These molecules mentioned above could potentially play a role in the regulation of MCTs expression.

T2DM rodents have been found to exhibit lower levels of BDNF in circulation and the hippocampus, along with decreased expression of TrkB, the receptor for BDNF, in the hippocampus [83–85]. This suggests that reduced BDNF signaling may contribute to the downregulation of hippocampal MCT2 levels in T2DM. Lowered BDNF levels in circulation have also been associated with cognitive dysfunction in T2DM patients [86]. Similarly, T2DM rodents show lower concentrations of IGF-1 and glutamate in the hippocampus compared to control rodents [87–89]. While insulin concentrations in the hippocampus may remain unchanged in T2DM [89], hippocampal insulin resistance could develop in T2DM. These alterations in biochemical profiles may be a result of the high-stress condition, including elevated levels of circulating corticosterone [85, 90]. On the other hand, T2DM is associated with decreased levels of FGF21 in circulation and peripheral organs, but the impact on hippocampal FGF21 levels is not [91]. Furthermore, there seems to be no difference in hippocampal noradrenaline levels between T2DM rats and control rats [92]. These findings suggest that FGF-21 and noradrenaline may not directly contribute to the deficit in hippocampal MCT2 levels associated with T2DM.

5 Effects of Exercise on Hippocampus in T2DM

Physical exercise is widely recognized as a therapeutic strategy for T2DM, as it has positive effects on peripheral organs. Interestingly, exercise also has beneficial effects on the brain, particularly on the hippocampus, not only in healthy mammals but also in those with T2DM [93, 94]. Indeed, regular exercise has been shown to enhance neurogenesis and reduce inflammation in the hippocampus of T2DM animals [95–97]. The American Diabetes Association (ADA) recommends that adults with diabetes engage in moderate-intensity aerobic physical activity (50–70% of maximum heart rate) to prevent diabetic complications and manage glycemic control [98]. A previous study demonstrated that a 4-week exercise regimen at a moderate intensity, based on the lactate threshold in rats (approximately 60% of maximal oxygen uptake [VO_{2max}]), improved hippocampal learning and memory dysfunction in advanced-stage T2DM rats (Fig. 2a-d; [56]). This suggests that the exercise therapy recommended by the ADA could effectively treat both peripheral and central complications in T2DM. Moreover, lowered hippocampal MCT2 expressions were restored with this moderate-intensity exercise in T2DM rats (Fig. 2e). Regular aerobic exercise has been shown to attenuate lactate accumulation in the hippocampus of T2DM animals [99], implying that the exercise-induced restoration of hippocampal MCT2 expression may contribute to the improvement of cognitive dysfunction by enhancing lactate transport in advanced-stage T2DM. However, it remains unclear whether the exercise-induced restoration of MCT2 contributes to enhanced neurogenesis and the amelioration of inflammation in the hippocampus of T2DM animals. Further studies are needed to investigate the specific mechanisms underlying these effects.

It is observed that regular exercise can improve BDNF levels in the hippocampus of T2DM animals [84]. However, in a previous study mentioned [56], the levels of hippocampal BDNF were not altered despite the improvement in hippocampal dysfunction and MCT2 levels with moderate-intensity exercise. This may be attributed to the short term of the exercise regimen or other factors. It is possible that restoring BDNF levels may not be necessary for the improvement of hippocampal dysfunction and MCT2 expressions in T2DM. Regarding IGF-1 levels, previous studies have shown that regular exercise does not alter hippocampal IGF-1 levels in T2DM rats [89]. Therefore, it appears that the molecular pathways underlying the regulation of hippocampal MCT2 expressions by exercise in T2DM are not fully understood. In the case of healthy rats, acute moderate-intensity exercise (50–70% VO_{2max}) has been shown to increase both MCT2 and MCT1 levels in the hippocampus [75]. In T2DM rats undergoing moderate-intensity exercise, on the other hand, hippocampal MCT1 expressions remained unchanged (Fig. 2f, g). This suggests that the lack of changes in hippocampal MCT1 expressions with moderate-intensity exercise may be specific to the T2DM condition. Overall, the molecular pathways and factors involved in the regulation of hippocampal MCT2 expressions by exercise in the context of T2DM are not fully elucidated and require further investigation.

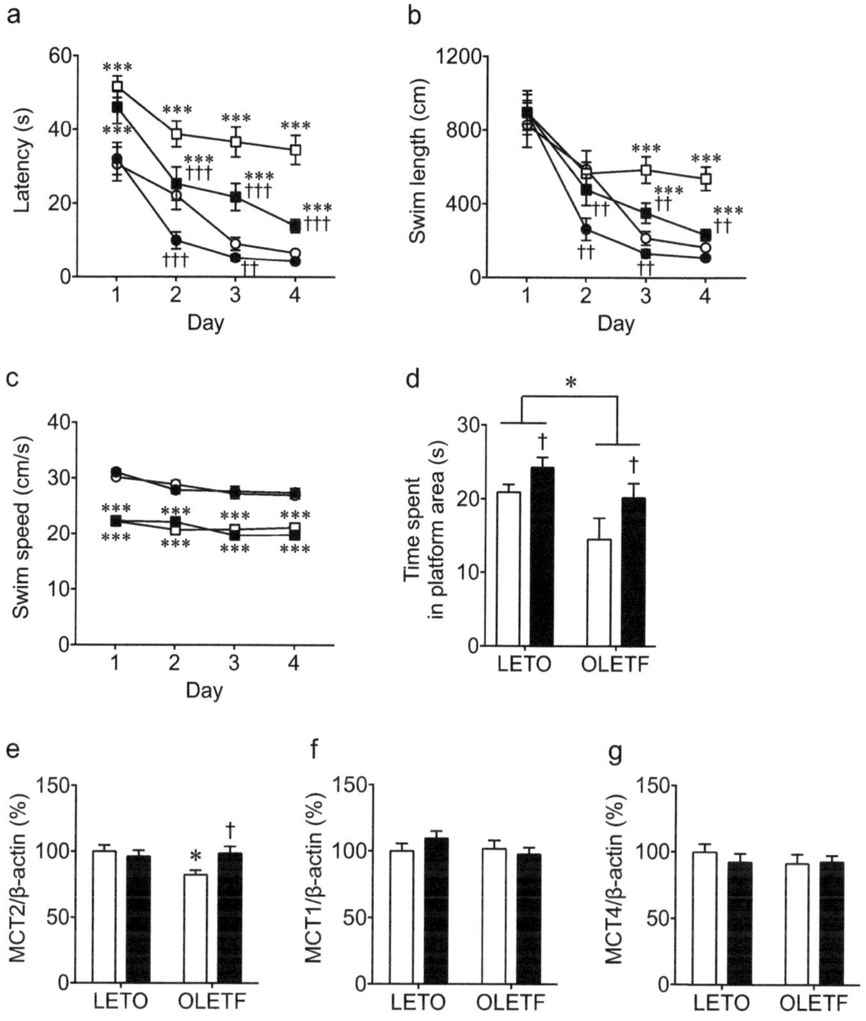

Fig. 2 The effects of moderate-intensity exercise on hippocampal cognitive function and MCTs expressions in T2DM rats. (**a**) Escape latency, (**b**) swim path length and (**c**) swim speed during learning test, and (**d**) time spent in platform area during probe test of Morris water maze. Moderate-intensity exercise improves learning and memory function in OLETF rats, a T2DM rat model. Additionally, (**e**) MCT2, (**f**) MCT1, and (**g**) MCT4 expressions in hippocampus of LETO (counterpart control) and OLETF rats. MCT2 expressions in OLETF rats are restored with moderate-intensity exercise. Values are mean ± SEM, white: sedentary groups, black: $^{*}p < 0.05$, $^{***}p < 0.001$ vs. LETO, $^{\dagger}p < 0.05$, $^{\dagger\dagger}p < 0.01$, $^{\dagger\dagger\dagger}p < 0.001$ vs. sedentary group. (Referenced by Shima et al., *Diabetologia*, 2017 [55])

Resistance exercise is also anticipated to serve as a beneficial intervention for managing T2DM [98]. Intriguingly, while aerobic exercise is effective in enhancing physical fitness parameters like VO_{2peak} and maximum heart rate, there exists a potential for resistance exercise to exert favorable effects on glycemic control in T2DM patients, similar to aerobic exercise [100]. Moreover, studies have reported that regular resistance exercise enhances cognitive function related to the hippocampus in T2DM rodents [101]. Consequently, resistance exercise employed may hold some significance in the treatment of cognitive dysfunction in T2DM. Although the amelioration of cognitive impairment induced by resistance exercise in T2DM may partly arise from the improvement of insulin resistance in the hippocampus, it remains unexplored whether resistance exercise contributes to the enhancement of hippocampal MCT2 expression. It is important to extensively contemplate and develop efficacious exercise regimens to address the treatment of T2DM.

6 How Exercise Is Effective on Hippocampus in T2DM

To develop the effective exercise regimens for the management of T2DM, careful consideration is given to the intensity, frequency, and timing of exercise [57, 102–104]. This section specifically focuses on exercise intensity. Previous research has demonstrated that engaging in light-intensity exercise ($<45\%\ VO_{2max}$) enhances hippocampal neuronal activity [105, 106] and cognitive function associated with the hippocampus in healthy mammals [107–110]. Similarly, regular light-intensity exercise has been shown to improve cognitive dysfunction in T2DM mice [111]. The amelioration of cognitive dysfunction induced by light-intensity exercise in T2DM mice would have been linked to the restoration of Mct2 mRNA levels. Conversely, the hippocampal mRNA levels of Mct1 and Mct4 remain unchanged with light-intensity exercise in T2DM mice. In a previous study, miRNA alterations were investigated to explore the underlying cause of the changes in Mct2 mRNA levels, resulting in the identification of nine reversed hippocampal miRNAs (miR-871-5p, miR-200a-3p, miR-200b-3p, miR-322-3p, miR-344d-1-5p, miR-351-5p, miR-429-3p, miR-542-3p, and miR-3474) associated with light exercise in T2DM (Fig. 3). While our findings hold promise for the development of novel clinical targets for addressing T2DM-induced memory dysfunction, this field of research is still evolving, necessitating further investigations. In the context of exercise intervention during the pre-symptomatic stage of T2DM, a four-month duration of both light- and moderate-intensity exercise demonstrated comparable efficacy in preventing the progression of memory dysfunction while enhancing hippocampal MCT2 expressions [57]. Considering that light-intensity exercise has shown potential for improving glycemic control not only during the pre-symptomatic stage but also in advanced stages of T2DM [111, 112], it could serve as a sufficient intervention for T2DM patients with reduced physical fitness.

Recently, high-intensity interval training (HIIT) has been noticed as an exercise regimen for enhancing various physiological functions. The effects of HIIT are

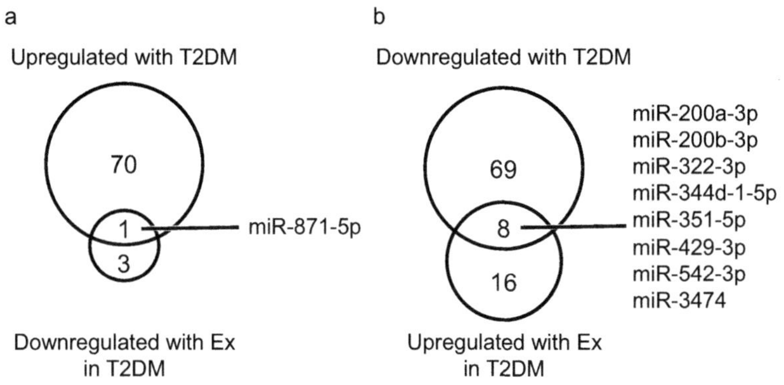

Fig. 3 Venn diagrams that depict the overlap of modulated hippocampal miRNA expressions between T2DM and light-intensity exercise in T2DM. (**a**) Hippocampal miR-871-5p level in T2DM mice is higher than that in control mice; however, 4 weeks of light-intensity exercise improves its levels. (**b**) Light-intensity exercise for 4 weeks restores T2DM hippocampal miRNAs, such as miR-200a-3p, miR-200b-3p, miR-322-3p, miR-344d-1-5p, miR-351-5p, miR-429-3p, miR-542-3p, and miR-3474. (Referenced by Shima et al. [111])

often compared to those of moderate-intensity exercise, with HIIT being deemed more beneficial for improving physical fitness [113]; conversely, moderate-intensity exercise could be more advantageous in terms of enhancing body composition in T2DM [114]. HIIT has also been found to enhance cognitive function in healthy mammals [115–117]. HIIT also leads to the upregulation of MCT1 and MCT4 expressions in the hippocampus of healthy mice [118]. Although HIIT may ameliorate hippocampal insulin resistance [119], its efficacy in improving hippocampus-based cognitive dysfunction in T2DM remains uncertain [120, 121]. Moreover, the specific impact of HIIT on the expression of MCTs in T2DM is not yet clearly understood. Consequently, a comprehensive investigation is warranted to elucidate the effects of HIIT on the hippocampus of individuals with T2DM.

7 Concluding Remarks and Future Direction

Hippocampus-based cognitive dysfunction represents a devastating complication associated with T2DM, which has implications for conditions such as dementia, Alzheimer's disease, and depression [2, 4, 5]. This chapter provided a comprehensive overview of previous studies examining the exercise-induced improvement in cognitive dysfunction and disruption of hippocampal MCT2 in T2DM. The disturbance of hippocampal MCT2-mediated lactate transport is posited as a potential underlying cause of hippocampus-based cognitive dysfunction in T2DM, thereby shedding light on the considering of novel and effective therapeutic targets. Although moderate-intensity exercise is commonly recommended as a standard exercise therapy for T2DM, it is worth considering that light-intensity exercise may also prove effective in the treatment of T2DM and its associated complications, including cognitive dysfunction. Further research is warranted to propose an optimal exercise regimen that can effectively treat hippocampal function and MCT2 expressions in T2DM.

Acknowledgements This work was supported by the Japan Society for the Promotion of Science (Grant-in-Aid for Early-Career Scientists, No. 22K17711). I thank the Laboratory for Exercise Physiology, and the Laboratory for Analytical Instruments, Education and Research Support Center, Gunma University.

References

1. Cukierman T, Gerstein HC, Williamson JD (2005) Cognitive decline and dementia in diabetes--systematic overview of prospective observational studies. Diabetologia 48:2460–2469
2. Whitmer R (2007) Type 2 diabetes and risk of cognitive impairment and dementia. Curr Neurol Neurosci Rep 7:373–380
3. McCrimmon RJ, Ryan CM, Frier BM (2012) Diabetes and cognitive dysfunction. Lancet 379:2291–2299
4. De Sousa RAL, Harmer AR, Freitas DA, Mendonça VA, Lacerda ACR, Leite HR (2020) An update on potential links between Type 2 diabetes mellitus and Alzheimer's disease. Mol Biol Rep 47:6347–6356
5. Farooqi A, Gillies C, Sathanapally H, Abner S, Seidu S, Davies MJ, Polonsky WH, Khunti K (2022) A systematic review and meta-analysis to compare the prevalence of depression between people with and without Type 1 and Type 2 diabetes. Prim Care Diabetes 16:1–10
6. Erickson KI, Hillman C, Stillman CM, Ballard RM, Bloodgood B, Conroy DE, Macko R, Marquez DX, Petruzzello SJ, Powell KE (2019) Physical activity, cognition, and brain outcomes: a review of the 2018 physical activity guidelines. Med Sci Sports Exerc 51:1242–1251
7. Morrison DJ, Kowalski GM, Grespan E, Mari A, Bruce CR, Wadley GD (2018) Measurement of postprandial glucose fluxes in response to acute and chronic endurance exercise in healthy humans. Am J Physiol Endocrinol Metab 314:E503–E511
8. Praet SFE, Van Loon LJC (2009) Exercise therapy in Type 2 diabetes. Acta Diabetol 46:263–278
9. Madden K (2013) Evidence for the benefit of exercise therapy in patients with Type 2 diabetes. Diabetes Metab Syndr Obes 6:233
10. Lundgren JR, Janus C, Jensen SBK et al (2021) Healthy weight loss maintenance with exercise, liraglutide, or both combined. N Engl J Med 384:1719–1730
11. Lupien SJ, Juster RP, Raymond C, Marin MF (2018) The effects of chronic stress on the human brain: from neurotoxicity, to vulnerability, to opportunity. Front Neuroendocrinol 49:91–105
12. Bartsch T, Wulff P (2015) The hippocampus in aging and disease: from plasticity to vulnerability. Neuroscience 309:1–16
13. Hu B, Yan LF, Sun Q et al (2019) Disturbed neurovascular coupling in Type 2 diabetes mellitus patients: evidence from a comprehensive fMRI analysis. Neuroimage Clin 22:101802
14. Li M, Li Y, Liu Y et al (2021) Altered hippocampal subfields volumes is associated with memory function in Type 2 diabetes mellitus. Front Neurol 12:1–10
15. Altman J, Das GD (1965) Autoradiographic and histological evidence of postnatal hippocampal neurogenesis in rats. J Comp Neurol 124:319–335
16. van Praag H, Schinder AF, Christie BR, Toni N, Palmer TD, Gage FH (2002) Functional neurogenesis in the adult hippocampus. Nature 415:1030–1034

17. Bracke A, Domanska G, Bracke K, Harzsch S, Van Den BJ, Bröker B, Von BO (2019) Obesity impairs mobility and adult hippocampal neurogenesis. J Exp Neurosci 13:1179069519883580
18. Stranahan AM, Arumugam TV, Cutler RG, Lee K, Egan JM, Mattson MP (2008) Diabetes impairs hippocampal function through glucocorticoid-mediated effects on new and mature neurons. Nat Neurosci 11:309–317
19. Bonds JA, Shetti A, Stephen TKL, Bonini MG, Minshall RD, Lazarov O (2020) Deficits in hippocampal neurogenesis in obesity-dependent and -independent Type-2 diabetes mellitus mouse models. Sci Rep 10:16368
20. Beauquis J, Homo-Delarche F, Giroix MH, Ehses J, Coulaud J, Roig P, Portha B, De Nicola AF, Saravia F (2010) Hippocampal neurovascular and hypothalamic-pituitary-adrenal axis alterations in spontaneously Type 2 diabetic GK rats. Exp Neurol 222:125–134
21. Zhou Z, Wang M, Huang C, Li Y, Gao L, Zhu Y, Ying C, Zhou X (2022) Treadmill exercise training alleviates diabetes-induced depressive-like behavior and cognitive impairment by improving hippocampal CA1 neurons injury in db/db mice. Brain Res Bull 190:84–96
22. Ma S, Bi W, Liu X, Li S, Qiu Y, Huang C, Lv R, Yin Q (2022) Single-cell sequencing analysis of the db/db mouse hippocampus reveals cell-type-specific insights into the pathobiology of diabetes-associated cognitive dysfunction. Front Endocrinol (Lausanne) 13:1–15
23. Taylor R (2012) Insulin resistance and Type 2 diabetes. Diabetes 61:778–779
24. Kadowaki T (2000) Insights into insulin resistance and Type 2 diabetes from knockout mouse models. J Clin Invest 106:459–465
25. Park S, Kim DS, Kang S, Moon NR (2013) β-Amyloid-induced cognitive dysfunction impairs glucose homeostasis by increasing insulin resistance and decreasing β-cell mass in non-diabetic and diabetic rats. Metabolism 62:1749–1760
26. Biessels GJ, Reagan LP (2015) Hippocampal insulin resistance and cognitive dysfunction. Nat Rev Neurosci 16:660–671
27. Grillo CA, Piroli GG, Lawrence RC et al (2015) Hippocampal insulin resistance impairs spatial learning and synaptic plasticity. Diabetes 64:3927–3936
28. Baker LD, Cross DJ, Minoshima S, Belongia D, Stennis Watson G, Craft S (2011) Insulin resistance and alzheimer-like reductions in regional cerebral glucose metabolism for cognitively normal adults with prediabetes or early Type 2 diabetes. Arch Neurol 68:51–57
29. Yang Y, Ma D, Wang Y, Jiang T, Hu S, Zhang M, Yu X, Gongb CX (2013) Intranasal insulin ameliorates tau hyperphosphorylation in a rat model of Type 2 diabetes. J Alzheimer's Dis 33:329–338
30. El Khoury NB, Gratuze M, Petry F et al (2016) Hypothermia mediates age-dependent increase of tau phosphorylation in db/db mice. Neurobiol Dis 88:55–65
31. Jolivalt CG, Lee CA, Beiswenger KK, Smith JL, Orlov M, Torrance MA, Masliah E (2008) Defective insulin signaling pathway and increased glycogen synthase kinase-3 activity in the brain of diabetic mice: parallels with Alzheimer's disease and correction by insulin. J Neurosci Res 86:3265–3274
32. Benarroch EE (2010) Glycogen metabolism: metabolic coupling between astrocytes and neurons. Neurology 74:919–923
33. Pellerin L, Bouzier-Sore A-K, Aubert A, Serres S, Merle M, Costalat R, Magistretti PJ (2007) Activity-dependent regulation of energy metabolism by astrocytes: an update. Glia 55:1251–1262
34. Pellerin L, Magistretti PJ (2012) Sweet sixteen for ANLS. J Cereb Blood Flow Metab 32:1152–1166
35. Pierre K, Pellerin L (2005) Monocarboxylate transporters in the central nervous system: distribution, regulation and function. J Neurochem 94:1–14
36. Tsacopoulos M, Magistretti P (1996) Metabolic coupling between glia and neurons. J Neurosci 16:877–885
37. Suzuki A, Stern SA, Bozdagi O, Huntley GW, Walker RH, Magistretti PJ, Alberini CM (2011) Astrocyte-neuron lactate transport is required for long-term memory formation. Cell 144:810–823

38. Descalzi G, Gao V, Steinman MQ, Suzuki A, Alberini CM (2019) Lactate from astrocytes fuels learning-induced mRNA translation in excitatory and inhibitory neurons. Commun Biol 2:247
39. Carrard A, Cassé F, Carron C, Burlet-Godinot S, Toni N, Magistretti PJ, Martin J (2021) Role of adult hippocampal neurogenesis in the antidepressant actions of lactate. Mol Psychiatry 26:6723–6735
40. Dembitskaya Y, Piette C, Perez S, Berry H, Magistretti PJ, Venance L (2022) Lactate supply overtakes glucose when neural computational and cognitive loads scale up. Proc Natl Acad Sci 119:e2212004119
41. Akter M, Ma H, Hasan M, Karim A, Zhu X, Zhang L, Li Y (2023) Exogenous L-lactate administration in rat hippocampus increases expression of key regulators of mitochondrial biogenesis and antioxidant defense. Front Mol Neurosci 16:1117146
42. Zhou J, Liu T, Guo H et al (2018) Lactate potentiates angiogenesis and neurogenesis in experimental intracerebral hemorrhage. Exp Mol Med 50:1–12
43. Karnib N, El-Ghandour R, El Hayek L et al (2019) Lactate is an antidepressant that mediates resilience to stress by modulating the hippocampal levels and activity of histone deacetylases. Neuropsychopharmacology 44:1152–1162
44. El Hayek L, Khalifeh M, Zibara V et al (2019) Lactate mediates the effects of exercise on learning and memory through SIRT1-dependent activation of hippocampal Brain-Derived Neurotrophic Factor (BDNF). J Neurosci 39(13):2369–2382
45. Lev-Vachnish Y, Cadury S, Rotter-Maskowitz A, Feldman N, Roichman A, Illouz T, Varvak A, Nicola R, Madar R, Okun E (2019) L-lactate promotes adult hippocampal neurogenesis. Front Neurosci 13:1–13
46. Won SJ, Jang BG, Yoo BH, Sohn M, Lee MW, Choi BY, Kim JH, Song HK, Suh SW (2012) Prevention of acute/severe hypoglycemia-induced neuron death by lactate administration. J Cereb Blood Flow Metab 32:1086–1096
47. Choi Y, Song JE, Lee JE, Kim E, Kim CH, Kim D, Song H-T (2019) Hyperpolarized [1-13C]lactate flux increased in the hippocampal region in diabetic mice. Mol Brain 12:88
48. Newman LA, Korol DL, Gold PE (2011) Lactate produced by glycogenolysis in astrocytes regulates memory processing. PLoS One 6:e28427
49. Duran J, Saez I, Gruart A, Guinovart JJ, Delgado-García JM (2013) Impairment in long-term memory formation and learning-dependent synaptic plasticity in mice lacking glycogen synthase in the brain. J Cereb Blood Flow Metab 33:550–556
50. Korol DL, Gardner RS, Tunur T, Gold PE (2019) Involvement of lactate transport in two object recognition tasks that require either the hippocampus or striatum. Behav Neurosci 133:176–187
51. Netzahualcoyotzi C, Pellerin L (2020) Neuronal and astroglial monocarboxylate transporters play key but distinct roles in hippocampus-dependent learning and memory formation. Prog Neurobiol 194:101888
52. Chen W, Sun X, Zhan L, Zhou W, Bi T (2021) Conditional knockout of Pdha1 in mouse hippocampus impairs cognitive function: the possible involvement of lactate. Front Neurosci 15:1–11
53. Zhou Z, Okamoto K, Onodera J, Hiragi T, Andoh M, Ikawa M, Tanaka KF, Ikegaya Y, Koyama R (2021) Astrocytic cAMP modulates memory via synaptic plasticity. Proc Natl Acad Sci 118:1–10
54. Herrera-López G, Griego E, Galván EJ (2020) Lactate induces synapse-specific potentiation on CA3 pyramidal cells of rat hippocampus. PLoS One 15:e0242309
55. Shima T, Jesmin S, Matsui T, Soya M, Soya H (2018) Differential effects of Type 2 diabetes on brain glycometabolism in rats: focus on glycogen and monocarboxylate transporter 2. J Physiol Sci 68:69–75
56. Shima T, Matsui T, Jesmin S, Okamoto M, Soya M, Inoue K, Liu Y, Torres-Aleman I, McEwen BS, Soya H (2017) Moderate exercise ameliorates dysregulated hippocampal glycometabolism and memory function in a rat model of Type 2 diabetes. Diabetologia 60:597–606

57. Jesmin S, Shima T, Soya M, Takahashi K, Omura K, Ogura K, Koizumi H, Soya H (2022) Long-term light and moderate exercise intervention similarly prevent both hippocampal and glycemic dysfunction in presymptomatic Type 2 diabetic rats. Am J Physiol Endocrinol Metab 322:E219–E230
58. Lu W, Sun S, Li Y et al (2019) Curcumin ameliorates memory deficits by enhancing lactate content and MCT2 expression in APP/PS1 transgenic mouse model of Alzheimer's disease. Anat Rec 302:332–338
59. Lu W, Huang J, Sun S, Huang S, Gan S, Xu J, Yang M, Xu S, Jiang X (2015) Changes in lactate content and monocarboxylate transporter 2 expression in Aβ25-35-treated rat model of Alzheimer's disease. Neurol Sci 36:871–876
60. Lauritzen KH, Morland C, Puchades M, Holm-Hansen S, Hagelin EM, Lauritzen F, Attramadal H, Storm-Mathisen J, Gjedde A, Bergersen LH (2014) Lactate receptor sites link neurotransmission, neurovascular coupling, and brain energy metabolism. Cereb Cortex 24:2784–2795
61. Morland C, Lauritzen KH, Puchades M, Holm-Hansen S, Andersson K, Gjedde A, Attramadal H, Storm-Mathisen J, Bergersen LH (2015) The lactate receptor, G-protein-coupled receptor 81/hydroxycarboxylic acid receptor 1: expression and action in brain. J Neurosci Res 93:1045–1055
62. Bergersen LH (2015) Lactate transport and signaling in the brain: potential therapeutic targets and roles in body-brain interaction. J Cereb Blood Flow Metab 35:176–185
63. Scavuzzo CJ, Rakotovao I, Dickson CT (2020) Differential effects of L- and D-lactate on memory encoding and consolidation: potential role of HCAR1 signaling. Neurobiol Learn Mem 168:107151
64. Lambertus M, Øverberg LT, Andersson KA, Hjelden MS, Hadzic A, Haugen ØP, Storm-Mathisen J, Bergersen LH, Geiseler S, Morland C (2021) L-lactate induces neurogenesis in the mouse ventricular-subventricular zone via the lactate receptor HCA1. Acta Physiol 231:1–15
65. Nicola R, Okun E (2021) Adult hippocampal neurogenesis: one lactate to rule them all. NeuroMolecular Med 23:445–448
66. Skwarzynska D, Sun H, Williamson J, Kasprzak I, Kapur J (2023) Glycolysis regulates neuronal excitability via lactate receptor, HCA1R. Brain 146:1888–1902
67. Boitsova EB, Morgun AV, Osipova ED et al (2018) The inhibitory effect of LPS on the expression of GPR81 lactate receptor in blood-brain barrier model in vitro. J Neuroinflammation 15:196
68. Feingold KR, Moser A, Shigenaga JK, Grunfeld C (2011) Inflammation inhibits GPR81 expression in adipose tissue. Inflamm Res 60:991–995
69. Chen L, Huang L, Gu Y, Cang W, Sun P, Xiang Y (2022) Lactate-Lactylation hands between metabolic reprogramming and immunosuppression. Int J Mol Sci 23:11943
70. Hagihara H, Shoji H, Otabi H, Toyoda A, Katoh K, Namihira M, Miyakawa T (2021) Protein lactylation induced by neural excitation. Cell Rep 37:109820
71. Maschari D, Saxena G, Law TD, Walsh E, Campbell MC, Consitt LA (2022) Lactate-induced lactylation in skeletal muscle is associated with insulin resistance in humans. Front Physiol 13:1–13
72. Robinet C, Pellerin L (2010) Brain-derived neurotrophic factor enhances the expression of the monocarboxylate transporter 2 through translational activation in mouse cultured cortical neurons. J Cereb Blood Flow Metab 30:286–298
73. Robinet C, Pellerin L (2011) Brain-derived neurotrophic factor enhances the hippocampal expression of key postsynaptic proteins in vivo including the monocarboxylate transporter MCT2. Neuroscience 192:155–163
74. Chenal J, Pierre K, Pellerin L (2008) Insulin and IGF-1 enhance the expression of the neuronal monocarboxylate transporter MCT2 by translational activation via stimulation of the phosphoinositide 3-kinase-Akt-mammalian target of rapamycin pathway. Eur J Neurosci 27:53–65

75. Takimoto M, Hamada T (2014) Acute exercise increases brain region-specific expression of MCT1, MCT2, MCT4, GLUT1, and COX IV proteins. J Appl Physiol 116:1238–1250
76. Chenal J, Pellerin L (2007) Noradrenaline enhances the expression of the neuronal monocarboxylate transporter MCT2 by translational activation via stimulation of PI3K/Akt and the mTOR/S6K pathway. J Neurochem 102:389–397
77. Pierre K, Debernardi R, Magistretti PJ, Pellerin L (2003) Noradrenaline enhances monocarboxylate transporter 2 expression in cultured mouse cortical neurons via a translational regulation. J Neurochem 86:1468–1476
78. Zhao L, Jiang H, Xie J, Shen D, Yi Q, Yan J, Li C, Zheng H, Gao H (2022) Effects of fibroblast growth factor 21 on lactate uptake and usage in mice with diabetes-associated cognitive decline. Mol Neurobiol 59:5656–5672
79. Tescarollo F, Covolan L, Pellerin L (2014) Glutamate reduces glucose utilization while concomitantly enhancing AQP9 and MCT2 expression in cultured rat hippocampal neurons. Front Neurosci 8:1–9
80. Ding R, Tan Y, Du A et al (2020) Redistribution of monocarboxylate 1 and 4 in hippocampus and spatial memory impairment induced by long-term ketamine administration. Front Behav Neurosci 14:1–10
81. Liu S, Ren C, Qu X, Wu X, Dong F, Chand YK, Fan H, Yao R, Geng D (2017) miR-219 attenuates demyelination in cuprizone-induced demyelinated mice by regulating monocarboxylate transporter 1. Eur J Neurosci 45:249–259
82. Ye XS, Li JX, Liu Q, Xu J, Huang J, Wei GS, Tian LW, Zhuo F, Yang M, Quan SS (2019) Role of rno-miR-124-3p in regulating MCT1 expression in rat brain after permanent focal cerebral ischemia. Genes Dis 6:398–406
83. Tomiga Y, Higaki Y, Anzai K, Takahashi H (2022) Behavioral defects and downregulation of hippocampal BDNF and nNOS expression in db/db mice did not improved by chronic TGF-β2 treatment. Front Physiol 13:1–11
84. Stranahan AM, Lee K, Martin B, Maudsley S, Golden E, Cutler RG, Mattson MP (2009) Voluntary exercise and caloric restriction enhance hippocampal dendritic spine density and BDNF levels in diabetic mice. Hippocampus 19:951–961
85. Stranahan AM, Arumugam TV, Mattson MP (2011) Lowering corticosterone levels reinstates hippocampal brain-derived neurotropic factor and Trkb expression without influencing deficits in hypothalamic brain-derived neurotropic factor expression in leptin receptor-deficient mice. Neuroendocrinology 93:58–64
86. Zhen YF, Zhang J, Liu XY, Fang H, Tian LB, Zhou DH, Kosten TR, Zhang XY (2013) Low BDNF is associated with cognitive deficits in patients with Type 2 diabetes. Psychopharmacology 227:93–100
87. Zheng Y, Yang Y, Dong B, Zheng H, Lin X, Du Y, Li X, Zhao L, Gao H (2016) Metabonomic profiles delineate potential role of glutamate-glutamine cycle in db/db mice with diabetes-associated cognitive decline. Mol Brain 9:40
88. Zhao L, Dong M, Ren M, Li C, Zheng H, Gao H (2018) Metabolomic analysis identifies lactate as an important pathogenic factor in diabetes-associated cognitive decline rats. Mol Cell Proteomics 17:2335–2346
89. Gomes RJ, de Oliveira CAM, Ribeiro C, Mota CSDA, Moura LP, Tognoli LMMC, Leme JACDA, Luciano E, de Mello MAR (2009) Effects of exercise training on hippocampus concentrations of insulin and IGF-1 in diabetic rats. Hippocampus 19:981–987
90. Piroli GG, Grillo CA, Reznikov LR, Adams S, McEwen BS, Charron MJ, Reagan LP (2007) Corticosterone impairs insulin-stimulated translocation of GLUT4 in the rat hippocampus. Neuroendocrinology 85:71–80
91. Lee H, Lim Y (2022) The potential role of myokines/hepatokines in the progression of neuronal damage in streptozotocin and high-fat diet-induced Type 2 diabetes mellitus mice. Biomedicine 10:1521

92. Sakimura K, Maekawa T, Sasagawa K, Ishii Y, Kume S, Ohta T (2018) Depression-related behavioural and neuroendocrine changes in the Spontaneously Diabetic Torii (SDT) fatty rat, an animal model of Type 2 diabetes mellitus. Clin Exp Pharmacol Physiol 45:927–933
93. De Sousa RAL, Improta-Caria AC, Cassilhas RC (2021) Effects of physical exercise on memory in Type 2 diabetes: a brief review. Metab Brain Dis 36:1559–1563
94. Yi SS (2015) Effects of exercise on brain functions in diabetic animal models. World J Diabetes 6:583
95. Yi SS, Hwang IK, Yoo K-Y et al (2009) Effects of treadmill exercise on cell proliferation and differentiation in the subgranular zone of the dentate gyrus in a rat model of Type II diabetes. Neurochem Res 34:1039–1046
96. Yoo DY, Chae J, Jung HY et al (2015) Treadmill exercise is associated with reduction of reactive microgliosis and pro-inflammatory cytokine levels in the hippocampus of Type 2 diabetic rats. Neurol Res 37:732–738
97. Hwang IK, Yi SS, Yoo K-Y, Park OK, Yan B, Song W, Won M-H, Yoon YS, Seong JK (2011) Differential effects of treadmill exercise in early and chronic diabetic stages on Parvalbumin Immunoreactivity in the hippocampus of a rat model of Type 2 diabetes. Neurochem Res 36:1526–1532
98. American Diabetes Association (2016) 3. Foundations of care and comprehensive medical evaluation. Diabetes Care 39:S23–S35
99. Li J, Liu B, Cai M, Lin X, Lou S (2019) Glucose metabolic alterations in hippocampus of diabetes mellitus rats and the regulation of aerobic exercise. Behav Brain Res 364:447–456
100. Yang Z, Scott CA, Mao C, Tang J, Farmer AJ (2014) Resistance exercise versus aerobic exercise for Type 2 diabetes: a systematic review and meta-analysis. Sports Med 44:487–499
101. De Sousa RAL, Improta-Caria AC, de Jesus-Silva FM, CODE M, Freitas DA, ACR L, Mendonça VA, Cassilhas RC, Leite HR (2020) High-intensity resistance training induces changes in cognitive function, but not in locomotor activity or anxious behavior in rats induced to Type 2 diabetes. Physiol Behav 223:112998
102. van der Velde JHPM, Boone SC, Winters-van Eekelen E, Hesselink MKC, Schrauwen-Hinderling VB, Schrauwen P, Lamb HJ, Rosendaal FR, de Mutsert R (2023) Timing of physical activity in relation to liver fat content and insulin resistance. Diabetologia 66:461–471
103. Savikj M, Stocks B, Sato S, Caidahl K, Krook A, Deshmukh AS, Zierath JR, Wallberg-Henriksson H (2022) Exercise timing influences multi-tissue metabolome and skeletal muscle proteome profiles in Type 2 diabetic patients—a randomized crossover trial. Metabolism 135:155268
104. Legaard GE, Lyngbæk MPP, Almdal TP et al (2023) Effects of different doses of exercise and diet-induced weight loss on beta-cell function in Type 2 diabetes (DOSE-EX): a randomized clinical trial. Nat Metab 5:880
105. Suwabe K, Byun K, Hyodo K et al (2018) Rapid stimulation of human dentate gyrus function with acute mild exercise. Proc Natl Acad Sci 115:10487–10492
106. Soya H, Nakamura T, Deocaris CC, Kimpara A, Iimura M, Fujikawa T, Chang H, McEwen BS, Nishijima T (2007) BDNF induction with mild exercise in the rat hippocampus. Biochem Biophys Res Commun 358:961–967
107. Inoue K, Hanaoka Y, Nishijima T, Okamoto M, Chang H, Saito T, Soya H (2014) Long-term mild exercise training enhances hippocampus-dependent memory in rats. Int J Sports Med 36:280–285
108. Inoue K, Okamoto M, Shibato J, Lee MC, Matsui T, Rakwal R, Soya H (2015) Long-term mild, rather than intense, exercise enhances adult hippocampal neurogenesis and greatly changes the transcriptomic profile of the hippocampus. PLoS One 10:e0128720
109. Okamoto M, Hojo Y, Inoue K, Matsui T, Kawato S, McEwen BS, Soya H (2012) Mild exercise increases dihydrotestosterone in hippocampus providing evidence for androgenic mediation of neurogenesis. Proc Natl Acad Sci 109:13100–13105
110. Okamoto M, Yamamura Y, Liu Y et al (2015) Hormetic effects by exercise on hippocampal neurogenesis with glucocorticoid signaling. Brain Plast 1:149–158

111. Shima T, Kawabata-Iwakawa R, Onishi H, Jesmin S, Yoshikawa T (2023) Light-intensity exercise improves memory dysfunction with the restoration of hippocampal MCT2 and miRNAs in Type 2 diabetic mice. Metab Brain Dis 38:245–254
112. Jesmin S, Shima T, Yamaguchi N et al (2021) Daily walking habit reduces fasting blood sugar level in newly diagnosed diabetic participants in rural Bangladesh. J Phys Fit Sports Med 10:85–97
113. De Nardi AT, Tolves T, Lenzi TL, Signori LU, da AMV S (2018) High-intensity interval training versus continuous training on physiological and metabolic variables in prediabetes and Type 2 diabetes: a meta-analysis. Diabetes Res Clin Pract 137:149–159
114. Magalhães JP, Júdice PB, Ribeiro R, Andrade R, Raposo J, Dores H, Bicho M, Sardinha LB (2019) Effectiveness of high-intensity interval training combined with resistance training versus continuous moderate-intensity training combined with resistance training in patients with Type 2 diabetes: a one-year randomized controlled trial. Diabetes Obes Metab 21:550–559
115. Okamoto M, Mizuuchi D, Omura K, Lee M, Oharazawa A, Yook JS, Inoue K, Soya H (2021) High-intensity intermittent training enhances spatial memory and hippocampal neurogenesis associated with BDNF signaling in rats. Cereb Cortex 31:4386–4397
116. Tsai CL, Pan CY, Tseng YT, Chen FC, Chang YC, Wang TC (2021) Acute effects of high-intensity interval training and moderate-intensity continuous exercise on BDNF and irisin levels and neurocognitive performance in late middle-aged and older adults. Behav Brain Res 413:113472
117. Mou H, Tian S, Fang Q, Qiu F (2022) The immediate and sustained effects of moderate-intensity continuous exercise and high-intensity interval exercise on working memory. Front Psychol 13:766679
118. Hu J, Cai M, Shang Q, Li Z, Feng Y, Liu B, Xue X, Lou S (2021) Elevated lactate by high-intensity interval training regulates the hippocampal BDNF expression and the mitochondrial quality control system. Front Physiol 12:1–17
119. Khoramipour K, Bejeshk MA, Rajizadeh MA, Najafipour H, Dehghan P, Farahmand F (2023) High-intensity interval training ameliorates molecular changes in the hippocampus of male rats with the diabetic brain: the role of adiponectin. Mol Neurobiol 60:3486–3495
120. Li X, He Q, Zhao N, Chen X, Li T, Cheng B (2021) High intensity interval training ameliorates cognitive impairment in T2DM mice possibly by improving PI3K/Akt/mTOR signaling-regulated autophagy in the hippocampus. Brain Res 1773:147703
121. Orumiyehei A, Khoramipour K, Rezaei MH, Madadizadeh E, Meymandi MS, Mohammadi F, Chamanara M, Bashiri H, Suzuki K (2022) High-intensity interval training-induced hippocampal molecular changes associated with improvement in anxiety-like behavior but not cognitive function in rats with Type 2 diabetes. Brain Sci 12:1280

Exercise as a Therapeutic Intervention for Alzheimer's Disease

Dong-Joo Hwang and Joon-Yong Cho

Abstract Alzheimer's disease (AD) is a common form of dementia characterized by cognitive decline and abnormal accumulation of proximate neurotoxins in older adults. It accounts for up to 80% of all dementia cases. AD is not exclusively attributed to aging; rather, it involves complex and multifactorial brain changes that can lead to severe functional dependence and ultimately death. Although there has been progress in the development of novel treatments for AD, they are yet to yield disease-modifying effects. Early detection and therapeutic interventions are critical for preventing or delaying the onset of AD. We aimed to provide an overview of emerging evidence on physical exercise as a therapeutic strategy for the prevention and treatment of AD. Studies have demonstrated the potential of exercise in improving cognitive function, reducing the risk of AD, and slowing disease progression by promoting various neuroplastic changes. Therefore, regular exercise should be considered as a disease-modifying intervention for AD and included in comprehensive treatment protocols. Further studies are warranted to establish the optimal exercise regimen for individuals with AD; nonetheless, incorporating exercise into daily routines may contribute toward the prevention and management of AD.

Keywords Alzheimer's disease · Cognitive Function · Exercise · Therapeutic intervention

---

D.-J. Hwang
Exercise Biochemistry Laboratory, Korea National Sport University,
Seoul, Republic of Korea

Sport Science Institute, Korea National Sport University, Seoul, Republic of Korea
e-mail: dongzoo87@knsu.ac.kr

J.-Y. Cho (✉)
Exercise Biochemistry Laboratory, Korea National Sport University,
Seoul, Republic of Korea
e-mail: chojy86@knsu.ac.kr

H. Soya (ed.), *Exercise Brain Stimulation for Cognitive Function and Mental Health*,
Advances in Neurobiology 44, https://doi.org/10.1007/978-981-95-0066-6_16

Abbreviations

AD	Alzheimer's disease
MCI	Mild Cognitive Impairment
EOAD	Early-Onset AD
LOAD	Late-Onset AD
APP	Amyloid Precursor Protein
Aβ	Amyloid-beta
NFTs	Neurofibrillary Tangles
BDNF	Brain-Derived Neurotrophic Factor
APOE	Apolipoprotein E
CBF	Cerebral Blood Flow
AHN	Adult Hippocampal Neurogenesis
TE	Treadmill Exercise
VWR	Voluntary Wheel Running

1 Introduction

Alzheimer's disease (AD) is the most common form of dementia and is characterized by cognitive deterioration in older adults. It accounts for approximately 60–80% of all cases. The global prevalence of AD is currently increasing, with the number of people living with dementia being expected to increase to 139 million by 2050 according to the World Health Organization [1].

The initial symptoms of AD vary according to the disease stage. Early-stage AD is characterized by memory and cognitive ability loss such as language and visual/spatial perception, which is termed as preclinical AD and mild cognitive impairment (MCI). In contrast, late-stage AD is characterized by severe cognitive decline and functional dependence along with an inability to hold a conversation and respond to environmental cues, which ultimately leads to death [2, 3]. Age is positively correlated with the risk of AD and other dementia types; moreover, nearly half of individuals aged >65 years experience some form of cognitive impairment. However, AD is not caused by aging only; rather, it results from complex brain changes. Although there has been tremendous progress in elucidating the mechanisms underlying AD and establishing potential novel treatments for AD, they remain unclear. In ~75% of individuals living with dementia, diagnosis is difficult, as the underlying pathological mechanisms might begin ≥20 years prior to the onset of noticeable warning signs or symptoms [4]. Therefore, early detection and therapeutic intervention methods are necessary for preventing or delaying the onset of AD.

There has been increasing research interest in AD. Several disease-modifying treatments for AD have been tested and approved in clinical trials. These treatments can slow the progression of AD symptoms by targeting the AD-related pathological changes, including deposition of amyloid plaque and neurofibrillary tangles (NFTs)

[5, 6]. However, none of these medications can achieve a definite cure; moreover, some of them are not cost-effective and there are no medications that can be prescribed to all patients in different AD stages [6, 7]. Therefore, there is a need for cost-effective and non-invasive interventions that can help reduce the risks of AD and dementia. Although there is no established intervention for preventing AD, there is increasing evidence suggesting that several lifestyle choices, including a healthy diet, physical exercise, and social engagement, may be potential interventions for preventing and delaying the onset of AD [8–11]. Notably, as a therapeutic strategy, physical exercise with a higher accessibility and affordability for individuals with AD has been proven to have multiple health benefits, including the improvement of cardiovascular health, insulin sensitivity, and immune response, as well as cognitive enhancement that have been linked to the development and progression of AD [9, 10]. Hence, this review aimed to provide an overview of emerging evidence on physical exercise as a therapeutic strategy for the prevention and treatment of AD. A comprehensive overview of the existing literature will contribute to a growing body of knowledge on the potential of physical exercise as a non-pharmacological intervention for the treatment of AD and help healthcare providers in a clinical setting to tailor exercise regime to individual patients' needs and preference.

2 Overview of Alzheimer's Disease

Extensive research spanning over 20 years has provided important insights into the nature of AD. However, the etiology and pathogenesis of AD remain unclear. Nonetheless, both genetic and environmental risk factors are known to be crucially involved in the initiation and progression of AD. AD can be classified according to the age at onset as follows: early (presenile)-onset AD (EOAD), which occurs at an age <65 years and accounts for a small number of cases, and late-onset AD (LOAD), which occurs at an age ≥65 years and accounts for approximately 95% of all cases. However, they are both basically considered as the same disease, as they do not differ in terms of pathological features. Almost all EOAD cases are familial because they usually result from causative genetic mutations or permanent changes in specific genes, including amyloid precursor protein (APP), presenilin 1, and presenilin; however, LOAD has no specific family link [6, 12]. Although the mechanisms that damage the brain cells remain unclear, abnormal accumulation of proteins that form amyloid-beta (Aβ) deposits and NFTs within and around brain cells is believed to damage neuronal cells. However, these are not the only pathologic features that contribute to AD symptoms; rather, inflammation, reactive oxygen species, dysregulation of neurotransmitters, and metabolic dysfunctions in the brain partly contribute to the disruption of neural network processes that cells require to survive [6, 13, 14] (Fig. 1).

As AD progresses, individuals tend to develop an increasing number of abnormal structures in an unpredictable pattern, which specifically begin in the hippocampus and entorhinal region, a neurogenic niche responsible for learning and memory, and

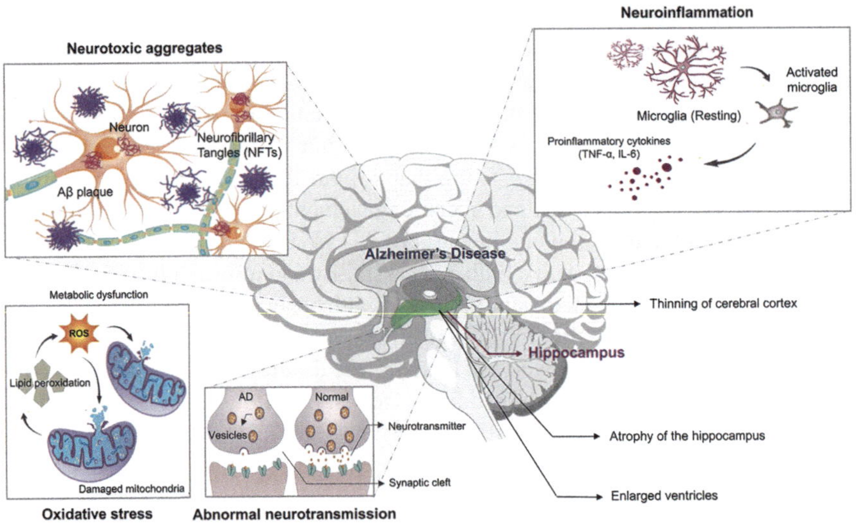

Fig. 1 A schematic representation of the pathophysiology of Alzheimer's Disease (AD) depicting the multifactorial features of AD

subsequently, spread to other brain regions [15, 16]. Objectively measurable pathological changes in the brain precede clinical AD symptoms by many years; accordingly, the onset of measurable cognitive decline may be preceded by the presence of aggregates in specific AD-related brain areas [4]. Taken together, these factors impede pre-symptomatic diagnosis and early therapeutic interventions for individuals with AD. However, some of the pathologic features are reversible, which suggests that several lifestyle choices and interventions can help reduce the risk of developing AD. Specifically, human and animal studies have suggested that different types of exercises can modulate AD-related pathologic conditions with concomitant cognitive changes. Therefore, cognitive enhancement mediated by physical exercise may be an alternative intervention for mitigating AD symptoms. Hence, this study aimed to review the literature on the application of exercise as a non-invasive intervention for AD.

3 Exercise as a Disease-Modifying Intervention

Exercise can allow an individual to live a longer and better life. Being both physically and mentally active provides a huge range of benefits. There is increasing evidence suggesting that regular exercise is important for preserving a healthy lifestyle and preventing diseases. Specifically, it promotes muscle growth and reduces the risk of cardiovascular disease; further, it can maintain brain health [17]. Human and animal studies have reported exercise-induced structural and functional changes

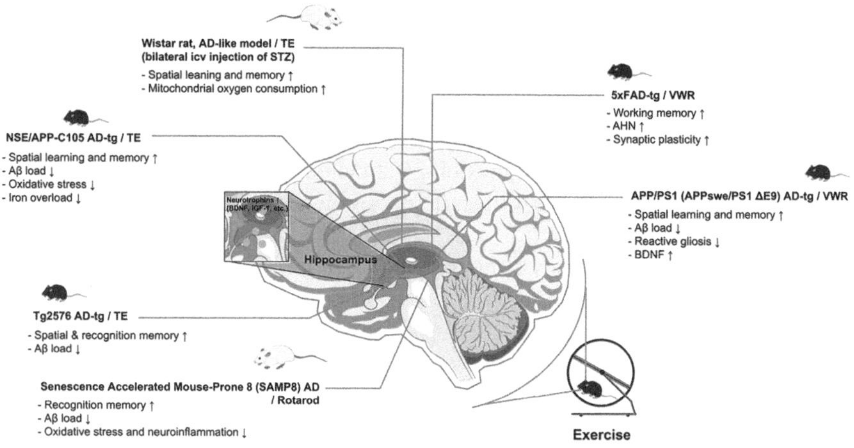

Fig. 2 A schematic representation of the potential role of exercise as a disease-modifying treatment for Alzheimer's disease

in the brains of patients with AD. Specifically, exercise has been shown to directly or indirectly reduce neuropathologic hallmarks (Aβ and hyperphosphorylated tau proteins) in the AD brain, which are strongly associated with cognitive deficits [17, 18]. Additionally, the benefits of exercise in maintaining cognitive functioning may be partly attributed to exercise-induced neuroplastic changes mediated by multiple molecular and signaling pathways, including anti-inflammation, antioxidation, and release of growth factors [17, 19–23]. However, as a disease-modifying treatment, it remains unclear how exercise mediates brain health and cognitive function in patients with AD. The subsequent sections mainly discuss how and what individuals with AD can achieve from regular exercise. Specifically, we discuss the following: (1) human studies on the effects of exercise on patients with dementia or AD and (2) animal studies on the cognitive benefits of exercise and comprehensive neurobiological changes occurring in AD brain, while performing different type and intensity of exercise (Fig. 2).

4 Physical Activity and Alzheimer's Disease: Clinical Evidence

Several pharmacological agents, including cholinesterase inhibitors (Donepezil, Galantamine, and Rivastigmine), Memantine, Aducanumab, and Lecanemab, have been used to treat AD symptoms, and studies trying to develop novel pharmacological agents are ongoing [24, 25]. However, as previously mentioned, these pharmacological agents have several limitations, including a lack of broad effectiveness, risk of side effects, and lack of cost-effectiveness [26–28]. Nonetheless, there is increasing evidence suggesting that regular physical activity, which is defined as movement that results in energy expenditure, has a positive effect on cognitive function, mental health, and overall well-being, with even a seemingly small activity exerting a large impact on the brain. Specifically, increased physical activity has a strong correlation with decreased risk and progression of AD. Physical activity can improve cognitive function by reducing the levels of Aβ deposition, neuroinflammation, and other AD-related pathophysiological factors. Therefore, promoting physical activity as part of a healthy lifestyle may reduce the risk and progression of AD. However, further research is warranted to determine the optimal regimen of physical activity for effectively reducing the risk of AD.

4.1 Physical Activity and Prevalence of AD

Several studies have revealed the relationship between physical activity and AD. Physical activity exerts positive effects on cognitive function in both healthy individuals and patients with AD and other dementia types. A meta-analysis of 12 studies reported that physical activity was associated with improved cognitive performance in older adults, with the greatest benefits being observed in patients with MCI and AD [29]. Similarly, a large-scale observational study reported that regular physical activity was associated with a decreased risk of MCI and AD in older adults [30, 31]. Physical activity stimulates the brain by increasing blood flow and oxygenation, which enhances the growth of newborn neurons and connections among existing neurons, eventually contributing to the overall brain function and cognitive processes [32]. Moreover, physical activity promotes the production of brain-derived neurotrophic factor (BDNF), which is a widely recognized protein known for its role in the central nervous system. BDNF is considered a key mediator of neuronal survival and synaptic plasticity to help protect and preserve brain function; additionally, it influences learning and memory processes in various neurological disorders, including depression, Parkinson's disease, and AD [33].

4.2 Physical Activity and AD Pathology

There has been extensive research on Aβ protein fragments in the context of AD. Abnormal Aβ accumulation in the brain is toxic to brain cells and has been implicated in AD-related cognitive decline. However, physical activity may reduce abnormal Aβ accumulation in the brain, indicating that physical exercise can reduce the risk of AD onset and slow down AD progression in diagnosed patients. Vemuri et al. [34] reported that older adults who engaged in regular physical activity had lower Aβ levels in their cerebrospinal fluid than those who were sedentary, indicating that exercise may be a possible therapeutic intervention. However, further research is warranted to elucidate the relationship between physical activity and Aβ accumulation. Nevertheless, regular physical activity is widely considered an effective intervention for maintaining brain health and reducing the risk of AD and other types of dementia.

Unregulated inflammatory responses in the brain may be a key factor in AD development. Several inflammatory cytokines and processes in the brain have been implicated in the death of brain cells and the progressive decline in cognitive function observed in AD. However, physical activity has anti-inflammatory effects on the brain, which may help reduce the risk of AD. Specifically, being physically active with occupational or leisure-time activities has been linked to decreased levels of inflammatory markers in the brain, including interleukin-6 (IL-6) and tumor necrosis factor-alpha (TNF), which are elevated in AD [35, 36]. Erickson et al. [30] reported that regular physical activity is associated with decreased inflammation levels in older adults, especially those with MCI and AD, highlighting that incorporating more physical activity into daily routines may serve as a potential non-pharmacological intervention for reducing neuroinflammation in AD.

5 Exercise as a Therapeutic Intervention for AD: Clinical Evidence

Although the terms exercise and physical activity are often used interchangeably, they are not synonymous. Exercise refers to planned, structured, and repetitive physical activity that is performed to improve one or more aspects of physical fitness, including cardiovascular endurance, muscle strength, and flexibility. Exercise has been shown to stimulate brain activity and promote neuroplasticity, which refers to the ability of the brain to change and adapt in response to new experiences and learning. This exerts positive effects on cognitive function and overall brain performance, which may provide benefits to individuals with dementia or AD.

5.1 Efficacy of Exercise Against AD

Exercise has been shown to exert beneficial effects, including enhancing memory, attention, and executive function, on cognitive function in patients with AD. The mechanisms underlying these positive effects are multifactorial, including the promotion of neuroplasticity and secretion of growth factors in the brain. Furthermore, exercise may improve cardiovascular health, which in turn may benefit brain health and cognitive function.

Recent studies investigating the clinical effectiveness of aerobic exercise for individuals with dementia or AD revealed that a 6-month aerobic exercise program improved memory performance with an increased cardiorespiratory fitness in patients with early-stage of AD [37]. Another study found that 1 year of moderate-to-vigorous aerobic exercise induced cognitive improvement mediated by redistribution of blood flow presumably accompanied by increased neural activation in patients with amnestic MCI [38]. In contrary, among cognitively intact older adults with elevated levels of cerebral amyloid, a 56-week aerobic exercise program was found to increase the amyloid burden, as well as the whole brain and hippocampal volume, with no cognitive benefits [39]. This suggests that the exercise-mediated benefits observed in the AD brain may not be associated with amyloid deposits. Moreover, exercise-induced neurotrophic changes were not observed in individuals who performed acute exercise and carry a specific variant of the apolipoprotein E gene (*APOE*) known as *APOE4*, which is a genetic risk factor for AD [40]. Taken together, as shown in Table 1, there have been inconsistent reports regarding the efficacy of exercise against AD. Specifically, the findings suggest that not all exercise-mediated beneficial effects in individuals with healthy brains necessarily extend to individuals with AD. Nevertheless, these findings are noteworthy and draw a significant degree of interest for determining factors influencing the effectiveness of exercise in individuals with AD.

Exercise may be a potential disease-modifying factor for the onset and progression of AD given its ability to affect various structural and functional aspects of the brain, including the volume and integrity of the brain, blood flow, and hallmark changes. Cerebral blood flow (CBF) in brain regions vulnerable to AD, including the hippocampus, is crucial for maintaining brain health, as it facilitates the delivery of oxygen and nutrients while simultaneously removing waste products. Both long-term and acute aerobic exercises increase CBF, which enhances brain function and decreases the risk of AD. This could be attributed to increased demand for oxygen during exercise, which may lead to enhancement of brain activity in a specific area, anti-inflammatory effects, and upregulation of nerve growth factor expression [40, 49, 53]. Exercise has been shown to downregulate amyloid accumulation in individuals with AD. Specifically, aerobic exercise may contribute to the clearance of neurotoxic Aβ species via its anti-inflammatory and antioxidant effects, leading to a concomitant cognitive improvement [10, 55, 56]. Although there have been studies on different types, durations, and intensities of exercise, the effectiveness of exercise in preventing and delaying cognitive impairment in individuals at risk of cognitive impairment or patients with dementia or AD remains unclear. Nonetheless, previous findings may contribute toward elucidating the role of exercise in individuals with dementia or AD.

5.2 *Type, Duration, and Intensity of Exercise for AD*

Most studies have selected moderate-intensity aerobic exercise as a potential therapeutic intervention (Table 1). This could be attributed to the fact that aerobic exercise is accessible and well-tolerated by individuals with AD regardless of sex, age, and physical ability, which makes it an attractive option for AD research.

Regarding accessibility, compared with other types of exercise, such as resistance exercise, aerobic exercise can be easily incorporated into the daily lives of

Table 1 Recent evidence for a role of exercise in individuals with AD or MCI

Author (year)	Research design	Subject	Type (intensity)	Duration	Primary outcomes
Steen Jensen et al. (2016) [41]	RCTs	AD	Aerobic exercise (moderate to high)	16 weeks	Aβ and tau in CSF →
Hoffmann et al. (2016) [42]	RCTs	Mild AD	Aerobic exercise (moderate to high)	16 weeks	Cognition → (MMSE and ADAS-Cog)
Morris et al. (2017) [37]	RCTs	Early AD	Aerobic exercise (moderate)	6 months	Memory ↑ (DAD) Hippocampal volume ↑
Van der Kleij et al. (2018) [43]	RCTs	AD	Aerobic exercise (moderate to high)	16 weeks	Cerebral blood flow →
Lamb et al. (2018) [44]	RCTs	Mild to moderate dementia	Cycling and strength training (moderate to hard)	4 months	No cognitive outcome
Tarumi et al. (2019) [45]	RCTs	aMCI	Aerobic exercise (moderate to high)	12 months	Memory and executive function ↑ (CVLT-II and D-KEFS) Hippocampal volume ↑ Aβ plaque →
Tsai et al. (2019) [21]	RCTs	aMCI	Aerobic or resistance exercise (moderate)	16 weeks	Neurocognitive performance ↑ Nuerotrophins ↑ Inflammatory cytokines ↓
Fonte et al. (2019) [46]	RCTs	AD	Endurance and resistance exercise (moderate)	6 months	Global cognition → (MMSE)
Tarumi et al. (2020) [47]	RCTs	aMCI	Aerobic exercise (moderate to high)	12 months	Memory and executive function ↑ (CVLT-II and D-KEFS) Hite matter integrity →
Sanders et al. (2020) [48]	RCTs	Dementia	Aerobic and strength training (low to high)	12 weeks	Global cognition → (MMSE)
Broadhouse et al. (2020) [22]	RCTs	MCI	Resistance exercise (high)	6 months	Global cognitive and executive function ↑ (ADAS-cog and executive domain) Hippocampal subfield plasticity ↑
Thomas et al. (2020) [49]	RCTs	aMCI	Aerobic exercise (moderate)	12 months	Memory function (WMS-R) ↑ Cerebral blood flow in ACC ↑

(continued)

Table 1 (continued)

Author (year)	Research design	Subject	Type (intensity)	Duration	Primary outcomes
Uijen et al. (2020) [50]	RCTs	Mild to moderate dementia	Aerobic exercise (not mentioned)	12 weeks	Global cognition ↑ (in a minority of individuals)
Sanders et al. (2020) [48]	RCTs	Dementia	Aerobic and strength training (low to high)	12 weeks	Global cognition → (MMSE)
Vidoni et al. (2021) [39]	RCTs	With cerebral amyloid	Aerobic exercise (moderate)	52 weeks	No cognitive outcome Amyloid burden ↑ Whole brain and hippocampal volume ↑
Yu et al. (2021) [51]	RCTs	AD	Cycling (moderate)	6 months	Decline in global cognition ↓ (ADAS-cog)
Tomoto et al. (2021) [38]	RCTs	aMCI	Aerobic exercise (moderate-to-vigorous)	12 months	Cognitive performance → CBF ↑ and central arterial stiffness and arterial pulsation ↓
Tomoto et al. (2021) [52]	RCTs	aMCI	Aerobic exercise (moderate-to-vigorous)	12 months	Episodic memory and executive function ↑ (CVLT-II and D-KEFS score) Cerebrovascular function ↑
yogev-Seligmann et al. (2021) [53]	RCTs	aMCI	Aerobic exercise (moderate-to-vigorous)	16 weeks	Brain activity in speicific area ↑ Cardiovascular fitness ↑
Vidoni et al. (2022) [40]	Non-RCTs	With APOE4 allele	Aerobic exercise (moderate)	Acute (15 min)	Cerebral blood flow in hippocampal region ↑ Neurotrophic factors →
Stein et al. (2023) [54]	RCTs	AD	Aerobic exercise (moderate)	12 weeks	Executive function → Neurotrophins →

individuals with dementia or AD and is a practical intervention for studies on AD research aimed to evaluate the effects of exercise on cognitive and brain health. Furthermore, aerobic exercise can be performed at low and moderate-to-vigorous intensity depending on the individual's condition, which makes it safe and tolerable for individuals with AD. This is important because individuals with AD may have physical or cognitive limitations that would impede them from performing various forms of exercise.

5.3 Potential Modalities of Exercise for AD

Other exercise modalities, including tai chi, yoga, dance, and qigong, have been suggested to attenuate cognitive impairment in patients with AD [27, 57, 58]. These modalities are mainly characterized by gentle movement, deep breathing, coordination, flexibility, etc., which may in turn benefit brain health and cognitive function. Although there remains limited evidence-based literature on the effectiveness of these modalities, they may be appropriate for individuals with dementia or AD, given their physical and cognitive limitations. Given the observed positive effects on the healthy brain, future studies are warranted to elucidate the beneficial effects of exercise on AD and to establish an optimal exercise regimen for relieving AD-related pathological symptoms.

6 Considerations for Evaluating the Effectiveness of Exercise for Patients with AD

As previously mentioned, there have been inconsistent findings regarding the effects of exercise on patients with dementia or AD (Table 1). However, many experts suggest that exercise may exert beneficial effects by improving pathological processes and clinical symptoms in AD. Several factors may have contributed to the aforementioned inconsistency in the findings.

First, the inconsistency may be attributed to the heterogeneity of the AD population. Patients with AD present with different symptoms and progression rates, which impedes the investigation of the effects of exercise on patients with AD [59, 60]. Moreover, the lack of standardized exercise regimens (type, duration, intensity, and follow-up period) across studies on AD impedes the generalization of the study findings. Second, confounding factors, such as diet, sleep pattern, and medication use, may contribute to the inconsistent findings regarding the effectiveness of exercise in patients with AD, affecting the validity and reliability of the findings. Third, the observed inconsistency may be attributed to limitations in the measurement methods of the effectiveness of exercise on cognition. In clinical settings, several tools are widely used to measure the cognitive state, including the Mini-Mental State Examination and the Alzheimer's Disease Assessment Scale-Cognitive Subscale, which assess memory, language, attention, and praxis. Although these tools remain crucial for providing valuable insights into the nature of AD and dementia, there are considerably limited in detecting and quantifying sensitive cognitive changes, especially in the assessment of the effects of exercise on individuals with AD or dementia. Taken together, there are limitations in the approaches used for measurement of the clinical effectiveness of exercise in patients with cognitive deficits. Nevertheless, further animal and human studies are warranted to establish exercise as a cost-effective and non-invasive therapeutic intervention for AD.

7 Effectiveness of Exercise in Animal Models of AD

Rodents, including mice and rats, have been used to establish valuable animal models for studying the effects of diverse therapeutic interventions in a controlled laboratory setting. Animal studies have yielded scientific evidence that has contributed toward the discovery of new medicine and therapeutics. However, given the complexities of biological systems, it is difficult to apply these findings to humans and use this evidence-based knowledge in clinical practice. There has been extensive research on mitigating these limitations to allow the application of findings in animal studies to go beyond the laboratory. This section briefly reviews evidence derived from animal studies on the effectiveness of exercise in AD and the underlying mechanisms. Studies using different AD animal models have demonstrated that aerobic exercise, including running on a treadmill, alleviates cognitive deficits. This alleviation is mediated by the clearance of neurotoxic molecules, restoration of neuroplasticity, anti-inflammatory effects, anti-oxidation effects, and effects on other AD-related pathologies.

7.1 Effects of Exercise on Potent Neurotoxins in AD

Exercise can reduce Aβ plaque accumulation by fine-tuning the amyloidogenic pathway, which is a key step in the production of Aβ peptides [19, 20, 61, 62]. Aβ peptides are prone to aggregation via the proteolytic processing of APP, a transmembrane protein, which disrupts neuronal function and promotes other neurotoxic mechanisms. Further, exercise promotes the activity of microglia, which are the immune cells in the brain that engulf and clear byproducts and thus enhances clearance of Aβ plaques and NFTs [63–65]. Additionally, exercise has been found to promote the production of α-secretase, which is an APP-cleaving enzyme that breaks down APP into non-toxic fragments, such as the C-terminal fragment-α and soluble ectodomain of APP [61, 63, 66]. Taken together, exercise might help tilt the balance of APP processing toward the non-amyloidogenic pathway, which could reduce the risk of developing AD. There has been controversy surrounding the amyloid cascade hypothesis, which has long been acknowledged as the primary pathogenic theory in AD, given the lack of success in clinical trials targeting Aβ for improving cognitive function as well as recent findings challenging the notion that Aβ accumulation is the sole cause of cognitive impairment in AD. Therefore, although exercise may be considered an effective preventive and therapeutic intervention, future animal studies on the beneficial effects of exercise on AD should consider other pathologic events, including tau pathology, inflammation, and synaptic dysfunction, contributing to learning and memory loss, while considering the exercise timing and magnitude of Aβ accumulation.

7.2 Exercise-Induced Enhancement of Neuroplastic Changes

Exercise enhances neuroplasticity in AD, especially in the hippocampal region, which is crucial for learning and memory. Exercise has been shown to increase the density and branching of dendrites, which are specialized branched threadlike structures on nerve cells that propagate electrochemical stimuli to neighboring cells [18, 75, 78]. This allows efficient processing of signals and information by the neuronal network, leading to cognitive enhancement in AD. Furthermore, exercise has been shown to promote adult hippocampal neurogenesis (AHN) in AD. Multiple evidence suggested that exercise increases blood flow to the brain, which improves the delivery of oxygen and nutrients to the brain and thus, may stimulate AHN in the AD brain. Even a various neurotransmitters and growth factors, including BDNF, insulin-like growth factor-1, and vascular endothelial growth factor, help create a favorable environment that supports AHN and promotes the survival of new neurons in AD [17, 18, 79]. Exercise-mediated reduction in the level of inflammation and oxidative stress, which negatively impact AHN, enhances the resiliency and recovery of AHN [80].

7.3 Exercise as a Potential Modulator of Neuroinflammation

Inflammation is a causative factor for numerous diseases, including AD. In the normal state, inflammation as a homeostatic mechanism is not necessarily undesirable; however, chronic inflammation in the brain damages the central nervous system and contributes to the development of AD. The exercise-induced inflammatory response is a complex process involving various molecular and cellular pathways. In the acute inflammation phase, an immediate immune response is normally triggered to release pro-inflammatory cytokines, including IL-6, TNF-α, and interleukin-1β (IL-1β) via mechanisms dependent on nuclear factor kappa B in the cortex and hippocampus [17, 81]. However, exercise can induce a comparatively sustained and controlled inflammatory response in the chronic phase, which involves the release of anti-inflammatory cytokines such as interleukin-10 and transforming growth factor-beta that help dampen the inflammatory response [82, 83]. These processes are considered beneficial for the AD brain. However, as AD is associated with chronic inflammation, exercise-induced inflammation may be potentially harmful by exacerbating neuroinflammation and contributing to cognitive decline in AD. These could be attributed to activated immune cells around pathologic aggregates, including Aβ or phosphorylated tau.

Microglia are innate immune cells that defend against infectious agents in the brain. Under pathological conditions, microglia are activated via morphological changes and altered protein expression to release pro-inflammatory cytokines. In

AD, microglial activation mediated by the accumulated Aβ and tau proteins can lead to chronic neuroinflammation, resulting in a vicious cycle [83–85]. Recent studies using advanced techniques (e.g., single-cell RNA-seq) have reported that microglia in AD have an altered gene expression profile, indicating that they functionally differ from normal microglia [86, 87]. These changes may contribute to the neuroinflammatory response in AD. Accordingly, interventions targeting microglia can have therapeutic benefits; moreover, exercise might contribute to reducing neuroinflammation and concomitant cognitive enhancement.

Taken together, it appears that not all exercise regimens beneficially regulate the inflammatory response, with a few studies reporting no improvement despite exercise administration. These findings indicate that the inflammation-mediated exercise-induced cognitive improvement in AD should be interpreted while considering the balance between pro- and anti-inflammatory responses according to the exercise regimen.

7.4 Type, Duration, and Intensity of Exercise in AD Animal Models

Unlike human patients with AD, after an appropriate adaptation period (approximately 7–14 days), rodents can perform exercise tasks, such as running on a treadmill or wheel, without any limitations. Nevertheless, it is important to consider the type, duration, and intensity of exercise when applying the findings in terms of the effectiveness and utility of exercise to human patients with AD. Accordingly, to minimize the neurobiological stress on rodents, numerous studies have applied voluntary and moderate-intensity treadmill running exercise at a frequency of 5 days/week for approximately 4 weeks–8 months (Table 2). Regarding the potential mechanisms underlying the exercise-induced benefits in AD, animal studies have shown that different exercise regimens may reverse several pathological processes, including abnormal protein aggregation, metabolic dysfunction, inflammatory response, oxidative stress, AHN, and epigenetic modification [20, 23, 64, 67, 69, 72]. However, the effectiveness of exercise in improving AD-related pathologies is dependent on the disease severity and other confounding factors; accordingly, further animal studies are warranted to help achieve desired exercise benefits. Taken together, although exercise as a disease-modifying treatment has shown considerable promise against AD, it is important to consider non-standard exercise regimens that are practical for patients with AD.

Table 2 Recent evidence for a role of different exercise regimes in AD models

Author (year)	Subject (model)	Type (intensity)	Frequency and duration	Primary outcomes
Choi et al. (2018) [18]	5xFAD-tg	Voluntary wheel running	2–3 months	Working memory ↑ AHN ↑ Synaptic plasticity ↑
Lamb et al. (2018) [67]	APP/PS1 (APPswe/PS1 ΔE9) AD-tg	Swimming exercise	60 min, 5 weeks (5 days per week)	Recognition memory ↑ Synaptic plasticity ↑ FNDC5/irisin ↑
Farzi et al. (2019) [68]	Wistar rat, AD-like model (bilateral icv injection of Aβ25–35 peptide)	Treadmill exercise (20 m/min for 40 min) and resistance exercise (10–100% of animal weight)	8 weeks	Recognition memory ↑ AChE activity ↓
Pang et al. (2019) [69]	APP/PS1 AD-tg	Swimming exercise (1 hour)	4 weeks (6 days per week)	Cognitive function ↑ Aβ load ↓ Synaptic plasticity ↑ Glucose utilization ↑
Broderick et al. (2020) [70]	3xFAD-tg	Treadmill exercise (10–15 m/min for 20–45 min)	5 months	Aβ load ↓ Neuroinflammation → Neurotrophin →
Wu et al. (2020) [62]	TgF344 (human APPswe and human PSEN1-dE9 mutation) AD-tg rat	Treadmill exercise (18 m/min for 45 min)	8 months (3 days per week)	Depressive-like behavior ↓ Aβ load ↓ Phosphorylated tau ↓ Inflammatory response ↓
Hashiguchi et al. (2020) [71]	APP/PS1 AD-tg	Resistance exercise (75, 90, and 100% of animal weight)	4 weeks (5 days per week)	Aβ load ↓ Neuroinflammation ↓ Hyperactivity ↓
Falkenhain et al. (2020) [72]	APP/PS1 (mouse/human APP695swe/PSE-dE9) AD-tg	Voluntary wheel running	N.M	Spatial memory ↑ Aβ load ↓ Neuroinflammation ↓ Capillary flow speeds →
Thomas et al. (2020) [73]	Tg2576 AD-tg	Treadmill exercise (low; 15 m/min for 60 min, high; 32 m/min for 60 min)	12 months (5 days per week)	Spatial & recognition memory ↑ Aβ load ↓
Bernardo et al. (2020) [23]	Wistar rat, AD-like model (bilateral icv injection of STZ)	Treadmill exercise (25 m/min for 60 min)	20 weeks (5 days per week)	Spatial leaning and memory ↑ Mitochondrial oxygen consumption ↑

(continued)

Table 2 (continued)

Author (year)	Subject (model)	Type (intensity)	Frequency and duration	Primary outcomes
Francis et al. (2020) [74]	Tg2576 (human APP harboring swedish familial AD mutations) AD-tg	Voluntary wheel running	4 months (5 days per week)	Spatial leaning and memory ↑ Aβ load ↓
Zhao et al. (2020) [75]	APP/PS1 (APPswe/ PS1dE9) AD-tg	Treadmill exercise (5–12 m/min for 45 min)	12 weeks (5 days per week)	Spatial leaning and memory ↑ Aβ load ↓ Mitochondrial function ↑ Mitophagy ↑ Synaptic plasticity ↑
Nakanishi et al. (2021) [64]	Senescence Accelerated Mouse-Prone 8 (SAMP8) AD	Rotarod-treadmill exercise (2.4 m/min for 15 min)	10 weeks (5 days per week)	Recognition memory ↑ Aβ load ↓ Oxidative stress ↓ Neuroinflammation ↓
Belaya et al. (2021) [76]	5xFAD (human APP; Swe, Flo, and Lon and human PSEN1; M146L and L286V) transgenes)-tg	Voluntary wheel running	6 months	Aβ load ↓ Iron homeostasis ↓
Choi et al. (2021) [19]	NSE/APP-C105 AD-tg	Treadmill exercise (8–14 m/min for 30 min)	8 weeks (5 days per week)	Spatial learning and memory ↑ Aβ load ↓ Oxidative stress ↓ Iron overload ↓
Hwang et al. (2022) [20]	NSE/PS2 mutant AD-tg	Treadmill exercise (10 m/min for 50 min)	6 weeks (5 days per week)	Recognition memory ↑ Spatial memory → Aβ plaque ↓ Cell death ↓
Liu et al. (2022) [65]	APP/PS1 (APPswe/ PS1 ΔE9) AD-tg	Voluntary wheel running	4 hours, 8 weeks (5 days per week)	Spatial learning and memory ↑ Aβ load ↓ Reactive gliosis ↓ BDNF ↑
Zhang et al. (2022) [77]	APP/PS1 (APPswe/ PS1 ΔE9) AD-tg	Voluntary wheel running	3 months	Spatial & recognition memory ↑ Hippocampal volume ↑ Aβ load ↓ Microglia glucose metabolism ↑

8 Perspectives and Future Direction of AD Research with Respect to Exercise as an Intervention

With increasing research interest in exercise, studies both on animals and humans have evaluated the benefits of exercise in terms of AD prevention and management, demonstrating that physical exercise can be a valuable tool for mitigating AD-related symptoms. However, there remains a considerable gap between preclinical and clinical research, which may be contributing to clinical failure of exercise treatment in individuals with AD and dementia. Further research is warranted to elucidate the mechanisms based on which exercise induces cognitive enhancement in AD as well as to establish the optimal exercise regimen in terms of the frequency, intensity, time (duration), and type of exercise.

Acknowledgments The authors declare no competing interests.

References

1. Global status report on the public health response to dementia. Geneva: World Health Organization, 2021. ISBN 978-92-4-003324-5 (electronic version)
2. Li TR et al (2022) Biomarkers and tools for predicting Alzheimer's disease in the preclinical stage. Curr Neuropharmacol 20(4):713–737
3. Langa KM, Levine DA (2014) The diagnosis and management of mild cognitive impairment: a clinical review. JAMA 312(23):2551–2561
4. Rabbito A et al (2020) Biochemical markers in Alzheimer's disease. Int J Mol Sci 21(6):1989
5. Zetterberg H, Bendlin BB (2021) Biomarkers for Alzheimer's disease-preparing for a new era of disease-modifying therapies. Mol Psychiatry 26(1):296–308
6. Khan S, Barve KH, Kumar MS (2020) Recent advancements in pathogenesis, diagnostics and treatment of Alzheimer's disease. Curr Neuropharmacol 18(11):1106–1125
7. Guzman-Martinez L et al (2021) New Frontiers in the prevention, diagnosis, and treatment of Alzheimer's disease. J Alzheimer's Dis 82(s1):S51–s63
8. Di Marco LY et al (2014) Modifiable lifestyle factors in dementia: a systematic review of longitudinal observational cohort studies. J Alzheimer's Dis 42(1):119–135
9. Serrano-Pozo A, Growdon JH (2019) Is Alzheimer's disease risk modifiable? J Alzheimer's Dis 67(3):795–819
10. López-Ortiz S et al (2021) Physical exercise and Alzheimer's disease: effects on pathophysiological molecular pathways of the disease. Int J Mol Sci 22(6):2897
11. Smith PJ (2019) Pathways of prevention: a scoping review of dietary and exercise interventions for Neurocognition. Brain Plast 5(1):3–38
12. Andrade-Guerrero J et al (2023) Alzheimer's disease: an updated overview of its genetics. Int J Mol Sci 24(4):3754
13. Scheltens P et al (2021) Alzheimer's disease. Lancet 397(10284):1577–1590
14. Abeysinghe A, Deshapriya R, Udawatte C (2020) Alzheimer's disease; a review of the pathophysiological basis and therapeutic interventions. Life Sci 256:117996
15. Sengoku R (2020) Aging and Alzheimer's disease pathology. Neuropathology 40(1):22–29
16. King-Robson J, Wilson H, Politis M (2021) Associations between amyloid and tau pathology, and connectome alterations, in Alzheimer's disease and mild cognitive impairment. J Alzheimer's Dis 82(2):541–560

17. Valenzuela PL et al (2020) Exercise benefits on Alzheimer's disease: state-of-the-science. Ageing Res Rev 62:101108
18. Choi SH et al (2018) Combined adult neurogenesis and BDNF mimic exercise effects on cognition in an Alzheimer's mouse model. Science 361(6406):eaan8821
19. Choi DH et al (2021) Treadmill exercise alleviates brain iron dyshomeostasis accelerating neuronal amyloid-β production, neuronal cell death, and cognitive impairment in transgenic mice model of Alzheimer's disease. Mol Neurobiol 58(7):3208–3223
20. Hwang DJ et al (2022) Exercise reverses amyloid β-peptide-mediated cognitive deficits in Alzheimer's disease mice expressing mutant Presenilin-2. Med Sci Sports Exerc 54(4):551–565
21. Tsai CL et al (2019) Distinctive effects of aerobic and resistance exercise modes on neurocognitive and biochemical changes in individuals with mild cognitive impairment. Curr Alzheimer Res 16(4):316–332
22. Broadhouse KM et al (2020) Hippocampal plasticity underpins long-term cognitive gains from resistance exercise in MCI. Neuroimage Clin 25:102182
23. Bernardo TC et al (2020) Physical exercise mitigates behavioral impairments in a rat model of sporadic Alzheimer's disease. Behav Brain Res 379:112358
24. Cummings J et al (2021) Alzheimer's disease drug development pipeline: 2021. Alzheimers Dement (N Y) 7(1):e12179
25. Cummings J et al (2022) Alzheimer's disease drug development pipeline: 2022. Alzheimers Dement (N Y) 8(1):e12295
26. Cummings J et al (2016) Drug development in Alzheimer's disease: the path to 2025. Alzheimer's Res Ther 8:39
27. Cui MY et al (2018) Exercise intervention associated with cognitive improvement in Alzheimer's disease. Neural Plast 2018:9234105
28. van Bokhoven P et al (2021) The Alzheimer's disease drug development landscape. Alzheimer's Res Ther 13(1):186
29. Lautenschlager NT et al (2008) Effect of physical activity on cognitive function in older adults at risk for Alzheimer disease: a randomized trial. JAMA 300(9):1027–1037
30. Erickson KI et al (2011) Exercise training increases size of hippocampus and improves memory. Proc Natl Acad Sci USA 108(7):3017–3022
31. Voss MW et al (2011) Exercise, brain, and cognition across the life span. J Appl Physiol (1985) 111(5):1505–1513
32. Erickson KI, Kramer AF (2009) Aerobic exercise effects on cognitive and neural plasticity in older adults. Br J Sports Med 43(1):22–24
33. Cotman CW, Berchtold NC (2002) Exercise: a behavioral intervention to enhance brain health and plasticity. Trends Neurosci 25(6):295–301
34. Vemuri P et al (2012) Effect of lifestyle activities on Alzheimer disease biomarkers and cognition. Ann Neurol 72(5):730–738
35. Ainsworth BE et al (2000) Compendium of physical activities: an update of activity codes and MET intensities. Med Sci Sports Exerc 32(9 Suppl):S498–S504
36. Shih IF et al (2019) The roles of physical activity and inflammation in mortality, cognition, and depressive symptoms among older Mexican Americans. Am J Epidemiol 188(11):1944–1952
37. Morris JK et al (2017) Aerobic exercise for Alzheimer's disease: a randomized controlled pilot trial. PLoS One 12(2):e0170547
38. Tomoto T et al (2021) One-year aerobic exercise reduced carotid arterial stiffness and increased cerebral blood flow in amnestic mild cognitive impairment. J Alzheimer's Dis 80(2):841–853
39. Vidoni ED et al (2021) Effect of aerobic exercise on amyloid accumulation in preclinical Alzheimer's: a 1-year randomized controlled trial. PLoS One 16(1):e0244893
40. Vidoni ED et al (2022) Dementia risk and dynamic response to exercise: a non-randomized clinical trial. PLoS One 17(7):e0265860
41. Steen Jensen C et al (2016) Cerebrospinal fluid amyloid beta and tau concentrations are not modulated by 16 weeks of moderate- to high-intensity physical exercise in patients with Alzheimer disease. Dement Geriatr Cogn Disord 42(3–4):146–158

42. Hoffmann K et al (2016) Moderate-to-high intensity physical exercise in patients with Alzheimer's disease: a randomized controlled trial. J Alzheimer's Dis 50(2):443–453
43. van der Kleij LA et al (2018) The effect of physical exercise on cerebral blood flow in Alzheimer's disease. Neuroimage Clin 20:650–654
44. Lamb SE et al (2018) Dementia And Physical Activity (DAPA) trial of moderate to high intensity exercise training for people with dementia: randomised controlled trial. BMJ 361:k1675
45. Tarumi T et al (2019) Exercise training in amnestic mild cognitive impairment: a one-year randomized controlled trial. J Alzheimer's Dis 71(2):421–433
46. Fonte C et al (2019) Comparison between physical and cognitive treatment in patients with MCI and Alzheimer's disease. Aging (Albany NY) 11(10):3138–3155
47. Tarumi T et al (2020) Cerebral white matter integrity in amnestic mild cognitive impairment: a 1-year randomized controlled trial of aerobic exercise training. J Alzheimer's Dis 73(2):489–501
48. Sanders LMJ et al (2020) Effects of low- and high-intensity physical exercise on physical and cognitive function in older persons with dementia: a randomized controlled trial. Alzheimer's Res Ther 12(1):28
49. Thomas BP et al (2020) Brain perfusion change in patients with mild cognitive impairment after 12 months of aerobic exercise training. J Alzheimer's Dis 75(2):617–631
50. Uijen IL et al (2020) Individual differences in the effects of physical activity on cognitive function in people with mild to moderate dementia. J Alzheimer's Dis 74(2):435–439
51. Yu F et al (2021) Cognitive effects of aerobic exercise in Alzheimer's disease: a pilot randomized controlled trial. J Alzheimer's Dis 80(1):233–244
52. Tomoto T et al (2021) One-year aerobic exercise altered cerebral vasomotor reactivity in mild cognitive impairment. J Appl Physiol (1985) 131(1):119–130
53. Yogev-Seligmann G et al (2021) Neurocognitive plasticity is associated with cardiorespiratory fitness following physical exercise in older adults with amnestic mild cognitive impairment. J Alzheimer's Dis 81(1):91–112
54. Stein AM et al (2023) Aerobic training and circulating neurotrophins in Alzheimer's disease patients: a controlled trial. Exp Aging Res 49(1):1–17
55. Baker LD et al (2012) High-intensity physical activity modulates diet effects on cerebrospinal amyloid-β levels in normal aging and mild cognitive impairment. J Alzheimer's Dis 28(1):137–146
56. Wang M et al (2023) Exercise suppresses neuroinflammation for alleviating Alzheimer's disease. J Neuroinflammation 20(1):76
57. Bisbe M et al (2020) Comparative cognitive effects of choreographed exercise and multimodal physical therapy in older adults with amnestic mild cognitive impairment: randomized clinical trial. J Alzheimer's Dis 73(2):769–783
58. Chen B et al (2022) Research trends and hotspots of exercise for Alzheimer's disease: a bibliometric analysis. Front Aging Neurosci 14:984705
59. Duara R, Barker W (2022) Heterogeneity in Alzheimer's disease diagnosis and progression rates: implications for therapeutic trials. Neurotherapeutics 19(1):8–25
60. Avelar-Pereira B et al (2022) Decoding the heterogeneity of Alzheimer's disease diagnosis and progression using multilayer networks. Mol Psychiatry 28:2423
61. Koo JH et al (2017) Treadmill exercise decreases amyloid-β burden possibly via activation of SIRT-1 signaling in a mouse model of Alzheimer's disease. Exp Neurol 288:142–152
62. Wu C et al (2020) Effects of exercise training on anxious-depressive-like behavior in Alzheimer rat. Med Sci Sports Exerc 52(7):1456–1469
63. Zhang J et al (2018) Long-term treadmill exercise attenuates Aβ burdens and astrocyte activation in APP/PS1 mouse model of Alzheimer's disease. Neurosci Lett 666:70–77
64. Nakanishi K et al (2021) Effect of low-intensity motor balance and coordination exercise on cognitive functions, hippocampal Aβ deposition, neuronal loss, neuroinflammation, and oxidative stress in a mouse model of Alzheimer's disease. Exp Neurol 337:113590

65. Liu Y et al (2022) Aquaporin 4 deficiency eliminates the beneficial effects of voluntary exercise in a mouse model of Alzheimer's disease. Neural Regen Res 17(9):2079–2088
66. Yu H et al (2021) Treadmill exercise ameliorates adult hippocampal neurogenesis possibly by adjusting the APP proteolytic pathway in APP/PS1 transgenic mice. Int J Mol Sci 22(17):9570
67. Lourenco MV et al (2019) Exercise-linked FNDC5/irisin rescues synaptic plasticity and memory defects in Alzheimer's models. Nat Med 25(1):165–175
68. Farzi MA et al (2019) Exercise improves recognition memory and acetylcholinesterase activity in the beta amyloid-induced rat model of Alzheimer's disease. Ann Neurosci 25(3):121–125
69. Pang R et al (2019) Regular exercise enhances cognitive function and Intracephalic GLUT expression in Alzheimer's disease model mice. J Alzheimer's Dis 72(1):83–96
70. Broderick TL et al (2020) Neuroprotective effects of chronic resveratrol treatment and exercise training in the 3xTg-AD mouse model of Alzheimer's disease. Int J Mol Sci 21(19):7337
71. Hashiguchi D et al (2020) Resistance exercise decreases amyloid load and modulates inflammatory responses in the APP/PS1 mouse model for Alzheimer's disease. J Alzheimer's Dis 73(4):1525–1539
72. Falkenhain K et al (2020) A pilot study investigating the effects of voluntary exercise on capillary stalling and cerebral blood flow in the APP/PS1 mouse model of Alzheimer's disease. PLoS One 15(8):e0235691
73. Thomas R et al (2020) Exercise training results in lower amyloid plaque load and greater cognitive function in an intensity dependent manner in the Tg2576 mouse model of Alzheimer's disease. Brain Sci 10(2):88
74. Francis N et al (2020) Voluntary wheel running reduces amyloid-β42 and rescues behavior in aged Tg2576 mouse model of Alzheimer's disease. J Alzheimer's Dis 73(1):359–374
75. Zhao N et al (2020) Treadmill exercise attenuates aβ-induced mitochondrial dysfunction and enhances mitophagy activity in APP/PS1 transgenic mice. Neurochem Res 45(5):1202–1214
76. Belaya I et al (2021) Regular physical exercise modulates iron homeostasis in the 5xFAD mouse model of Alzheimer's disease. Int J Mol Sci 22(16):8715
77. Zhang SS et al (2022) Long-term running exercise improves cognitive function and promotes microglial glucose metabolism and morphological plasticity in the hippocampus of APP/PS1 mice. J Neuroinflammation 19(1):34
78. Yang L et al (2022) Long-term exercise pre-training attenuates Alzheimer's disease-related pathology in a transgenic rat model of Alzheimer's disease. Geroscience 44(3):1457–1477
79. Horowitz AM et al (2020) Blood factors transfer beneficial effects of exercise on neurogenesis and cognition to the aged brain. Science 369(6500):167–173
80. Joshi R, Salton SRJ (2022) Neurotrophin crosstalk in the etiology and treatment of neuropsychiatric and neurodegenerative disease. Front Mol Neurosci 15:932497
81. Chow LS et al (2022) Exerkines in health, resilience and disease. Nat Rev Endocrinol 18(5):273–289
82. De Miguel Z et al (2021) Exercise plasma boosts memory and dampens brain inflammation via clusterin. Nature 600(7889):494–499
83. Liu Y et al (2020) Short-term resistance exercise inhibits neuroinflammation and attenuates neuropathological changes in 3xTg Alzheimer's disease mice. J Neuroinflammation 17(1):4
84. Heneka MT et al (2015) Neuroinflammation in Alzheimer's disease. Lancet Neurol 14(4):388–405
85. Arranz AM, De Strooper B (2019) The role of astroglia in Alzheimer's disease: pathophysiology and clinical implications. Lancet Neurol 18(4):406–414
86. Mathys H et al (2019) Single-cell transcriptomic analysis of Alzheimer's disease. Nature 570(7761):332–337
87. Belonwu SA et al (2022) Sex-stratified single-cell RNA-Seq analysis identifies sex-specific and cell type-specific transcriptional responses in Alzheimer's disease across two brain regions. Mol Neurobiol 59(1):276–293

Neuroprotective and Neurotrophic Effects of Astaxanthin on the Brain

Jang Soo Yook and Hideaki Soya

Abstract Oxidative stress in the brain is associated with the development and progression of neurological disorders, posing antioxidant nutrients as an effective strategy for protecting neuronal cells and potentially slowing cognitive decline. Bioactive compounds from natural sources with antioxidant effects promote brain health. Among various natural compounds, astaxanthin (ASX), a potent red-pigment carotenoid found in various microorganisms and marine animals, is well recognized for its potential health benefits. In this review, we highlight the promising neuroprotective effects of ASX through cellular experiments and animal models. This review can provide novel insights on the therapeutic potential of ASX through its antioxidant, anti-inflammatory, and anti-apoptotic effects against neurodegenerative diseases such as Alzheimer's disease and Parkinson's disease. In addition to demonstrating the neurotrophic effects of ASX on structural and functional changes in hippocampal neuronal plasticity, this review also discusses its synergistic potential with other interventions, such as exercise.

Keywords Astaxanthin · Antioxidant · Neuroprotection · Adult hippocampal neurogenesis · Exercise

J. S. Yook (✉)
Division of Humanities and Social Sciences, Pohang University of Science and Technology (POSTECH), Pohang, Gyeongsangbuk-do, South Korea
e-mail: yookjs@postech.ac.kr

H. Soya
Laboratory of Exercise Biochemistry and Neuroendocrinology, Institute of Health and Sport Sciences, University of Tsukuba, Tsukuba, Ibaraki, Japan

Sport Neuroscience Division, Kokoro Division, Advanced Research Initiative for Human High Performance (ARIHHP), Institute of Health and Sport Sciences, University of Tsukuba, Tsukuba, Ibaraki, Japan

Center for Cybernics Research, University of Tsukuba, Tsukuba, Ibaraki, Japan

H. Soya (ed.), *Exercise Brain Stimulation for Cognitive Function and Mental Health*, Advances in Neurobiology 44, https://doi.org/10.1007/978-981-95-0066-6_17

1 Introduction

Unhealthy lifestyle-related factors, including physical inactivity, poor diet, and sleep deprivation, are associated with cognitive impairment and dementia, which adversely affect the quality of independent living. Exploring the beneficial effects of changes in modifiable lifestyle factors as a non-pharmacological treatment in maintaining and improving cognitive health is crucial, owing to the lack of effective pharmacological modalities. Among a number of identified non-pharmacological factors, dietary supplements and physical activity have been shown to improve cognitive impairment in neurological disorders, such as Alzheimer's disease (AD) and Parkinson's disease (PD), through multidimensional cellular and molecular changes.

Oxidative stress in the brain is associated with the onset and progression of neurodegenerative diseases, such as AD, which involve cognitive dysfunction [1]. Therefore, attenuating oxidative stress is an effective approach for protecting neuronal cells implicated in the development of neurodegenerative diseases. Consumption of antioxidant nutrients induces potential neuroprotective effects that may suppress the decline in cognitive function. Currently, the search and discovery of novel bioactive compounds from natural sources in the food and pharmaceutical industry for application in the brain health are challenging.

Among effective dietary compounds for brain health, astaxanthin (ASX), a natural carotenoid, has gained attention as a promising antioxidant ingredient with higher efficacy than other carotenoids such as lutein and β-carotene [2, 3]. Notably, ASX plays a role in protecting neuronal cells by attenuating oxidative damage and inhibiting neuroinflammation, as it can readily cross the blood–brain barrier owing to its unique chemical structure [4, 5]. Consequently, ASX may induce neuronal and synaptic plasticity, including neurogenesis and long-term potentiation (LTP), which is a cellular mechanism underlying ASX-regulated hippocampal functions, including learning and memory, through its antioxidant effect [6, 7]. Moreover, ASX also stimulates various molecular pathways involved in hippocampal neurogenesis, enhancing cell proliferation and differentiation and reducing oxidative stress and neuroinflammation [8]. Notably, numerous studies have reported the potential therapeutic role of ASX in neurological disorders, including AD and PD, as a modifiable lifestyle behavior.

Exercise is also an important modifiable lifestyle behavior that is an essential component of brain health because of its influence on brain structure and function. Exercise improves hippocampus-related cogntivie functions, such as learning and memory, in humans and rodents [9], probably by enhancing neuronal plasticity, including neurogenesis and LTP [10, 11]. Considering the brain's capacity for activity, which is partly a response to extrinsic factors, combining exercise and dietary supplements may potentiate the induction of neuronal plasticity involving all the aspects that these factors influence individually. In human studies, dietary factors combined with physical activity exhibited a greater preventive effect on the onset of dementia and depression [105, 106]. A previous study in rodents demonstrated that ASX intake potentiates the effects of exercise on neurogenesis and memory function through a specific molecular factor in the hippocampus [12], demonstrating the synergistic benefits of dietary supplements with antioxidants and exercise on cognitive enhancement. Therefore, this review focuses on the neuroprotective properties of ASX, including its antioxidant, anti-inflammatory, and anti-apoptotic effects, against neurological disorders, such as AD and PD, based on cellular experiments and animal models. In addition, this review underscores the role of ASX in neurotrophic effects to promote neuronal plasticity as well as its synergistic potential with exercise.

2 Biological Properties of ASX in Brain Health

2.1 Source of ASX

ASX is a naturally occurring marine carotenoid with potent antioxidant activity. It appears as a red–orange pigment in various marine animals, including salmonids, shrimp, and crayfish [13]. The primary natural source of ASX is the microalga *Haematococcus pluvialis*, which is the most prolific ASX producer [14]. Although synthetic forms of ASX have been developed recently, safety concerns exist regarding their use for human consumption. Therefore, natural ASX extracted from *Haematococcus pluvialis* is a primary source for human applications, including dietary supplements, cosmetics, and food products.

The chemical structure of ASX comprises a unique keto and hydroxyl group at each end of the molecule [15], imparting it stronger bioactivity and antioxidant capacity than zeaxanthin, lutein, and β-carotene. Carotenoids, as precursors of vitamin A, can be converted to retinoic acid, contributing to its associated health benefits and risks. However, unlike other carotenoids, ASX cannot be converted to retinoic acid [16]. Owing to its unique chemical characteristics, ASX exhibits strong potential health-promoting biological properties, including antioxidant, anti-inflammatory, and anti-apoptotic properties (Fig. 1).

2.2 Antioxidant Effects of ASX

Reactive oxygen species (ROS) are generated during the oxidation of metabolic and physiological byproducts. Notably, excessive cellular energy metabolism and oxygen consumption coupled with ROS overproduction induce oxidative stress in cells [17]. Oxidative stress and excessive ROS levels induce deleterious effects on cell

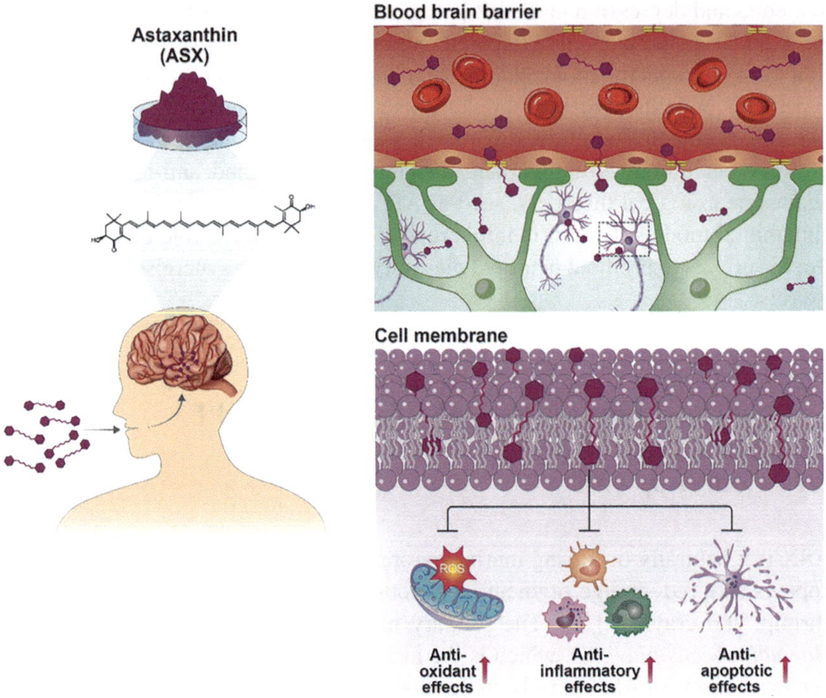

Fig. 1 Effects of astaxanthin in central nerve system. Dietary astaxanthin (ASX) can cross the blood-brain barrier (BBB) and act both inside and on the surface of the double-layered cell membrane because of its unique chemical structure. ASX enhances anti-oxidant, anti-inflammatory, and anti-apoptotic capacity, providing neuroprotective effect in the treatment of neurological disease

structure and function, leading to the development of several metabolic and chronic disorders and neurodegenerative diseases [18, 19]. Notably, excessive ROS levels result in lipid membrane disruption, increased membrane fluidity and permeability, and damage to proteins and DNA [20]. Owing to its unique chemical features, ASX can neutralize ROS, scavenge free radicals, and prevent lipid peroxidation, thereby maintaining cell membrane integrity and subsequently protecting cells against oxidative damage [21]. ASX exhibits a stronger antioxidant capacity than other nutrients, including vitamin C, coenzyme Q, and catechins [3, 22].

Oxidative stress is strongly associated with the development of multiple neurodegenerative diseases, such as AD and PD, as evidenced by increased free radicals, lipid peroxidation, and DNA damage during neurodegenerative diseases [19, 23, 24]. ASX counteracted oxidative stress in an in vitro model of Parkinsonism induced by 1-methyl-4-phenylpyridinium (MPP+) and paraquat (PQ) by regulating several molecular pathways, including N-methyl-D-aspartate (NMDA) receptors, mitogen-activated protein kinase (MAPK)/extracellular signal-regulated protein kinase (ERK), and heme oxygenase-1 (HO-1), associated with neurodegenerative diseases [25–27]. Moreover, the mouse model of PD fed ASX exhibited increased efficacy of endogenous antioxidant enzymes, as evidenced by the reduced glutathione (GSH) to oxidized glutathione ratio [28]. These studies suggest that ASX alleviates motor impairments and the loss of dopaminergic (DA) neurons and axonal projections in the substantia nigra and striatum, respectively.

In addition to the antioxidant effects in PD, ASX ameliorated cognitive decline and reduced AD pathophysiology in vivo and in vitro. In neuroblastoma cells (SH-SY5Y), ASX administration induced a neuroprotective effect against amyloid beta (Aβ) toxicity by increasing HO-1 expression, an antioxidative enzyme, through the ERK signaling pathway [29]. Wen et al. [30] replicated this finding and demonstrated an ASX-induced dose- and time-dependent increase in HO-1 expression, thereby protecting mouse hippocampal neurons against glutamate toxicity-induced ROS [30]. Notably, ASX induced HO-1 expression through the nuclear factor erythroid 2-related factor 2 (NRF2)-mediated antioxidant response element signaling pathway. Further, Hafez et al. reported that ASX treatment significantly increases NRF2 expression in a dose-dependent manner in the hippocampus and cortex of hydrated aluminum chloride-induced AD rat model [31]. Another study demonstrated that ASX supplementation in App knock-in AD mice ($App^{NL\text{-}G\text{-}F}$) protected hippocampal parvalbumin interneurons, which are antioxidant regulators of oxidative stress. This effect may be attributed to an ASX-induced increase in endogenous antioxidants such as GSH [32]. In a recent study that used AD model mice, constructed through injecting $A\beta_{(25\text{–}35)}$ into the lateral ventricle, ASX intake inhibited oxidative stress in the hippocampus, as evidenced by a decrease in ROS and malondialdehyde levels and an increase in (superoxide dismutase) SOD and GSH levels through the silent information regulator 1 (SIRT1) signaling pathway [33].

2.3 *Anti-Inflammatory Effects of ASX*

The immune system plays a crucial role in response to infection, ischemia, and injury in the human body. However, abnormal inflammatory reactions can cause numerous diseases, such as neurodegenerative disorders, chronic obstructive pulmonary disorder, atherosclerosis, and rheumatoid arthritis [34]. The balance between pro- and anti-inflammatory cytokines is crucial in inducing the appropriate immune response during homeostasis and disease-associated inflammation [35]. The aberrant production of pro-inflammatory cytokines, such as interleukin (IL)-1β, IL-6, IL-8, and tumor necrosis factor-α (TNF-α), in the central nervous system serves as etiological factors for neuroinflammation, which is a key pathological driver of many neurological diseases. In addition, transcription factors and several pathways, including nuclear factor kappa-light-chain enhancer of activated B cells

(NF-κB), MAPK, and the NRF2 signaling pathway, contribute to the persistent neuroinflammation [36]. These transcription factors and pathways are involved in microglia and astrocyte activation and play pivotal roles in the progression of many neurological diseases, including AD, PD, depression, and cerebral vascular diseases [37].

In AD, chronically activated microglia and astrocytes in the brain that interact with Aβ plaques and tau protein overexpression contribute to aberrant neuroinflammation, resulting in neuronal damage [38]. Regulation of neuroinflammation by suppressing microglial and astroglial activation using natural compounds with anti-inflammatory properties represents a potentially promising treatment strategy for AD [39]. Several in vivo studies in AD animal models have demonstrated that ASX reduces neuroinflammation in the brain, thereby suppressing Aβ accumulation and Tau hyperphosphorylation. In $A\beta_{(1-42)}$-infused AD rats, oral ASX administration for 4 weeks suppressed pro-inflammatory TNF-α in the hippocampus [40]. Further, the ASX diet decreased ionized calcium-binding adaptor molecule 1 (Iba-1) expression, a marker for activated microglia, in the hippocampus of $App^{NL\text{-}G\text{-}F}$ mice [32]. In another AD transgenic mouse (APP/PS1), ASX and docosahexaenoic acid-acylated ASX reduced the expression of both Iba-1 and glial fibrillary acidic protein (GFAP), a marker of astrocyte activation, accompanied by a decrease in TNF-α and IL-1β levels [41]. Additionally, ASX treatment decreased Iba-1 and GFAP expression in the hippocampus and ameliorated deficits in spatial learning and memory in a ferrous amyloid buthionine (FAB)-infused sporadic late-onset AD rat model [42]. Several molecular mechanisms underlying the anti-neuroinflammatory effects of ASX have been proposed. Microglia-mediated neuroinflammation can be categorized into either M1 pro- or M2 anti-inflammatory phenotypes, and M1 to M2 shifting is regulated by several important signaling pathways, including NF-κB, MAPK, and Janus kinase/signal transducer and activator of transcription (JAK/STAT) [43]. In lipopolysaccharide (LPS)-treated BV1 microglia cells, ASX promoted M2 anti-inflammatory polarization by inhibiting NF-κB and JNK signaling pathways [44] and suppressed M1 pro-inflammatory factors by down-regulating inflammation-related miRNA (miR-31-5p) [45]. Overall, ASX may decelerate the progression of AD by reducing neuroinflammation.

Neuroinflammatory processes play a crucial role in PD pathophysiology. In patients with PD, the levels of pro-inflammatory cytokines are elevated in response to activated microglia in the cerebrospinal fluid, serum, and brain, which are further correlated with the severity and disability of motor and non-motor dysfunction [46–48]. Pro-inflammatory cytokines overexpression affects DA neuron degeneration in rodent models of PD induced by neurotoxins such as 6-hydroxydopamine and 1-methyl-4-phenyl-1,2,3,6 tetrahydropyridine (MPTP) [49]. As chronic neuroinflammation in microglia considerably contributes to the loss of DA neurons, targeting inflammatory pathways mediated by activated glial cells could be a desirable approach for therapies aimed at preventing PD progression [50]. ASX exerts anti-inflammatory effects by suppressing pro-inflammatory cytokines [51]. In activated microglia cells, ASX inhibited LPS-induced IL-6 mRNA and protein levels through MAPK- and NF-κB-mediated pathways [52]. Additionally, ASX pre-treatment of PC12 neuronal cells induced neuroprotective effects against H_2O_2 or MPP + -induced damage by reducing IL-1, IL-6, and TNF-α production [53]. These results are also supported by further animal studies, indicating that ASX-induced anti-neuroinflammation may have positive effects in PD models. Dietary ASX reduced the microglial expression of Iba-1 and prevented the neurodegeneration of DA neurons in the substantia nigra pars compacta (SNpc) and striatum in MPTP-induced young PD mice [28], which also contributed to the alleviation of motor impairment in PQ-induced PD mice [27]. A recent study demonstrated that ASX attenuates MPTP-induced motor impairment and DA neuron damage by suppressing α-synuclein, a major gene implicated in the onset of PD, and regulates microRNA-7 [54]. Moreover, a DNA omics study revealed that the transcription of pro- (IL-6 and TNF) and anti-inflammatory (S100A8 and S100A9) factors in the hippocampus contributed to ASX-enhanced learning and memory function in wild-type ASX-treated mice [7]. Considering the role of inflammatory in non-motor dysfunction of PD, ASX could be a potential dietary compound for the presentation of mental disorder, such as depress and anxiety, in PD.

2.4 *Anti-Apoptotic Effects of ASX*

Apoptosis, a programmed cell death, typically occurs during tissue development and aging as a homeostatic process to maintain the cell population. Apoptosis is involved in various defense processes, such as immune reactions and cellular damage [55]. However, dysregulated, aberrant apoptosis can lead to deleterious effects at the cellular level, such as neuronal cell death via oxidative stress, disruption of calcium homeostasis, and mitochondrial dysfunction, contributing to the pathology of neurodegenerative diseases, including AD and PD [56]. At the molecular level, apoptosis is regulated by anti-apoptotic molecules, such as the B-cell lymphoma-2 (BCL −2) family proteins, and pro-apoptotic molecules, including BCL-2-associated X protein (BAX), caspases, and p53 [57]. As oxidative stress triggers apoptosis, intervention with antioxidants may offer potential anti-apoptotic neuroprotective effects.

Apoptosis contributes to the neuronal death in AD. Upregulated pro-apoptotic proteins, including the BCL-2 family proteins, have been observed in the brains of patients with AD and animal models [58, 59]. ASX, owing to its antioxidant effects, attenuates glutamate- or H_2O_2-induced cytotoxicity in cell lines that serve as mimic cell models of AD by modulating the BAX/BCL-2 pathway, suggesting that ASX-induced neuroprotective effect could be attributed to its anti-apoptotic capacity [60, 61]. A previous study also demonstrated that ASX administration mitigates cytotoxic damage in PC12 cells induced by Aβ25–35 peptide by reducing apoptosis rate.

In addition to the in vitro study, oral ASX intake increased *Bcl-2* expression and decreased *Bax* expression, demonstrating that ASX prevents learning and memory impairment. Accordingly, ASX may exert neuroprotective effects against Aβ toxicity-induced oxidative stress through the SIRT1/PGC-1α pathway, which is involved in mitochondrial function [33]. Considering another molecular factor involved in ASX-regulated apoptosis, Yook et al. found that ASX treatment of neuroblastoma cells (SH-SY5Y) increased leptin (LEP) expression in a dose-dependent manner. Further, the ASX diet upregulated the *Lep* gene in the hippocampus of young wild-type mice and its receptor, including LEPRα, and partly the downstream pathway, including the AKT pathway [12]. Notably, the hippocampal LEP enhances neuroplasticity and memory function, making it a potential therapeutic target for AD [62, 63]. Therefore, ASX exerts a neuroprotective effect against neuronal apoptosis and prevents cognitive impairment in AD.

Apoptosis is a crucial mechanism of neuronal loss in PD, as evidenced by the presence of apoptotic markers, including caspase-3 and -9, in the SN of postmortem patients and animal models [64, 65]. Furthermore, the roles of anti- and pro-apoptotic proteins, including BCL-2 and BAX, in PD pathogenesis have been confirmed in post-mortem, in vivo, and in vitro models [66]. Although PD pathogenesis is not fully understood, apoptosis plays an important role in its pathogenesis as a mediator. In a study of MPTP/MPP + -induced neuronal death in neuroblastoma cells (SH-SY5Y) and a mouse model of PD, ASX exerted anti-apoptotic effects by upregulating the BCL-1/BAX ratio and downregulating caspase-3 activation [67]. Additionally, a recent study demonstrated that ASX treatment mitigates apoptotic neuronal death in PQ-treated SH-SY5Y cells by decreasing cleaved caspase-9 expression [27]. Considering its potential anti-apoptotic efficacy, ASX may play a significant role in the alleviation and treatment of PD.

3 Neuroplasticity Induction by ASX in Brain

3.1 Effects of ASX on Adult Hippocampal Neurogenesis

Adult neurogenesis is a process through which new neurons are generated from neural progenitor cells in specific areas of the adult brain. Cell proliferation and neuronal differentiation occur continuously throughout their lifespan in restricted regions of the brain, including the subgranular zone (SGZ) of the dentate gyrus (DG) in the hippocampus and the subventricular zone adjacent to the lateral ventricles, where the integration of new neurons into neural circuits contributes to neuroplasticity [68]. Neurogenesis in the SGZ plays a crucial role in hippocampus-dependent learning and memory [69]. Importantly, hippocampal neurogenesis is involved in mood disorders, such as depression, and neurodegenerative diseases, such as AD, and its reduction is associated with cognitive decline during aging [70]. Therefore, preservation and enhancement of adult neurogenesis may be crucial for maintaining brain health.

Adult hippocampal neurogenesis is a highly sensitive process regulated by both extrinsic and intrinsic factors [71]. The most effective modulators of neuronal activity, such as exercise and dietary factors, positively regulate adult mammalian neurogenesis [72, 73]. Furthermore, dietary compounds with antioxidant effects, including carotenoids, polyphenols, and vitamin E, induce neurogenesis in the adult brain. Therefore, these compounds may play a role in improving cognitive function and attenuating cognitive decline [74–76]. Among carotenoids, ASX, through its neurogenic effects, can increase neurogenesis in rodent models. In an in vitro study, 1, 5, and 10 ng/mL of ASX enhanced the proliferation of mouse neural progenitor cells in a dose- and time-dependent manner [77]. Particularly, the colony-forming cell assay, which discriminates proliferation efficiency, demonstrated a dynamic increase in colony formation in ASX-treated neural progenitor cells. This research group also revealed that ASX treatment improved the osteogenic and adipogenic differentiation potential of the neural stem cells [78]. Yook et al. suggested that ASX is a potential neurotrophic modulator for enhancing cognitive function [7]. Young adult mice (C57BL6/J) were fed an ASX-supplemented diet in a dose-dependent manner (0.02, 0.1, and 0.5%) for four weeks to investigate its effects on cell proliferation in the DG of the hippocampus, as confirmed by an increase in Ki67-positive neurons, a useful endogenous marker of cell proliferation in adult neurogenesis. In addition to affecting the initial stages of neurogenesis, the diet containing 0.5% ASX enhanced the maturation of new neurons, as evidenced by the co-labeling of newly generated neurons with BrdU and the neuronal marker NeuN in the DG. Moreover, the mice demonstrated superior learning and memory using the Morris water maze test, which evaluates hippocampus-dependent cognition. Therefore, ASX promotes adult hippocampal neurogenesis, contributing to the enhancement of hippocampal function.

Adult neurogenesis is regulated through many molecular pathways, including those involving neurotrophic and growth factors. Neurotrophic factors are involved in the maintenance of neuroplasticity. Among the identified neurotrophins, brain-derived neurotrophic factor (BDNF) is most extensively studied because of its role in regulating the differentiation and survival of hippocampal neurons. Importantly, BDNF expression is strongly associated with extrinsic modulators of neurogenesis, including exercise and dietary factors [79, 80]. In an in vivo study, oral ASX administration increased hippocampal BDNF expression and attenuated oxidative stress in D-galactose-induced aged rats [81]. As BDNF stimulates several intracellular pathways that regulate neurogenesis [82], PI3K and MEK signaling pathway activation may be involved in the ASX-induced proliferation of neural progenitor cells [77]. Although ASX exerts neurotrophic effects, the underlying molecular mechanism remains unknown. Yook et al. provided a comprehensive molecular understanding of ASX-induced neurogenesis based on high-throughput DNA microarray analysis [7, 83]. Subsequent bioinformatics analyses, including ingenuity pathway analysis, revealed molecular networks, including prolactin (*Prl*), integrin alpha 4 (*Itga4*), and interleukin-4 (*Il4*) genes, related to neurogenesis and memory function. Additionally, correlation analysis revealed the involvement of downstream factors such as *Mapk 7*, focal adhesion kinase (*Fak*), *Jak1*, and *Akt* mRNAs in ASX-enhanced memory function. Consequently, the promotion of adult hippocampal neurogenesis by ASX owing to its neurotrophic effect is beneficial for brain health.

3.2 Effects of ASX on Synaptic Plasticity

Synaptic plasticity, which refers to the interconnected capacity of neural circuits at synapses, plays an important role in cognitive functions such as learning and memory in the mammalian brain. Many neurons can modify the activity-dependent synaptic connections through different forms of functional synaptic plasticity, including LTP and long-term depression (LTD) [84]. Notably, the pathological disruption of LTP or LTD in the hippocampus contributes to most neurodegenerative diseases, such as AD [85]. Therefore, restoring hippocampal synaptic deficits may be valuable to maintaining cognitive function. Several in vivo studies using AD or aged rodent models have revealed the positive effects of ASX on synaptic plasticity. Electrophysiological experiments conducted using ASX-supplemented young and aged mice exhibited an increased LTP in Schaffer collateral–CA1 synapses [86]. As a form of synaptic structural plasticity, dynamic changes in dendritic spine morphology and density are associated with the efficiency of synaptic connectivity and memory formation [87]. A recent study found that ASX administration to FAB-induced sporadic AD-like rats restored the loss of spines in the dendrites of hippocampal CA1 pyramidal neurons [42]. Therefore, ASX may affect the structural and functional plasticity of synapses, particularly within the hippocampal neuronal circuitry.

Synaptic proteins play an important role in regulating synaptic plasticity and memory [88]. Among the identified synaptic proteins, postsynaptic density protein-95 is associated with synaptic maturation and signal transduction. Synapsin-I (SYN-1), a presynaptic protein, is involved in the regulation of synaptic vesicles and neurotransmitter release [89]. Further, growth-associated protein 43 and synaptophysin (SYP), localized in the presynaptic nerve terminal, play central roles in hippocampal synaptic plasticity [90]. In AD-like mice induced by Aβ (25–35), ASX supplementation restored the reduced SYN expression in the hippocampus via the SIRT1/PGC-1α pathway, which is associated with synaptic plasticity [91]. Another recent study using an APP/PS1 transgenic mouse model of AD showed that ASX could induce neuroprotective effects by regulating the expression of SYP, GAP43, and PSD95 in the cerebral cortex via the mammalian target of the rapamycin (mTOR) pathway [92], thereby improving spatial learning and memory, as evidenced by the Morris water maze test.

4 Neuroplasticity Induction by Combined ASX and Exercise Intervention

4.1 Effects of ASX and Exercise on Adult Hippocampal Neurogenesis

Dietary factors and exercise are key modifiable lifestyle elements that optimize cognitive function [93], and recent studies have indicated their promising benefit. Notably, certain dietary compounds may potentiate the health effects of exercise on hippocampal neuroplasticity and function. For example, a systematic review suggested that the combination of docosahexaenoic acid and exercise exerts greater behavioral, neuroplastic, and neurotrophic benefits [94, 95]. Considering the boosting effects of exercise combined with antioxidants, epicatechin (EGCG) enhanced the effects of voluntary exercise on hippocampal function. However, several studies have failed to demonstrate the additive effects of EGCG and voluntary exercise on adult hippocampal neurogenesis and memory [96–98]. Nevertheless, these studies highlight the importance of selecting nutritional compounds and exercise regimens that may evoke promising additive or synergistic effects. Importantly, both interventional factors must be involved in the same molecular pathways [99].

In a recent study, we investigated the extent of the additive/synergistic effects of ASX supplementation and exercise on adult neurogenesis and memory function in young mice [12]. Specifically, we applied light-intensity running exercise, which exerts beneficial effects on adult hippocampal neurogenesis and spatial memory in young rodents [100, 101], as part of a combined intervention involving ASX. The Morris water maze test revealed that exercise alone increased spatial memory performance, which was further enhanced with ASX. Notably, the combined ASX and exercise intervention synergistically increased cell proliferation by 62.2% and maturation of new cells by 116.9% during adult hippocampal neurogenesis.

A systematic review by Schättin et al. suggested that the possible additive or synergistic effects of dietary factors and exercise may require the involvement of common molecular pathways to promote neuronal plasticity [99]. Therefore, we explored the molecular mechanisms underlying the combined effects of ASX and exercise using DNA microarray and bioinformatics [12]. Subsequently, the omics analysis of the hippocampus revealed that the upregulation of the *Lep* gene, which was upregulated by exercise alone, was further enhanced by ASX. Therefore, hippocampal LEP may be involved in the common pathway between ASX and exercise. LEP, a hormone derived from adipose tissue, plays a crucial role in the regulation of food intake. In addition to its role in energy homeostasis, hippocampal LEP promotes adult neurogenesis and memory function [102, 103]. The combination of ASX and exercise synergistically increased in LEP expression in the hippocampus but not in the plasma. Additionally, LEP deficiency in mice (*ob/ob* mice) did not increase hippocampal LEP expression as observed with the combination of ASX and exercise compared to exercise alone. However, the synergistic enhancement by the combined intervention was restored by the exogenous administration of

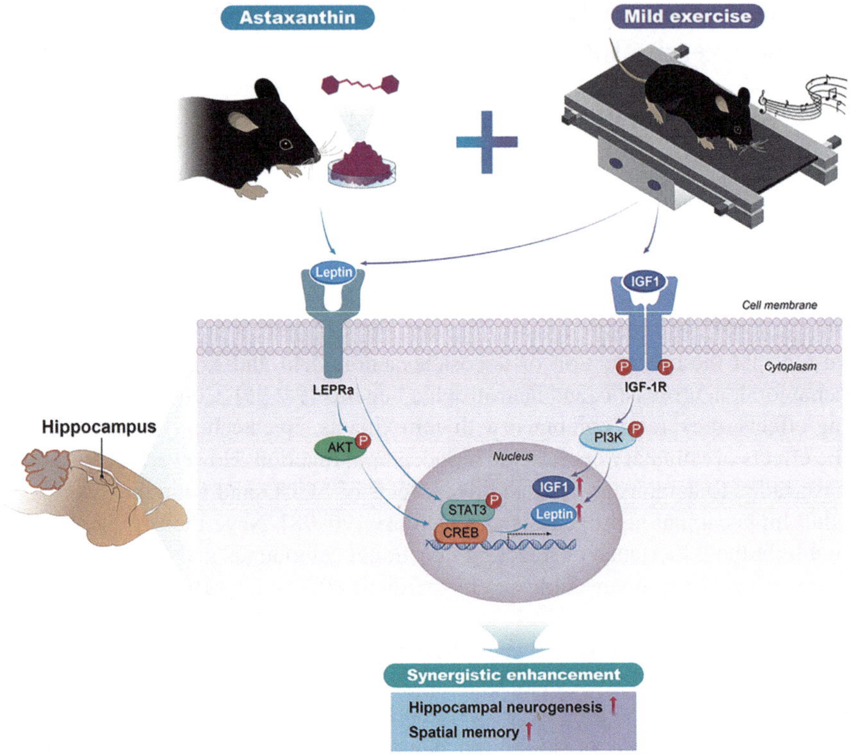

Fig. 2 Potential molecular pathways of hippocampal function induced by astaxanthin combined with exercise. Dietary astaxanthin (ASX) combined with light-intensity exercise may enhance adult neurogenesis and spatial memory function through the leptin and IGF-1 signaling pathways. In particular, brain-derived leptin expression in the hippocampus may mediate the synergistic enhancement of spatial memory function [12]. AKT protein kinase B, CREB cAMP response element-binding, IGF1 insulin-like growth factor-1, IGF-1R insulin-like growth factor-1 receptor, LEPRa shot isoform of leptin receptor, PI3K phosphoinositide 3-kinases, and STAT3 signal transducer and activator of transcription 3

LEP into the brain, indicating that brain-derived hippocampal LEP and its downstream cellular pathway are implicated in the synergistic enhancement of learning and memory with combined ASX and exercise intervention (Fig. 2).

5 Conclusion

ASX exerts promising beneficial effects against neuropathology in addition to its antioxidant, anti-inflammatory, and anti-apoptotic effects. In addition to its neuroprotective ability, ASX enhances adult hippocampal neurogenesis and hippocampal-dependent memory function by interacting with various molecular factors associated with the proliferation and maturation of neurons in the DG. The neurotropic effects of ASX, involving an increase in hippocampal LEP expression, synergistically potentiate the effects of exercise on adult neurogenesis and memory. Although further research is needed to understand the cellular and molecular mechanisms underlying the combined effects of ASX and exercise on brain dysfunction, this strategy represents a powerful modifiable lifestyle intervention for improving cognitive function and attenuating cognitive decline throughout life.

Acknowledgment This work was supported by the Ministry of Education of the Republic of Korea and the National Research Foundation of Korea (NRF-2024S1A5A8024477).

Conflict of Interest The authors declare no financial interests or potential conflicts of interest.

References

1. Ionescu-Tucker A, Cotman CW (2021) Emerging roles of oxidative stress in brain aging and Alzheimer's disease. Neurobiol Aging 107:86–95
2. Miki W (1991) Biological functions and activities of animal carotenoids. Pure Appl Chem 63:141–146
3. Naguib YMA (2000) Antioxidant activities of astaxanthin and related carotenoids. J Agric Food Chem 48:1150–1154
4. Grimmig B, Kim S-H, Nash K, Bickford PC, Shytle RD (2017) Neuroprotective mechanisms of astaxanthin: a potential therapeutic role in preserving cognitive function in age and neurodegeneration. GeroScience 39:19–32
5. Tso MOM, Lam T-T (1996) Method of retarding and ameliorating central nervous system and eye damage. US Patent Office
6. Medoro A, Davinelli S, Milella L, Willcox BJ, Allsopp RC, Scapagnini G, Willcox DC (2023) Dietary astaxanthin: a promising antioxidant and anti-inflammatory agent for brain aging and adult neurogenesis. Mar Drugs 21:643
7. Yook JS, Okamoto M, Rakwal R, Shibato J, Lee MC, Matsui T, Chang H, Cho J-Y, Soya H (2016) Astaxanthin supplementation enhances adult hippocampal neurogenesis and spatial memory in mice. Mol Nutr Food Res 60:589–599
8. Bahbah EI, Ghozy S, Attia MS, Negida A, Emran TB, Mitra S, Albadrani GM, Abdel-Daim MM, Uddin MS, Simal-Gandara J (2021) Molecular mechanisms of astaxanthin as a potential neurotherapeutic agent. Mar Drugs 19:201
9. Erickson KI, Voss MW, Prakash RS et al (2011) Exercise training increases size of hippocampus and improves memory. Proc Natl Acad Sci USA 108:3017–3022
10. van PH, Shubert T, Zhao C, Gage FH (2005) Exercise enhances learning and hippocampal neurogenesis in aged mice. J Neurosci 25:8680–8685
11. van PH, Christie BR, Sejnowski TJ, Gage FH (1999) Running enhances neurogenesis, learning, and long-term potentiation in mice. Proc Natl Acad Sci 96:13427–13431
12. Yook JS, Rakwal R, Shibato J, Takahashi K, Koizumi H, Shima T, Ikemoto MJ, Oharomari LK, McEwen BS, Soya H (2019) Leptin in hippocampus mediates benefits of mild exercise by an antioxidant on neurogenesis and memory. Proc Natl Acad Sci 116:10988–10993
13. Davinelli S, Nielsen ME, Scapagnini G (2018) Astaxanthin in skin health, repair, and disease: a comprehensive review. Nutrients 10:522

14. Boussiba S (2000) Carotenogenesis in the green alga Haematococcus pluvialis: cellular physiology and stress response. Physiol Plant 108:111–117
15. Higuera-Ciapara I, Félix-Valenzuela L, Goycoolea FM (2006) Astaxanthin: a review of its chemistry and applications. Crit Rev Food Sci Nutr 46:185–196
16. Guerin M, Huntley ME, Olaizola M (2003) Haematococcus astaxanthin: applications for human health and nutrition. Trends Biotechnol 21:210–216
17. Nathan C, Cunningham-Bussel A (2013) Beyond oxidative stress: an immunologist's guide to reactive oxygen species. Nat Rev Immunol 13:349–361
18. Manzoor MF, Arif Z, Kabir A et al (2022) Oxidative stress and metabolic diseases: relevance and therapeutic strategies. Front Nutr 9:994309
19. Niedzielska E, Smaga I, Gawlik M, Moniczewski A, Stankowicz P, Pera J, Filip M (2016) Oxidative stress in neurodegenerative diseases. Mol Neurobiol 53:4094–4125
20. Hammarlund EU, Flashman E, Mohlin S, Licausi F (2020) Oxygen-sensing mechanisms across eukaryotic kingdoms and their roles in complex multicellularity. Science 370(6515):eaba3512.
21. Kamath BS, Srikanta BM, Dharmesh SM, Sarada R, Ravishankar GA (2008) Ulcer preventive and antioxidative properties of astaxanthin from Haematococcus pluvialis. Eur J Pharmacol 590:387–395
22. Shimidzu N, Goto M, Miki W (1996) Carotenoids as singlet oxygen quenchers in marine organisms. Fish Sci 62:134–137
23. Meraz-Ríos MA, Franco-Bocanegra D, Rios DT, Campos-Peña V (2014) Early onset Alzheimer's disease and oxidative stress. Oxid Med Cell Longev 2014:375968
24. Manoharan S, Guillemin GJ, Abiramasundari RS, Essa MM, Akbar M, Akbar MD (2016) The role of reactive oxygen species in the pathogenesis of Alzheimer's disease, Parkinson's disease, and Huntington's disease: a mini review. Oxid Med Cell Longev 2016:8590578
25. Ye Q, Zhang X, Huang B, Zhu Y, Chen X (2013) Astaxanthin suppresses MPP(+)-induced oxidative damage in PC12 cells through a Sp1/NR1 signaling pathway. Mar Drugs 11:1019–1034
26. Ye Q, Huang B, Zhang X, Zhu Y, Chen X (2012) Astaxanthin protects against MPP+-induced oxidative stress in PC12 cells via the HO-1/NOX2 axis. BMC Neurosci 13:156
27. Wang L, Lu K, Lou X, Zhang S, Song W, Li R, Geng L, Cheng B (2023) Astaxanthin ameliorates dopaminergic neuron damage in paraquat-induced SH-SY5Y cells and mouse models of Parkinson's disease. Brain Res Bull 202:110762
28. Grimmig B, Daly L, Hudson C, Nash KR, Bickford PC (2017) Astaxanthin attenuates neurotoxicity in a mouse model of Parkinson's disease. Funct Foods Health Dis 7:562–576
29. Wang H-Q, Sun X-B, Xu Y-X, Zhao H, Zhu Q-Y, Zhu C-Q (2010) Astaxanthin upregulates heme oxygenase-1 expression through ERK1/2 pathway and its protective effect against beta-amyloid-induced cytotoxicity in SH-SY5Y cells. Brain Res 1360:159–167
30. Wen X, Huang A, Hu J, Zhong Z, Liu Y, Li Z, Pan X, Liu Z (2015) Neuroprotective effect of astaxanthin against glutamate-induced cytotoxicity in HT22 cells: involvement of the Akt/GSK-3β pathway. Neuroscience 303:558–568
31. Hafez HA, Kamel MA, Osman MY, Osman HMY, Elblehi SS, Mahmoud SA (2021) Ameliorative effects of astaxanthin on brain tissues of Alzheimer's disease-like model: cross talk between neuronal-specific microRNA-124 and related pathways. Mol Cell Biochem 476:2233–2249
32. Hongo N, Takamura Y, Nishimaru H, Matsumoto J, Tobe K, Saito T, Saido TC, Nishijo H (2020) Astaxanthin ameliorated parvalbumin-positive neuron deficits and Alzheimer's disease-related pathological progression in the hippocampus of AppNL-G-F/NL-G-F mice. Front Pharmacol 11:307.
33. Liu N, Lyu X, Zhang X, Zhang F, Chen Y, Li G (2023) Astaxanthin attenuates cognitive deficits in Alzheimer's disease models by reducing oxidative stress via the SIRT1/PGC-1α signaling pathway. Cell Biosci 13:173

34. McInnes IB, Gravallese EM (2021) Immune-mediated inflammatory disease therapeutics: past, present and future. Nat Rev Immunol 21:680–686
35. Boshtam M, Asgary S, Kouhpayeh S, Shariati L, Khanahmad H (2017) Aptamers against pro- and anti-inflammatory cytokines: a review. Inflammation 40:340–349
36. Kempuraj D, Thangavel R, Natteru PA, Selvakumar GP, Saeed D, Zahoor H, Zaheer S, Iyer SS, Zaheer A (2016) Neuroinflammation induces neurodegeneration. J Neurol Neurosurg Spine 1:1003
37. Kwon HS, Koh S-H (2020) Neuroinflammation in neurodegenerative disorders: the roles of microglia and astrocytes. Transl Neurodegener 9:42
38. Leng F, Edison P (2021) Neuroinflammation and microglial activation in Alzheimer disease: where do we go from here? Nat Rev Neurol 17:157–172
39. Wang C, Zong S, Cui X, Wang X, Wu S, Wang L, Liu Y, Lu Z (2023) The effects of microglia-associated neuroinflammation on Alzheimer's disease. Front Immunol 14:1117172
40. Rahman SO, Panda BP, Parvez S, Kaundal M, Hussain S, Mohd A, Najmi AK (2019) Neuroprotective role of astaxanthin in hippocampal insulin resistance induced by Aβ peptides in animal model of Alzheimer's disease. Biomed Pharmacother 110:47–58
41. Che H, Li Q, Zhang T, Wang D, Yang L, Xu J, Yanagita T, Xue C, Chang Y, Wang Y (2018) Effects of astaxanthin and docosahexaenoic-acid-acylated astaxanthin on Alzheimer's disease in APP/PS1 double-transgenic mice. J Agric Food Chem 66:4948–4957
42. Chen M-H, Wang T-J, Chen L-J, Jiang M-Y, Wang Y-J, Tseng G-F, Chen J-R (2021) The effects of astaxanthin treatment on a rat model of Alzheimer's disease. Brain Res Bull 172:151–163
43. Guo S, Wang H, Yin Y (2022) Microglia polarization from M1 to M2 in neurodegenerative diseases. Front Aging Neurosci 14:815347
44. Wen X, Xiao L, Zhong Z, Wang L, Li Z, Pan X, Liu Z (2017) Astaxanthin acts via LRP-1 to inhibit inflammation and reverse lipopolysaccharide-induced M1/M2 polarization of microglial cells. Oncotarget 8:69370–69385
45. Zhou X, Zhang J, Li Y, Cui L, Wu K, Luo H (2021) Astaxanthin inhibits microglia M1 activation against inflammatory injury triggered by lipopolysaccharide through down-regulating miR-31-5p. Life Sci 267:118943
46. Reale M, Iarlori C, Thomas A, Gambi D, Perfetti B, Nicola MD, Onofrj M (2009) Peripheral cytokines profile in Parkinson's disease. Brain Behav Immun 23:55–63
47. Harms AS, Ferreira SA, Romero-Ramos M (2021) Periphery and brain, innate and adaptive immunity in Parkinson's disease. Acta Neuropathol 141:527–545
48. Lindqvist D, Kaufman E, Brundin L, Hall S, Surova Y, Hansson O (2012) Non-motor symptoms in patients with Parkinson's disease—correlations with inflammatory cytokines in serum. PLoS One 7:e47387
49. Grotemeyer A, McFleder RL, Wu J, Wischhusen J, Ip CW (2022) Neuroinflammation in Parkinson's disease—putative pathomechanisms and targets for disease-modification. Front Immunol 13:878771
50. Hirsch EC, Hunot S (2009) Neuroinflammation in Parkinson's disease: a target for neuroprotection? Lancet Neurol 8:382–397
51. Chang MX, Xiong F (2020) Astaxanthin and its effects in inflammatory responses and inflammation-associated diseases: recent advances and future directions. Molecules 25:5342
52. Kim YH, Koh H-K, Kim D-S (2010) Down-regulation of IL-6 production by astaxanthin via ERK-, MSK-, and NF-κB-mediated signals in activated microglia. Int Immunopharmacol 10:1560–1572
53. Chan K, Mong M, Yin M (2009) Antioxidative and anti-inflammatory neuroprotective effects of astaxanthin and canthaxanthin in nerve growth factor differentiated PC12 cells. J Food Sci 74:H225–H231
54. Shen D-F, Qi H-P, Ma C, Chang M-X, Zhang W-N, Song R-R (2021) Astaxanthin suppresses endoplasmic reticulum stress and protects against neuron damage in Parkinson's disease by regulating miR-7/SNCA axis. Neurosci Res 165:51–60

55. Elmore S (2007) Apoptosis: a review of programmed cell death. Toxicol Pathol 35:495–516
56. Radi E, Formichi P, Battisti C, Federico A (2014) Apoptosis and oxidative stress in neurodegenerative diseases. J Alzheimers Dis 42:S125–S152
57. Bedoui S, Herold MJ, Strasser A (2020) Emerging connectivity of programmed cell death pathways and its physiological implications. Nat Rev Mol Cell Biol 21:678–695
58. Kitamura Y, Shimohama S, Kamoshima W, Ota T, Matsuoka Y, Nomura Y, Smith MA, Perry G, Whitehouse PJ, Taniguchi T (1998) Alteration of proteins regulating apoptosis, Bcl-2, Bcl-x, Bax, Bak, Bad, ICH-1 and CPP32, in Alzheimer's disease. Brain Res 780:260–269
59. Goel P, Chakrabarti S, Goel K, Bhutani K, Chopra T, Bali S (2022) Neuronal cell death mechanisms in Alzheimer's disease: an insight. Front Mol Neurosci 15:937133
60. Zhang Y, Wang W, Hao C, Mao X, Zhang L (2015) Astaxanthin protects PC12 cells from glutamate-induced neurotoxicity through multiple signaling pathways. J Funct Foods 16:137–151
61. Yan T, Ding F, Zhang Y et al (2024) Astaxanthin inhibits H2O2-induced excessive mitophagy and apoptosis in SH-SY5Y cells by regulation of Akt/mTOR activation. Mar Drugs 22:57
62. McGregor G, Harvey J (2017) Food for thought: leptin regulation of hippocampal function and its role in Alzheimer's disease. Neuropharmacology 136:298–306
63. Hamilton K, Harvey J (2021) The neuronal actions of leptin and the implications for treating Alzheimer's Disease. Pharmaceuticals 14:52
64. Viswanath V, Wu Y, Boonplueang R, Chen S, Stevenson FF, Yantiri F, Yang L, Beal MF, Andersen JK (2001) Caspase-9 activation results in downstream caspase-8 activation and bid cleavage in 1-Methyl-4-Phenyl-1,2,3,6-Tetrahydropyridine-induced Parkinson's disease. J Neurosci 21:9519–9528
65. Mogi M, Togari A, Kondo T, Mizuno Y, Komure O, Kuno S, Ichinose H, Nagatsu T (2000) Caspase activities and tumor necrosis factor receptor R1 (p55) level are elevated in the substantia nigra from Parkinsonian brain. J Neural Transm 107:335–341
66. Erekat NS. (2018) Apoptosis and its Role in Parkinson's Disease. In: Stoker TB, Greenland JC (eds). Parkinson's Disease: Pathogenesis and Clinical Aspects: Brisbane (AU), Codon Publications; Chapter 4. 11:307
67. Lee D-H, Kim C-S, Lee YJ (2011) Astaxanthin protects against MPTP/MPP+-induced mitochondrial dysfunction and ROS production in vivo and in vitro. Food Chem Toxicol 49:271–280
68. Denoth-Lippuner A, Jessberger S (2021) Formation and integration of new neurons in the adult hippocampus. Nat Rev Neurosci 22:223–236
69. Jurkowski MP, Bettio L, Woo EK, Patten A, Yau S-Y, Gil-Mohapel J (2020) Beyond the Hippocampus and the SVZ: adult neurogenesis throughout the brain. Front Cell Neurosci 14:576444
70. Berger T, Lee H, Young AH, Aarsland D, Thuret S (2020) Adult hippocampal neurogenesis in major depressive disorder and Alzheimer's disease. Trends Mol Med 26:803–818
71. Aimone JB, Li Y, Lee SW, Clemenson GD, Deng W, Gage FH (2014) Regulation and function of adult neurogenesis: from genes to cognition. Physiol Rev 94:991–1026
72. Poulose SM, Miller MG, Scott T, Shukitt-Hale B (2017) Nutritional factors affecting adult neurogenesis and cognitive function. Adv Nutr 8:804–811
73. Ma C-L, Ma X-T, Wang J-J, Liu H, Chen Y-F, Yang Y (2017) Physical exercise induces hippocampal neurogenesis and prevents cognitive decline. Behav Brain Res 317:332–339
74. Pietrasik S, Cichon N, Bijak M, Gorniak L, Saluk-Bijak J (2022) Carotenoids from marine sources as a new approach in neuroplasticity enhancement. Int J Mol Sci 23:1990
75. Dias GP, Cavegn N, Nix A, do N BMC, Stangl D, MSA Z, Nardi AE, Gardino PF, Thuret S (2012) The role of dietary polyphenols on adult hippocampal neurogenesis: molecular mechanisms and behavioural effects on depression and anxiety. Oxid Med Cell Longev 2012:541971
76. Roy-Chowdhury S, Bhavsar D, Singh J, Mardon A (2021) Role of antioxidant vitamins in neurogenesis. J Undergrad Life Sci 15:11–11

77. Kim J-H, Nam S-W, Kim B-W, Choi W, Lee J-H, Kim W-J, Choi Y-H (2010) Astaxanthin improves stem cell potency via an increase in the proliferation of neural progenitor cells. Int J Mol Sci 11:5109–5119
78. Kim JH, Nam S-W, Kim BW, Kim W-J, Choi YH (2010) Astaxanthin improves the proliferative capacity as well as the osteogenic and adipogenic differentiation potential in neural stem cells. Food Chem Toxicol 48:1741–1745
79. Gomez-Pinilla F (2008) Brain foods: the effects of nutrients on brain function. Nat Rev Neurosci 9:568–578
80. Liu PZ, Nusslock R (2018) Exercise-mediated neurogenesis in the hippocampus via BDNF. Front Neurosci 12:52
81. Wu W, Wang X, Xiang Q, Meng X, Peng Y, Du N, Liu Z, Sun Q, Wang C, Liu X (2013) Astaxanthin alleviates brain aging in rats by attenuating oxidative stress and increasing BDNF levels. Food Funct 5:158–166
82. Ohira K, Hayashi M (2009) A new aspect of the TrkB signaling pathway in neural plasticity. Curr Neuropharmacol 7:276–285
83. Yook JS, Shibato J, Rakwal R, Soya H (2016) DNA microarray-based experimental strategy for trustworthy expression profiling of the hippocampal genes by astaxanthin supplementation in adult mouse. GDATA 7:32–37
84. Citri A, Malenka RC (2008) Synaptic plasticity: multiple forms, functions, and mechanisms. Neuropsychopharmacology 33:18–41
85. Balschun D, Rowan MJ (2018) Hippocampal synaptic plasticity in neurodegenerative diseases: Aß, tau and beyond. Neuroforum 24:A133–A141
86. Grimmig B, Hudson C, Moss L, Peters M, Subbarayan M, Weeber EJ, Bickford PC (2019) Astaxanthin supplementation modulates cognitive function and synaptic plasticity in young and aged mice. GeroScience 777:60–11
87. Gipson CD, Olive MF (2017) Structural and functional plasticity of dendritic spines—root or result of behavior? Genes Brain Behav 16:101–117
88. Rosenberg T, Gal-Ben-Ari S, Dieterich DC, Kreutz MR, Ziv NE, Gundelfinger ED, Rosenblum K (2014) The roles of protein expression in synaptic plasticity and memory consolidation. Front Mol Neurosci 7:86
89. Castejón OJ, Fuller L, Dailey ME (2004) Localization of synapsin-I and PSD-95 in developing postnatal rat cerebellar cortex. Dev Brain Res 151:25–32
90. Webster MJ, Weickert CS, Herman MM, Hyde TM, Kleinman JE (2001) Synaptophysin and GAP-43 mRNA levels in the hippocampus of subjects with schizophrenia. Schizophr Res 49:89–98
91. Liu N, Zeng L, Zhang Y-M, Pan W, Lai H (2020) Astaxanthin alleviates pathological brain aging through the upregulation of hippocampal synaptic proteins. Neural Regen Res 16:1062–1067
92. Huang C, Wen C, Yang M, Li A, Fan C, Gan D, Li Q, Zhao J, Zhu L, Lu D (2021) Astaxanthin improved the cognitive deficits in APP/PS1 transgenic mice via selective activation of mTOR. J Neuroimmune Pharmacol 16:609–619
93. Meeusen R (2014) Exercise, nutrition and the brain. Sports Med 44:47–56
94. Gomez-Pinilla F (2011) The combined effects of exercise and foods in preventing neurological and cognitive disorders. Prev Med 52:S75–S80
95. Wu A, Ying Z, Gomez-Pinilla F (2008) Docosahexaenoic acid dietary supplementation enhances the effects of exercise on synaptic plasticity and cognition. Neuroscience 155:751–759
96. van Praag H, Lucero MJ, Yeo GW et al (2007) Plant-derived flavanol (−)epicatechin enhances angiogenesis and retention of spatial memory in mice. J Neurosci 27:5869–5878
97. Gibbons TE, Pence BD, Petr G et al (2014) Voluntary wheel running, but not a diet containing (−)-epigallocatechin-3-gallate and β-alanine, improves learning, memory and hippocampal neurogenesis in aged mice. Behav Brain Res 272:131–140

98. Bhattacharya TK, Pence BD, Ossyra JM, Gibbons TE, Perez S, McCusker RH, Kelley KW, Johnson RW, Woods JA, Rhodes JS (2015) Exercise but not (−)-epigallocatechin-3-gallate or β-alanine enhances physical fitness, brain plasticity, and behavioral performance in mice. Physiol Behav 145:29–37
99. Schättin A, Baur K, Stutz J, Wolf P, de BED (2016) Effects of physical exercise combined with nutritional supplements on aging brain related structures and functions: a systematic review. Front Aging Neurosci 8:161
100. Inoue K, Hanaoka Y, Nishijima T, Okamoto M, Chang H, Saito T, Soya H (2014) Long-term mild exercise training enhances hippocampus-dependent memory in rats. Int J Sports Med 36:280–285
101. Inoue K, Okamoto M, Shibato J, Lee MC, Matsui T, Rakwal R, Soya H (2015) Long-term mild, rather than intense, exercise enhances adult hippocampal neurogenesis and greatly changes the transcriptomic profile of the hippocampus. PLoS One 10:e0128720
102. Garza JC, Guo M, Zhang W, Lu X-Y (2008) Leptin increases adult hippocampal neurogenesis in vivo and in vitro. J Biol Chem 283:18238–18247
103. Harvey J (2003) Novel actions of leptin in the hippocampus. Ann Med 35:197–206
104. Nour S., Erekat Thomas B., Stoker Julia C., Greenland Parkinson's Disease: Pathogenesis and Clinical Aspects Apoptosis and its Role in Parkinson's Disease Codon Publications 65-82
105. Walker JG, Mackinnon AJ, Batterham P, Jorm AF, Hickie I, McCarthy A, Fenech M, Christensen H (2010) Mental health literacy, folic acid and vitamin B12, and physical activity for the prevention of depression in older adults: randomised controlled trial. Br J Psychiatry 197(1):45–54
106. Wang RS, Wang B-L, Huang Y-N, Wan TTH (2022) The combined effect of physical activity and fruit and vegetable intake on decreasing cognitive decline in older Taiwanese adults. Sci Rep 12(1):9825

HIT Your Brain: Neuron and New Run

Min-Chul Lee and Hideaki Soya

Abstract The importance of physical activity in neuroscience is gaining increasing recognition. The question arises: What is the specific focus of exercise, and what factors contribute to the observed benefits of exercise in neuroscience? Various forms of exercise have been examined across physiological, psychological, and biochemical experiments within neuroscience. Still, there is a need for greater clarity to identify optimal exercise conditions, including the FITT-VP variables (frequency, intensity, type, and time).

This chapter aims to shed light on the positive impacts of high-intensity training (HIT) exercises in facilitating physiological adaptation and exploring the newfound role in brain functions. Key areas explored include (1) exercise neuroscience at the structural level involving synaptic plasticity and neurogenesis; (2) functional level concerning behavioral development; and (3) molecular level addressing potential mechanisms underlying exercise-induced brain plasticity.

Overall, high-intensity training emerges as a more cost-effective method for enhancing physiological adaptations, including improvements in aerobic capacity. Additionally, it has been shown to influence brain functions such as hippocampus-dependent learning and memory positively. These findings offer valuable insights into the practicality of high-intensity training for performance improvement and suggest directions for future research.

M.-C. Lee (✉)
Department of Sports Medicine, College of Health Science, CHA University, Pocheon, South Korea
e-mail: mclee@cha.ac.kr

H. Soya
Laboratory of Exercise Biochemistry and Neuroendocrinology, Institute of Health and Sport Sciences, University of Tsukuba, Tsukuba, Ibaraki, Japan

Sport Neuroscience Division, Kokoro Division, Advanced Research Initiative for Human High Performance (ARIHHP), Institute of Health and Sport Sciences, University of Tsukuba, Tsukuba, Ibaraki, Japan

Center for Cybernics Research, University of Tsukuba, Tsukuba, Ibaraki, Japan

H. Soya (ed.), *Exercise Brain Stimulation for Cognitive Function and Mental Health*, Advances in Neurobiology 44, https://doi.org/10.1007/978-981-95-0066-6_18

Keywords High-intensity interval training (HIT) · Resistant wheel running (RWR) · Brain functions · Hippocampus · Brain-derived neurotrophic factor (BDNF)

1 Introduction

The effectiveness of physical activity and exercise training in enhancing physiological adaptations, encompassing the cardiovascular and musculoskeletal systems, as well as brain function, is widely acknowledged. The field of sports neuroscience is actively evolving, involving a diverse range of multidisciplinary studies to gain a deeper understanding of the interplay between the brain and brawn (skeletal muscle). Numerous studies have provided evidence that exercise contributes to improvements in various indicators of cardiovascular and skeletal muscle adaptation [1–3]. Furthermore, it has been demonstrated to have a positive impact on the central nervous system, particularly in promoting adult hippocampal neurogenesis (AHN) [4, 5]. AHN is closely associated with hippocampus-dependent functions such as spatial learning and memory [6, 7]. This body of evidence suggests that exercise might play a role in enhancing synaptic transmission and plasticity. Despite the wealth of studies indicating the link between exercise and central as well as peripheral changes, the optimal type and quantity required to elicit these beneficial effects continue to be a topic of debate.

As per the American College of Sports Medicine, high-intensity interval training (HIT) continues to maintain its status as a consistently top-ranking trend [6]. Notably, HIT has secured the number one position and has consistently held a spot within the top 10 trends for a decade. HIT workouts involve brief bursts of high-intensity, strenuous training (typically lasting 20–90 seconds), followed by short rest periods or low-intensity recovery [7]. The popularity of HIT has soared, making it one of the most favored exercise trends globally. It has gained widespread recognition in gyms worldwide, despite cautionary notes from industry experts highlighting its relatively higher risk and potential for causing injuries.

In this overview, we examine the mechanisms underlying the improved metabolic control of skeletal muscles and alterations in cardiovascular function resulting from HIT. Furthermore, we speculate about the impact of implementing HIT in practical terms for performance. While the underlying mechanisms might vary in individuals with less training, studying the responses in athletes may provide insights into why HIT serves as a powerful exercise stimulus for brain function.

2 Characteristics of HIT and Physiological Adaptations

While HIT has consistently ranked among the top fitness trends, feedback from ACSM surveys often notes that clients may enjoy the program for a brief period before seeking alternative exercises. Concerns have been raised about the potential for a high risk of injury despite its popularity. Some professionals working with clinical populations in medical fitness centers express interest in trying HIT with their patients but consider substituting high-intensity with moderate-intensity interval training.

HIT workouts are consist of short bursts of maximum effort, typically lasting approximately 20–90 seconds, followed by brief periods of rest or low-intensity recovery. These exercise programs typically last less than 30 minutes, although some may extend beyond that timeframe. The primary goal is to allow for sufficient recovery, enabling the trainee to exert maximum effort in the subsequent work interval. Thus, HIT exercises can vary in terms of the duration of low-intensity recovery, but not in the duration of high-intensity effort.

HIT exhibits extensive variability, with the specific physiological adaptations elicited by this training method are influenced by various factors. These factors include the precise nature of the exercise stimulus, encompassing intensity, duration, and the number of intervals performed. Additionally, the duration and activity patterns during recovery play a crucial role. Research has demonstrated that HIT can impact aerobic capacity, mitochondrial fatty acid oxidation, and cardiovascular risk as effectively as traditional endurance training (ET) [7–9]. When energy expenditure is comparable, HIT proves to be a viable alternative to traditional endurance training, generating similar or even superior changes in various physiological adaptations, performance, and health-related markers in both healthy individuals and those with health conditions [10, 11].

While much remains unknown about the effects of HIT, emerging evidence suggests that it stimulates physiological remodeling similar to moderate-intensity continuous training, despite requiring substantially less time and a reduced total exercise volume [12]. The Wingate test, a common model used in low-volume HIT studies, involves a 30-second "all-out" cycling effort against a supra-maximal workload. Typically, subjects perform four to six work bouts separated by approximately 4 minutes of recovery, totaling 2–3 minutes of intense exercise during a 20-minute training session. Remarkably, as few as six sessions of this type of training, totaling approximately 15 minutes of all-out cycling exercise over 2 weeks, increase skeletal muscle oxidative capacity, reflected by the maximal activity and/or protein content of mitochondrial enzymes [13]. Furthermore, direct comparisons have shown that 6 weeks of Wingate-based HIT is equivalent to traditional endurance training.

3 Potential Underlying Molecular Mechanisms Adaptations to HIT

Recent investigations have delved into the molecular mechanisms that govern the metabolic adaptations of skeletal muscles to HIT. Given HIT's ability to enhance mitochondrial capacity, researchers have naturally explored its impact on the activation of peroxisome-proliferator-activated receptor γ co-activator (PGC)-1α [14]. A pivotal breakthrough in understanding the cellular processes promoting mitochondrial biogenesis was the identification of PGC-1α, an inducible coactivator that regulates the coordinated expression of mitochondrial proteins encoded in both nuclear and mitochondrial genomes [15].

In skeletal muscle, PGC-1α has emerged as a crucial regulator of mitochondrial biogenesis, responsive to neuromuscular input and prevailing contractile activity. A single bout of endurance exercise rapidly and sustainably increases the PGC-1α gene and protein levels in skeletal muscles [16], Additionally, muscle-specific overexpression of PGC-1α leads to a substantial increase in functional mitochondria, improvements in whole-body VO_{2max}, a shift from carbohydrate to fat as fuel during submaximal exercise, and enhanced endurance performance [17]. Two vital signaling cascades, AMPK and p38 MAPK, converge on the regulation of PGC-1α and, consequently, the control of mitochondrial biogenesis [18]. AMPK and p38 MAPK phosphorylate and activate PGC-1α expression, a known regulator of PGC-1α, which, in turn, increase PGC-1α protein by binding to and activating the CREB site on the PGC-1α promoter [19].

Evidence suggests that exercise intensity significantly influences PGC-1α activation in human skeletal muscle. Loss-of-function studies have questioned the absolute necessity of PGC-1α during training-induced changes in muscle mitochondrial biogenesis, angiogenesis, and fiber type alterations [20]. Recent findings position PGC-1α as a central player in orchestrating many oxidative adaptations to exercise. Acute HIT, similar to endurance exercise, increases PGC-1α mRNA post-exercise [21]. Moreover, acute HIT may activate PGC-1α by enhancing its nuclear translocation [22]. The rise in nuclear PGC-1α following low-volume HIT aligns with increased mRNA expression of several mitochondrial genes, suggesting the initiation of a program for mitochondrial adaptation with these brief bursts of intense exercise [23].

4 Adaptation to the High-Intensity Training Model in Animals

Recent insights have expanded our understanding of the myriad effects of HIT, positioning it as a time-efficient alternative to traditional endurance training that triggers similar metabolic responses in both humans [13, 21] and animals [7, 9]. Despite these known effects, the impact of HIT on cognitive functions and the underlying mechanisms remains unexplored. Our selection of HIT as a model from our earlier observations suggested the potential of voluntary resistance wheel running (RWR) to enhance brain functions. To investigate further, our goal was to establish an animal model for HIT and explore its effectiveness in enhancing hippocampal-related spatial learning and memory. This exploration was grounded in the potential mechanisms associated with hippocampal brain-derived neurotrophic factor (BDNF) signaling.

While acute and chronic voluntary resistance wheel running did not increase running distance (remaining at half of the control) despite the high-energy expenditure with load, it mimicked the effects of traditional wheel running on learning and memory associated with BDNF signaling. It also induced hippocampal adaptations, including neurogenesis in rats [24, 25]. Additionally, it revealed a spectrum of newly altered gene expressions and highlighted changes in several transcriptional pathways related to RWR rat hippocampal plasticity [26].

RWR offers a controlled load on a running wheel, providing a valuable exercise model to elevate energy expenditure without imposing physical and psychological stressors. The relatively high intensity and short duration of intermittent exercise associated with RWR allow for enhanced activities of oxidative enzyme citrate synthase and hypertrophy in fast-twitch plantaris muscles [27, 28]. This aligns with prior findings indicating that HIT has more pronounced effects on fast-twitch muscle fibers [13]. Therefore, HIT emerges as a distinct form of high-intensity exercise, adapting to high-energy demands by increasing strength, in contrast to the traditional ET model. Moreover, the present HIT approach in rats appears to mirror adaptations observed in studies conducted in humans.

The role of the BDNF is crucial in supporting the growth, development, and survival of neurons. Physical activity has been shown to increase hippocampal BDNF mRNA and protein levels [29], leading to enhanced hippocampal-dependent cognitive function and neurogenesis [30, 31]. Changes in BDNF signaling are essential for exercise to influence hippocampal plasticity in rodents; inhibiting BDNF signaling hampers exercise-induced learning, memory [32], and neurogenesis [33]. These findings, consistent with earlier studies, suggest that RWR serves as an effective model for enhancing spatial cognitive functions through increased hippocampal BDNF signaling, even with a lower exercise volume and time commitment.

Data from RWR indicate that a lower volume and shorter duration of exercise, similar to HIT, are more effective in improving hippocampus-dependent learning and memory associated with hippocampal BDNF signaling compared to endurance training. This supports the idea that HIT is a viable alternative to traditional endurance training, eliciting comparable or even superior changes in various physiological, performance, and health-related markers. Similar to RWR, intermittent exercise like HIT, which covers shorter distances but involves higher work levels, proves to be an efficient exercise method that positively impacts brain functions. In summary, intermittent high-intensity exercise, such as HIT, not only enhances muscle adaptation and aerobic capacity but also holds promise for improving brain function.

5 Conclusion

In summary, this information is invaluable for guiding critical decisions in the health and fitness industry, influencing both activity programming and business strategies. The outcomes of this annual survey can significantly inform investment decisions crucial for the industry's future growth and development. In our discussion, we have explored the mechanisms responsible for enhancing physiological responses and influencing cardiovascular function in response to HIT. We have also delved into the potential health-related implications for athletes and individuals with chronic diseases.

Additionally, we have pondered the practical applications of HIT in enhancing performance, a consideration of utmost importance in public health promotion. This is particularly relevant given that the lack of time remains a commonly cited barrier to regular exercise. While recognizing that the underlying mechanisms of HIT may vary among individuals with different training backgrounds, insights from elite athletes can contribute significantly to understanding why HIT is such a powerful exercise stimulus. Furthermore, recent studies have illuminated how HIT plays a role in training methods that may prevent and/or treat brain functions. Future investigations are warranted to uncover the specific impact of HIT parameters on brain function.

References

1. Kobayashi YM et al (2008) Sarcolemma-localized nNOS is required to maintain activity after mild exercise. Nature 456(7221):511–515
2. Hawley JA et al (2014) Integrative biology of exercise. Cell 159(4):738–749
3. Ploutz-Snyder LL et al (2014) Integrated resistance and aerobic exercise protects fitness during bed rest. Med Sci Sports Exerc 46(2):358–368
4. Kempermann G, Wiskott L, Gage FH (2004) Functional significance of adult neurogenesis. Curr Opin Neurobiol 14(2):186–191
5. van Praag H, Kempermann G, Gage FH (1999) Running increases cell proliferation and neurogenesis in the adult mouse dentate gyrus. Nat Neurosci 2(3):266–270
6. Thompson WR (2017) World wide survey of fitness trends for 2018: the CREP Edition. ACSMs Health Fit J 21(6):10–19
7. Afzalpour ME et al (2015) Comparing interval and continuous exercise training regimens on neurotrophic factors in rat brain. Physiol Behav 147:78–83
8. Haram PM et al (2009) Aerobic interval training vs. continuous moderate exercise in the metabolic syndrome of rats artificially selected for low aerobic capacity. Cardiovasc Res 81(4):723–732
9. Holloway TM et al (2015) High intensity interval and endurance training have opposing effects on markers of heart failure and cardiac remodeling in hypertensive rats. PLoS One 10(3):e0121138
10. Wisloff U et al (2007) Superior cardiovascular effect of aerobic interval training versus moderate continuous training in heart failure patients: a randomized study. Circulation 115(24):3086–3094

11. Hwang CL, Wu YT, Chou CH (2011) Effect of aerobic interval training on exercise capacity and metabolic risk factors in people with cardiometabolic disorders: a meta-analysis. J Cardiopulm Rehabil Prev 31(6):378–385
12. Gibala MJ, McGee SL (2008) Metabolic adaptations to short-term high-intensity interval training: a little pain for a lot of gain? Exerc Sport Sci Rev 36(2):58–63
13. Burgomaster KA et al (2008) Similar metabolic adaptations during exercise after low volume sprint interval and traditional endurance training in humans. J Physiol 586(1):151–160
14. Wu Z et al (1999) Mechanisms controlling mitochondrial biogenesis and respiration through the thermogenic coactivator PGC-1. Cell 98(1):115–124
15. Lin J, Handschin C, Spiegelman BM (2005) Metabolic control through the PGC-1 family of transcription coactivators. Cell Metab 1(6):361–370
16. Mathai AS et al (2008) Rapid exercise-induced changes in PGC-1alpha mRNA and protein in human skeletal muscle. J Appl Physiol (1985) 105(4):1098–1105
17. Calvo JA et al (2008) Muscle-specific expression of PPARgamma coactivator-1alpha improves exercise performance and increases peak oxygen uptake. J Appl Physiol (1985) 104(5):1304–1312
18. Puigserver P et al (2001) Cytokine stimulation of energy expenditure through p38 MAP kinase activation of PPARgamma coactivator-1. Mol Cell 8(5):971–982
19. Akimoto T et al (2005) Exercise stimulates Pgc-1alpha transcription in skeletal muscle through activation of the p38 MAPK pathway. J Biol Chem 280(20):19587–19593
20. Geng T et al (2010) PGC-1alpha plays a functional role in exercise-induced mitochondrial biogenesis and angiogenesis but not fiber-type transformation in mouse skeletal muscle. Am J Physiol Cell Physiol 298(3):C572–C579
21. Gibala MJ et al (2009) Brief intense interval exercise activates AMPK and p38 MAPK signaling and increases the expression of PGC-1alpha in human skeletal muscle. J Appl Physiol (1985) 106(3):929–934
22. Little JP et al (2011) An acute bout of high-intensity interval training increases the nuclear abundance of PGC-1alpha and activates mitochondrial biogenesis in human skeletal muscle. Am J Physiol Regul Integr Comp Physiol 300(6):R1303–R1310
23. Gibala MJ et al (2012) Physiological adaptations to low-volume, high-intensity interval training in health and disease. J Physiol 590(5):1077–1084
24. Lee MC et al (2012) Voluntary resistance running with short distance enhances spatial memory related to hippocampal BDNF signaling. J Appl Physiol (1985) 113(8):1260–1266
25. Lee M, Soya H (2017) Effects of acute voluntary loaded wheel running on BDNF expression in the rat hippocampus. J Exerc Nutr Biochem 21(4):52–57
26. Lee MC et al (2014) DNA microarray-based analysis of voluntary resistance wheel running reveals novel transcriptome leading robust hippocampal plasticity. Physiol Rep 2(11):e12206
27. Ishihara A et al (1998) Hypertrophy of rat plantaris muscle fibers after voluntary running with increasing loads. J Appl Physiol (1985) 84(6):2183–2189
28. Legerlotz K et al (2008) Voluntary resistance running wheel activity pattern and skeletal muscle growth in rats. Exp Physiol 93(6):754–762
29. Neeper SA et al (1995) Exercise and brain neurotrophins. Nature 373(6510):109
30. Pencea V et al (2001) Infusion of brain-derived neurotrophic factor into the lateral ventricle of the adult rat leads to new neurons in the parenchyma of the striatum, septum, thalamus, and hypothalamus. J Neurosci 21(17):6706–6717
31. Lee MC et al (2013) Voluntary resistance running induces increased hippocampal neurogenesis in rats comparable to load-free running. Neurosci Lett 537:6–10
32. Vaynman S, Ying Z, Gomez-Pinilla F (2004) Hippocampal BDNF mediates the efficacy of exercise on synaptic plasticity and cognition. Eur J Neurosci 20(10):2580–2590
33. Li Y et al (2008) TrkB regulates hippocampal neurogenesis and governs sensitivity to antidepressive treatment. Neuron 59(3):399–412

Enriched Exercise Environment Boosting Exercise Effects on the Brain: Beneficial Effects of Music

Kazuya Suwabe, Takemune Fukuie, and Hideaki Soya

Abstract Physical exercise is a potential medicine for cognitive function and mental health; however, regular exercise is more difficult than taking a pill every day. Developing an exercise environment that promotes a positive affective response to exercise and exercise benefits on the brain may encourage people to participate in physical activities. Listening to music while exercising is a promising candidate for such an "enriched exercise environment." This chapter reviews the studies demonstrating the beneficial effects of music on enhancing mood and cognitive function in both acute and chronic settings. Furthermore, the underlying neural mechanisms involved in the effects of exercise with music are discussed from the following three perspectives: (1) musical reward and pleasure, (2) rhythmic entrainment, and (3) sensory distraction. In addition, the concept of groove, which is "the pleasurable sensation of wanting to move the body to music" was used to explore the characteristics of music that are compatible with exercise. Finally, individual variations on

K. Suwabe (✉)
Faculty of Health and Sport Sciences, Ryutsu Keizai University, Ryugasaki, Ibaraki, Japan

Laboratory of Exercise Biochemistry and Neuroendocrinology, Institute of Health and Sport Sciences, University of Tsukuba, Tsukuba, Ibaraki, Japan

Center for Cybernics Research, University of Tsukuba, Tsukuba, Ibaraki, Japan
e-mail: ksuwabe@rku.ac.jp

T. Fukuie
School of Nursing and Social Services, Health Sciences University of Hokkaido, Tobetsu-cho, Ishikari-gun, Hokkaido, Japan

H. Soya
Laboratory of Exercise Biochemistry and Neuroendocrinology, Institute of Health and Sport Sciences, University of Tsukuba, Tsukuba, Ibaraki, Japan

Center for Cybernics Research, University of Tsukuba, Tsukuba, Ibaraki, Japan

Sport Neuroscience Division, Kokoro Division, Advanced Research Initiative for Human High Performance (ARIHHP), Institute of Health and Sport Sciences, University of Tsukuba, Tsukuba, Ibaraki, Japan

H. Soya (ed.), *Exercise Brain Stimulation for Cognitive Function and Mental Health*, Advances in Neurobiology 44, https://doi.org/10.1007/978-981-95-0066-6_19

whether combining exercises with music are appropriate and the considerations that should be addressed for implementation in the field are discussed.

Keywords Motivation · Positive affect · Cognition · Groove · Dopamine

1 Introduction

As previously described in other chapters, several animal and human studies have demonstrated that physical exercise and an active lifestyle help improve cognitive function and mental health. Based on these widely recognized findings, various international organizations recommend exercise and physical activity not only for physical fitness and cardiometabolic health but also for brain health. The World Health Organization (WHO) guidelines on physical activity and sedentary behavior state that physical activity has a positive impact on mental and cognitive health across the life span [1]. In addition, from the perspective of dementia care, engaging in regular exercise and physical activity has a preventive effect on dementia [2]. The "Global action plan on the public health response to dementia 2017–2025" by WHO strongly recommends increasing physical activity to reduce the risk of cognitive decline and dementia [3].

Despite these benefits, adopting a physically active lifestyle is not simple. This is evident from the fact that in 2016, 27.5% of adults (aged $>$ 18 years) worldwide did not meet the WHO Global Recommendations on Physical Activity for Health criteria of a minimum of 150 min of moderate-intensity, 75 min of vigorous-intensity physical activity per week, or any equivalent combination of both [4]. Moreover, the proportion of adolescents (11–17 years) who did not meet the WHO recommendations for physical activity (with $<$ 60 min of moderate-to-vigorous intensity daily physical activity) was significantly higher (81.0%) [5].

Potential barriers to participation in exercise and physical activity include biological, psychological, behavioral, social, and cultural factors [6, 7]. Similarly, factors that promote exercise and physical activity have been identified. Recently, the affective response to physical exercise has been identified as a key factor for enhancing motivation toward exercise and physical activity [8–12]. Affective states can be assessed according to a two-dimensional model of mood states consisting of pleasantness (pleasant-unpleasant) and arousal (excitement-sedation) [13]. Theoretical models from a psychological perspective have proposed that hedonic pleasure and rewards improve adherence to exercise [9–12]. A systematic review of actual surveys concluded that positive affective responses (pleasure) during and after exercise were reliably correlated with affective judgments about future physical activity [14].

Furthermore, the exercise-induced positive affective valence may be beneficial not only with exercise adherence but also with cognitive function [15, 16]. Exposure to pleasant stimuli can elicit increased activation of the prefrontal cortex (PFC) [17–19], whereas aversive stimuli have the opposite effects [20, 21]. Based on these findings, we postulated that exercise-induced mood enhancement can amplify the cognitive effects of exercise.

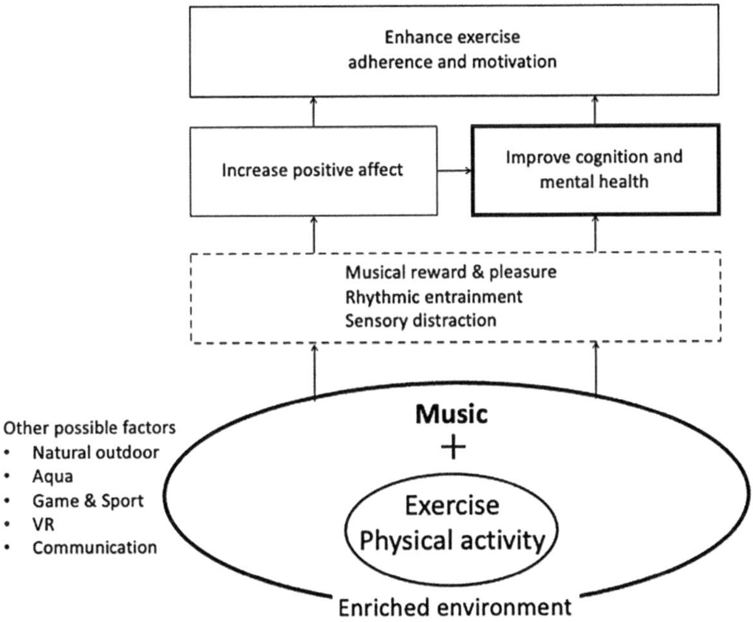

Fig. 1 A theoretical framework of enriched exercise environment using music boosts exercise effects on the brain and its potential mechanism

Therefore, to encourage people to continue exercising and to implement the benefits of exercise in society, it may be useful to develop an exercise environment that can 1) induce a positive affective response to exercise and 2) potentiate the beneficial effects of exercise. We defined an exercise environment that meets these conditions as an "enriched exercise environment." Previous studies, particularly those conducted on animals, have defined "enriched environment" as those environments with high levels of stimulation (i.e., sensory enrichment), such as physical exercise [22, 23]. Therefore, we focused not only on the stimulatory factors but also on identifying the environmental conditions that are compatible with exercise, thereby facilitating high levels of adherence, with a view toward social implementation.

This chapter focuses on the use of music as a potential factor for an enriched exercise environment (Fig. 1) and current literature review.

2 Why Music? : Strong Connection Between Exercise and Music

Music is one of the most compatible environmental factors for exercising, from various perspectives. Rhythmic body movements to music are a highly instinctive and cross-specific phenomenon. Even neonates, as young as 48 h after birth, can detect a regular beat [24], and 5-month-old infants exhibit rhythmic body movements accompanied by smiles when listening to Mozart's classical music and drum rhythms [25]. In addition to humans, birds and other mammals can sense rhythms and move their bodies. Vocal-mimicking birds such as parrots and macaws [26, 27] and even California sea lions [28] dance to music. Furthermore, a recent study using electrophysiological techniques demonstrated that rats can also synchronize their body movements to beats, with this synchronization observed not only in head movements but also in cortical activity in the auditory region [29].

Music is advantageous for social implementation. The widespread use of digital audio equipment has made it easier for anyone to access and enjoy music in their daily lives. Music is also used in a variety of situations in physical exercise and sports and is an essential part of exercise and sports that are engaged in music, such as dance and gymnastics. Therefore, music is closely related to exercise and physical activity, not only from an evolutionary and biological perspective but also from a sociological perspective.

Listening to music has a beneficial effect on positive affective responses. A recent meta-analysis of 139 studies revealed that listening to music prior to or during acute physical activity reduces ratings of perceived exertion (RPE) and increases positive affective valence compared to those without music [30]. Music is also a powerful tool for boosting motivation toward exercise in an exercise setting [31], which aligns with the theoretical model named "Music as an Affective Stimulant for Physical Activity," indicating that exercise-induced affective valence promotes exercise adherence [32].

3 Music Boosts Exercise Effects on the Brain: Empirical Evidence

Does listening to music potentiate the benefits of exercise on the brain? Because exercise itself has beneficial effects on cognitive and mental health, an experimental design that compares the effects of exercise with and without music is needed to explore the additive and/or synergistic effects of combining exercise and music. More specifically, it is necessary to have a control group/condition in which they perform exercises without music or with controlled auditory stimuli. Furthermore, the control group/condition should be allowed to exercise under the same conditions as the experimental group/condition preferably. To determine whether listening to music during exercise enhances exercise effects on cognition and mood, that is, whether it can meet the criteria as an enriched environment, it can only be made feasible with an appropriate experimental design. Comprehensive studies on acute and chronic interventions are discussed below.

3.1 Acute Effects

Acute effects have been analyzed using acute exercise models in healthy young adults. Tanaka et al. recruited 15 students, and the participants completed 30 min of moderate cycling exercise under two experimental conditions: listening to self-selected music (non-matched with pedaling tempo) and no music. Music listening reduces RPE; however, executive performance, as measured by the Stroop task, does not change [33]. Chen et al. conducted an experiment that focused on the music tempos, in which 90 young adults performed 20 min of cycling exercise at 50–60 bpm with one of three music tempos: slow (60–65 bpm), moderate (120–140 bpm), and fast (155–165 bpm). The moderate-tempo group showed a higher positive mood (happy and excited) and a lower negative mood (anger and hate). In addition, the moderate-tempo group outperformed the inhibition task (Stroop task) and working memory task (N-back task) [34]. These results suggest that music tempo is important, and a suitable tempo is imperative for combination with exercise.

Suwabe et al. conducted a similar experiment and explored the neural substrates underlying these effects using neuroimaging techniques. Thirty-three young adults completed 10 min of moderate cycling exercise with music (matched with pedaling tempo) and metronome beeps at a regular pace. Exercise with music elicited a lower RPE and greater enhancement of vitality (high arousal, high pleasure) than exercise with beeps. Although there was no interaction with ANOVA in the pre/post differences by condition, listening to music positively influenced executive performance (Stroop interference effect), and a small effect size (Cohen's $d = 0.46$) was observed in the music condition. An interesting finding of this study was that the magnitude of enhanced vitality predicted the extent of executive performance improvement and cortical activation in the left dorsolateral prefrontal cortex (DLPFC) measured by functional near-infrared spectroscopy (fNIRS) at the individual participant level [15] (Fig. 2). These results support the hypothesis that music potentiates the acute exercise effects on positive mood and cognition; however, the magnitude of the effect on mood and cognitive function varied widely among individuals. Therefore, demonstrating the effect of exercise with music was difficult.

A few studies have examined the effects of music addition on the effects of acute exercise in populations other than young adults. Emery et al. recruited older adults with coronary artery disease, and the participants performed two sessions of approximately 20 min of exercise on a treadmill: with music (Vivaldi's Four Seasons) and without music. The exercise intensity was gradually increased until it reached an equivalent of 85% of the VO_{2max}. No differences in mood changes were observed, as measured by the Profile of Mood States, second edition (POMS); however, performance on the verbal fluency task improved after the music condition [35]. Mezghani et al. explored this topic among children in school settings and included 27 children aged between 11 and 12. The children completed 10 min of four break activity sessions: music only, playing only, music and playing, and no-activity sessions. During the playing-only and music and playing sessions, the children played a hunter and fishing game. Children listened to "Mozart's Sonata KV.448" during music only

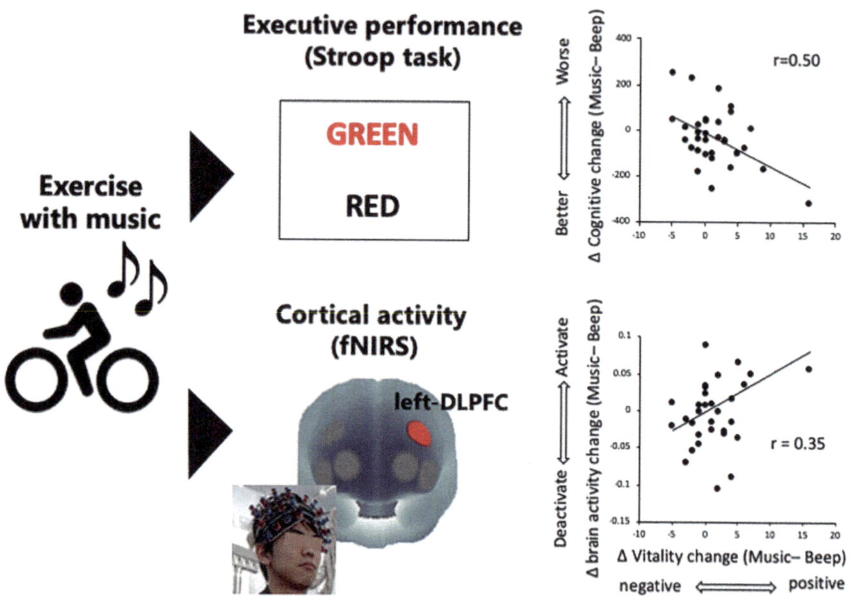

Fig. 2 Exercising while listening to music increases vitality, and the more positive the mood, the greater the improvement in executive performance and cortical activity [15]

and music and playing sessions. During the no-activity session, the participants were instructed to rest without activity. The results indicated that attention and simple reaction time were better in the three activity sessions (music, playing, and music and playing) than in the non-activity session; however, no significant differences were observed among the three activity groups [36].

Although strong evidence to indicate significant additive/synergistic effects of music on acute cognitive improvement with a single session of exercise is lacking, few studies have elucidated that music potentially improves the effects of exercise on mood and cognition.

3.2 *Chronic Effects*

The chronic effects of exercise with music on cognitive function have been extensively studied in older adults; however, well-controlled studies are limited [37–39]. Satoh et al. addressed the additive/synergistic effects of music on exercise benefits on cognition through a carefully controlled experimental design in a research project named "The Mihama-Kiho Project." Healthy older adults were divided into three groups: exercise with music, exercise without music, and non-exercise waiting control. The exercise with music group performed an exercise program with musical accompaniment composed of movements such as aerobic exercise by marching and clapping in various rhythms and resistance exercises of the lumbar, limb, and truncal muscles. The exercise-alone group performed the same exercise program without music. After the 1-year intervention, the visuospatial functioning (copying figures) of the exercise with music group, but not of the exercise alone group, was better than that of the control group. Structural MRI revealed that the exercise with music group showed greater gray matter volume in several brain regions, including the superior frontal gyrus, anterior cingulate gyrus, superior temporal gyrus, and insula [40, 41].

Several studies focused on ballroom dancing. Merom et al. attempted to determine the effect of ballroom dancing in a two-arm randomized controlled trial involving community-dwelling older adults. The dance group performed ballroom dancing for 1 h twice weekly for 8 months. A collection of ballroom dances, Rock and Roll, Foxtrot, Waltz, and some Latin styles (Salsa and Rumba) was delivered. The control group underwent a combination of home- and group-based walking to facilitate socialization. The results suggest dancing has a higher potential for improving spatial memory than walking, as assessed by total learning and delayed recall in geometric visual designs [42]. Moreover, Kosmat and Vranic examined the effects of ballroom dancing on older adults without cognitive impairment. They were randomly assigned to either a dance group or an active control group. The dance group attended a dance program with Slow Waltz music, whereas the active control group was involved in a group-based, non-dance program. Beneficial effects were observed in the dance group in terms of memory (total learning in a modified version of the RVALT) and executive functioning (error rate in the Wisconsin Card Sorting Test) but not for the active control group, and these benefits were maintained at the 5-month follow-up [43]. In these ballroom dance studies, exercise conditions, such as intensity and type of exercise, were not standardized among the groups, and the dance group had additional influencing factors, such as imitating and learning the choreography. Therefore, the observed positive effects cannot simply be attributed to the on/off effect of music, implying a positive effect of music.

A few studies have examined the effects of exercise and music on cognitively impaired older adults. In a study conducted in South Korea, nursing home residents aged 65–96 years (mean Mini-Mental Status Examination [MMSE] score: 21.0) were divided into two groups: music and the control groups. The music group participated in a rhythmic exercise program comprising hand clapping, stretching, dancing, and percussion of various rhythms. The control group was only provided regular activity therapy without music, which included recreation, physical exercise, and/or counseling. Cognition (MMSE), depression (geriatric depression scale [GDS]), anxiety (geriatric anxiety inventory [GAI]), and life satisfaction improved in the music intervention with rhythmic exercise program group than in the control group [44]. Existing comparative studies on patients with Alzheimer's disease (AD) with or without music demonstrating the combined effects of physical activity and music are lacking [45]. Nevertheless, a study by Sakamoto et al. assigned 39 individuals with severe AD to three groups (passive music, interactive music, and no-music control) and compared the effects of listening to music passively or interactively. Music was individually selected and related to special memories for each participant to evoke positive emotions, such as pleasure or joy. Participants in the passive music group listened passively to selected music. The interactive music group participants listened to the selected music and participated in interactive activities (e.g., clapping, singing, and dancing) guided by a facilitator. The interactive intervention caused not only the greatest short-term improvement in emotional state but also a greater long-term reduction in behavioral and psychological symptoms of dementia compared with passive music intervention and a no-music control condition [46].

Not many, but a limited number of studies have examined the effects of musical exercise on children. Children aged 6–11 years were enrolled in one of two programs for 7 weeks: a rhythmic program and a physical education program. The rhythmic program emphasized moving to the beat of music in various rhythmic patterns, including whole-body movements, clapping, and drumsticks. The physical education group engaged in ball skills, locomotor patterns, team sports, and moving through stations in small groups, with no emphasis on rhythm. The physical activity levels of the two groups were comparable. Children in both groups showed improvements in executive function (Flanker task) with no difference between the groups [47]. Takehara et al. conducted a randomized controlled trial with 2301 fourth-grade students (aged 9–10 years) in Mongolia. The intervention group received a music-based, 3-minute, high-intensity interval exercise program composed of music during a physical exercise class that included jumps, squats, and various steps. The control group received the usual physical education classes. Compared with the control group, students in the intervention group showed a significant improvement not only in physical fitness tests (20 m shuttle run and side-to-side jump) but also in academic performance [48]. Researchers from Korea reported similar findings among children in later adolescence. This study included 66 girls in middle school (13–16 years old) who were divided into three groups: rhythm step training with music group, step training without music group, and non-exercise group (program duration: 12 weeks). The rhythm step training program involved performing various step patterns by moving to a space divided into a line or step ladder for 45 min. The rhythm step training with music group outperformed not only in a series of physical fitness tests but also in cognitive tests, including verbal memory (word list memory test) and attention (trail-making test) [49]. These findings in childhood and adolescence suggest that music may promote the effects of exercise on physical fitness, which, in turn, may have a positive effect on the brain.

To summarize the literature on the long-term effects of exercise with music, music tends to positively influence the effects of exercise on the brain, although it should be noted that findings on cognitive performance are limited. There are a limited number of well-controlled studies, most of them have been focused mainly on older adults, and only a few have examined neural substrates using neuroimaging techniques.

4 Neural Mechanisms of Music Facilitate the Effects of Exercise on Cognition and Mental Health

Exercise is executed by multiple brain regions working together to activate a multi-step motor control system (Fig. 3a network a–g) [50–54]. Exercise stimulates cognition-related brain regions, including the PFC and hippocampus, possibly by the modulation of the ascending arousal system (e.g. catecholaminergic projections from the locus coeruleus) [55, 56] (Fig. 3a network h). Therefore, the possible mechanisms underlying the facilitative effect of music on the effect of exercise on cognitive function have been discussed.

The most plausible explanation for this mechanism is that increased sensory stimulation and cognitive demands activate the brain and increase plasticity; however, empirical evidence supporting this idea in humans is limited [57]. Generally, combined physical and cognitive training has a positive impact on cognitive improvement; however, the impact was confined to trained cognitive functions rather than generalizing to other cognitive functions [58]. In addition, excessive cognitive demands lead to the depletion of finite cognitive resources from multiple perspectives, such as metabolic energy in the brain, working memory capacity, and information processing pathways [59]. This may lead to cognitive fatigue and impair the beneficial effects of acute exercise on cognition [60–63].

Music may enhance the effects of exercise on cognition and mental health through other mechanisms, rather than simply increasing sensory stimulation and cognitive demands. Park et al. [32] proposed a theoretical model of Music as an Affective Stimulant for Physical Activity (MASPA) to explain the presumed mechanisms underlying the use of music to promote physical activity. In this model, three putative mechanisms underlie affective and ergogenic responses to physical activity with music: (1) musical pleasure and reward, (2) rhythmic entrainment, and (3) sensory distraction from physical exertion. Herein, we extend this theory to investigate the mechanism by which music promotes the effects of exercise on the brain (Fig. 1).

4.1 *Musical Reward and Pleasure*

Music elicits positive emotions, increases arousal, and activates the brain areas responsible for pleasure and rewards. Sound stimuli converted into electrical signals at the eardrum and auditory nerves were perceived as music through the inferior colliculus, medial geniculate body, auditory cortex, and PFC (Fig. 3b network i, j, k). Music evokes a variety of emotions that involve brain activity in various regions, including the amygdala, nucleus accumbens (NAcc), ventral globus pallidus, hippocampus, insula, anterior cingulate cortex (ACC), and orbitofrontal cortex (OFC) [64] (Fig. 3b network n). Listening to pleasurable music is often accompanied by measurable bodily reactions such as goose bumps or shivers down the spine. This

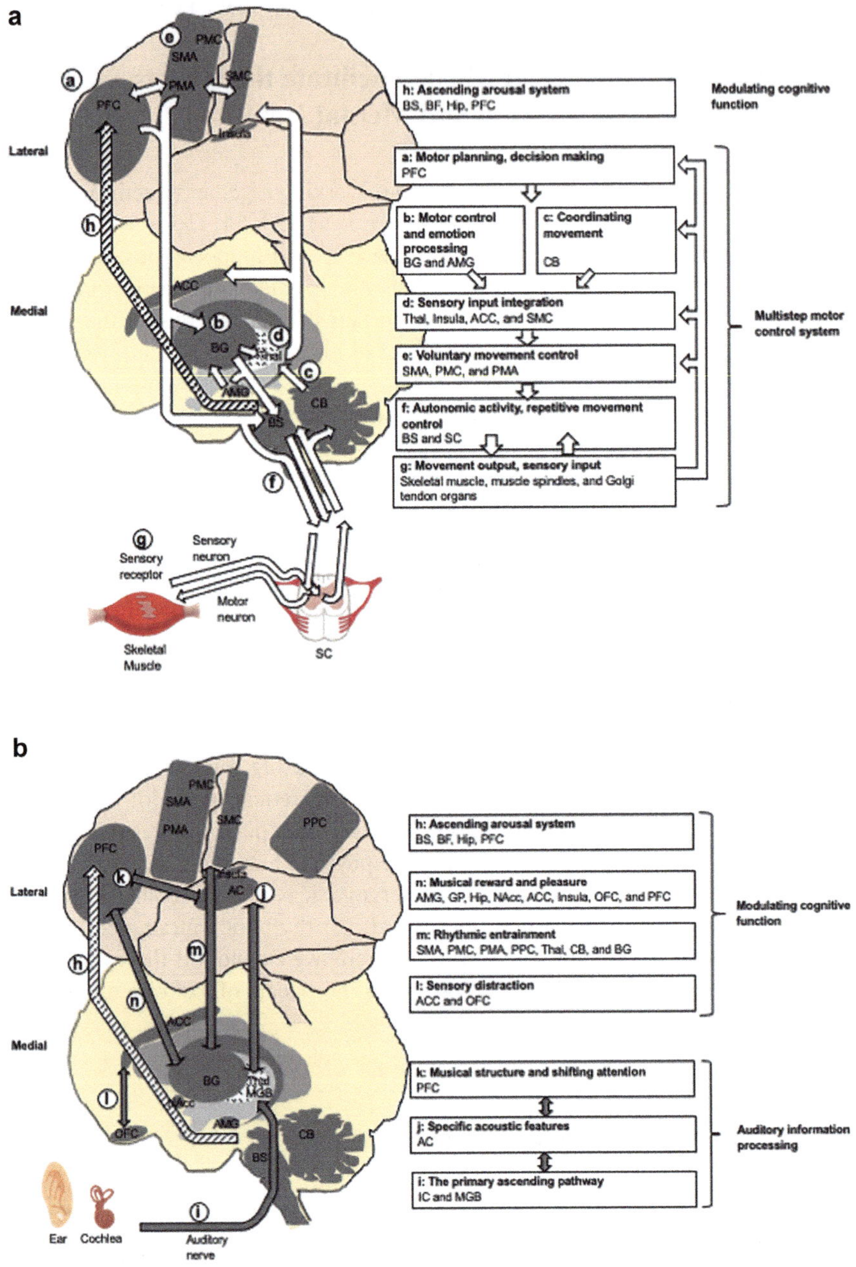

Fig. 3 (**a**) Possible neural mechanism of motor control and cognitive improvement by exercise. (**b**) Possible neural mechanism of cognitive modulation by exercise combined with music. The figure includes four elements of mechanisms, musical pleasure and reward, rhythmic entrainment, sensory distraction, cognitive load and neuroplasticity. Abbreviation. BS, brain stem, BF basal forebrain, SC spinal cord, MGB medial geniculate body, IC inferior colliculus, CB cerebellum, AMG amygdala, NAcc nucleus accumbens, GP globus pallidus, Thal thalamus, Hip hippocampus, OFC orbitofrontal cortex, ACC anterior cingulate cortex, AC auditory cortex, PPC posterior parietal cortex, SMC somatosensory cortex, PMA premotor area, PMC primary motor cortex, SMA supplementary motor area, DLPFC dorsolateral prefrontal cortex

phenomenon is commonly called "chills" or "frissons," and this could be attributed to the activation of the dopaminergic reward system based on the brain regions including the midbrain, striatum, and NAcc [65–68], and activation of these regions is associated with subjective arousal [69]. Transcranial magnetic stimulation (TMS)-induced activation of the DLPFC increases NAcc activity and enhances reward sensitivity to music through top-down control [70]. One hypothesis that explains the biological significance of music activation in the reward system is reward prediction error theory. The idea is that the accumulation of information that enhances the accuracy of predictions, i.e., predictions that are met or marginally erroneous, is rewarding [71]. Listening to music produces predictions about the flow of music, and the reward system is activated not only at the timing of emotional peaks but also during the prediction phase [72].

Music-induced positive emotions may contribute to improved cognitive functions. The ascending arousal system, which includes the dopaminergic neuromodulatory system from the ventral tegmental area, controls arousal levels and cognitive function. The activation of the system at a medium level is beneficial for higher cognitive performance (an inverted-U relationship) [73, 74]. Since each of moderate exercise [75] and listening to music [68] activates the ascending arousal system, it is postulated that exercise with music may further potentiate the effect (Fig. 3b, network h).

Several studies support this hypothesis by showing that improved cognitive function with exercise and music listening is associated with improved mood. Higher subjective pleasure and arousal when listening to music enhanced music memory function, and dopamine receptor antagonists reduced this memory performance [76]. Increases in executive function and DLPFC activity induced by acute moderate-intensity exercise following preferred music are correlated with an increase in positive mood [15]. A long-term intervention study in older patients with MCI also showed that moderate-intensity square dance exercise improved cognitive function, which was mediated by a reduction in depressive symptoms and quality of life [77].

4.2 Rhythmic Entrainment

Music can support movement and elicit smoother and more active movements, thereby enhancing positive affect and modulating brain activation through changes in the quality of movement and sensory stimulus feedback. A noteworthy element is the synchronization of rhythm and body movement (audiomotor entrainment). In the MASPA model, this factor is called rhythmic entrainment [32]. Providing sound stimuli or music that support the innate cyclic rhythm in patients with Parkinson's disease who exhibit gait difficulties could improve gait performance, positive emotions, and motivation [78, 79]. This effect has also been observed in healthy individuals [80]. Rhythmic entrainment can bring about pleasant emotions and neural activation of motor and cognitive systems [81–84]. The basal ganglia are involved in this process, as they are responsible for both motor execution and emotional control [83] (Fig. 3b network m, n). Rhythmic entrainment requires activation and coordination of many brain regions involved in temporal attentional processes and motor planning, such as the supplementary motor area, premotor area, primary motor cortex, posterior parietal cortex, thalamus, and cerebellum [85, 86] (Fig. 3b, network m). In addition, DLPFC activity is recruited to perceive rhythm and achieve entrainment with the rhythm [84]. Taken together, music externally supports the innate rhythm of movement, and when entrainment between the internal rhythm and external stimuli occurs, pleasant feelings and arousal increase, leading to improved cognitive function.

Several empirical studies support this idea. The long-term intervention of exercise followed by music with percussion instruments, which requires rhythmic entrainment, had a positive impact on the test battery for prefrontal cognitive function in older patients with MCI [87]. A series of studies from Germany have attempted to determine the effect of dance exercises with music on brain structure. Healthy older individuals in the dance group constantly learned new movement sequences that required whole-body coordination with music. After 6–18 months of dance intervention, individuals in the dance group exhibited an increase in brain volume, including the precentral gyrus, cingulate cortex, insula, corpus callosum, sensorimotor cortex, and parahippocampus, compared to the control sports training group that matched exercise conditions to the dance group, including music listening [88–90]. A more detailed analysis of the hippocampus revealed that dance training alone increased the volume of the dentate gyrus [89]. In addition, interpersonal rhythmic entrainment can activate the basal ganglia, which is the hub of the reward and motor network, and the prefrontal regions [91, 92], suggesting that various aspects of entrainment, such as perceptual, physiological, motor, and social entrainment, can stimulate the brain [83].

4.3 Sensory Distraction

Moderate- or high-intensity exercise reduces physical discomfort, referred to as sensory distraction in the MASPA model [32]. Moderate-intensity exercises considerably increase the internal discomfort [93], which is relieved by music by diverting attention from the inside of the body to the outside [94]. Applying music to moderate-intensity exercise reduces the RPE [30]. This distraction effect is caused by the limited capacity of the neural signals from the afferent nervous system to reach the brain [30]. Additionally, the activities of the ACC and OFC increased, whereas the activities of the thalamus and insula involved in the pain matrix decreased when attention was diverted from internal pain to external stimuli [95] (Fig. 3b network l). A similar neural mechanism may be involved in music-induced distraction. The distraction effect was greater when music was preferable, whereas discomfort increased when music was not preferable [96]. Healthy young adults may not obtain the distraction effect when the exercise intensity is too high, concretely more than 75% of their aerobic capacity [97]. However, in exercisers and athletes, RPE can be reduced even at high intensities [98, 99]. This may be because exercisers tend to direct their attention to internal sensations and let music help them (such as rhythmic entrainment) rather than distract them; the manner in which practitioners direct attention also seems to influence the distraction effect [30]. The distraction effect was also observed in exercise types other than aerobic exercises. When the preferred music is used for resistance exercises, a distraction effect is observed even at high intensities of 75% of the maximal load [100]. The distraction effect enables exercise to be performed by reducing discomfort while fully achieving the arousing effects of exercise, especially above a moderate intensity, which may lead to an improvement in cognitive function.

5 Potential of Groove Music and Combination with Exercise

The potential of music to promote exercise effects has been examined; however, the characteristic features of music that are compatible with exercise and effective for enhancing cognitive function are unclear. In particular, the detailed specifications of music and exercise conditions and the optimal combination to elicit enhancing effects on cognitive function have not yet been fully investigated. Considering the mechanisms of these effects, we summarized the literature on the optimal musical characteristics and exercise conditions for enhancing cognitive function.

The three major musical elements were rhythm, melody, and harmony. As a more intuitive concept induced by the combination of these elements, groove sensation may be useful when considering compatibility with exercise. Although there is variation in the definition of groove sensation, a common feature of groove across cultures is "the pleasurable sensation of wanting to move the body to music [81, 101, 102]." Moving the body to music and dancing is a universal phenomenon in humans, and groove is inextricably linked to physical exercise and sports. Groove is also considered to be a good subject for research to reveal the role of "anticipation" as the basis for reward in dance and music [103].

Groove sensation is influenced by various musical factors, such as musical genre, beat salience, micro-timing, strength of bass sound, and tempo [104–106]. Among the three major musical elements, rhythm primarily influences groove sensation [107]. Periodic rhythm is also involved in the execution of movement, and its influence is significant during musical exercise [31]. Rhythmic syncopation, the shifting of rhythmic emphasis, is an important component of groove [108]. Psychological models of groove have demonstrated the influence of individual characteristics, scenes, and various other factors on groove sensation [109].

Notably, the sensation of groove was accompanied by positive emotions. Listen to groove music activates motor, reward, and prefrontal neural activity [110]. Furthermore, the intensity of neural activity in the NAcc, which is a part of the reward system, correlates with both subjective groove sensations and pleasure [110]. As previously explained, the neural activity associated with positive emotion induction can produce positive effects on cognitive performance.

We addressed the effect of groove rhythms (GR) on cognitive performance using drum patterns (bass-drum, snare-drum, and hi-hat). Compared to white noise metronome, listening to the GR for 3 min induced a groove sensation and positive affective states. Although GR did not enhance executive function and l-DLPFC activity in all participants, cluster analysis revealed that executive function improvement and l-DLPFC activation were found in groove enjoiners; both "Good *Nori*" (Japanese term for similar concepts with groove sensation) and "Clear headed" (cognitive alertness) were significantly enhanced in response to GR. In addition, groove-related sensations were correlated with l-DLPFC activity and executive function [111] (Fig. 3b, network h, Fig. 4). These results suggest that subjective groove and associated arousal are important modulators of the effects of GR on executive function. Furthermore, we examined the effects of very-light-intensity bicycle exercise combined with GR on PFC function in a similar experiment. Similar to the GR listening effect, 3 min of very-light-intensity exercise combined with GR (GREX) induced positive psychological responses compared to exercise with a white noise metronome; however, it did not enhance executive function and l-DLPFC activity. After excluding participants who exhibit negative practice effect in a pre-practice for following the beat with bicycle exercise, cluster analysis

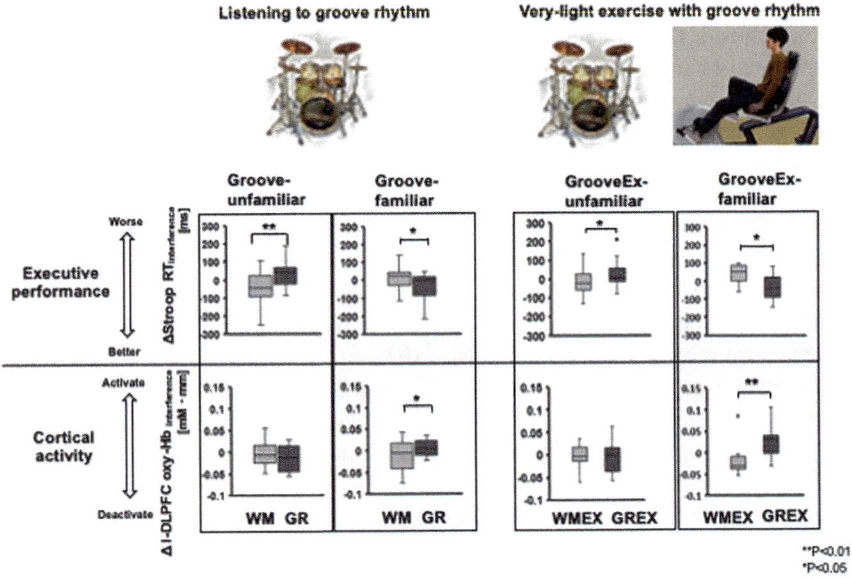

Fig. 4 Listening to groove rhythm (GR) enhanced EF and l-DLPFC activity compared to white-noise metronome (WM) only in groove enjoiner who significantly felt both "Good Nori" (Japanese term means similar concepts with groove sensation) and "Clear headed" (cognitive alertness) in response to GR [111]. Very-light exercise with groove rhythm (GREX) enhanced EF and l-DLPFC activity compared to exercise with white-noise metronome (WMEX) only in groove enjoiner who significantly felt both "Feeling like my body is resonating with the rhythm" (subjective audiomotor entrainment) and "excited" in response to GR [112]

revealed that enhancement of executive function and l-DLPFC activity was found only in those in which both "feeling like my body is resonating with the rhythm" (subjective feeling of audiomotor entrainment) and "excited" were significantly enhanced, and these were correlated with l-DLPFC activity and executive function [112] (Fig. 4). These results suggest that there are substantial individual variations and that audiomotor entrainment is a critical factor when exercising with GR. On comparing the effect sizes of GREX and GR, focusing on the familiar group that showed a benefit in prefrontal executive function, the effect size of GREX (d=1.17) was larger than that of GR (d=0.69). This suggests that exercise with music is more effective than music alone for the same intervention time if the combination of music and exercise matches personal characteristics and enhances psychological response. However, of the total participants, GR and GREX were beneficial to 31.3% (16/51) and 21.9% (9/41) in the familiar group, indicating that the percentage of people who benefited from the intervention was lower in the GREX experiment. The addition of an element of physical exercise using large muscle groups and the exercise type of bicycling may have limited the success of eliciting audiomotor entrainment and positive affect. These findings conclude that high-groove music, which is compatible with exercise, effectively enhances motivation and enjoyment during physical exercise; however, its effects on cognitive function are mixed, and individual differences in psychological responsiveness to music may be related to its modifying factors. Therefore, exercise conditions that are compatible with groove music should be considered.

Groove music induces body movements, especially of the feet and head, which are associated with pleasure [81]. Groove music stirs both the urge to move and the accompanying pleasure because physical movement promotes the perception and anticipation of rhythm and beat, and physical movement reduces the prediction error of the beat and makes it a reward [102]. The desire to move and positive affect generated by groove music are inseparable and a key factor in combination of groove music and exercise could be audiomotor entrainment. This indicates that adjusting the exercise conditions to facilitate entrainment is imperative to improve cognitive function through exercise with groove music.

The intensity of exercise should be mainly considered. As previously mentioned, listening to music is recommended when the exercise intensity is greater than moderate to divert attention from the discomfort caused to outside the body [94]. However, music that reduces boredom and enhances positive emotions is expected to be effective in light-intensity exercises, where such discomfort does not occur; groove music with positive emotions can meet this characteristic. In addition, the cognitive demands for exercise execution are relatively small in light-intensity exercises [113], enabling more attention to the body and entrainment to the rhythm, which is considered a suitable intensity for groove music. Our preliminary experiment revealed that, compared to moderate-intensity (50% VO_{2peak}) exercise, GR tended to increase groove sensation and excitement more than control stimuli (too complex a rhythm) when combined with very light-intensity (30% VO_{2peak}) exercise.

Exercise type is also an important factor for audiomotor entrainment. Exercise that allows the body to be more easily attuned to rhythm is desirable. Rhythmic body movements are easier to perform when the economy of movement is high (e.g., contraction rather than extension) and move toward gravity than against gravity [137]. Rhythmic entrainment is easier to achieve when the movements are discontinuous than when they are continuous [114, 115]. Clapping movements are characterized by discontinuous movements, and sound feedback facilitates rhythmic entrainment [116], which is likely to be useful as a type of exercise performed in conjunction with groove music. "Slow aerobic dance exercise" is a concrete example of exercise that is familiar with audiomotor entrainment. Slow aerobic dance was developed by revising aerobic dance exercises into an easy, simple, light-intensity exercise program with music. Online exercise classes for older adults, including slow aerobic dance for 20 min per session, five times a week for 8 weeks, showed a high participation rate of 97.4% [117]. Slow aerobic dancing for 10 min enhanced mood and cognitive function, with greater increases in mood (vitality and pleasure levels) than cycling exercises did [118]. In addition to dance-like movements, familiar movement types, such as walking, are also suitable for audiomotor entrainment because this exercise type can be automatically performed and has low cognitive demand in the execution of exercise [119]. Future research should consider and set compatible exercise conditions for audiomotor entrainment, such as light intensity, discontinuous, and familiar exercise types, when assessing the effects of exercise following groove music on cognitive function.

The compatibility between exercise conditions and mechanisms can be verified by focusing on specific musical characteristics. Groove music that "makes the body want to move," or, conversely, increases the motivation to exercise, is a musical characteristic that should be focused on for a maximized overall effect, if the overall effect is expressed as the frequency of implementation multiplied by single bout effect. Future research should verify the effects of exercise with groove music on executive function.

6 Considering the Suitability of the Exercise with Music for Individuals

To verify the effects of exercise with music on executive function, examining the optimal combination of exercise conditions and specific music characteristics is essential. In addition, individual characteristics are an important factor and will be a major concern for future research.

As previously mentioned, the effect of GR on cognitive function is influenced by the level of the subjective psychological response to music [111], and audiomotor entrainment is considered a key influential factor. Therefore, individual variations in the factors that affect affective states and audiomotor entrainment could consequently influence the effects of exercise with music on cognitive function. Individual factors, such as pitch sensitivity, beat processing ability, musical reward sensitivity, cultural background, listening experience (familiarity), and music and dance training, influence the effects of music, exercise with music, and dance on various outcomes, including emotion and body movements.

Congenital inability to perceive differences in musical pitch and melody is referred to as amusia (tone deafness) [120]. Perceiving tone accuracy and precision were measured using the Montreal Battery of Evaluation of Amusia (MBEA) and singing accuracy and precision [121, 122]. The impact of amusia on recognition and emotional responses to music was mixed, impairing, or negligible [123, 124].

Beat deafness is a condition that exists in addition to tone deafness, where individuals who are not tone deaf are unable to extract beats from music [125]. Beat-processing ability is the ability to extract (perceive) beats, which are rhythms at regular intervals from music, or to internally generate rhythms and reproduce them with body movements, as measured by the beat alignment test (BAT) and the Harvard beat alignment test (H-BAT) [126, 127]. This ability varies greatly among individuals and between non-musicians and musicians. The ability to perceive rhythm probably has a greater impact on the effects of exercise with music because the rhythm in exercise with music is a major factor influencing the effect [31]. Neural activity and networks in the putamen, caudate, supplementary motor cortex, premotor cortex, and auditory cortex form the neural basis for this ability [128, 129]. Beat-processing ability modulates the effects of groove music on gait performance [79, 130], which could be attributed to an increased beat salience or other factors. The ability to synchronize with the rhythm during whole-body movements is strictly distinct from the ability to internally generate rhythm and take the beat with partial body movements (e.g., tapping), which is part of the beat processing ability. Dancers performed better than musicians on the task of moving the whole body to music [131], and musicians performed better than dancers on the rhythm-taking task, which also included tapping with the fingers [131, 132]. Assessment of rhythmic ability through whole-body movement is preferable for evaluating beat processing ability as a factor influencing the effects of exercise with music.

In addition to the aforementioned individual differences in the level of perception of music as a sensory stimulus, individual differences in sensitivity to pleasure from listening to music do exist [133]. The inability to perceive pleasure and reward for music is called musical anhedonia, whereas monetary rewards are normal [134] and are characterized by a neural basis of network dysfunction in the auditory cortex and mesolimbic reward system [135]. Individual differences in susceptibility have been attributed to differences in the neural connections of the NAcc, OFC, and superior temporal gyrus (STG), which are part of the reward system [136]. In addition, the PFC controls these networks in a top-down fashion based on reports that stimulation and activation of the DLPFC with TMS increase music-reward sensitivity [70]. Music reward sensitivity influences subjective emotional responses to music, particularly the degree of positive emotional elicitation [138]. The fact that there are individual differences in the degree of rewarding pleasure, even when listening to the same music, could show that music reward sensitivity could also influence individual differences in cognitive function. This is because the activation of the reward system and increment of arousal due to positive emotions could be the mechanisms underlying the effects of exercise with music on cognitive function.

Social and cultural differences can potentially influence music preferences and psychological responses to music [139]. For instance, the perception of rhythmic coherence differs depending on whether the native language is Japanese or English [140], and Westerners who are familiar with regular meters of Western music have difficulty perceiving the exact irregular meters of music in the Balkan, Asian, and African regions [141]. The listening experience, which varies among individuals with different demographic factors, leads to the accumulation of structural information about music, which contributes to the predictive framework. Familiarity with music explicitly affects preferences and psychological responses. This can explain the cultural or generational similarities in these preferences and psychological responses. In a single listening experience, the complex features of music, such as ever-changing melodies, harmonies, and rhythms, cannot be fully understood, and the memory of musical features is incomplete. Each successive listening session brings greater pleasure as errors become smaller. Furthermore, even after multiple listenings, there may be a margin of error that continues to produce positive prediction errors [139]. Familiarity increases groove sensation when listening to groovy music [108, 142], increases neural activation in the limbic regions involved in the reward system [143], and affects walking speed by increasing the enjoyment of the music [144].

Music and dance training can alter the factors described above and have a wide range of effects on brain structure and function, psychological responses to music, and cognitive functions. Musicians are more likely to exhibit groove sensations when listening to groove music than non-musicians, and they have greater neural activity in areas related to beat perception and audiomotor entrainment, such as the supplementary motor cortex and premotor cortex [110]. Listening to high-groove music activated the corticospinal excitability responsible for motor control in musicians, whereas suppressed in non-musicians [145]. Differences in brain structure and function were observed, with increased brain volume and enhanced functional connectivity between regions in the auditory cortex as well as in brain regions associated with music and movement, such as the motor cortex, sensorimotor integration areas, and cerebellum [146]. These structural and functional changes may have beneficial effects on language, social and cognitive functions, and musical ability [147]. The experience of dance, a typical example of exercise with music, also influences brain structure and function. Long-term training in dance leads to gray and white matter neuroplasticity related to movement and hearing, which is also correlated with dance performance [148]. Although both music and dance experience increase beat processing ability and whole-body rhythmic entrainment compared to those in controls, differences between music and dance experience (including differences in the influence of artistic and motor components) and differences by music and dance genre experience have not yet been validated, warranting further research [132, 148]. Owing to their wide-ranging influences, music and dance experiences should be considered when examining the effects of music-enhanced exercises on cognitive function.

In summary, individual variations can modulate the effects of combining music and exercise on cognitive function; however, the mechanism is complex. Considering the influence of individual characteristics can be helpful when setting up optimal exercise programs for individuals. For example, for individuals with low beat processing ability, it may be possible to adjust music to emphasize the beat using low-frequency sounds (e.g., bass drum). For exercise conditions, adjustments could be made to use exercise types (clapping, downward movement, and walking) that are more easily synchronized with the beat. Further studies on the individual characteristics add potential insights to the transformation of individual characteristics through music/dance training and education, thereby increasing the effects of exercise to music on cognitive function, and the benefits of training will be recognized.

7 Possible Combination with Other "Enriched Exercise" Factors

The promotion of exercise effects by combining music with other environmental factors can be maximized. The effects of various environmental factors on cognitive function and mental health have been demonstrated. Outdoor environments are enriched with sensory stimuli, such as sunshine, birdsong, green scents, and wind. Attentional performance and event-related neural responses (P300 amplitude) improved during a 2-km walk in the natural outdoors (green exercise), as opposed to indoor walking [149]. Moreover, green exercises induce more positive emotions and brain theta activity than did virtual green exercise and control [150]. Aquatic environments have higher pressure, resistance, buoyancy, and thermal conductivity, leading to different physiological and psychological responses than on land. A pilot study suggested that aquatic walking enhanced cortical activation and executive function more than land walking did [151]. Virtual reality (VR) environments provide immersive experiences. Compared with control cycling, VR-based cycling was beneficial in reducing tiredness and increasing enjoyment and self-efficacy [152]. Boxing motion exercises in a VR environment are more effective in enhancing positive affect than the same exercises with a 2D display [153]. Exercises under 360° video with music elicited the most positive affect and dissociative thoughts than those of exercises under music, video, music-video, or 360° video [154]. Group exercise is a component of social interaction. Longitudinal studies have documented that, compared with individual exercises, exercise with others is beneficial for preventing the risk of cognitive impairment and depression [155, 156]. Games and competitive sports provide cognitive load and excitement. Badminton enhances executive function, but not running, compared to seated rest [157]. Future research should consider effective combinations that consider the compatibility and context of the environmental stimuli.

8 Conclusion

This chapter focuses on music as an "enriched exercise environment" that enhances positive affects during exercise and promotes the exercise effects on the brain. To apply these effects in the field, we discussed the possibility of groovy music, which is compatible with exercise, and individual variations in the suitability and effectiveness of exercise with music.

Finally, we summarized the future directions for this promising research area. First, existing studies examining the effect of adding music to exercise in an appropriate experimental design are scarce. Future studies should control for various factors, including the exercise itself, across experimental conditions and groups. Second, few studies have examined the neural mechanisms of musical exercise in the brain. Further studies using cutting-edge neuroimaging techniques and biomarkers are warranted. Third, careful attention should be paid to the environmental setting, as it is likely that the suitability of the exercise conditions and music selected for the individual will modify the effects of the program.

Although "exercise is medicine" is a household catchline, exercise also has an aspect as a "hobby" [158]. The key to promoting high levels of adherence to exercise and physical activity is to allow individuals to design the exercise environment according to their goals, preferences, timing and location, and practice exercise enjoyably. Music is a potential tool to make physical exercise more enjoyable and effective. Further evidence may facilitate its social implementation.

References

1. World Health Organization (2020) WHO guidelines on physical activity and sedentary behaviour. ISBN: 9789240015128
2. Livingston G, Sommerlad A, Orgeta V et al (2017) Dementia prevention, intervention, and care. Lancet 390:2673–2734
3. World Health Organization (2017) Global action plan on the public health response to dementia 2017–2025. ISBN: 978-92-4-151348-7
4. Guthold R, Stevens GA, Riley LM, Bull FC (2018) Worldwide trends in insufficient physical activity from 2001 to 2016: a pooled analysis of 358 population-based surveys with 1·9 million participants. Lancet Glob Health 6:e1077–e1086
5. Guthold R, Stevens GA, Riley LM, Bull FC (2020) Global trends in insufficient physical activity among adolescents: a pooled analysis of 298 population-based surveys with 1·6 million participants. Lancet Child Adolesc Health 4:23–35
6. Choi J, Lee M, Lee JK, Kang D, Choi JY (2017) Correlates associated with participation in physical activity among adults: a systematic review of reviews and update. BMC Public Health 17:1–13
7. Stojanovic M, Babulal GM, Head D (2023) Determinants of physical activity engagement in older adults. J Behav Med 46:757–769
8. Ekkekakis P (2017) People have feelings! exercise psychology in paradigmatic transition. Curr Opin Psychol 16:84–88
9. Lee HH, Emerson JA, Williams DM (2016) The exercise–affect–adherence pathway: an evolutionary perspective. Front Psychol. 7:1285

10. Murphy SL, Eaves DL (2016) Exercising for the pleasure and for the pain of it: the implications of different forms of hedonistic thinking in theories of physical activity behavior. Front Psychol 7:1–3
11. Stevens CJ, Baldwin AS, Bryan AD, Conner M, Rhodes RE, Williams DM (2020) Affective determinants of physical activity: a conceptual framework and narrative review. Front Psychol. 11:568331
12. Stults-Kolehmainen MA, Blacutt M, Bartholomew JB, Gilson TA, Ash GI, McKee PC, Sinha R (2020) Motivation states for physical activity and sedentary behavior: desire, urge, wanting, and craving. Front Psychol 11:1–17
13. Russell JA, Barrett LF (1999) Core affect, prototypical emotional episodes, and other things called emotion: dissecting the elephant. J Pers Soc Psychol 76:805–819
14. Rhodes RE, Kates A (2015) Can the affective response to exercise predict future motives and physical activity behavior? a systematic review of published evidence. Ann Behav Med 49:715–731
15. Suwabe K, Hyodo K, Fukuie T, Ochi G, Inagaki K, Sakairi Y, Soya H (2021) Positive mood while exercising influences beneficial effects of exercise with music on prefrontal executive function: a functional NIRS study. Neuroscience 454:61–71
16. Nouchi R, Nouchi H, Kawashima R (2020) A Single 30 Minutes Bout of Combination Physical Exercises Improved Inhibition and Vigor-Mood in Middle-Aged and Older Females: Evidence From a Randomized Controlled Trial. Front Aging Neurosci 12:1–12
17. Herrington JD, Mohanty A, Koven NS, Fisher JE, Stewart JL, Banich MT, Webb AG, Miller GA, Heller W (2005) Emotion-modulated performance and activity in left dorsolateral prefrontal cortex. Emotion 5:200–207
18. Perlstein WM, Elbert T, Stenger VA (2002) Dissociation in human prefrontal cortex of affective influences on working memory-related activity. Proc Natl Acad Sci 99:1736–1741
19. Dolcos F, Labar KS, Cabeza R (2004) Dissociable effects of arousal and valence on prefrontal activity indexing emotional evaluation and subsequent memory: an event-related fMRI study. Neuroimage 23:64–74
20. Dolcos F, McCarthy G (2006) Brain systems mediating cognitive interference by emotional distraction. J Neurosci 26:2072
21. Qin S, Hermans EJ, Van MHJF, Luo J, Fernández G (2009) Acute psychological stress reduces working memory- related activity in the dorsolateral prefrontal cortex. BPS 66:25–32
22. Kempermann G, Kuhn HG, Gage FH (1997) More hippocampal neurons in adult mice living in an enriched environment. Nature 386:493–495
23. Hattori S, Hashimoto R, Miyakawa T, Yamanaka H, Maeno H, Wada K, Kunugi H (2007) Enriched environments influence depression-related behavior in adult mice and the survival of newborn cells in their hippocampi. Behav Brain Res 180:69–76
24. Winkler I, Háden GP, Ladinig O, Sziller I, Honing H (2009) Newborn infants detect the beat in music. Proc Natl Acad Sci 106:2468–2471
25. Zentner M, Eerola T (2010) Rhythmic engagement with music in infancy. Proc Natl Acad Sci 107:5768–5773
26. Patel AD, Iversen JR, Bregman MR, Schulz I (2009) Experimental evidence for synchronization to a musical beat in a nonhuman animal. Curr Biol 19:827–830
27. Schachner A, Brady TF, Pepperberg IM, Hauser MD (2009) Spontaneous motor entrainment to music in multiple vocal mimicking species. Curr Biol 19:831–836
28. Cook P, Rouse A, Wilson M, Reichmuth C (2013) A california sea lion (Zalophus californianus) can keep the beat: motor entrainment to rhythmic auditory stimuli in a non vocal mimic. J Comp Psychol 127:412–427
29. Ito Y, Shiramatsu TI, Ishida N, Oshima K, Magami K, Takahashi H (2022) Spontaneous beat synchronization in rats: neural dynamics and motor entrainment. Sci Adv. 8:eabo7019
30. Terry PC, Karageorghis CI, Curran ML, Martin OV, Parsons-Smith RL (2020) Effects of music in exercise and sport: a meta-analytic review. Psychol Bull 146:91–117

31. Karageorghis CI, Priest D (2012) Music in the exercise domain: a review and synthesis (Part I). Int Rev Sport Exerc Psychol 5:44–66
32. Park KS, Williams DM, Etnier JL (2023) Exploring the use of music to promote physical activity: from the viewpoint of psychological hedonism. Front Psychol 14:1–12
33. Tanaka D, Tsukamoto H, Suga T, Takenaka S, Hamaoka T, Hashimoto T, Isaka T (2018) Self-selected music-induced reduction of perceived exertion during moderate-intensity exercise does not interfere with post-exercise improvements in inhibitory control. Physiol Behav 194:170–176
34. Chen J, Su R, Lv Z, Xiao J, Zhao Y, Wang D, Jiang E (2021) The effect of acute aerobic exercise with music on executive function: the major role of tempo matching. Phys Act Health 5:31–44
35. Emery CF, Hsiao ET, Hill SM, Frid DJ (2003) Short-term effects of exercise and music on cognitive performance among participants in a cardiac rehabilitation program. Hear Lung J Acute Crit Care 32:368–373
36. Mezghani N, Ammar A, Alzahrani TM et al (2022) Listening to music and playing activities during recreation between lessons regenerate children's cognitive performance at different times of day. Children 9:1–10
37. Hewston P, Kennedy CC, Borhan S et al (2021) Effects of dance on cognitive function in older adults: a systematic review and meta-analysis. Age Ageing 50:1084–1092
38. Ma C, Li M, Li R, Wu C (2023) The effect of rhythmic movement on physical and cognitive functions among cognitively healthy older adults: a systematic review and meta-analysis. Arch Gerontol Geriatr 104:104837
39. Muiños M, Ballesteros S (2021) Does dance counteract age-related cognitive and brain declines in middle-aged and older adults? a systematic review. Neurosci Biobehav Rev 121:259–276
40. Satoh M, Ogawa J-II, Tokita T, Nakaguchi N, Nakao K, Kida H, Tomimoto H (2014) The effects of physical exercise with music on cognitive function of elderly people: Mihama-Kiho Project. PLoS One 9:e95230
41. Tabei KI, Satoh M, Ogawa JI, Tokita T, Foster PP, Nakaguchi N, Nakao K, Kida H, Tomimoto H (2017) Physical exercise with music reduces gray and white matter loss in the frontal cortex of elderly people : the Mihama-Kiho Scan Project. Front Aging Neurosci 9:1–12
42. Merom D, Mathieu E, Cerin E, Morton RL, Simpson JM, Rissel C, Anstey KJ, Sherrington C, Lord SR, Cumming RG (2016) Social dancing and incidence of falls in older adults: a cluster randomised controlled trial. PLoS Med 13:1–19
43. Kosmat H, Vranic A (2017) The efficacy of a dance intervention as cognitive training for the old-old. J Aging Phys Act 25:32–40
44. Kim HS, Kang JS (2021) Effect of a group music intervention on cognitive function and mental health outcomes among nursing home residents: a randomized controlled pilot study. Geriatr Nurs (Minneap) 42:650–656
45. Li K, Cui CC, Zhang H, Jia L, Li R, Hu HY (2022) Exploration of combined physical activity and music for patients with Alzheimer's disease: a systematic review. Front Aging Neurosci. 14:962475
46. Sakamoto M, Ando H, Tsutou A (2013) Comparing the effects of different individualized music interventions for elderly individuals with severe dementia. Int Psychogeriatr 25:775–784
47. Vazou S, Klesel B, Lakes KD, Smiley A (2020) Rhythmic physical activity intervention: exploring feasibility and effectiveness in improving motor and executive function skills in children. Front Psychol 11:1–14
48. Takehara K, Togoobaatar G, Kikuchi A et al (2021) Exercise intervention for academic achievement among children: a randomized controlled trial. Pediatrics. 148:e2021052808
49. Park SK, Jee YS (2022) Effects of rhythm step training on physical and cognitive functions in adolescents: a prospective randomized controlled trial. Health. 10:712
50. Kandel ER, James HS, Thomas MJ, Siegelbaum SA, H AJ (2012) Principles of neural science, 5th edn

51. Thaut MH, Donald AH (2019) The oxford handbook of music and the brain. 90-144, 166-234
52. Gage NM, Baars BJ (2018) Fundamentals of cognitive neuroscience. Acad Press. 18-52, 143-319
53. Kenney WL, Wilmore JH, David LC (2022) Physiology of sport and exercise, 8th edn. Human Kinetic, Champaign
54. Schmidt RF (1978) Fundamentals of neurophysiology. 158-204
55. Kuwamizu R, Yamazaki Y, Aoike N, Hiraga T (2023) Pupil dynamics during very light exercise predict benefits to prefrontal cognition. NeuroImage 277:120244
56. Suwabe K, Kuwamizu R, Hyodo K, Yoshikawa T, Otsuki T, Zempo-Miyaki A, Yassa MA, Soya H (2024) Improvement of mnemonic discrimination with acute light exercise is mediated by pupil-linked arousal in healthy older adults. Neurobiol Aging 133:107–114
57. Sale A, Berardi N, Maffei L (2014) Environment and brain plasticity: towards an endogenous pharmacotherapy. Physiol Rev 94:189–234
58. Lauenroth A, Ioannidis AE, Teichmann B (2016) Influence of combined physical and cognitive training on cognition: a systematic review. BMC Geriatr 16:21–23
59. Shenhav A, Musslick S, Lieder F, Kool W, Griffiths TL, Cohen JD, Botvinick MM (2017) Toward a rational and mechanistic account of mental effort. Annu Rev Neurosci 40:99–124
60. Kamijo K, Abe R (2019) Aftereffects of cognitively demanding acute aerobic exercise on working memory. Med Sci Sports Exerc 51:153–159
61. Egger F, Conzelmann A, Schmidt M (2018) The effect of acute cognitively engaging physical activity breaks on children's executive functions: too much of a good thing? Psychol Sport Exerc 36:178–186
62. Gallotta MC, Guidetti L, Franciosi E, Pietro EG, Bonavolontà V, Baldari C (2012) Effects of varying type of exertion on children's attention capacity. Med Sci Sports Exerc 44:550–555
63. O'Leary KC, Pontifex MB, Scudder MR, Brown ML, Hillman CH (2011) The effects of single bouts of aerobic exercise, exergaming, and videogame play on cognitive control. Clin Neurophysiol 122:1518–1525
64. Koelsch S (2014) Brain correlates of music-evoked emotions. Nat Rev Neurosci 15:170–180
65. Grewe O, Kopiez R, Altenmüller E (2009) Chills as an indicator of individual emotional peaks. Ann N Y Acad Sci 1169:351–354
66. Salimpoor VN, Benovoy M, Longo G, Cooperstock JR, Zatorre RJ (2009) The rewarding aspects of music listening are related to degree of emotional arousal. PLoS One 4:e7487
67. Ferreri L, Mas-Herrero E, Zatorre RJ et al (2019) Dopamine modulates the reward experiences elicited by music. Proc Natl Acad Sci 116:3793–3798
68. Salimpoor VN, van den Bosch I, Kovacevic N, McIntosh AR, Dagher A, Zatorre RJ (2013) Interactions between the nucleus accumbens and auditory cortices predict music reward value. Science 80-) 340:216–219
69. Blood AJ, Zatorre RJ (2001) Intensely pleasurable responses to music correlate with activity in brain regions implicated in reward and emotion. Proc Natl Acad Sci USA 98:11818–11823
70. Mas-Herrero E, Dagher A, Zatorre RJ (2018) Modulating musical reward sensitivity up and down with transcranial magnetic stimulation. Nat Hum Behav 2:27–32
71. Zatorre RJ, Salimpoor VN (2013) From perception to pleasure: music and its neural substrates. Proc Natl Acad Sci 110:10430–10437
72. Salimpoor VN, Benovoy M, Larcher K, Dagher A, Zatorre RJ (2011) Anatomically distinct dopamine release during anticipation and experience of peak emotion to music. Nat Neurosci 14:257–262
73. Berridge CW, Arnsten AFT (2013) Neuroscience and biobehavioral reviews psychostimulants and motivated behavior : arousal and cognition. Neurosci Biobehav Rev 37:1976–1984
74. Ashby FG, Isen AM, Turken AU (1999) A neuropsychological theory of positive affect and its influence on cognition. Psychol Rev 106:529–550
75. Goekint M, Bos I, Heyman E, Meeusen R, Michotte Y, Sarre S (2012) Acute running stimulates hippocampal dopaminergic neurotransmission in rats, but has no influence on brain-derived neurotrophic factor. J Appl Physiol 112:535–541

76. Ferreri L, Mas-Herrero E, Cardona G, Zatorre RJ, Antonijoan RM, Valle M, Riba J, Ripollés P, Rodriguez-Fornells A (2021) Dopamine modulations of reward-driven music memory consolidation. Ann N Y Acad Sci nyas.14656
77. Chang J, Zhu W, Zhang J, Yong L, Yang M, Wang J, Yan J (2021) The effect of Chinese square dance exercise on cognitive function in older women with mild cognitive impairment: the mediating effect of mood status and quality of life. Front Psych. 12:711079
78. Cochen De Cock V, Dotov D, Damm L et al (2021) BeatWalk: personalized music-based gait rehabilitation in Parkinson's disease. Front Psychol. 12:655121
79. Leow LA, Watson S, Prete D, Waclawik K, Grahn JA (2021) How groove in music affects gait. Exp Brain Res 239:2419–2433
80. Ready EA, Holmes JD, Grahn JA (2022) Gait in younger and older adults during rhythmic auditory stimulation is influenced by groove, familiarity, beat perception, and synchronization demands. Hum Mov Sci 84:102972
81. Janata P, Tomic ST, Haberman JM (2012) Sensorimotor coupling in music and the psychology of the groove. J Exp Psychol Gen 141:54–75
82. Vuilleumier P, Trost W (2015) Music and emotions: from enchantment to entrainment. Ann N Y Acad Sci 1337:212–222
83. W JT, Labbé C, Grandjean D (2017) Rhythmic entrainment as a musical affect induction mechanism. Neuropsychologia 96:96–110
84. Chen JL, Penhune VB, Zatorre RJ (2008) Moving on time: brain network for auditory-motor synchronization is modulated by rhythm complexity and musical training. J Cogn Neurosci 20:226–239
85. Krause V, Schnitzler A, Pollok B (2010) NeuroImage functional network interactions during sensorimotor synchronization in musicians and non-musicians. NeuroImage 52:245–251
86. Coull JT (2004) fMRI studies of temporal attention: allocating attention within, or towards, time. Cogn Brain Res 21:216–226
87. Shimizu N, Umemura T, Matsunaga M, Hirai T (2018) Effects of movement music therapy with a percussion instrument on physical and frontal lobe function in older adults with mild cognitive impairment: a randomized controlled trial. Aging Ment Health 22:1614–1626
88. Rehfeld K, Lüders A, Hökelmann A, Lessmann V, Kaufmann J, Brigadski T, Müller P, Müller NG (2018) Dance training is superior to repetitive physical exercise in inducing brain plasticity in the elderly. PLoS One 13:e0196636
89. Rehfeld K, Müller P, Aye N, Schmicker M, Dordevic M, Kaufmann J, Hökelmann A, Müller NG (2017) Dancing or fitness sport? The effects of two training programs on hippocampal plasticity and balance abilities in healthy seniors. Front Hum Neurosci. 11:305
90. Müller P, Rehfeld K, Schmicker M, Hökelmann A, Dordevic M, Lessmann V, Brigadski T, Kaufmann J, Müller NG (2017) Evolution of neuroplasticity in response to physical activity in old age: the case for dancing. Front Aging Neurosci 9:1–8
91. Engel A, Hoefle S, Monteiro MC, Moll J, Keller PE (2022) Neural correlates of listening to varying synchrony between beats in samba percussion and relations to feeling the groove. Front Neurosci. 16:779964
92. Kokal I, Engel A, Kirschner S, Keysers C (2011) Synchronized drumming enhances activity in the caudate and facilitates prosocial commitment – if the rhythm comes easily. PLoS One. 6:e27272
93. Ekkekakis P, Hall EE, Petruzzello SJ (2005) Variation and homogeneity in affective responses to physical activity of varying intensities: an alternative perspective on dose-response based on evolutionary considerations. J Sports Sci 23:477–500
94. Razon S, Basevitch I, Land W, Thompson B, Tenenbaum G (2009) Perception of exertion and attention allocation as a function of visual and auditory conditions. Psychol Sport Exerc 10:636–643
95. Bantick SJ, Wise RG, Ploghaus A, Clare S, Smith SM, Tracey I (2002) Imaging how attention modulates pain in humans using functional MRI. Brain 125:310–319
96. Nakamura PM, Pereira G, Papini CB, Nakamura FY, Kokubun E (2010) Effects of preferred and nonpreferred music on continuous cycling exercise performance. Percept Mot Skills 110:257–264

97. Karageorghis CI, a MD, Priest D-L, a ST, Morrish DJ, CLJ W (2009) Psychophysical and ergogenic effects of synchronous music during treadmill walking. J Sport Exerc Psychol 31:18–36
98. Jarraya M, Chtourou H, Aloui A, Hammouda O, Chamari K, Chaouachi A, Souissi N (2012) The effects of music on high-intensity short-term exercise in well trained athletes. Asian J Sports Med. 3:34543
99. Terry PC, Karageorghis CI, Saha AM, D'Auria S (2012) Effects of synchronous music on treadmill running among elite triathletes. J Sci Med Sport 15:52–57
100. Ballmann CG, McCullum MJ, Rogers RR, Marshall MM, Williams TD (2018) Effects of Preferred vs. nonpreferred music on resistance exercise performance. J Strength Cond Res. 35:1650-1655
101. Madison G (2006) Experiencing groove induced by music: consistency and phenomenology. Music Percept 24:201–208
102. Etani T, Miura A, Kawase S, Fujii S, Keller PE, Vuust P, Kudo K (2023) A review of psychological and neuroscientific research on musical groove. Neurosci Biobehav Rev:105522
103. Foster Vander Elst O, Vuust P, Kringelbach ML (2021) Sweet anticipation and positive emotions in music, groove, and dance. Curr Opin Behav Sci 39:79–84
104. Madison G, Gouyon F, Ullén F, Hörnström K (2011) Modeling the tendency for music to induce movement in humans: first correlations with low-level audio descriptors across music genres. J Exp Psychol Hum Percept Perform 37:1578–1594
105. Stupacher J, Hove MJ, Janata P (2016) Audio features underlying perceived groove and sensorimotor synchronization in music. Music Percept *33*(5):571–589
106. Etani T, Marui A, Kawase S, Keller PE (2018) Optimal tempo for groove: its relation to directions of body movement and japanese nori. Front Psychol 9:462
107. Matthews TE, Witek MAGG, Heggli OA, Penhune VB, Vuust P (2019) The sensation of groove is affected by the interaction of rhythmic and harmonic complexity. PLoS One 14:e0204539
108. G WM a, Clarke EF, Wallentin M, Kringelbach ML, Vuust P (2014) Syncopation, body-movement and pleasure in groove music. PLoS One 9:e94446
109. Senn O, Rose D, Bechtold T, Kilchenmann L, Hoesl F, Jerjen R, Baldassarre A, Alessandri E (2019) Preliminaries to a psychological model of musical groove. Front Psychol 10:1–5
110. Matthews TE, Witek MAG, Lund T, Vuust P, Penhune VB (2020) The sensation of groove engages motor and reward networks. NeuroImage 214:116768
111. Fukuie T, Suwabe K, Kawase S, Shimizu T, Ochi G, Kuwamizu R, Sakairi Y, Soya H (2022) Groove rhythm stimulates prefrontal cortex function in groove enjoyers. Sci Rep 12:7377
112. Fukuie T, Suwabe K, Kawase S, Shimizu T, Ochi G, Kuwamizu R, Sakairi Y, Soya H (2023) Groove rhythm enhances exercise impact on prefrontal cortex function in groove enjoyers. Neuroscience. 531:117-129
113. Dietrich A, Audiffren M (2011) The reticular-activating hypofrontality (RAH) model of acute exercise. Neurosci Biobehav Rev 35:1305–1325
114. Elliott MT, Welchman AE, Wing AM (2009) Being discrete helps keep to the beat. Exp Brain Res 192:731–737
115. Torre K, Balasubramaniam R (2009) Two different processes for sensorimotor synchronization in continuous and discontinuous rhythmic movements. Exp Brain Res 199:157–166
116. Tranchant P, Vuvan DT, Peretz I (2016) Keeping the beat: a large sample study of bouncing and clapping to music. PLoS One 11:e0160178
117. Hyodo K, Kidokoro T, Yamaguchi D et al (2023) Feasibility, safety, enjoyment, and system usability of web-based aerobic dance exercise program in older adults: single-arm pilot study. JMIR Aging 6:e39898
118. Hyodo K, Takashi J, Suwabe K, Soya H, Nagamatsu T (2019) Acute effects of light-intensity, slow-tempo aerobic dance exercise on mood and executive function in older adults. Bull Phys Fit Res Inst 117:8–16
119. MacKay-Lyons M (2002) Central pattern generation of locomotion: a review of the evidence. Phys Ther 82:69–83

120. PERETZ I (2001) Brain specialization for music new evidence from congenital amusia. Ann N Y Acad Sci 930:153–165
121. Bella SD, Berkowska M, Sowiński J (2011) Disorders of pitch production in tone deafness. Front Psychol 2:1–11
122. Peretz I, Champod AS, Hyde K (2003) Varieties of musical disorders: the montreal battery of evaluation of amusia. Ann N Y Acad Sci 999:58–75
123. Lévêque Y, Teyssier P, Bouchet P, Bigand E, Caclin A, Tillmann B (2018) Musical emotions in congenital amusia: impaired recognition, but preserved emotional intensity. Neuropsychology 32:880–894
124. Gosselin N, Paquette S, Peretz I (2015) Sensitivity to musical emotions in congenital amusia. Cortex 71:171–182
125. Phillips-Silver J, Toiviainen P, Gosselin N, Piché O, Nozaradan S, Palmer C, Peretz I (2011) Born to dance but beat deaf: a new form of congenital amusia. Neuropsychologia 49:961–969
126. Fujii S, Schlaug G (2013) The Harvard beat assessment test (H-BAT): a battery for assessing beat perception and production and their dissociation. Front Hum Neurosci 7:1–16
127. Iversen JR, Patel AD (2008) The beat alignment test (BAT): surveying beat processing abilities in the general population. In: Proceedings of the 10th International Conference Music Perception and Cognition, Sapporo, pp 465–468
128. Grahn JA, Rowe JB (2009) Feeling the beat: premotor and striatal interactions in musicians and nonmusicians during beat perception. J Neurosci 29:7540–7548
129. Noda Y, Matsushita K (2023) Glutamatergic neurometabolite levels in the caudate are associated with the ability of rhythm production. 1–7
130. Leow L-A, Parrott T, Grahn JA (2014) Individual differences in beat perception affect gait responses to low- and high-groove music. Front Hum Neurosci 8:811
131. Karpati FJ, Giacosa C, Foster NEV, Penhune VB, Hyde KL (2016) Sensorimotor integration is enhanced in dancers and musicians. Exp Brain Res 234:893–903
132. Nguyen T, Sidhu RK, Everling JC, Wickett MC, Gibbings A, Grahn JA (2022) Beat perception and production in musicians and dancers. Music Percept 39:229–248
133. Mas-Herrero E, Marco-Pallares J, Lorenzo-Seva U, Zatorre RJ, Rodriguez-Fornells A (2013) Individual differences in music reward experiences. Music Percept 31:118–138
134. Mas-Herrero E, Zatorre RJ, Rodriguez-Fornells A, Marco-Pallarés J (2014) Dissociation between musical and monetary reward responses in specific musical anhedonia. Curr Biol 24:699–704
135. Martínez-Molina N, Mas-Herrero E, Rodríguez-Fornells A, Zatorre RJ, Marco-Pallarés J (2016) Neural correlates of specific musical anhedonia. Proc Natl Acad Sci USA 113:E7337–E7345
136. Martínez-Molina N, Mas-Herrero E, Rodríguez-Fornells A, Zatorre RJ, Marco-Pallarés J (2019) White matter microstructure reflects individual differences in music reward sensitivity. J Neurosci 39:5018–5027
137. Miura A, Kudo K, Ohtsuki T, Kanehisa H (2011) Coordination modes in sensorimotor synchronization of whole-body movement: A study of street dancers and non-dancers. Hum Mov Sci 30:1260–1271
138. Fuentes-Sánchez N, Pastor MC, Eerola T, Pastor R (2023) Individual differences in music reward sensitivity influence the perception of emotions represented by music. Music Sci 27:313–331
139. Salimpoor VN, Zald DH, Zatorre RJ, Dagher A, McIntosh AR (2015) Predictions and the brain: how musical sounds become rewarding. Trends Cogn Sci 19:86–91
140. Iversen JR, Patel AD, Ohgushi K (2008) Perception of rhythmic grouping depends on auditory experience. J Acoust Soc Am 124:2263–2271
141. Stevens CJ (2012) Music perception and cognition: a review of recent cross-cultural research. Top Cogn Sci 4:653–667
142. O'Connell SR, Nave-Blodgett JE, Wilson GE, Hannon EE, Snyder JS (2022) Elements of musical and dance sophistication predict musical groove perception. Front Psychol 13:1–16

143. Pereira CS, Teixeira J, Figueiredo P, Xavier J, Castro SL, Brattico E (2011) Music and emotions in the brain: familiarity matters. PLoS One 6:e27241
144. Leow LA, Rinchon C, Grahn J (2015) Familiarity with music increases walking speed in rhythmic auditory cuing. Ann N Y Acad Sci. 1337:53-61
145. Stupacher J, Hove MJ, Novembre G, Schütz-Bosbach S, Keller PE (2013) Musical groove modulates motor cortex excitability: a TMS investigation. Brain Cogn 82:127–136
146. Barrett KC, Ashley R, Strait DL, Kraus N (2013) Art and science: how musical training shapes the brain. Front Psychol 4:1–13
147. Miendlarzewska EA, Trost WJ, Sammler D, Planck M (2014) How musical training affects cognitive development: rhythm, reward and other modulating variables. Front Neurosci 7:1–18
148. Karpati FJ, Giacosa C, Foster NEV, Penhune VB, Hyde KL (2015) Dance and the brain: a review. Ann N Y Acad Sci 1337:140–146
149. Boere K, Lloyd K, Binsted G, Krigolson OE (2023) Exercising is good for the brain but exercising outside is potentially better. Sci Rep 13:1–8
150. Mavrantza AM, Bigliassi M, Calogiuri G (2023) Psychophysiological mechanisms underlying the effects of outdoor green and virtual green exercise during self-paced walking. Int J Psychophysiol 184:39–50
151. Hashitomi T, Hoshi D, Tarumi T, Sugawara J, Watanabe K (2023) Effect of aquatic walking on prefrontal activity and executive function in healthy middle- and old-aged adults: a pilot study. J Phys Fit Sport Med 12:59–67
152. Zeng N, Pope Z, Lee JE, Gao Z (2018) Virtual reality exercise for anxiety and depression: a preliminary review of current research in an emerging field. J Clin Med 7:1–13
153. Ochi G, Kuwamizu R, Fujimoto T, Ikarashi K, Yamashiro K, Sato D (2022) The effects of acute virtual reality exergaming on mood and executive function: exploratory crossover trial. JMIR Serious Games 10:1–13
154. Bird JM, Karageorghis CI, Baker SJ, Brookes DA (2019) Effects of music, video, and 360-degree video on cycle ergometer exercise at the ventilatory threshold. Scand J Med Sci Sports 29:1161–1173
155. Nagata K, Tsunoda K, Fujii Y, Jindo T, Okura T (2023) Impact of exercising alone and exercising with others on the risk of cognitive impairment among older Japanese adults. Arch Gerontol Geriatr 107:104908
156. Kanamori S, Takamiya T, Inoue S, Kai Y, Tsuji T, Kondo K (2018) Frequency and pattern of exercise and depression after two years in older Japanese adults: the JAGES longitudinal study. Sci Rep 8:1–9
157. Takahashi S, Grove PM (2019) Comparison of the effects of running and badminton on executive function: a within-subjects design. PLoS One 14:1–11
158. Mak HW, Noguchi T, Bone JK, Wels J, Gao Q, Kondo K, Saito T, Fancourt D (2023) Hobby engagement and mental wellbeing among people aged 65 years and older in 16 countries. Nat Med 29:2233–2240

Application of Minimum Exercise Model to the Hypoxic Environment

Genta Ochi, Yuhki Yamada, and Hideaki Soya

Abstract While moderate exercise has been demonstrated to enhance executive function, this beneficial effect may vary depending on the exercise environment. For instance, the decline in blood oxygen levels (hypoxemia) associated with ascent to high altitude has been shown not only to induce acute mountain sickness but also to potentially cause decreased cognitive performance. Therefore, exercise under hypoxic conditions may reduce oxygen delivery to various tissues, thereby attenuating the executive function-enhancing effects of exercise. Previous studies have examined the impact of exercise in hypoxic environments on cognitive function using cognitive task paradigms; however, a consensus has not been reached. One contributing factor to this lack of consensus is the insufficient investigation of how exercise in hypoxic environments affects neural activity in brain regions specific to cognitive function tasks. This limitation stems from the practical difficulties of utilizing positron emission tomography (PET) and magnetic resonance imaging (MRI) systems in hypoxic environments. We addressed these challenges by employing functional near-infrared spectroscopy (fNIRS), which requires only a compact experimental system, is portable, and can be readily installed in gym settings. Our

G. Ochi (✉)
Faculty of Health Sciences, Department of Health and Sports, Niigata University of Health and Welfare, Niigata, Niigata, Japan
e-mail: ochi@nuhw.ac.jp

Y. Yamada
Laboratory of Exercise Biochemistry and Neuroendocrinology, Institute of Health and Sport Sciences, University of Tsukuba, Tsukuba, Ibaraki, Japan

H. Soya
Laboratory of Exercise Biochemistry and Neuroendocrinology, Institute of Health and Sport Sciences, University of Tsukuba, Tsukuba, Ibaraki, Japan

Sport Neuroscience Division, Kokoro Division, Advanced Research Initiative for Human High Performance (ARIHHP), Institute of Health and Sport Sciences, University of Tsukuba, Tsukuba, Ibaraki, Japan

Center for Cybernics Research, University of Tsukuba, Tsukuba, Ibaraki, Japan

H. Soya (ed.), *Exercise Brain Stimulation for Cognitive Function and Mental Health*, Advances in Neurobiology 44, https://doi.org/10.1007/978-981-95-0066-6_20

findings revealed that exercise in hypoxic environments induces decreasing cognitive performance, specifically cognitive fatigue, by reducing task-specific neural activity. This chapter provides an overview of our research methodology and results.

Keywords Cognitive fatigue · Hypoxic exercise · fNIRS · DLPFC · Executive function

1 Introduction

The effects of acute hypoxia on brain function, particularly the impact of altitude exposure on cerebral function, represent a critical concern for maintaining safe activities during recreational and occupational exposure to high-altitude environments, including skiing, trekking, mountaineering, mining, mountain guiding, and aviation. Several studies have demonstrated that acute exercise improves executive function in the lateral prefrontal cortex [1–4]. However, in hypoxic environments, exercise may adversely affect not only physical performance [5–7] but also executive function [8]. Executive function is associated with motor performance, including motor coordination [9] and behavioral inhibition and decision-making [10, 11], and impairments may lead not only to decreased physical performance but also to accidents. Nevertheless, the fundamental physiological and neurological mechanisms underlying decreased cognitive performance (cognitive fatigue) under such conditions remain unclear.

Cerebral hypoxia induced by hypoxemia (decreased arterial oxygen saturation, SpO_2) may represent a potential mechanism for cognitive fatigue. SpO_2 demonstrates a strong positive correlation with prefrontal cortex oxygen saturation [12, 13], and reduced SpO_2 during exercise may induce cerebral hypoxia [14]. These systemic reductions in oxygen levels affect cerebral oxygen dynamics even during light to moderate-intensity exercise under moderate hypoxic conditions [14, 15]. Cerebral hypoxia is considered a contributing factor to central fatigue by limiting motor execution through suppression of motor cortex activation [16, 17]. Interestingly, 2 weeks of acclimatization to hypoxic environments can suppress hypoxemia and cerebral hypoxia during exercise under hypoxic conditions, reducing the central drive from the motor cortex (central fatigue) [18]. Thus, hypoxemia characterized by decreased SpO_2 may constitute a critical factor in brain function impairment and may contribute to executive function decline during exercise in hypoxic environments. However, the specific neural substrates of executive function decline during hypoxic exercise remain unclear. To elucidate the neural basis of executive function impairment due to exercise in hypoxic environments, it is necessary to establish methods for evaluating neural activity following exercise under such conditions. This chapter provides an overview of hypoxic environment exposure and exercise models and clarifies the effects of exercise in hypoxic environments on executive function and its neural substrates.

2 Effects of Acute Exposure to the Hypoxic Environment on Executive Function

To clarify the combined effects of hypoxic environments and exercise, it is necessary to establish the impact of acute exposure to hypoxic environments alone on executive function. Traditionally, the effects of hypoxic environment exposure have been considered to arise from hypoxemia, as hypoxemia is associated with the induction of acute mountain sickness, including symptoms such as headache, nausea, and insomnia [19]. Furthermore, neuropsychological adverse effects may be induced, particularly executive function impairment that affects the prefrontal cortex [20–23]. Multiple studies have demonstrated that when SpO_2 falls below 80% (corresponding to altitudes exceeding 5000 m with inspired oxygen fraction (FiO_2) = 0.100), cognitive task performance is significantly affected [24, 25]. Although cognitive performance is generally considered to decline markedly in a severity-dependent manner under hypoxic conditions, direct empirical evidence examining the relationship between blood oxygen levels and executive function decline does not exist. This gap exists because while there are many studies conducting meta-analyses on the effects of hypoxic environment exposure on executive function, there are few studies that have examined executive function tasks with well-established neural correlates under hypoxic environments with different oxygen levels to determine whether cognitive performance declines in a severity-dependent manner with hypoxic condition severity. Therefore, we decided to address this issue using the Color-Word Stroop Task (CWST), which has been used for many years as an executive function assessment task.

The CWST is widely used in experimental and clinical settings to measure prefrontal cortex-based executive function [26, 27] and has frequently been employed in previous studies using hypoxic conditions [28–30]. Behavioral performance in the CWST is typically measured by comparing the delay in reaction time (RT) that occurs when subjects name colors in incongruent trials (e.g., the word "yellow" printed in green ink) versus neutral trials (e.g., "XXXX" printed in green ink). This time delay is called Stroop interference. We used the CWST to assess executive function by measuring Stroop interference. We hypothesized that exposure to hypoxic conditions would increase Stroop interference, manifesting as slower reaction times reflecting executive function impairment.

We examined whether exposing subjects to acute normobaric hypoxic conditions (FiO_2 = 0.165, 0.135, 0.105; corresponding to altitudes of 2000 m, 3500 m, and 5000 m, respectively) would impair CWST performance, and if performance declined, whether the decline in CWST performance was associated with SpO_2 (Fig. 1). The results revealed that under severe hypoxic conditions, Stroop interference in RT increased compared to normoxic, mild, and moderate hypoxic conditions. These findings demonstrate that severe hypoxic environments equivalent to 5000 m altitude impair executive function regardless of the presence or absence of exercise, suggesting that this effect may occur in a manner dependent on the decline in SpO_2 levels.

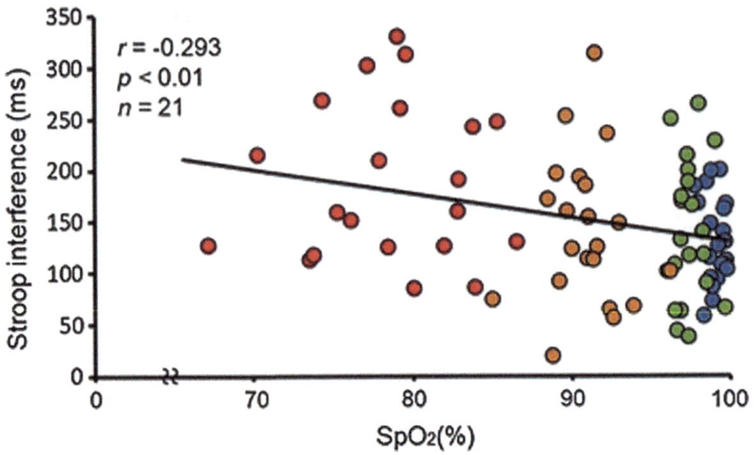

Fig. 1 Relationship between arterial blood oxygen saturation and Stroop interference. When CWST was performed during exposure to four different concentrations of inspired oxygen, Stroop interference, a measure of executive function, was significantly delayed under severe hypoxia equivalent to 5000 m elevation. Furthermore, a significant negative correlation was found between SpO_2 and Stroop interference, suggesting that reduced arterial blood oxygen levels may be associated with reduced executive function. (This figure is taken from [31])

Furthermore, there was a significant negative correlation between Stroop interference in RT and SpO_2. These results suggest that executive function decreases in hypoxic environments in a manner dependent on reduced SpO_2. Exercise in a hypoxic environment results in a significant drop in SpO_2. This is due to increased oxygen demand in skeletal muscles and the brain compared to resting conditions. Therefore, even in a moderately hypoxic environment where no executive function decline occurs at rest, it is conceivable that exercise may cause a significant drop in SpO_2 and a decline in executive function.

3 Neural Function Assessment Method Using fNIRS in the Hypoxic Environment

To elucidate the neural mechanisms underlying executive dysfunction, researchers employ techniques for non-invasively evaluating neural activity in the human brain. In numerous prior studies, PET, MRI, and fNIRS have been utilized to assess neural activity in the human brain non-invasively. fNIRS, in particular, offers the advantage of radiation-free measurement and enables recordings in natural, stress-free postures such as seated positions. Consequently, our laboratory employs this technology to investigate the neural mechanisms through which exercise modulates executive function. fNIRS can evaluate the tight coupling between neural activity and local cerebral blood flow (neurovascular coupling), using changes in hemodynamic responses during task performance as an indirect indicator of cortical activation [32]. Therefore, we employ this approach to assess the effects of exercise by implementing executive function tasks during and after exercise and measuring neural activity during task execution using fNIRS.

However, fNIRS is a technology that measures light transmission through the scalp surface, making it strongly influenced by factors such as skin blood flow, global cerebral blood flow, and baseline hemoglobin concentrations [33, 34]. In fMRI studies, it has been reported that despite similar levels of neural activity occurring under both normoxic and hypoxic conditions, there is a reduction in hemodynamic responses in the hypoxic environment compared to the normoxic environment [35]. These findings suggest that decreases in SpO_2 and prefrontal cortex oxygen saturation (rSO_2) attenuate local blood flow responses to neural activity, potentially complicating the evaluation of fNIRS data. Due to these technical challenges, the effects of exercise under hypoxic conditions on prefrontal cortex neural activity have remained unclear.

To verify the effects of exercise intervention under hypoxia on executive function, we deemed it necessary to confirm that conditions such as skin blood flow, whole-brain blood flow, and hemoglobin concentration during the task remained unchanged before and after the exercise intervention. Therefore, our first step was to identify the time required for physiological parameters altered by exercise under hypoxia to return to resting levels. We aimed to create an experimental protocol that would allow us to assess neural activity during the task using fNIRS both before and after exercise. To measure the neural basis of hypoxic exercise-induced executive function impairment using fNIRS, we performed 10 minutes of moderate-intensity exercise in a hypoxic environment equivalent to an elevation of 3,500 m. We examined the time required for the mean velocity of the middle cerebral artery (MCA V_{mean}), skin blood flow (SBF), SpO_2, rSO_2, end-tidal carbon dioxide ($ETCO_2$), and heart rate to recover to resting levels, as these factors are known to affect fNIRS data.

Previous research demonstrated that the MCA V_{mean} peaked 3–4 minutes into exercise at low- (30% VO_2peak), moderate- (50% VO_2peak), and high-intensity (70% VO_2peak) exercise and then decreased throughout the exercise [36]. However, this effect was not observed in the hypoxic environment. One possible reason for this is the involvement of a decrease in blood carbon dioxide concentration due to increased ventilation during exercise. Indeed, $ETCO_2$, which reflects the arterial blood carbon dioxide concentration that typically increases during moderate-intensity exercise in normoxic environments, was not found to increase in hypoxic environments. Blood carbon dioxide concentration is involved in vascular contraction and dilation, and cerebral blood flow is reduced as the brain undergoes vasoconstriction when $ETCO_2$ is low [37, 38]. Therefore, during hypoxic exercise, increased ventilation may have limited the rise in blood carbon dioxide concentration, thereby attenuating the expected increase in cerebral blood flow during exercise. Figure 2 shows the effects of hypoxic exercise on physiological parameters that can generally affect fNIRS data. SBF increased at the end of exercise and decreased quickly after the end of exercise. Heart rate also decreased quickly after exercise and was within 10% of the resting level within 5 minutes. SpO_2 and rSO_2 decreased immediately after the start of exercise and recovered to resting levels

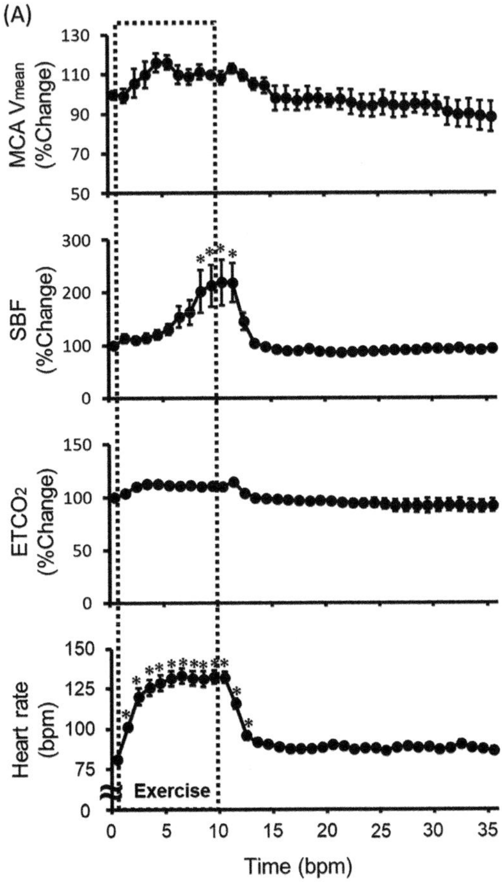

Fig. 2 Physiological parameters including MCA V_{mean}, SBF, $ETCO_2$, and HR. Time course of physiological measures during rest, exercise, and recovery phases. Data points represent group averages for each physiological variable across measurement periods. Standard deviations are shown as error bars. Asterisks denote time points showing statistically significant differences from baseline values ($P < 0.05$, Dunnett's test). MCA V_{mean} middle cerebral artery blood velocity, SBF skin blood flow, HR: heart rate, $ETCO_2$ end-tidal carbon dioxide

soon after the end of exercise. These results are consistent with a previous study [39]. These results confirmed that even in a hypoxic environment at an altitude equivalent to 3,500 m, physiological parameters affecting fNIRS returned to pre-exercise baseline levels 15 minutes after the end of exercise, and that factors affecting fNIRS measurements could be excluded.

The decline in SpO_2 and reduction in cerebral blood flow (CBF) induce cerebral hypoxia, which serves as a major contributing factor to the deterioration of executive function. SpO_2 levels have been shown to exhibit a positive correlation with prefrontal cortex oxygenation [12, 13, 40], and our data confirmed that rSO_2 levels decreased when SpO_2 levels declined during exercise under hypoxic conditions. In addition to decreased arterial oxygen saturation, hyperventilation-induced cerebral vasoconstriction may affect prefrontal cortex oxygenation through the limitation of CBF [41, 42]. While moderate-intensity exercise in normoxic conditions increases end-tidal CO_2 partial pressure ($PEtCO_2$) associated with changes in CBF [43], $PEtCO_2$ did not increase during moderate-intensity exercise under hypoxic conditions, and no increase in CBF was observed. These findings suggest that during moderate-intensity exercise in hypoxic environments, the decline in

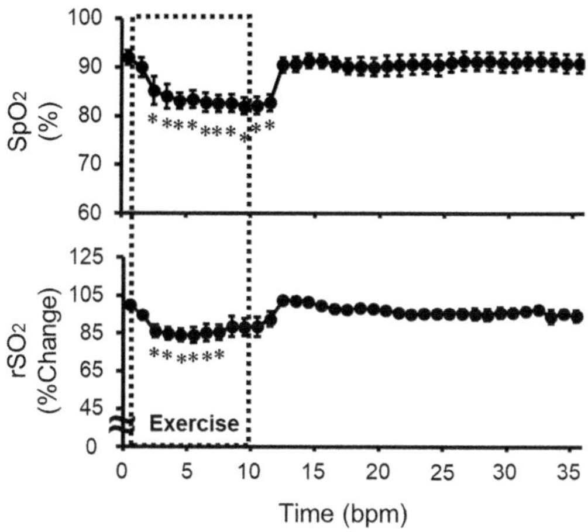

Fig. 3 Physiological parameters, including SpO_2 and rSO_2. Time course of physiological measures across rest, exercise, and recovery phases of an acute exercise session. Group averages for each physiological variable are displayed at all measurement time points. Standard deviations are represented by error bars. Asterisks indicate measurement points showing statistically significant differences from baseline values ($P < 0.05$, Dunnett's test). SpO_2 percutaneous arterial oxygen saturation, rSO_2 regional cerebral oxygen saturation measured at baseline, throughout, and following 10 minutes of moderate-intensity exercise (50% peak oxygen uptake) under hypoxic conditions ($FiO_2 = 0.135$)

SpO_2, rSO_2levels and reduction in CBF may induce cerebral hypoxia, potentially contributing to cognitive fatigue (Fig. 3).

4 Effects of Exercise in a Hypoxic Environment on Executive Function

From the previous chapters, it is evident that executive function is impaired when SpO_2 is significantly reduced. Even in a moderately hypoxic environment below 3500 m elevation, where no decrease in executive function occurred at rest, SpO_2 was found to decrease significantly during exercise. This suggests that the combination of moderate hypoxia and exercise may reduce executive function. Furthermore, 15 minutes after hypoxic exercise, physiological parameters that may affect fNIRS were found to recover to pre-exercise levels. These results suggest that exercise in a moderate hypoxic environment (equivalent to 3500 m) impairs executive function, and that the neural basis of this impairment can be verified using fNIRS 15 minutes after the end of the exercise.

The primary objective of this study was to determine whether exercise under hypoxic conditions impairs executive function. On the initial visit, participants underwent graded exercise testing to assess their VO_2peak. Following the American College of Sports Medicine's physical activity intensity classification, individual exercise intensities were determined to correspond to moderate-intensity exercise, defined as 50% of participants' VO_2peak. Several days after the initial visit, participants completed two experimental conditions: exercise under normoxic (NO) conditions and exercise under normobaric hypoxic (HY) conditions. In both conditions, participants performed 10 minutes of moderate-intensity exercise using a recumbent ergometer. Pedaling cadence was fixed at 60 revolutions per minute, synchronized with metronome beats. The CWST described above was administered before exercise and 15 minutes post-exercise. The timing of post-exercise CWST administration was based on previous findings indicating that physiological parameters of non-cortical origin, such as MCA V_{mean} and forehead SBF, return to resting levels within 15 minutes following moderate-intensity exercise under hypoxic conditions. In both NO and HY conditions, participants demonstrated reduced reaction times and lower error rates during neutral trials compared to incongruent trials, confirming that Stroop interference was successfully induced during all CWST administrations. Using this Stroop interference as an indicator of executive function, we first compared the effects of acute moderate-intensity exercise on executive function under hypoxic versus normoxic conditions. Results demonstrated that 10 minutes of moderate-intensity exercise under normoxic conditions reduced Stroop interference, thereby improving executive function. This finding aligns with previous research demonstrating that moderate-intensity exercise enhances inhibitory control function [1]. Conversely, 10 minutes of moderate-intensity exercise under hypoxic conditions eliminated this Stroop interference reduction effect. Since hypoxic exposure alone did not influence Stroop interference (Fig. 4), this finding likely represents a combined effect of hypoxic conditions and moderate-intensity exercise. These results suggest that acute moderate-intensity exercise under hypoxic conditions may induce impairment in executive function.

Furthermore, we examined whether moderate-intensity exercise in hypoxic environments would increase Stroop interference by comparing a 10-minute moderate-intensity exercise condition with a rest condition under hypoxic exposure. Participants engaged in two primary conditions under hypoxic exposure: CTL (control) or EX (exercise), which were implemented on separate days using a counterbalanced design. In the EX condition, similar to the HY condition, participants performed moderate-intensity exercise on a cycle ergometer at 60 rpm under moderate hypoxic conditions ($FiO_2 = 0.135$). In the CTL condition, participants maintained rest for 10 minutes under the same hypoxic environment as the EX condition. The CWST was administered before and 15 minutes after the exercise or rest period. During CWST performance, cortical hemodynamic changes in regions of interest within the lateral prefrontal cortex (dorsolateral: DLPFC; ventrolateral: VLPFC; frontal pole areas: FPA) were monitored using fNIRS.

The results demonstrated that the EX condition produced increased Stroop interference following exercise compared to the CTL condition, revealing that moderate-intensity exercise in hypoxic environments exerts negative effects on executive function (Fig. 5). These findings replicated the results from previous experiments, indicating that the experimental model employed in this study reliably demonstrates executive function impairment.

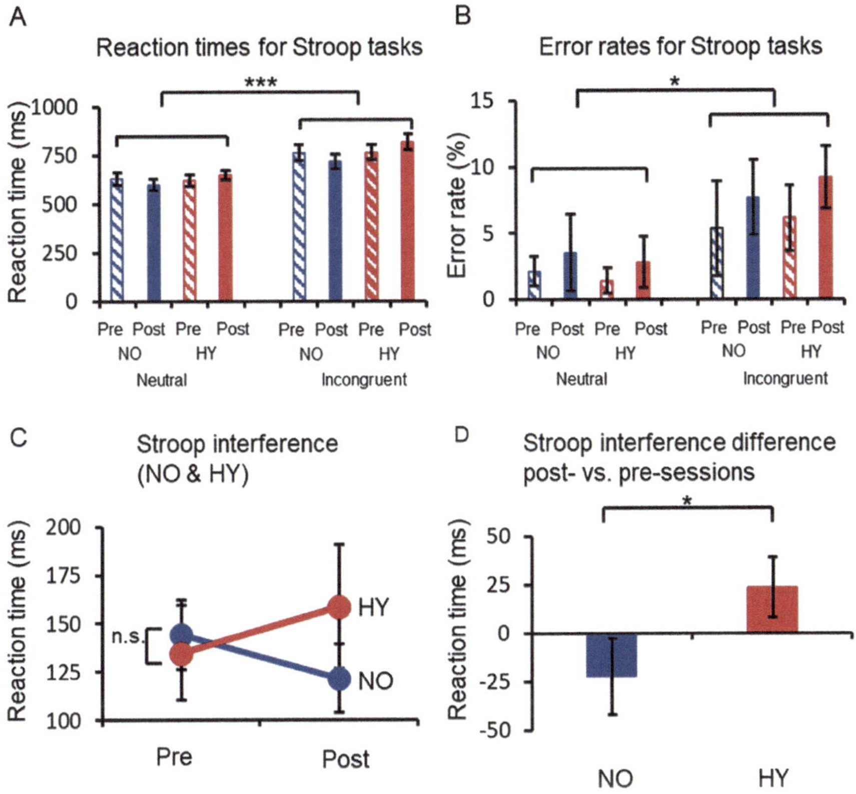

Fig. 4 Effects of exercise in normoxic and hypoxic environments on executive function. (**a**) Reaction time (RT) across incongruent versus neutral task conditions. Incongruent trials demonstrated markedly prolonged response times (***$P < 0.001$). (**b**) Error frequency comparison for incongruent versus neutral task conditions. Clear Stroop interference patterns were evident (*$P < 0.05$). (**c**) Average response time differential between incongruent and neutral trials representing Stroop interference magnitude for each experimental condition. (**d**) Changes in Stroop interference (response time) from pre- to post-exercise sessions across conditions. The hypoxic condition showed significantly greater increases in interference compared to the normoxic condition ($P < 0.05$). Standard errors are displayed as error bars. (This figure is adapted from [44])

Next, we examined the neural mechanisms underlying exercise-induced executive function decline observed under hypoxic conditions. During the CWST administered before exercise, no significant increase in oxy-Hb signals related to Stroop interference (incongruent-neutral condition contrast) was observed, although small effect sizes for signal increases were detected in the DLPFC (left: 0.16, right: 0.20), VLPFC (left: 0.58, right: 0.40), and FPA (left: 0.19, right: 0.32). Our recent findings demonstrate that CWST under normoxic conditions significantly increases activation in bilateral DLPFC, VLPFC, and FPA [3]. Animal fMRI studies have reported that mild hypoxic conditions reduce hemodynamic responses to electrical stimulation of the forelimb, while electroencephalographic (EEG) responses remain

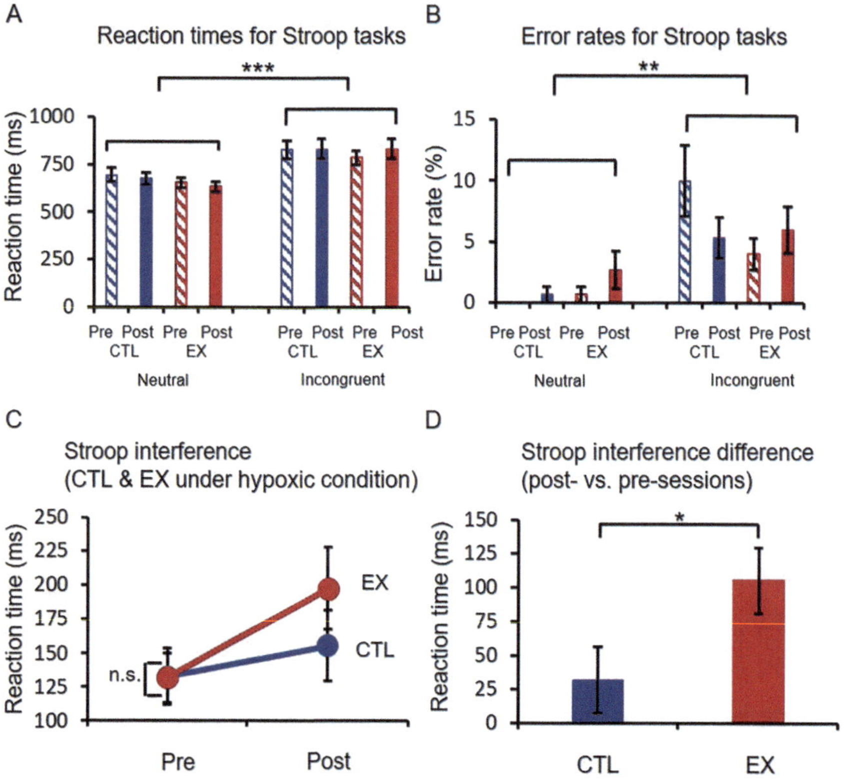

Fig. 5 Effects of exercise and rest in a hypoxic environment on executive function. (**a**) Reaction time (RT) analysis across incongruent versus neutral task conditions. Incongruent trials showed markedly prolonged RT (***$P < 0.001$). (**b**) Error frequency analysis for incongruent versus neutral task trials. Clear Stroop interference patterns emerged (**$P < 0.01$). (**c**) Average RT differential between incongruent and neutral conditions representing Stroop interference magnitude across experimental conditions. (**d**) Pre-to-post session changes in Stroop interference for each experimental condition. The exercise (EX) condition demonstrated significantly greater interference increases compared to the control (CTL) condition ($P < 0.05$). Standard errors are shown as error bars. (This figure is adapted from [44])

unchanged compared to normoxic conditions [35]. In our preliminary experiments, we confirmed that prefrontal cortex activation related to Stroop interference was reduced under hypoxic conditions compared to normoxic conditions across DLPFC, VLPFC, and FPA, despite no changes in CWST performance. This suggests that while neural activity in response to CWST may increase under hypoxic environments, oxy-Hb signals may be detected to a lesser extent in hypoxic conditions.

In subsequent analyses, we examined brain regions activated after exercise. Since exercise under hypoxic conditions prolonged Stroop interference RT, we evaluated prefrontal cortical activation recruited for Stroop interference. Results showed that only left DLPFC activation was significantly decreased, which was consistent

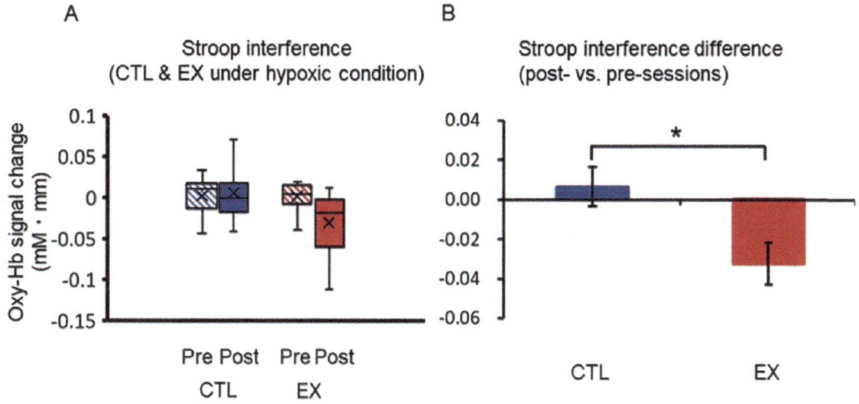

Fig. 6 Effect of exercise in a hypoxic environment on task-specific left DLPFC activity. (**a**) Box plot distributions showing Stroop interference in oxy-Hb signal across rest (CTL) and exercise (EX) experimental conditions. Box boundaries represent first and third quartiles (bottom and top, respectively). Whisker extensions indicate maximum data points within 1.5 interquartile ranges above the upper quartile and minimum data points within 1.5 interquartile ranges below the lower quartile. Median values are displayed as bands within each box. Mean values are marked with x symbols. (**b**) Post-exercise minus pre-exercise oxy-Hb signal changes in the EX condition show significantly reduced values compared to the CTL condition ($P < 0.05$). Standard errors are represented by error bars. (This figure is adapted from [44])

with the RT prolongation. These findings provide experimental evidence that the left DLPFC represents the neural basis for executive function decline following acute moderate-intensity exercise under hypoxic conditions (Fig. 6).

During hypoxemia, CBF increases to maintain brain activity [45–47]. Under moderate hypoxic conditions ($FiO_2 = 0.135$), the absence of changes in CWST performance suggests that executive function was preserved through increased CBF, even when SpO_2 levels declined to approximately 90%. Conversely, during moderate-intensity exercise in hypoxic environments, SpO_2 levels decreased below 80%, and executive function impairment was observed following exercise. Given that SpO_2 levels recovered to pre-exercise baseline levels during post-exercise CWST testing, the severe SpO_2 reduction during exercise may have contributed to the subsequent executive function decline.

Since SpO_2 and prefrontal cortex oxygenation exhibit a positive correlation [12], cerebral oxygenation during exercise under hypoxic conditions may be lower than during rest under identical conditions. One potential explanation for this phenomenon involves competition for blood flow between skeletal muscle and the brain during exercise in hypoxic environments [48]. This competition may result in reduced oxygen delivery to the prefrontal cortex, potentially inducing cerebral hypoxia. Furthermore, the CBF increase observed in hypoxic environments demonstrates greater magnitude in basal ganglia and diencephalic structures (including the nucleus accumbens, putamen, globus pallidus, caudate nucleus, and thalamus) compared to the cerebral cortex, presumably to maintain essential homeostatic functions even at the expense of cognitive performance [49]. During severe SpO_2 reduction, regional disparities in CBF increases may emerge, resulting in insufficient oxygen delivery to the prefrontal cortex and consequent executive function impairment.

The persistence of executive function decline despite post-exercise SpO_2 recovery remains a subject of debate. Previous research has demonstrated that moderate-intensity exercise under normobaric hypoxic conditions ($FiO_2 = 0.10$; equivalent to approximately 5500 m altitude) reduces cerebral metabolic rate of oxygen consumption ($CMRO_2$) [50], and that $CMRO_2$ reduction induced by exposure to normobaric hypoxic conditions ($FiO_2 = 0.08$; equivalent to approximately 7500 m altitude) fails to return to baseline following reoxygenation [51]. These findings suggest that the reduced prefrontal cortex activation resulting from severe hypoxemia during exercise under hypoxic conditions may persist beyond the exercise period.

5 Physiological Mechanisms of Hypoxic Exercise-Induced Decrease in Executive Function

As previously mentioned, SpO_2 decreased to 80% during moderate-intensity exercise in hypoxic environments, and this exercise-induced SpO_2 decline was suggested to potentially contribute to post-exercise executive function impairment. The SpO_2 reduction during hypoxic exercise is thought to decrease oxygen supply to the brain, thereby reducing neural activity in the motor cortex following exercise [16, 50]. Two weeks of exposure to hypoxic environments promotes adaptation to such conditions, and hypoxic states have the effect of mitigating exercise-induced SpO_2 decline, which improves cerebral hypoxia and ameliorates the reduction in central commands from the motor cortex (central fatigue) [18]. Prevention of exercise-induced hypoxemia through acclimatization to hypoxic environments may also be effective for executive function decline; however, whether improvements in exercise-induced SpO_2 reduction are related to post-exercise executive function decline remains insufficiently investigated.

Therefore, to clarify whether exercise-induced executive function decline in hypoxic environments is caused by SpO_2 reduction during exercise, we employed a condition in which participants inhaled hypoxic gas adjusted to maintain SpO_2 at resting levels during exercise (mild hypoxic exercise [ME]) and examined the effects of exercise-induced SpO_2 decline on executive performance and activation of the dorsolateral prefrontal cortex, which is associated with such performance. Specifically, in the HE condition involving inhalation of normobaric hypoxic gas ($FiO_2 = 0.135$), participants inhaled hypoxic gas (13.5% oxygen and 0.03% carbon dioxide mixed in nitrogen, $FiO_2 = 0.135$, equivalent to approximately 3500 m altitude) through a mask connected to a Douglas bag, similar to our previous study [44]. In the ME condition, participants inhaled moderate hypoxic gas during the CWST, but during exercise only, they inhaled milder hypoxic gas ($FiO_2 = 0.161 \pm 0.018$) to prevent SpO_2 from declining below resting levels. To adjust oxygen concentration, oxygen was directly injected into the hose while monitoring SpO_2 during exercise. Similar to [44], both conditions involved moderate-intensity cycling exercise, with CWST and neural activity of the left DLPFC evaluated using fNIRS before and after exercise.

Consequently, in the HE condition, the reaction time for Stroop interference in executive function was delayed post-exercise compared to pre-exercise, demonstrating results consistent with previous studies [44]. Conversely, in the ME condition, no delay in Stroop interference was observed. Subsequently, we focused on the decreased left DLPFC activity corresponding to the increased Stroop interference. As a result, reduced neural activity in the left DLPFC was observed post-exercise in the HE condition, which was similar to effects reported in previous research. In contrast, the ME condition showed an attenuated reduction in neural activation post-exercise. These findings suggest that the prevention of behavioral performance deterioration post-exercise in the ME condition may be associated with the suppression of task-specific activity reduction in the left DLPFC. This supports the hypothesis that exercise-induced cognitive fatigue under hypoxic conditions is attributable to exercise-induced hypoxemia and suggests that amelioration of exercise-induced hypoxemia may be beneficial for preventing cognitive fatigue.

During CWST implementation, SpO_2, minute ventilation (VE), and $ETCO_2$, which are involved in cerebrovascular dilation, showed no significant differences between pre- and post-exercise or between conditions. This finding suggests that the negative effects on CWST performance and left DLPFC activity are related to changes in physiological parameters during exercise rather than changes in physiological parameters during the task. Given that CWST performance is impaired in severe hypoxic environments where SpO_2 falls below 80% [31], the executive function impairment in the present study is suggested to be attributable to low SpO_2 during exercise. Previous research has reported that hypoxemia during exercise in hypoxic environments is associated with decreased neural activity in the motor cortex and reduced exercise performance [18]. Additionally, moderate exercise under normobaric hypoxic conditions ($FiO_2 = 0.10$) caused a reduction in $CMRO_2$ [17], and the decrease in $CMRO_2$ due to exposure to normobaric hypoxic conditions ($FiO_2 = 0.08$) did not return to baseline even after reoxygenation [51]. The present study supports these research findings and demonstrates that the effects of severe hypoxemia-induced prefrontal cortex activity reduction may persist even after blood oxygen levels have improved.

During exercise in hypoxic environments, increased cardiac output shortens red blood cell transit time within pulmonary capillaries, resulting in ventilation-perfusion mismatch and diffusion limitation, which leads to decreased SpO_2 [52]. Although the decline in SpO_2 restricts oxygen delivery to the brain, it is considered that neural activity is maintained through increased cerebral blood flow [53–56]. However, this vascular response does not occur uniformly throughout the brain. Paleocortical regions exhibit greater changes than neocortical regions, suggesting that cerebral blood flow changes during SpO_2 decline may serve to protect brain regions that play crucial roles in homeostatic maintenance [49]. The mechanisms underlying the decline in neural activity under cerebral hypoxia remain unclear, but one potential cause is neural inactivation related to inflammatory responses. Animal studies have demonstrated that hypoxia triggers inflammatory responses mediated by hypoxia-inducible factor and nuclear factor-κB [57], and the combination of hypoxia and inflammation has been reported to rapidly reduce neural excitability in hippocampal CA1 neurons [58]. Acute exercise also induces acute inflammatory markers and elevates inflammatory cytokine levels in the prefrontal cortex. Therefore, it is possible that the combination of hypoxia and inflammation in the human brain similarly reduces the excitability of prefrontal cortical neurons, thereby inducing cognitive fatigue, although detailed verification of this hypothesis is required in future studies.

6 Future Directions

We demonstrated that moderate-intensity exercise in hypoxic environments impairs executive function through reduced task-specific neural activity in the prefrontal cortex. This represents one of the few studies examining the role of the prefrontal cortex in executive function decline following hypoxic exposure in healthy individuals. Furthermore, we propose that severe hypoxemia during exercise serves as a physiological condition underlying this dysfunction. Prevention of exercise-induced hypoxemia, such as through hypoxic acclimatization, may be beneficial for mitigating exercise-induced executive function decline. However, the physiological mechanisms leading to post-exercise executive dysfunction in exercise-induced hypoxic states remain unclear. Given that prefrontal cortex oxygenation decreases when SpO_2 levels decline [12, 13, 40], we hypothesized that reduced SpO_2 levels are associated with executive function decline. However, our previous studies found no significant correlation between SpO_2 decline during exercise and executive function impairment, leaving the relationship between low SpO_2 and negative effects on executive function unclear. Since executive function decline occurred under severe conditions where SpO_2 fell below 80%, there may be a threshold level of hypoxemia at which executive function decline occurs, with compensatory mechanisms such as increased cerebral blood flow maintaining function until this threshold is reached. The fMRI and fNIRS techniques employed in this study face challenges in evaluating neural activity during exercise due to factors such as body movement and skin blood flow effects, making it difficult to elucidate the relationship between post-exercise executive function decline and exercise-induced cerebral hypoxia. Transcranial magnetic stimulation (TMS) has been utilized to evaluate motor cortex neural activity during exercise under hypoxic conditions. Assessment of prefrontal cortex neural activity and oxygen metabolism during hypoxic exercise using TMS is expected to contribute to elucidating the mechanisms of cognitive fatigue.

Individual differences between subjects represent an intriguing area for future research. Indeed, compared to non-climbers, elite climbers are known to exhibit smaller SpO_2 decreases in the quadriceps muscle under hypoxic conditions ($FiO_2 = 0.125$) [59]. Climbers may maintain better oxygen delivery to tissues during exercise in hypoxic environments compared to non-climbers, potentially preventing cognitive fatigue. Additionally, ventilatory responses to hypoxic environments are higher in women than in men [60], suggesting that women may possess more robust protective mechanisms against acute hypoxic exposure [61]. Cognitive fatigue from exercise under hypoxic conditions may also be reduced in women, though this hypothesis requires future validation. Furthermore, several previous studies have reported that acclimatization to hypoxic environments increases oxygen delivery to the brain during both rest [62] and exercise [18] in hypoxic environments compared to pre-acclimatization conditions, improving hypoxia-induced motor cortex activity decline. Based on this research, further studies examining whether acclimatization to hypoxic environments improves cognitive fatigue in climbers and trail runners who have adapted to hypoxic environments through hypoxic training could significantly contribute to elucidating the physiological mechanisms of exercise-induced executive function decline in hypoxic environments and developing countermeasures. This would lead to the realization of safe, high-performance activities at altitude.

References

1. Yanagisawa H, Dan I, Tsuzuki D, Kato M, Okamoto M, Kyutoku Y, Soya H (2010) Acute moderate exercise elicits increased dorsolateral prefrontal activation and improves cognitive performance with Stroop test. NeuroImage 50:1702–1710
2. Hyodo K, Dan I, Suwabe K, Kyutoku Y, Yamada Y, Akahori M, Byun K, Kato M, Soya H (2012) Acute moderate exercise enhances compensatory brain activation in older adults. Neurobiol Aging 33:2621–2632
3. Byun K, Hyodo K, Suwabe K, Ochi G, Sakairi Y, Kato M, Dan I, Soya H (2014) Positive effect of acute mild exercise on executive function via arousal-related prefrontal activations: an fNIRS study. NeuroImage 98:336–345
4. Kujach S, Byun K, Hyodo K, Suwabe K, Fukuie T, Laskowski R, Dan I, Soya H (2018) A transferable high-intensity intermittent exercise improves executive performance in association with dorsolateral prefrontal activation in young adults. NeuroImage 169:117–125
5. Chapman RF (2013) The individual response to training and competition at altitude. Br J Sports Med 47(Suppl 1):i40–i44
6. Fulco CS, Rock PB, Cymerman A (1998) Maximal and submaximal exercise performance at altitude. Aviat Space Environ Med 69:793–801
7. Wehrlin JP, Hallén J (2006) Linear decrease in.VO2max and performance with increasing altitude in endurance athletes. Eur J Appl Physiol 96:404–412
8. Lefferts WK, Babcock MC, Tiss MJ, Ives SJ, White CN, Brutsaert TD, Heffernan KS (2016) Effect of hypoxia on cerebrovascular and cognitive function during moderate intensity exercise. Physiol Behav 165:108–118
9. Piek JP, Dyck MJ, Nieman A, Anderson M, Hay D, Smith LM, McCoy M, Hallmayer J (2004) The relationship between motor coordination, executive functioning and attention in school aged children. Arch Clin Neuropsychol 19:1063–1076
10. Vestberg T, Gustafson R, Maurex L, Ingvar M, Petrovic P (2012) Executive functions predict the success of top-soccer players. PLoS One 7:e34731
11. Vestberg T, Reinebo G, Maurex L, Ingvar M, Petrovic P (2017) Core executive functions are associated with success in young elite soccer players. PLoS One 12:e0170845
12. Eichhorn L, Erdfelder F, Kessler F, Doerner J, Thudium MO, Meyer R, Ellerkmann RK (2015) Evaluation of near-infrared spectroscopy under apnea-dependent hypoxia in humans. J Clin Monit Comput 29:749–757

13. Ricci M, Lombardi P, Schultz S, Galindo A, Coscarella E, Vasquez A, Rosenkranz E (2006) Near-infrared spectroscopy to monitor cerebral oxygen saturation in single-ventricle physiology. J Thorac Cardiovasc Surg 131:395–402
14. Subudhi AW, Dimmen AC, Roach RC (2007) Effects of acute hypoxia on cerebral and muscle oxygenation during incremental exercise. J Appl Physiol 103:177–183
15. Ando S, Yamada Y, Kokubu M (2010) Reaction time to peripheral visual stimuli during exercise under hypoxia. J Appl Physiol 108:1210–1216
16. Goodall S, González-Alonso J, Ali L, Ross EZ, Romer LM (2012) Supraspinal fatigue after normoxic and hypoxic exercise in humans. J Physiol 590:2767–2782
17. Rasmussen P, Nybo L, Volianitis S, Møller K, Secher NH, Gjedde A (2010) Cerebral oxygenation is reduced during hyperthermic exercise in humans. Acta Physiol 199:63–70
18. Goodall S, Twomey R, Amann M, Ross EZ, Lovering AT, Romer LM, Subudhi AW, Roach RC (2014) AltitudeOmics: exercise-induced supraspinal fatigue is attenuated in healthy humans after acclimatization to high altitude. Acta Physiol 210:875–888
19. Mandolesi G, Avancini G, Bartesaghi M, Bernardi E, Pomidori L, Cogo A (2014) Long-term monitoring of oxygen saturation at altitude can be useful in predicting the subsequent development of moderate-to-severe acute mountain sickness. Wilderness Environ Med 25:384–391
20. McMorris T, Hale BJ, Barwood M, Costello J, Corbett J (2017) Effect of acute hypoxia on cognition: a systematic review and meta-regression analysis. Neurosci Biobehav Rev 74:225–232
21. Taylor L, Watkins SL, Marshall H, Dascombe BJ, Foster J (2015) The impact of different environmental conditions on cognitive function: a focused review. Front Physiol 6:372
22. Yan X (2014) Cognitive impairments at high altitudes and adaptation. High Alt Med Biol 15:141–145
23. Virués-Ortega J, Buela-Casal G, Garrido E, Alcázar B (2004) Neuropsychological functioning associated with high-altitude exposure. Neuropsychol Rev 14:197–224
24. Phillips JB, Hørning D, Funke ME (2015) Cognitive and perceptual deficits of normobaric hypoxia and the time course to performance recovery. Aerosp Med Hum Perform 86:357–365
25. Turner CE, Barker-Collo SL, Connell CJW, Gant N (2015) Acute hypoxic gas breathing severely impairs cognition and task learning in humans. Physiol Behav 142:104–110
26. MacLeod CM (1991) Half a century of research on the Stroop effect: an integrative review. Psychol Bull 109:163–203
27. Stroop JR (1935) Studies of interference in serial verbal reactions. J Exp Psychol 18:643–662
28. Asmaro D, Mayall J, Ferguson S (2013) Cognition at altitude: impairment in executive and memory processes under hypoxic conditions. Aviat Space Environ Med 84:1159–1165
29. Issa AN, Herman NM, Wentz RJ, Taylor BJ, Summerfield DC, Johnson BD (2016) Association of cognitive performance with time at altitude, sleep quality, and acute mountain sickness symptoms. Wilderness Environ Med 27:371–378
30. de Aquino LV, Antunes HKM, dos Santos RVT, Lira FS, Tufik S, de Mello MT (2012) High altitude exposure impairs sleep patterns, mood, and cognitive functions. Psychophysiology 49:1298–1306
31. Ochi G, Kanazawa Y, Hyodo K, Suwabe K, Shimizu T, Fukuie T, Byun K, Soya H (2018) Hypoxia-induced lowered executive function depends on arterial oxygen desaturation. J Physiol Sci 68:847–853
32. Obrig H, Villringer A (2003) Beyond the visible--imaging the human brain with light. J Cereb Blood Flow Metab 23:1–18
33. Kirilina E, Jelzow A, Heine A, Niessing M, Wabnitz H, Brühl R, Ittermann B, Jacobs AM, Tachtsidis I (2012) The physiological origin of task-evoked systemic artefacts in functional near infrared spectroscopy. NeuroImage 61:70–81
34. Takahashi T, Takikawa Y, Kawagoe R, Shibuya S, Iwano T, Kitazawa S (2011) Influence of skin blood flow on near-infrared spectroscopy signals measured on the forehead during a verbal fluency task. NeuroImage 57:991–1002
35. Sumiyoshi A, Suzuki H, Shimokawa H, Kawashima R (2012) Neurovascular uncoupling under mild hypoxic hypoxia: an EEG-fMRI study in rats. J Cereb Blood Flow Metab 32:1853–1858

36. Byun K, Hyodo K, SuwabeFfniras K, Kujach S, Kato M, Soya H (2014) Possible influences of exercise-intensity-dependent increases in non-cortical hemodynamic variables on NIRS-based neuroimaging analysis during cognitive tasks: technical note. J Exerc Nutr Biochem 18:327–332
37. Karsli C, Luginbuehl I, Bissonnette B (2004) The cerebrovascular response to hypocapnia in children receiving propofol. Anesth Analg 99:1049–1052
38. Rostrup E, Law I, Pott F, Ide K, Knudsen GM (2002) Cerebral hemodynamics measured with simultaneous PET and near-infrared spectroscopy in humans. Brain Res 954:183–193
39. Oguri K, Fujimoto H, Sugimori H, Miyamoto K, Tachi T, Nagasaki S, Kato Y, Matsuoka T (2008) Pronounced muscle deoxygenation during supramaximal exercise under simulated hypoxia in sprint athletes. J Sports Sci Med 7:512–519
40. Kusaka T, Isobe K, Nagano K, Okubo K, Yasuda S, Kondo M, Itoh S, Hirao K, Onishi S (2002) Quantification of cerebral oxygenation by full-spectrum near-infrared spectroscopy using a two-point method. Comp Biochem Physiol A Mol Integr Physiol 132:121–132
41. Teppema LJ, Dahan A (2010) The ventilatory response to hypoxia in mammals: mechanisms, measurement, and analysis. Physiol Rev 90:675–754
42. Steinback CD, Poulin MJ (2008) Cardiovascular and cerebrovascular responses to acute isocapnic and poikilocapnic hypoxia in humans. J Appl Physiol 104:482–489
43. Ide K, Eliasziw M, Poulin MJ (2003) Relationship between middle cerebral artery blood velocity and end-tidal PCO2 in the hypocapnic-hypercapnic range in humans. J Appl Physiol 95:129–137
44. Ochi G, Yamada Y, Hyodo K, Suwabe K, Fukuie T, Byun K, Dan I, Soya H (2018) Neural basis for reduced executive performance with hypoxic exercise. NeuroImage 171:75–83
45. Julien-Dolbec C, Tropres I, Montigon O, Reutenauer H, Ziegler A, Decorps M, Payen JF (2002) Regional response of cerebral blood volume to graded hypoxic hypoxia in rat brain. Br J Anaesth 89:287–293
46. Koehler RC, Traystman RJ, Zeger S, Rogers MC, Jones MD Jr (1984) Comparison of cerebrovascular response to hypoxic and carbon monoxide hypoxia in newborn and adult sheep. J Cereb Blood Flow Metab 4:115–122
47. Ulatowski JA, Bucci E, Razynska A, Traystman RJ, Koehler RC (1998) Cerebral blood flow during hypoxic hypoxia with plasma-based hemoglobin at reduced hematocrit. Am J Phys 274:H1933–H1942
48. Verges S, Rupp T, Jubeau M, Wuyam B, Esteve F, Levy P, Perrey S, Millet GY (2012) Cerebral perturbations during exercise in hypoxia. Am J Physiol Regul Integr Comp Physiol 302:R903–R916
49. Binks AP, Cunningham VJ, Adams L, Banzett RB (2008) Gray matter blood flow change is unevenly distributed during moderate isocapnic hypoxia in humans. J Appl Physiol 104:212–217
50. Rasmussen P, Nielsen J, Overgaard M, Krogh-Madsen R, Gjedde A, Secher NH, Petersen NC (2010) Reduced muscle activation during exercise related to brain oxygenation and metabolism in humans. J Physiol 588:1985–1995
51. Tichauer KM, Brown DW, Hadway J, Lee T-Y, St Lawrence K (2006) Near-infrared spectroscopy measurements of cerebral blood flow and oxygen consumption following hypoxia-ischemia in newborn piglets. J Appl Physiol 100:850–857
52. Calbet JAL, Lundby C (2009) Air to muscle O2 delivery during exercise at altitude. High Alt Med Biol 10:123–134
53. Miyamoto O, Auer RN (2000) Hypoxia, hyperoxia, ischemia, and brain necrosis. Neurology 54:362–371
54. Ellingsen I, Hauge A, Nicolaysen G, Thoresen M, Walløe L (1987) Changes in human cerebral blood flow due to step changes in PAO2 and PACO2. Acta Physiol Scand 129:157–163
55. Cohen PJ, Alexander SC, Smith TC, Reivich M, Wollman H (1967) Effects of hypoxia and normocarbia on cerebral blood flow and metabolism in conscious man. J Appl Physiol 23:183–189

56. Kety SS, Schmidt CF (1948) The effects of altered arterial tensions of carbon dioxide and oxygen on cerebral blood flow and cerebral oxygen consumption of normal young men. J Clin Invest 27:484–492
57. Taylor CT (2008) Interdependent roles for hypoxia inducible factor and nuclear factor-kappaB in hypoxic inflammation. J Physiol 586:4055–4059
58. Yang Y-S, Son SJ, Choi JH, Rah J-C (2018) Synaptic transmission and excitability during hypoxia with inflammation and reoxygenation in hippocampal CA1 neurons. Neuropharmacology 138:20–31
59. Puthon L, Bouzat P, Rupp T, Robach P, Favre-Juvin A, Verges S (2016) Physiological characteristics of elite high-altitude climbers. Scand J Med Sci Sports 26:1052–1059
60. Paterson DJ, Pinnington H, Pearce AR, Morton AR (1987) Maximal exercise cardiorespiratory responses of men and women during acute exposure to hypoxia. Aviat Space Environ Med 58:243–247
61. Woorons X, Mollard P, Lamberto C, Letournel M, Richalet J-P (2005) Effect of acute hypoxia on maximal exercise in trained and sedentary women. Med Sci Sports Exerc 37:147–154
62. Subudhi AW, Fan J-L, Evero O, Bourdillon N, Kayser B, Julian CG, Lovering AT, Roach RC (2014) AltitudeOmics: effect of ascent and acclimatization to 5260 m on regional cerebral oxygen delivery. Exp Physiol 99:772–781

Healthy Cognitive Aging Through Movement: A Practical Approach of Light-Intensity Aerobic Dance for Older Adults

Kazuki Hyodo and Hideaki Soya

Abstract Cognitive decline is a natural part of aging, though its progression varies significantly among individuals. There is a great deal of evidence showing that exercise is one of the most promising lifestyle factors that can both improve cognitive function and reduce the risk of dementia by causing molecular, cellular, structural, and functional changes. In this chapter, based on observational studies and intervention studies, we will briefly outline the effects of physical activity and exercise on cognitive function in older adults, focusing on the potential of light-intensity dance as a practical and enjoyable exercise that many older adults can do.

Keywords Aging · Executive function · Cardiovascular fitness · Physical activity · Dance exercise · fNIRS

1 Introduction

Due to improvements in healthcare, living environments, and hygiene, life expectancies are increasing in many countries, and aging rates are continuing to rise in tandem with the decline in birth rates [1]. In this context, preventing age-related

K. Hyodo (✉)
Physical Fitness Research Institute, Meiji Yasuda Life Foundation of Health and Welfare, Tokyo, Japan
e-mail: k-hyodo@my-zaidan.or.jp

H. Soya
Laboratory of Exercise Biochemistry and Neuroendocrinology, Institute of Health and Sport Sciences, University of Tsukuba, Tsukuba, Ibaraki, Japan

Sport Neuroscience Division, Kokoro Division, Advanced Research Initiative for Human High Performance (ARIHHP), Institute of Health and Sport Sciences, University of Tsukuba, Tsukuba, Ibaraki, Japan

Center for Cybernics Research, University of Tsukuba, Tsukuba, Ibaraki, Japan

H. Soya (ed.), *Exercise Brain Stimulation for Cognitive Function and Mental Health*, Advances in Neurobiology 44, https://doi.org/10.1007/978-981-95-0066-6_21

cognitive decline is extremely important from a social welfare perspective in terms of curbing healthcare and nursing care costs, and from a personal perspective in terms of maintaining quality of life and enabling older adults to lead independent and healthy social lives [2]. Many studies have reported that regular exercise and increasing the amount of physical activity in daily life are not only effective for physical function, but also for improving cognitive function and preventing dementia. However, even if people recognize that exercise or physical activity is beneficial for both physical and cognitive health, not many older adults are able to do or maintain it [3]. In this chapter, we will outline recent research, together with our own research, on the effects of physical activity and exercise on cognitive function and introduce the potential of light-intensity dance exercise, which is easy for many older people to engage in.

2 Cognitive Aging

The impact of aging on cognitive function differs in each cognitive domain, and it is known that some functions are maintained even in old age, while others decline easily with age. For example, crystallized intelligence, which refers to things such as the meaning of words and knowledge acquired through experience and learning, improves until around the age of 60 and is maintained until around the age of 80 in people without cognitive impairment [4]. On the other hand, cognitive functions classified as fluid intelligence, such as episodic memory, information processing speed, selective attention, and executive function, all of which are related to quickly processing and manipulating information according to a given situation or context, are more susceptible to the deleterious effects of aging [5, 6]. The age at which these functions begin to decline is still under debate, partly because such decline differs depending on the cognitive domain. While some cross-sectional studies have reported that memory, reasoning, and processing speed begin to decline from the 20s and 30s [4, 7], longitudinal studies that have tracked changes in individual cognitive function have reported that memory begins to decline from around the 50s [8]. As cognitive functions such as memory and executive function are closely related to everyday activities such as managing money, driving, cooking, and conversation, it is important to suppress functional decline as much as possible in order to maintain the quality of life of older adults and to allow them to safely live independently [9–11].

2.1 Age-Related Structural Brain Changes

One potential factor contributing to age-related cognitive decline is the structural change, such as decline in gray matter or white matter volume, in brain regions responsible for cognitive functions. In pathological aging, such as that associated with neurodegenerative diseases like Alzheimer's, the number of nerve cells in the brain decreases significantly. In contrast, in normal aging, the number of neurons in the neocortex remains relatively stable, but reduction in dendritic length and number, loss of dendritic spines, synaptic loss, and decreased myelination play a role in the observed decrease in gray matter and white matter volume [12, 13]. The degree of brain atrophy due to aging differs depending on the region. Several longitudinal and cross-sectional studies have reported that the occipital and part of temporal lobes, for example, show little atrophy with normal aging, while the volume of the frontal and parietal lobes, which are mainly responsible for executive function, and of the hippocampus and entorhinal cortex, which play important roles in memory function, tends to decrease even with normal aging [14–18]. Furthermore, in research using diffusion tensor imaging, multiple studies have indicated that the anterior brain regions, including the frontal white matter and the anterior corpus callosum (genu), exhibit the most significant decline in white matter integrity with aging [19, 20]. These age-related structural changes in the brain are associated with a decline in various cognitive functions [19, 21–23].

2.2 Cognitive Reserve

In contrast to the cognitive decline described above, it has been reported that the relationship between structural changes and cognitive function is heterogeneous [24]. The concept of "cognitive reserve" has been proposed to explain this [25, 26]. Cognitive reserve refers to the brain's ability to optimize performance through the use of available brain resources, even in the presence of age-related changes or neurological damage. It is a concept that helps explain why some individuals can maintain cognitive function despite brain aging or pathology, while others may experience significant cognitive decline. Cognitive reserve operates through several mechanisms that can be categorized into three main perspectives: neural capacity, neural maintenance, and neural compensation [27–30]. For example, neural capacity refers to the degree to which a task-related brain network can be maximally activated to maintain performance under increasing demands. Neural efficiency refers to the brain's ability to use fewer resources to achieve the same cognitive performance. Individuals with a higher cognitive reserve may exhibit greater neural efficiency, allowing them to perform cognitive tasks with less effort or fewer neural resources. Neural compensation refers to the brain's ability to recruit alternative neural pathways or networks to compensate for deficits in areas that may have been affected by aging or pathology. For example, if one area of the brain responsible for a specific cognitive function is compromised, individuals with a higher cognitive reserve may activate other regions or utilize different cognitive strategies to achieve similar performance levels. It has been reported that these compensatory brain activations are often seen in the prefrontal cortex (PFC) [31]. This flexibility in neural function allows for continued cognitive engagement and performance despite underlying challenges. Therefore, in order to prevent cognitive decline due to aging, it is important not only to suppress brain atrophy throughout life, but also to increase cognitive reserve in order to use the brain efficiently and effectively, even if there is atrophy.

3 The Role of ole of Physical Activity and Exercise in Brain Health

So, how is it possible to maintain the brain structure and improve cognitive reserve in old age? Cognitive reserve is reported to be influenced by a combination of factors including genetics, education, occupational complexity, health, and lifestyle, for example, the presence or absence of mentally stimulating activities or the rate of alcohol consumption [29, 32]. Among these factors, physical activity is one that is not only modifiable, but has accumulated the most evidence of being beneficial to brain health.

3.1 Observational Studies

Cross-sectional and longitudinal studies have shown that physical activity levels and physical fitness levels, particularly cardiovascular fitness, are related to cognitive function, brain structure and function, and the risk of developing dementia in the future. For example, several studies have reported that older adults with a higher peak oxygen uptake (VO_{2peak}) or a moderate- to vigorous-intensity physical activity level have PFCs and hippocampi with greater volumes [33–37], higher white matter integrity [38], and higher resting-state functional connectivity [39, 40]. Task-related brain activation studies suggest that older people with high aerobic capacity may flexibly recruit their cerebral resources depending on the type and difficulty of the cognitive task [41–44]. For example, using functional near-infrared spectroscopy (fNIRS), we examined the mediation effects of prefrontal activation pattern between ventilatory threshold (VT), an indicator of cardiovascular fitness, and executive function in older adults. We showed that older adults with higher VT scores exhibited left-hemisphere-dominant prefrontal activation and high-level Stroop task performance, as might be observed in a younger individual [43], and that they recruited the PFC in a compensatory manner and exhibited high-level performance in a high working memory load task [42] (Fig. 1). In addition, prospective research has also reported that older people with high levels of physical activity and cardiovascular fitness have a lower risk of developing dementia [45, 46] and a lower rate of cognitive decline [47].

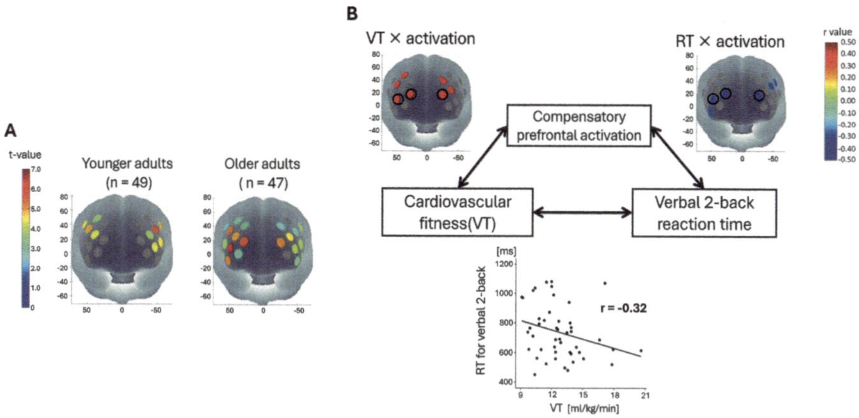

Fig. 1 Mediating effect of prefrontal activation on the relationship between cardiovascular fitness and n-back task performance in older adults. (**a**) Prefrontal activation pattern during 2-back condition in younger and older adults. (**b**) Relationships between prefrontal activation, cardiovascular fitness, and reaction time (RT) for the verbal 2-back condition. The areas circled in black were positively correlated with cardiovascular fitness and negatively correlated with RT for the verbal 2-back task

3.2 *Intervention Studies*

In addition to observational studies, many intervention studies have shown that both acute and chronic exercise interventions can improve cognitive function with functional or structural changes to the brain. The type of exercise that has been most widely studied for its effects on cognitive function is moderate- to high-intensity aerobic exercise (such as brisk walking, jogging, and cycling) [48]. Regarding the acute effects of exercise, several fNIRS studies have reported that moderate-intensity aerobic exercise improves inhibitory function and working memory and increases task-related prefrontal activation [49–51]. We found that only 10 minutes of moderate-intensity cycling exercise reduced Stroop interference and increased activation in the right frontopolar area (FPA), which is interpreted as compensatory recruitment in older adults [49].

Regarding the effects of chronic exercise, meta-analyses and reviews have generally confirmed the beneficial effects of aerobic exercise on several cognitive functions such as executive function and memory [48, 52, 53], though there is some variation among the studies. As structural and functional changes underlie the effects of exercise on cognitive function, neuroimaging studies using magnetic resonance imaging (MRI) and functional MRI (fMRI) have found that moderate-intensity aerobic exercise leads to increased volume of the PFC and hippocampus [54–57], strengthens the functional and structural networks [58, 59], increases efficiency in task-related brain activation [41, 60–62], and improves cerebrovascular function [63].

4 Exploring the Benefits of Light-Intensity Dance Exercise

As mentioned above, many studies have reported positive effects of moderate- to vigorous-intensity exercise on cognitive function. However, it has been argued that it is difficult for many older people to actually practice and continue such exercise in their respective real-life settings [3]. Several studies examining the barriers to exercise among older adults have identified factors that may contribute to this: environmental and social factors, such as lack of facilities and lack of social support, and physical and psychological factors, such as chronic disease, low physical capabilities, fear of injury, low self-efficacy, and lack of enjoyment [64–66]. Therefore, when considering the practicality and continuity of exercise for older people with chronic diseases, low physical strength, or low motivation, it is necessary to identify exercise conditions that are easy and enjoyable to do with little physical burden and that also have a positive effect on the brain. In this section, we outline the potential of light-intensity aerobic dance exercise from the perspective of exercise intensity and mode.

4.1 Effect of Light-Intensity Physical Activity and Exercise

Light-intensity physical activity or exercise can be defined using both relative exercise intensity (VO_{2peak} or %heart rate reserve (HRR)) and absolute exercise intensity (metabolic equivalents (METs)). In terms of VO_{2peak}, light-intensity (including very-light-intensity) exercise corresponds to less than 40% of an individual's VO_{2peak} or HRR [67]. For METs, 1.5–3 METS of activity (including light housework, slow walking, etc.) is defined as light-intensity physical activity.

Light-intensity exercise is attracting attention as a feasible form of exercise with health benefits, as it has been shown to reduce mortality rates and have a positive effect on cardiometabolic health [68, 69]. For the brain, animal studies have shown that light-intensity exercise can increase brain-derived neurotrophic factor (BDNF) and promote neurogenesis (see chapter "Hippocampal Neurogenesis via Light-Intensity Running and Its Mechanism" for details). Human studies have also shown some evidence of the beneficial effects of light-intensity physical activity and exercise [70, 71].

Observational studies suggest that light-intensity physical activity in daily life has a positive effect on the cognitive function and brain structure of older adults. For example, cross-sectional studies have reported that light-intensity physical activity assessed using an accelerometer was associated with beneficial effects on working memory, cognitive flexibility, and cognitive inhibition in older adults [72–76]. Hyodo et al. [77] used compositional isotemporal substitution analysis to reveal that replacing a portion of sedentary time (sedentary behavior: SB) or sleep time with light physical activity (LPA) could improve inhibitory control as evaluated by Stroop task performance (Fig. 2). Furthermore, several neuroimaging studies have revealed a positive relationship between light-intensity walking and the volume of

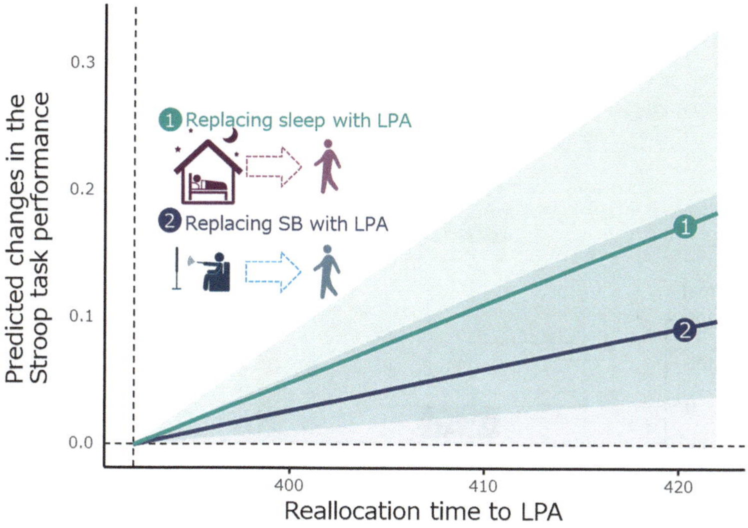

Fig. 2 Differences in predicted changes in Stroop task performance when fixed amounts of time were reallocated from Sleep or Sedentary behavior to LPA while keeping the remaining components constant at compositional means. Adjusted for age, gender, years of education, living conditions, household composition, frequency of social participation, depression, and medical history. SB sedentary behavior, LPA light-intensity physical activity

the hippocampus in healthy older adults [78] as well as a positive relationship between the volume of the whole brain and the amount of light-intensity physical activity in middle-aged and older adults [70, 79]. In a longitudinal study, Volders et al. [80] reported that changes in light-intensity physical activity over a 6-month period were associated with significant improvements in executive function in 432 older adults with chronic diseases. A 22-month follow-up study of 274 older adults found that LPA, measured by an accelerometer, was associated with reduced cognitive decline independent of moderate-to-vigorous physical activity [81].

Although the amount of research is small compared to observational studies, similar effects have also been reported in intervention studies. Regarding the effects of acute exercise, Suwabe et al. [82] reported that 10 minutes of light-intensity bicycle exercise enhanced memory function in older adults. Regarding the effects of chronic exercise, Ruscheweyh et al. [83] randomly divided healthy older adults into a moderate-intensity (Nordic walking) group, a light-intensity (stretching, limbering, and toning) group, and a control group, and conducted a 6-month intervention program of 50 minutes of exercise (excluding the control group) three times a week. They reported that light-intensity exercise improves memory function as well as moderate-intensity exercise does. In addition, we reported that a group that performed very-light-intensity cycling exercise for 40–50 minutes three times a week for 3 months had improved Stroop task performance compared to a wait-list control group, and that the older-old exercise group, which was divided into two groups based on the median age of the participants, had an increase in their neural

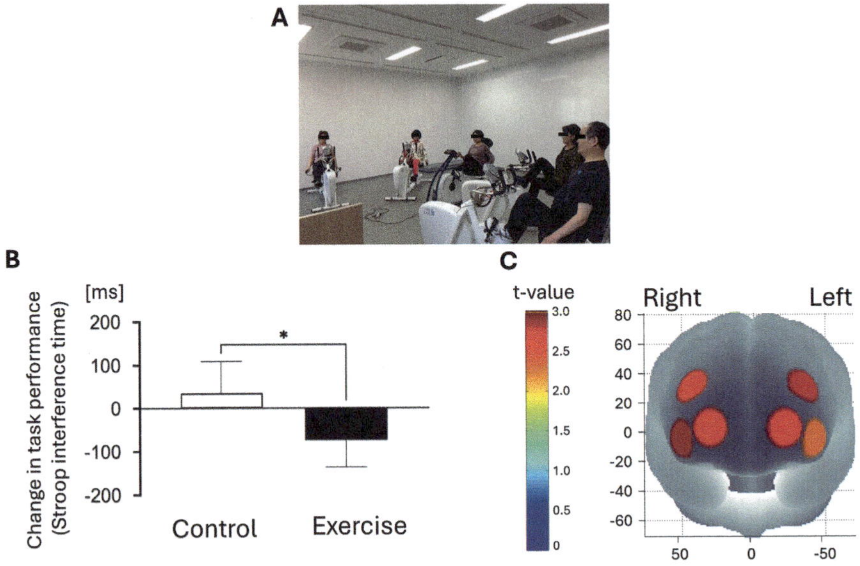

Fig. 3 Three-month light-intensity exercise effect on cognitive function and prefrontal activation. (**a**) Exercise class; (**b**) Change in Stroop task performance after 3 months from baseline for each group; (**c**) t-map of group comparison for the change in the neural efficiency score of the prefrontal cortex during the Stroop task. A positive t-value indicates that the change in the neural efficiency score is greater in the exercise group than in the control group

efficiency score related to the PFC during a task, as measured by fNIRS (Fig. 3) [71]. Although more research is needed, these observational and intervention studies suggest that increasing light-intensity physical activity in daily life and light-intensity aerobic exercise may be beneficial for brain health in older adults.

4.2 *Effect of Dance Exercise*

Aerobic exercise is the most common intervention reported to have beneficial effects on cognitive function, but other exercise modes (i.e., resistance exercise, tai chi, yoga, dance exercise) have, in recent years, also been demonstrated to have positive effects on the brain [84–86]. Among these, dance exercise may be an exercise mode that is highly effective for cognitive function. For example, in some meta-analyses that compared the effects of various types of exercise interventions on cognitive function, it was reported that dance and the kind of coordinated exercise required for dance were more effective than aerobic exercise [87, 88]. Hewston [89] conducted a meta-analysis of 11 randomized control studies (RCTs) that examined the effects of dance on cognitive function in healthy older adults and older adults with mild cognitive impairment (MCI), and reported that dance has a positive effect on executive function, memory, and global cognitive function. Neuroimaging studies have also reported positive structural and functional changes due to dance exercise [90–93]. Along with the benefits of aerobic exercise, the mechanisms by which dance has a positive effect on brain function include learning and executing complex sequences of steps and arm movements, applying spatial navigation, and coordinating movements while keeping time with the rhythm and music, all of which involve various brain regions and have the potential to stimulate the brain even more than less complex forms of light-intensity exercise [94–96]. Dance movements with rhythm and music may also enhance the enjoyment of exercise and boost the effect of improving cognitive function (for details, see chapter "Enriched Exercise Environment Boosting Exercise Effects on the Brain: Beneficial Effects of Music").

4.3 *Potential Effects of Light-Intensity Aerobic Dance Exercise*

Considering the benefits of both light-intensity exercise and dance, light-intensity aerobic dance exercise holds significant potential as a practical and enjoyable activity for older adults. Not only is it easy to adopt and sustain, facilitating its broader implementation in society, but it is also expected to have notable positive effects on cognitive function. Thus, we have developed a light-intensity and low-impact aerobic dance program called Slow Aerobic®. This program features simple movements that require minimal space and primarily involve the upper limbs and trunk of the body, making it suitable for both group settings and at-home practice. This dance exercise program mainly comprises three dynamic upper-body movements: (1) twisting the upper and lower body in opposite directions; (2) pulling the elbows back and then clapping the hands while shaking the hips from side to side; and (3) waving the arms like wiping the windows while shaking the hips from side to side (Fig. 4). Laboratory-based research has confirmed the transient improvement of

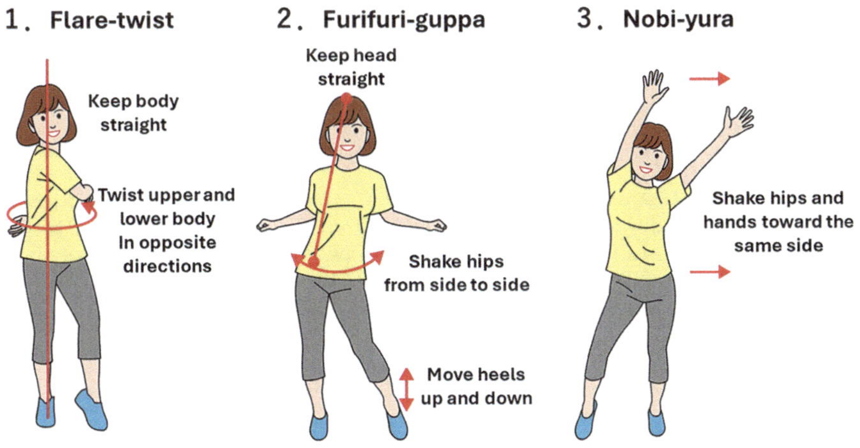

Fig. 4 Three components of the Slow-Aerobic® program

cognitive function and mood via exercise in older adults [97]. Furthermore, a 2-year light-intensity dance exercise intervention led to the maintenance of gray matter volume in the prefrontal cortex and improved performance on cognitive tasks [98]. This form of exercise, being easily performed at home, presents a valuable opportunity for physical activity among older adults who may face environmental barriers limiting their ability to exercise. During the COVID-19 pandemic, we held an 8-week web-based Slow Aerobic program for community-dwelling older adults, with 20-minute sessions held every weekday morning via a web-conference platform. We achieved a high adherence rate (97.4%) and a high level of enjoyment [99]. These results suggest the promising effects of light-intensity dance exercise and its applicability for many older adults. Further research is needed on the long-term effects on brain health, its efficacy for older adults with cognitive decline, and the effect of social implementation in the real world.

5 Conclusion

Cognitive decline with aging often prevents older people from living independently and has a significant impact on their quality of life. In this chapter, we have summarized the role of exercise and physical activity in alleviating age-related cognitive decline and in promoting brain health, focusing on light-intensity exercise and dance exercise. The number of studies on the effect of physical activity or exercise on cognitive function is increasing year by year, but it is still unclear which type of exercise is most effective for brain health. Although it is important to pursue exercise conditions that maximize positive effects on the brain, it is also necessary to pursue exercise programs that both have positive cognitive effects and are easy to practice and maintain, while at the same time considering their social dissemination effects. In this respect, we believe that light-intensity aerobic dance is a promising form of exercise. Further research is needed in order to clarify the mechanisms and long-term effects of light-intensity dance exercise on brain function.

References

1. Kyu HH, Abate D, Abate KH, Abay SM, Abbafati C, Abbasi N et al (2018) Global, regional, and national disability-adjusted life-years (DALYs) for 359 diseases and injuries and healthy life expectancy (HALE) for 195 countries and territories, 1990–2017: a systematic analysis for the Global Burden of Disease Study 2017. Lancet 392:1859–1922
2. Taniguchi Y, Kitamura A, Ishizaki T, Fujiwara Y, Shinozaki T, Seino S et al (2019) Association of trajectories of cognitive function with cause-specific mortality and medical and long-term care costs. Geriatr Gerontol Int 19:1236–1242
3. Sparling PB, Howard BJ, Dunstan DW, Owen N (2015) Recommendations for physical activity in older adults. BMJ 350:h100

4. Salthouse T (2012) Consequences of age-related cognitive declines. Annu Rev Psychol 63:201–226
5. Park DC, Lautenschlager G, Hedden T, Davidson NS, Smith AD, Smith PK (2002) Models of visuospatial and verbal memory across the adult life span. Psychol Aging 17:299–320
6. Salthouse TA (2010) Selective review of cognitive aging. J Int Neuropsychol Soc 16:754–760
7. Salthouse TA (2009) When does age-related cognitive decline begin? Neurobiol Aging 30:507–514
8. Nyberg L, Lovden M, Riklund K, Lindenberger U, Backman L (2012) Memory aging and brain maintenance. Trends Cogn Sci 16:292–305
9. Davis JC, Marra CA, Najafzadeh M, Liu-Ambrose T (2010) The independent contribution of executive functions to health related quality of life in older women. BMC Geriatr 10:16
10. Lee M-T, Jang Y, Chang W-Y (2019) How do impairments in cognitive functions affect activities of daily living functions in older adults? PLoS One 14:e0218112
11. Marshall GA, Rentz DM, Frey MT, Locascio JJ, Johnson KA, Sperling RA et al (2011) Executive function and instrumental activities of daily living in mild cognitive impairment and Alzheimer's disease. Alzheimers Dement 7:300–308
12. Fjell AM, Walhovd KB (2010) Structural brain changes in aging: courses, causes and cognitive consequences. Rev Neurosci 21:187–222
13. Pannese E (2011) Morphological changes in nerve cells during normal aging. Brain Struct Funct 216:85–89
14. Bagarinao E, Watanabe H, Maesawa S, Kawabata K, Hara K, Ohdake R et al (2022) Reserve and maintenance in the aging brain: a longitudinal study of healthy older adults. eNeuro 9:ENEURO.0455-21
15. Fjell AM, Walhovd KB, Fennema-Notestine C, McEvoy LK, Hagler DJ, Holland D et al (2009) One-year brain atrophy evident in healthy aging. J Neurosci 29:15223–15231
16. Pfefferbaum A, Rohlfing T, Rosenbloom MJ, Chu W, Colrain IM, Sullivan EV (2013) Variation in longitudinal trajectories of regional brain volumes of healthy men and women (ages 10 to 85 years) measured with atlas-based parcellation of MRI. NeuroImage 65:176–193
17. Raz N, Lindenberger U, Rodrigue KM, Kennedy KM, Head D, Williamson A et al (2005) Regional brain changes in aging healthy adults: general trends, individual differences and modifiers. Cereb Cortex 15:1676–1689
18. Ziegler G, Dahnke R, Jäncke L, Yotter RA, May A, Gaser C (2012) Brain structural trajectories over the adult lifespan. Hum Brain Mapp 33:2377–2389
19. Kennedy KM, Raz N (2009) Aging white matter and cognition: differential effects of regional variations in diffusion properties on memory, executive functions, and speed. Neuropsychologia 47:916–927
20. Sullivan EV, Pfefferbaum A (2006) Diffusion tensor imaging and aging. Neurosci Biobehav Rev 30:749–761
21. Fjell AM, Sneve MH, Grydeland H, Storsve AB, Amlien IK, Yendiki A et al (2017) Relationship between structural and functional connectivity change across the adult lifespan: a longitudinal investigation. Hum Brain Mapp 38:561–573
22. Jessen F, Feyen L, Freymann K, Tepest R, Maier W, Heun R et al (2006) Volume reduction of the entorhinal cortex in subjective memory impairment. Neurobiol Aging 27:1751–1756
23. Raz N, Rodrigue KM (2006) Differential aging of the brain: patterns, cognitive correlates and modifiers. Neurosci Biobehav Rev 30:730–748
24. Oschwald J, Guye S, Liem F, Rast P, Willis S, Rocke C et al (2019) Brain structure and cognitive ability in healthy aging: a review on longitudinal correlated change. Rev Neurosci 31:1–57
25. Stern Y (2009) Cognitive reserve. Neuropsychologia 47:2015–2028
26. Stern Y (2002) What is cognitive reserve? Theory and research application of the reserve concept. J Int Neuropsychol Soc 8:448–460
27. Barulli D, Stern Y (2013) Efficiency, capacity, compensation, maintenance, plasticity: emerging concepts in cognitive reserve. Trends Cogn Sci 17:502–509

28. Cabeza R, Albert M, Belleville S, Craik FIM, Duarte A, Grady CL et al (2018) Maintenance, reserve and compensation: the cognitive neuroscience of healthy ageing. Nat Rev Neurosci 19:701–710
29. Pappalettera C, Carrarini C, Miraglia F, Vecchio F, Rossini PM (2024) Cognitive resilience/reserve: myth or reality? A review of definitions and measurement methods. Alzheimers Dement 20:3567–3586
30. Steffener J, Stern Y (2012) Exploring the neural basis of cognitive reserve in aging. Biochim Biophys Acta 1822:467–473
31. Cabeza R, Dennis NA (2013) Frontal lobes and aging: deterioration and compensation. In: Stuss DT, Knight RT (eds) Principles of frontal lobe function. Oxford University Press, pp 628–652
32. Clare L, Wu Y-T, Teale JC, MacLeod C, Matthews F, Brayne C et al (2017) Potentially modifiable lifestyle factors, cognitive reserve, and cognitive function in later life: a cross-sectional study. PLoS Med 14:e1002259
33. Erickson KI, Kramer AF (2009) Aerobic exercise effects on cognitive and neural plasticity in older adults. Br J Sports Med 43:22–24
34. Hamer M, Sharma N, Batty GD (2018) Association of objectively measured physical activity with brain structure: UK Biobank study. J Intern Med 284:439–443
35. Melo Neves L, Ritti-Dias R, Juday V, Marquesini R, Mendes Gerage A, Cândido Laurentino G et al (2023) Objective physical activity accumulation and brain volume in older adults: an MRI and whole-brain volume study. J Gerontol A Biol Sci Med Sci 78:902–910
36. Raji CA, Meysami S, Hashemi S, Garg S, Akbari N, Gouda A et al (2024) Exercise-related physical activity relates to brain volumes in 10,125 individuals. J Alzheimers Dis 97:829–839
37. Weinstein AM, Voss MW, Prakash RS, Chaddock L, Szabo A, White SM et al (2012) The association between aerobic fitness and executive function is mediated by prefrontal cortex volume. Brain Behav Immun 26:811–819
38. Oberlin LE, Verstynen TD, Burzynska AZ, Voss MW, Prakash RS, Chaddock-Heyman L et al (2016) White matter microstructure mediates the relationship between cardiorespiratory fitness and spatial working memory in older adults. NeuroImage 131:91–101
39. Kawagoe T, Onoda K, Yamaguchi S (2017) Associations among executive function, cardiorespiratory fitness, and brain network properties in older adults. Sci Rep 7:40107
40. Voss MW, Erickson KI, Prakash RS, Chaddock L, Malkowski E, Alves H et al (2010) Functional connectivity: a source of variance in the association between cardiorespiratory fitness and cognition? Neuropsychologia 48:1394–1406
41. Colcombe SJ, Kramer AF, Erickson KI, Scalf P, McAuley E, Cohen NJ et al (2004) Cardiovascular fitness, cortical plasticity, and aging. Proc Natl Acad Sci USA 101:3316–3321
42. Hyodo K, Dan I, Jindo T, Niioka K, Naganawa S, Mukoyama A et al (2024) Neural mechanisms of the relationship between aerobic fitness and working memory in older adults: an fNIRS study. Imaging Neurosci 2:1–19
43. Hyodo K, Dan I, Kyutoku Y, Suwabe K, Byun K, Ochi G et al (2016) The association between aerobic fitness and cognitive function in older men mediated by frontal lateralization. NeuroImage 125:291–300
44. Voelcker-Rehage C, Godde B, Staudinger UM (2010) Physical and motor fitness are both related to cognition in old age. Eur J Neurosci 31:167–176
45. Iso-Markku P, Kujala UM, Knittle K, Polet J, Vuoksimaa E, Waller K (2022) Physical activity as a protective factor for dementia and Alzheimer's disease: systematic review, meta-analysis and quality assessment of cohort and case–control studies. Br J Sports Med 56:701–709
46. Wang S, Xu L, Yang W, Wang J, Dove A, Qi X et al (2025) Association of cardiorespiratory fitness with dementia risk across different levels of genetic predisposition: a large community-based longitudinal study. Br J Sports Med 59:150–158
47. Iso-Markku P, Aaltonen S, Kujala UM, Halme H-L, Phipps D, Knittle K et al (2024) Physical activity and cognitive decline among older adults: a systematic review and meta-analysis. JAMA Netw Open 7:e2354285
48. Erickson KI, Donofry SD, Sewell KR, Brown BM, Stillman CM (2022) Cognitive aging and the promise of physical activity. Annu Rev Clin Psychol 18:417–442

49. Hyodo K, Dan I, Suwabe K, Kyutoku Y, Yamada Y, Akahori M et al (2012) Acute moderate exercise enhances compensatory brain activation in older adults. Neurobiol Aging 33:2621–2632
50. Ji Z, Feng T, Mei L, Li A, Zhang C (2019) Influence of acute combined physical and cognitive exercise on cognitive function: an NIRS study. PeerJ 7:e7418
51. Stute K, Hudl N, Stojan R, Voelcker-Rehage C (2020) Shedding light on the effects of moderate acute exercise on working memory performance in healthy older adults: an fNIRS study. Brain Sci 10:813
52. Colcombe SJ, Kramer AF (2003) Fitness effects on the cognitive function of older adults: a meta-analytic study. Psychol Sci 14:125–130
53. Hoffmann CM, Petrov ME, Lee RE (2021) Aerobic physical activity to improve memory and executive function in sedentary adults without cognitive impairment: a systematic review and meta-analysis. Prev Med Rep 23:101496
54. Colcombe SJ, Erickson KI, Scalf PE, Kim JS, Prakash R, McAuley E et al (2006) Aerobic exercise training increases brain volume in aging humans. J Gerontol A Biol Sci Med Sci 61:1166–1170
55. Erickson KI, Voss MW, Prakash RS, Basak C, Szabo A, Chaddock L et al (2011) Exercise training increases size of hippocampus and improves memory. Proc Natl Acad Sci USA 108:3017–3022
56. Soshi T, Andersson M, Kawagoe T, Nishiguchi S, Yamada M, Otsuka Y et al (2021) Prefrontal plasticity after a 3-month exercise intervention in older adults relates to enhanced cognitive performance. Cereb Cortex 31:4501–4517
57. Thomas AG, Dennis A, Rawlings NB, Stagg CJ, Matthews L, Morris M et al (2016) Multimodal characterization of rapid anterior hippocampal volume increase associated with aerobic exercise. NeuroImage 131:162–170
58. Voss MW, Heo S, Prakash RS, Erickson KI, Alves H, Chaddock L et al (2013) The influence of aerobic fitness on cerebral white matter integrity and cognitive function in older adults: results of a one-year exercise intervention. Hum Brain Mapp 34:2972–2985
59. Voss MW, Prakash RS, Erickson KI, Basak C, Chaddock L, Kim JS et al (2010) Plasticity of brain networks in a randomized intervention trial of exercise training in older adults. Front Aging Neurosci 2:32
60. Hsu CL, Best JR, Davis JC, Nagamatsu LS, Wang S, Boyd LA et al (2018) Aerobic exercise promotes executive functions and impacts functional neural activity among older adults with vascular cognitive impairment. Br J Sports Med 52:184–191
61. Nocera J, Crosson B, Mammino K, McGregor KM (2017) Changes in cortical activation patterns in language areas following an aerobic exercise intervention in older adults. Neural Plast 2017:6340302
62. Voelcker-Rehage C, Godde B, Staudinger UM (2011) Cardiovascular and coordination training differentially improve cognitive performance and neural processing in older adults. Front Hum Neurosci 5:26
63. Maaß A, Düzel S, Goerke M, Becke A, Sobieray U, Neumann K et al (2015) Vascular hippocampal plasticity after aerobic exercise in older adults. Mol Psychiatry 20:585–593
64. Bethancourt HJ, Rosenberg DE, Beatty T, Arterburn DE (2014) Barriers to and facilitators of physical activity program use among older adults. Clin Med Res 12:10–20
65. Moschny A, Platen P, Klaaßen-Mielke R, Trampisch U, Hinrichs T (2011) Barriers to physical activity in older adults in Germany: a cross-sectional study. Int J Behav Nutr Phys Act 8:121
66. Spiteri K, Broom D, Bekhet AH, De Caro JX, Laventure B, Grafton K (2019) Barriers and motivators of physical activity participation in middle-aged and older adults—a systematic review. J Aging Phys Act 27:929–944
67. American College of Sports Medicine (2013) ACSM's guidelines for exercise testing and prescription. Lippincott Williams & W
68. Chastin SFM, De Craemer M, De Cocker K, Powell L, Van Cauwenberg J, Dall P et al (2019) How does light-intensity physical activity associate with adult cardiometabolic health and mortality? Systematic review with meta-analysis of experimental and observational studies. Br J Sports Med 53:370–376

69. Hamer M, de Oliveira C, Demakakos P (2014) Non-exercise physical activity and survival: English longitudinal study of ageing. Am J Prev Med 47:452–460
70. Spartano NL, Davis-Plourde KL, Himali JJ, Andersson C, Pase MP, Maillard P et al (2019) Association of accelerometer-measured light-intensity physical activity with brain volume: the Framingham heart study. JAMA Netw Open 2:e192745
71. Byun K, Hyodo K, Suwabe K, Fukuie T, Ha M, Damrongthai C et al (2023) Mild exercise improves executive function with increasing neural efficiency in the prefrontal cortex of older adults. GeroScience 46:309–325
72. Gerten S, Engeroff T, Fleckenstein J, Füzéki E, Matura S, Pilatus U et al (2022) Deducing the impact of physical activity, sedentary behavior, and physical performance on cognitive function in healthy older adults. Front Aging Neurosci 13:777490
73. Gothe NP (2021) Examining the effects of light versus moderate to vigorous physical activity on cognitive function in African American adults. Aging Ment Health 25:1659–1665
74. Johnson LG, Butson ML, Polman RC, Raj IS, Borkoles E, Scott D et al (2016) Light physical activity is positively associated with cognitive performance in older community dwelling adults. J Sci Med Sport 19:877–882
75. Umegaki H, Makino T, Uemura K, Shimada H, Cheng XW, Kuzuya M (2018) Objectively measured physical activity and cognitive function in urban-dwelling older adults. Geriatr Gerontol Int 18:922–928
76. Wilbur J, Marquez DX, Fogg L, Wilson RS, Staffileno BA, Hoyem RL et al (2012) The relationship between physical activity and cognition in older Latinos. J Gerontol B Psychol Sci Soc Sci 67:525–534
77. Hyodo K, Kitano N, Ueno A, Yamaguchi D, Watanabe Y, Noda T et al (2022) Association between intensity or accumulating pattern of physical activity and executive function in community-dwelling older adults: a cross-sectional study with compositional data analysis. Front Hum Neurosci 16:1018087
78. Varma VR, Chuang YF, Harris GC, Tan EJ, Carlson MC (2015) Low-intensity daily walking activity is associated with hippocampal volume in older adults. Hippocampus 25:605–615
79. Gu Y, Beato JM, Amarante E, Chesebro AG, Manly JJ, Schupf N et al (2020) Assessment of leisure time physical activity and brain health in a multiethnic cohort of older adults. JAMA Netw Open 3:e2026506
80. Volders E, de Groot RHM, Bolman CAW, Lechner L (2021) The longitudinal associations between change in physical activity and cognitive functioning in older adults with chronic illness (es). BMC Geriatr 21:478
81. Stubbs B, Chen LJ, Chang CY, Sun WJ, Ku PW (2017) Accelerometer-assessed light physical activity is protective of future cognitive ability: a longitudinal study among community dwelling older adults. Exp Gerontol 91:104–109
82. Suwabe K, Kuwamizu R, Hyodo K, Yoshikawa T, Otsuki T, Zempo-Miyaki A et al (2024) Improvement of mnemonic discrimination with acute light exercise is mediated by pupil-linked arousal in healthy older adults. Neurobiol Aging 133:107–114
83. Ruscheweyh R, Willemer C, Krüger K, Duning T, Warnecke T, Sommer J et al (2011) Physical activity and memory functions: an interventional study. Neurobiol Aging 32:1304–1319
84. Bhattacharyya KK, Andel R, Small BJ (2021) Effects of yoga-related mind-body therapies on cognitive function in older adults: a systematic review with meta-analysis. Arch Gerontol Geriatr 93:104319
85. Liu-Ambrose T, Nagamatsu LS, Graf P, Beattie BL, Ashe MC, Handy TC (2010) Resistance training and executive functions: a 12-month randomized controlled trial. Arch Intern Med 170:170–178
86. Wayne PM, Walsh JN, Taylor-Piliae RE, Wells RE, Papp KV, Donovan NJ et al (2014) Effect of tai chi on cognitive performance in older adults: systematic review and meta-analysis. J Am Geriatr Soc 62:25–39
87. Chen FT, Etnier JL, Chan KH, Chiu PK, Hung TM, Chang YK (2020) Effects of exercise training interventions on executive function in older adults: a systematic review and meta-analysis. Sports Med 50:1451–1467

88. Ludyga S, Gerber M, Puhse U, Looser VN, Kamijo K (2020) Systematic review and meta-analysis investigating moderators of long-term effects of exercise on cognition in healthy individuals. Nat Hum Behav 4:603–612
89. Hewston P, Kennedy CC, Borhan S, Merom D, Santaguida P, Ioannidis G et al (2021) Effects of dance on cognitive function in older adults: a systematic review and meta-analysis. Age Ageing 50:1084–1092
90. Muller P, Rehfeld K, Schmicker M, Hokelmann A, Dordevic M, Lessmann V et al (2017) Evolution of neuroplasticity in response to physical activity in old age: the case for dancing. Front Aging Neurosci 9:56
91. Rehfeld K, Luders A, Hokelmann A, Lessmann V, Kaufmann J, Brigadski T et al (2018) Dance training is superior to repetitive physical exercise in inducing brain plasticity in the elderly. PLoS One 13:e0196636
92. Tabei KI, Satoh M, Ogawa JI, Tokita T, Nakaguchi N, Nakao K et al (2017) Physical exercise with music reduces gray and white matter loss in the frontal cortex of elderly people: the Mihama-Kiho scan project. Front Aging Neurosci 9:174
93. Teixeira-Machado L, Arida RM, de Jesus MJ (2019) Dance for neuroplasticity: a descriptive systematic review. Neurosci Biobehav Rev 96:232–240
94. Brown S, Martinez MJ, Parsons LM (2006) The neural basis of human dance. Cereb Cortex 16:1157–1167
95. Kimura K, Hozumi N (2012) Investigating the acute effect of an aerobic dance exercise program on neuro-cognitive function in the elderly. Psychol Sport Exerc 13:623–629
96. Miyazaki A, Okuyama T, Mori H, Sato K, Kumamoto K, Hiyama A (2022) Effects of two short-term aerobic exercises on cognitive function in healthy older adults during COVID-19 confinement in Japan: a pilot randomized controlled trial. Int J Environ Res Public Health 19:6202
97. Hyodo K, Suwabe K, Yamaguchi D, Soya H, Arao T (2021) Comparison between the effects of continuous and intermittent light-intensity aerobic dance exercise on mood and executive functions in older adults. Front Aging Neurosci 13:723243
98. Tamura M, Nemoto K, Kawaguchi A, Kato M, Arai T, Kakuma T et al (2014) Long-term mild-intensity exercise regimen preserves prefrontal cortical volume against aging. Int J Geriatr Psychiatry 30:686–694
99. Hyodo K, Kidokoro T, Yamaguchi D, Iida M, Watanabe Y, Ueno A et al (2023) Feasibility, safety, enjoyment, and system usability of web-based aerobic dance exercise program in older adults: single-arm pilot study. JMIR Aging 6:e39898

Gating of Neuroplasticity: Effects and Mechanisms of Acute Aerobic Exercise as a Brain Stimulation for the Sensorimotor Cortex

Yudai Yamazaki, Daisuke Sato, and Koya Yamashiro

Abstract Aerobic exercise works as a "brain stimulation" or "medicine" to improve executive function, memory, and mental health. In addition to this evidence, recent studies have demonstrated that aerobic exercise positively affects neuroplasticity in the sensorimotor cortex, which is essential for motor skill learning, or rehabilitation in patients with central nervous system disorders. A shift in the excitatory/inhibitory balance within the sensorimotor cortex through the modulation of intracortical excitability elicited by aerobic exercise has been postulated as the underlying mechanism. In this chapter, we focus on acute aerobic exercise and summarize the effects of aerobic exercise on neuroplasticity in the sensorimotor cortex. Additionally, we describe the effects of acute aerobic exercise on changes in intracortical excitability and shifts in excitatory/inhibitory balance, which are considered mechanisms of enhanced neuroplasticity, with a brief explanation of the methods used to evaluate them. We provide important insights into the potential benefits of exercise beyond cognition and memory, thus expanding the role of aerobic exercise as a "brain stimulation" and "medicine."

Keywords Acute aerobic exercise · Sensorimotor cortex · Transcranial magnetic stimulation · Somatosensory-evoked potentials

Y. Yamazaki (✉)
Laboratory of Exercise Biochemistry and Neuroendocrinology, Institute of Health and Sport Sciences, University of Tsukuba, Tsukuba, Ibaraki, Japan
e-mail: yamazaki.yudai.gn@u.tsukuba.ac.jp

D. Sato
Advanced Research Initiative for Human High Performance (ARIHHP), Institute of Health and Sport Sciences, University of Tsukuba, Tsukuba, Ibaraki, Japan

K. Yamashiro
Institute for Human Movement and Medical Sciences, Niigata University of Health and Welfare, Niigata, Japan

Department of Health and Sports, Niigata University of Health and Welfare, Niigata, Japan

H. Soya (ed.), *Exercise Brain Stimulation for Cognitive Function and Mental Health*, Advances in Neurobiology 44, https://doi.org/10.1007/978-981-95-0066-6_22

1 Introduction

Long-term physical activity or exercise enhances neuroplasticity and maintains brain health. This topic has often been investigated in the hippocampus and prefrontal cortex with accumulated evidence. Additionally, acute aerobic exercise is well known to be beneficial for hippocampal-related memory and prefrontal-related cognitive function. In the last decade, a new potential for exercise has been identified. Aerobic exercise rapidly enhances the neuroplasticity of the sensorimotor cortex. Non-invasive brain stimulation (NIBS), a neurophysiological approach, has remarkably advanced our knowledge of the effects of aerobic exercise on neuroplasticity in the sensorimotor cortex. Previous studies, including ours, have demonstrated that the modulation of intracortical circuits in the sensorimotor cortex may be a key mechanism for enhancing neuroplasticity. These findings have increased our focus on the potential of aerobic exercise as a neuromodulatory tool capable of facilitating neural plasticity in the sensorimotor cortex.

This chapter briefly summarizes the effects of aerobic exercise on neuroplasticity in the sensorimotor cortex and excitability of the sensorimotor cortex, which is an essential component for plasticity modulation. Additionally, we discuss potential candidates for exercise-induced modulation of sensorimotor cortex excitability and their potential influence on behaviors, such as motor learning.

2 Exercise-Induced Enhancement of Sensorimotor Plasticity and Motor Learning

Maintaining high neuroplasticity in the sensorimotor cortex is essential for acquiring fine and outstanding sensorimotor functions through motor learning. For example, the long-term potentiation (LTP) of horizontal intracortical connections in layers II and III of the primary motor cortex (M1), with a decrease in GABAergic inhibition, is a well-known key mechanism critical for use-dependent plasticity and motor learning [1–5]. Additionally, because plastic changes occur in the primary somatosensory cortex (S1) during motor learning [6–10], and lesions or decreased activity of S1 impair motor learning [6, 11–13], S1 neuroplasticity is necessary for motor learning.

The magnitude of the effect of NIBS can be used to assess neuroplasticity in the human cortex. Thus, the higher the cortical response to NIBS (e.g., modulation of corticospinal excitability [CSE] and somatosensory-evoked potentials [SEPs]), the higher the neuroplasticity. The effects of aerobic exercise on sensorimotor plasticity were investigated using a paradigm combining exercise and NIBS. An early study has reported that LTP-like plastic changes induced by paired associative stimulation (PAS), an NIBS method that repeatedly induces paired transcranial magnetic stimulation (TMS) on M1 and peripheral nerve stimulation at certain intervals [14, 15], were more remarkable in individuals with higher physical activity levels than in those with lower physical activity levels [16]. Initially, the effects of acute aerobic exercise were examined in studies using theta burst stimulation (TBS), a repetitive TMS method that applies three bursts of 50 Hz at 5 Hz intervals (i.e., theta rhythm) and induces plastic changes in stimulated regions [17, 18]. McDonnell and colleagues demonstrated inhibitory plastic changes (i.e., a long-term depression [LTD]-like effect) induced by continuous TBS on M1 facilitated by prior low-intensity exercise [19]. Subsequent studies have demonstrated that acute aerobic exercise enhances M1 plasticity using PAS and intermittent TBS (iTBS; facilitatory effect on M1) [20–23]. Interestingly, individuals who did not fully benefit from iTBS alone achieved the expected effects when performing high-intensity interval training (HIIT) before iTBS [23]. A problem with NIBS is large interindividual variability [24–26]. Performing exercise before NIBS may overcome this interindividual variability and enable the achievement of the expected NIBS effect. Exercise-induced modulation of sensorimotor plasticity depends on exercise intensity and the direction of plasticity (i.e., LTP- or LTD-like effects). For example, above moderate-intensity exercise facilitates the LTP-like effect [20–23] but does not affect or abolish the LTD-like effect [19, 27]. The common finding in these studies is that aerobic exercise using the lower legs enhances neuroplasticity in the M1 non-exercised hand regions, indicating that aerobic exercise using the lower limb affects neuroplasticity throughout M1 rather than locally in the exercised regions.

2.1 Potential Mechanisms that Explain the Enhancement of Sensorimotor Plasticity Subsequent Exercise—Metaplasticity

The facilitation of neural plasticity following exercise may be interpreted based on the metaplasticity theory. Metaplasticity, in other words, the plasticity of plasticity, is a theory that the direction, magnitude, and duration of neural plasticity are influenced by prior neural activity [28–31]. Metaplasticity comprises two concepts: homeostatic metaplasticity and gating [30]. Homeostatic metaplasticity is consistent with the Bienenstock–Cooper–Munro theory [32] that the plasticity threshold is not constant, but varies and is influenced by previous neural activity. Hence, LTP and LTD are homeostatically regulated, depending on prior LTP and LTD, respectively. For example, prior stimulation that induces LTP-like plasticity attenuates or reverses the effects of subsequent stimulation that induces LTP-like plasticity. By contrast, prior LTD-like plastic changes facilitated subsequent LTP-like plastic changes. This phenomenon may be a protective mechanism for maintaining synaptic activity homeostasis. Gating implies that prior neural activity that does not induce plastic changes facilitates subsequent neural plasticity. An important fact regarding gating is that prior neural activity alters the degree of synaptic response, even if the stimulus remains at the same level. Gating could contribute to the priming effect of aerobic exercise since aerobic exercise itself does not cause LTP- or LTD-like plastic changes (i.e., modulation of CSE, as described in the next section) and facilitates subsequent plastic changes induced by NIBS in humans.

Gating of M1 in humans is represented as modulating intracortical inhibition [30, 33]. A previous study has reported that deafferentation-induced M1 disinhibition facilitates motor learning and induces plastic changes, and that the induction of cortical disinhibition in M1 can enhance plasticity via a gating mechanism. A gating mechanism that enhances neuroplasticity in the sensorimotor cortex has been observed in studies using deafferentation-induced M1 disinhibition and NIBS [34–36]. Additionally, repetitive TMS (rTMS), which increases intracortical inhibition, abolishes bidirectional plastic changes induced by subsequent PAS [37]. According to the gating mechanism, the exercise-induced facilitation of plasticity may be represented by changes in the excitatory/inhibitory (E/I) balance involved in the reduction of GABAergic inhibition. The following section summarizes the studies that have investigated cortical excitability, particularly intracortical inhibitory mechanisms, using neurophysiological techniques.

3 Aerobic Exercise Modulates the Intracortical Inhibitory Circuits in the Primary Motor Cortex

In this section, we summarize previous studies using TMS, including ours, to elucidate exercise-induced changes in the E/I balance in the M1, with a particular focus on intracortical inhibitory circuits.

3.1 TMS

TMS is a non-invasive technique that uses a powerful magnetic field to stimulate the human brain through the scalp. According to Faraday's law, a fluctuating magnetic field is induced by passing a current pulse through a coil orthogonal to the coil, generating an electric field in the brain. The electric field generated by TMS reaches the brain from the scalp through the skull and cerebrospinal fluid. Because electric fields diminish proportionally to the square of the distance, TMS can only stimulate the cortex. When TMS is applied to M1, a series of high-frequency waves are recorded from an electrode placed in the medullary pyramid or on the dorsolateral surface of the cervical spinal cord [38]. This series of descending volleys evokes a twitch in the targeted muscle on the contralateral side to the stimulus. Muscle twitch can be recorded by electromyography, and this recorded response is called motor-evoked potential (MEP), which is an index of excitability of corticospinal tract. Single- and paired-pulse stimulations with a specific intensity and inter-stimulus interval (ISI) can also induce MEP inhibition or facilitation and assess intracortical excitability in M1.

3.2 *Effects of Aerobic Exercise on CSE*

Corticospinal tract excitability can be assessed using motor thresholds and single-pulse motor-evoked pulse MEP amplitudes. The former includes the resting and active motor thresholds (RMT and AMT, respectively) measured at rest and during weak contractions of the target muscle. RMT may be dependent on the excitability of neuronal substrates that are excited by TMS, including corticocortical axons, excitatory synaptic contacts with corticospinal neurons, and the initial axon segments of corticospinal neurons [38, 39]. The MEP amplitude elicited by single-pulse TMS provided the CSE. Two methods can be used to assess CSE: one is stimulation with a single intensity (many studies use intensity to evoke 1 mV or 120% RMT), and the other stimulation with multiple intensities and assessment of intensity–response relationships (i.e., input–output [I–O] curve). In the latter case, the MEP amplitude increased with increasing stimulus intensity in a sigmoid curve. The slope of the I–O curve has a strong relationship with the glutamate concentration in M1 [40]. In addition, as stimulus intensity increased, the influence of GABAergic control increased. Hence, suprathreshold stimulus-evoked MEPs are reflected in multiple synaptic activities of corticospinal neurons, which occur because of a complex network of excitatory circuits controlled by inhibitory circuits [39].

Many studies have reported no changes in CSE or motor thresholds induced by a single bout of aerobic exercise [19, 21, 41–46]. Our research has also determined that 30 min of low-intensity pedaling exercise (30%VO_{2peak}) produced no changes in RMT and CSE (i.e., MEP amplitude induced by single-pulse stimulation) in the M1 leg (exercised limb) and hand (non-exercised limb) regions [47]. By contrast, several studies have reported that high-intensity exercise decreases RMT [43] and increases CSE [44–47], and this is also observed in older adults [48]. Among studies on light-to-moderate-intensity exercise, one study has reported that individuals' physical activity levels affect CSE after exercise [49], whereas another study has indicated no influence of physical fitness levels [50]. Altogether, CSE modulation after exercise is often not observed; however, it may be affected by exercise intensity or physical activity levels.

3.3 *Effects of Aerobic Exercise on Intracortical Circuits*

3.3.1 Measurement of Intracortical Circuits by TMS

Short-Interval Intracortical Inhibition

As described above, paired-pulse stimulation can be used to assess the E/I balance of intracortical circuits in the M1. Subthreshold-conditioned stimuli preceded by suprathreshold test stimuli inhibited or facilitated the MEP induced by the test stimulus, depending on the ISI of both stimuli. The former is called short-interval intracortical inhibition (SICI) and is observed when the ISI is set at 1–5 ms [51]. Conversely, the latter is called intracortical facilitation and is observed when the ISI is set at 7–20 ms [51, 52]. MEP inhibition by SICI is thought to be reflected in the inhibitory postsynaptic potentials (IPSP) of corticospinal neurons induced by low-threshold inhibitory interneurons activated by conditioned stimuli [39]. Meanwhile, the inhibitory phenomenon caused by ISI below 1.5 ms is thought to reflect the refractoriness of cortico-cortical axons rather than synaptic inhibition [53]. Since the administration of lorazepam and diazepam, GABAA receptor agonists, strengthens the inhibition of SICI, but not zolpidem, SICI is believed to reflect the activity of postsynaptic GABAA receptor, particularly α2- and α3-receptors, in the M1 [54].

Long-Interval Intracortical Inhibition

A suprathreshold-conditioned stimulus 50–200 ms before the suprathreshold test stimulus inhibits test stimulus MEPs, a phenomenon called long-interval intracortical inhibition (LICI) [55]. Inhibition by the LICI paradigm is increased by baclofen, a GABAB receptor agonist [56], but is not modulated by benzodiazepines [57, 58], suggesting that the inhibitory phenomenon of LICI reflects long-latency inhibition through the GABABergic mechanism [39, 59].

3.3.2 Effects of Exercise on Intracortical Inhibitory Circuits

Regarding intracortical excitability, many studies have demonstrated that SICI decreases after a single bout of aerobic exercise, irrespective of the exercise intensity. Exercise-induced SICI reduction was first reported by Yamaguchi et al. in 2012, who investigated the effect of low-intensity pedaling for 7 min on SICI in the M1 leg region [41]. Following this study, other studies have demonstrated a reduction in SICI in the M1 hand region after moderate-intensity exercise or HIIT [42, 43, 46, 60, 61]. Moreover, our research has demonstrated that 30 min of low-intensity exercise decreased SICI in the M1 exercised leg region and non-exercised hand regions (Fig. 1A, B) [47]. Although a reduction in SICI after a single bout of aerobic exercise has often been observed in many studies, some have not confirmed this observation [44, 62]. This discrepancy may be attributed to the difference in the stimulus intensity of the conditioning pulse, degree of inhibition at the baseline level, and target muscle for stimulation. Changes in LICI reflecting GABABergic inhibition by exercise have been inconsistent compared to those in SICI (Fig. 1C, D) [44, 47].

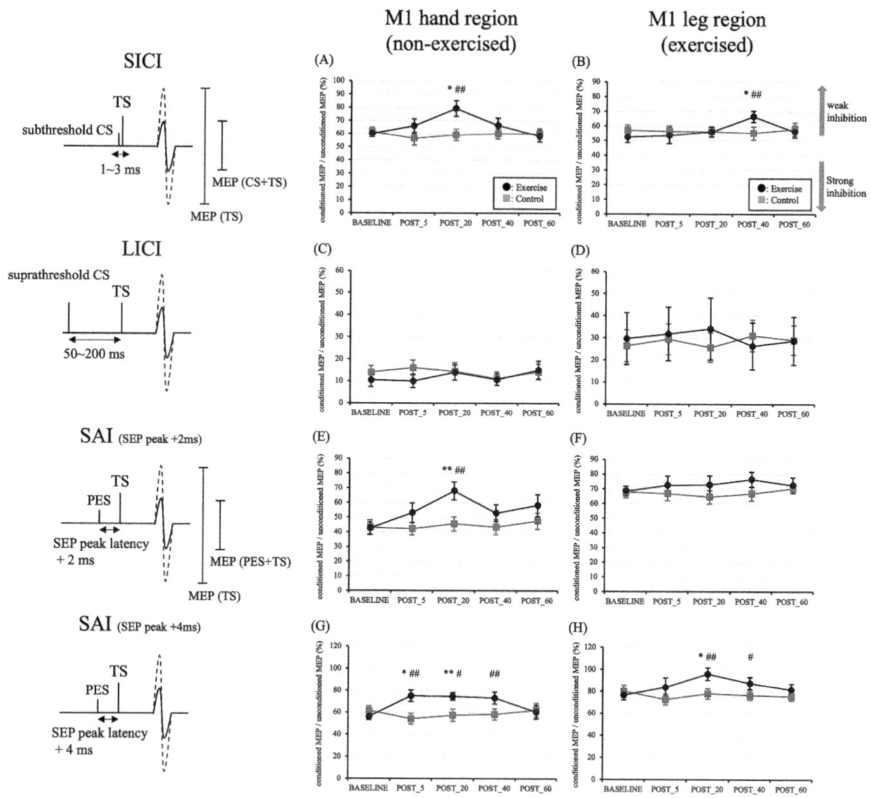

Fig. 1 Exercise-induced changes in M1 intracortical inhibitory function. Left panel: Conceptual illustration of the methodology of intracortical inhibitory circuits and short-latency afferent inhibition. Right panel indicates a changes in intracortical inhibition and short-latency afferent inhibition before and after each conditions. These parameters are calculated by conditioned MEP amplitude (with conditioning stimulus)/single MEP amplitude (without conditioning stimulus). The dark gray circle and light gray rectangle indicate low-intensity exercise (30%VO_{2peak}, 30 min) and control condition, respectively. *$p < 0.05$ compared with baseline, **$p < 0.01$ compared with baseline, #$p < 0.05$ compared with control condition, ##$p < 0.01$ compared with control condition. SICI short-interval intracortical inhibition, LICI long-interval intracortical inhibition, SAI short-latency afferent inhibition,TS test stimulus, CS conditioning stimulus, MEP motor-evoked potentials, PES peripheral electrical stimulation, SEP somatosensory-evoked potential

3.4 *Effects of Acute Aerobic Exercise on Sensorimotor Integration*

3.4.1 Measurement of Sensorimotor Integration by TMS

The peripheral somatosensory input before TMS for M1 on the contralateral side modulated the MEP amplitude depending on the ISI of peripheral stimulation and TMS. In the case of the M1 hand region, electrical stimulation of the peripheral nerve at the wrist (the median nerve is commonly stimulated) was approximately 20 ms before TMS inhibits MEP; this phenomenon is called short-latency afferent inhibition (SAI) [63]. The ISI of peripheral nerve stimulation and TMS corresponds to the conductance time from the stimulated peripheral nerve to the contralateral S1 (i.e., the latency of N20, which is one of the components of SEPs), and the ISI of SAI is often set based on the latency of N20. The inhibitory phenomenon of SAI is thought to be caused by thalamocortical inputs of somatosensory information directly to corticospinal neurons in M1 or via corticocortical projections from S1. Because SAI is reduced in patients with Alzheimer's disease and is strengthened by acetylcholine (Ach) esterase inhibitors or nicotine [64–66], it is modulated by Ach, making it a physiological marker of the central cholinergic pathway. Although SAI may also be controlled by the GABAA receptor, the response to benzodiazepines differs between SICI and SAI [54]. This difference may be due to the different inhibitory mechanisms of SICI and SAI. Another possibility is that they have common inhibitory mechanisms at the M1 level; however, SAI has an additional factor, GABAergic control of Ach release, which may produce a different benzodiazepine response. Several studies have suggested that the M1 and S1, as well as the thalamus [67] and cerebellum [68, 69], are involved in SAI. Beyond the SAI interval, MEPs increase when the ISI of peripheral nerve stimulation and TMS is set at 50–80 ms [70], and MEP suppression is again observed when the ISI is further lengthened (100–200 ms) [71–73]. These phenomena are known as afferent facilitation (AF) and long-latency afferent inhibition (LAI), respectively. Although the detailed mechanisms of these phenomena remain unknown, corticocortical connections may play a role in peripheral input-induced MEP modulations.

3.4.2 Effects of Exercise on Sensorimotor Integration

To the best of our knowledge, our research group was the first to investigate the effects of acute aerobic exercise on SAI [47]. Low-intensity cycling exercise lasting 30 min (30%VO_{2peak}) decreased the SAI in the upper limb 20 min after the exercise (Fig. 1E, G). In addition, exercise decreased SAI in the lower limb (peripheral electrical stimulation [PES] at the tibial nerve and TMS at the M1 tibialis anterior dominance area) 20–40 min after exercise (Fig. 1F, H). Brown et al. investigated the effect of moderate-intensity exercise on afferent modulation of M1 using SAI, LAI, and AF [74]. In contrast to our study, 20 min of moderate-intensity cycling exercise (65–70% age-predicted maximal heart rate) increased LAI, without changes in SAI and AF. LAI and SICI respond inversely to GABAA agonists; a single dose of lorazepam increases SICI but decreases LAI [75]. Based on the evidence that acute aerobic exercise tends to decrease SICI, aerobic exercise may attenuate GABAAergic inhibition and enhance LAI. Although the mechanisms of LAI are unclear, inhibition through the cortico-cortical pathway, such as the parietal cortex, is expected as the candidate based on the ISI length (approximately 200 ms).

3.5 Summary

Since mid-2010, the number of studies investigating the modulation of corticospinal and intracortical excitability in M1 by acute aerobic exercise has drastically increased. Previous studies have generally reported that acute aerobic exercise decreases intracortical inhibition in the M1 (i.e., disinhibition), which is regulated by the GABAA receptor, regardless of exercise intensity. The consistently observed reduction in SICI after acute aerobic exercise suggests a shift in the E/I balance toward excitation, which is involved in the gating mechanism. By contrast, CSE often does not change with exercise, although this may depend on the exercise intensity or participants' physical activity level. Therefore, acute aerobic exercise locally modulates intracortical excitability without altering corticospinal tract excitability.

4 Aerobic Exercise Modulates the Cortical Excitability in the Primary Somatosensory Cortex

In the previous section, we summarized the modulation of E/I balance in M1 intracortical circuits induced by acute aerobic exercise. For M1, we assessed the E/I balance using TMS-evoked MEPs. Conversely, in S1, the E/I balance can be assessed using evoked potentials measured by electroencephalography (EEG).

4.1 Measurement of the Excitability in the Somatosensory Cortex

4.1.1 SEPs

The neural activity generated by somatosensory input from the peripheral site was recorded as SEPs from electrodes placed on the scalp above S1 on the contralateral side of the stimulus. When the median nerve at the wrist is stimulated, electrodes placed in the S1 region exhibit remarkable and stable negative potentials approximately 20 ms after the stimulus, a phenomenon called N20, which is thought to be reflected in the excitatory postsynaptic potentials in Brodmann's area (BA) 3b [76]. This is followed by P25, the positive potentials thought to originate from the neural processing of BAs 1, 2, and 4 [77, 78]. These components are often used to assess the S1 activity.

4.1.2 Paired-Pulse Inhibition (PPI)

A paired pulse with a specific ISI to the peripheral sensory nerve attenuated the amplitude of the second pulse-evoked potentials. This phenomenon is called PPI which indicates the inhibitory function of the sensory cortex. The PPI is calculated as the ratio of the amplitude of the second pulse to that of the first pulse, and lower values indicate potent inhibition. PPI occurs in various sensory regions (visual, auditory, and somatosensory cortices) and has been extensively investigated; however, its neural basis is unclear. Although the detailed mechanisms of PPI remain unclear, inhibition of pyramidal cells by feedforward and feedback inhibition via various inhibitory interneurons may be associated with this phenomenon [79]. However, PPI may be affected by GABAergic mechanisms [80, 81], which are related to the activity of inhibitory interneurons, and glutamatergic mechanisms, which are related to the excitatory activity of thalamocortical fibers, pyramidal cells, and spiny neurons in the layer IV, which synapse onto the interneurons.

Generally, the ISI of the PPI for S1 ranges from 5 to 100 ms, and potent inhibition is observed at an ISI of 40–60 ms. A PPI with a long ISI (>100 ms) may be associated with synaptic depression but not forward suppression [82]. Many studies have demonstrated that a PPI of 30 ms ISI is abnormal in patients with neurological disorders and is modulated by NIBS to S1. A study using multichannel EEG recordings has demonstrated that the inhibitory phenomenon by PPI of 30 ms ISI originates from at least the rostral part of the brainstem [83].

4.2 Effects of Aerobic Exercise on the Excitability in the S1

4.2.1 Effects of Exercise on SEPs

Fewer studies have investigated the effects of acute aerobic exercise on S1 excitability than on M1 excitability. Typically, a single bout of aerobic exercise does not alter the amplitude of SEPs obtained by stimulating exercised and non-exercised limbs [74, 84–86]. By contrast, the latency of SEPs decreased when the exercise was performed above a certain intensity. The latency of SEPs obtained from the exercised lower limb area was shortened after moderate-intensity exercise [85] and high-intensity exercise to exhaustion [84]. In addition, we investigated changes in SEP latency in the non-exercised upper area using moderate-intensity (20 min, 50%VO_{2peak}) and low-intensity (30 min, 30%VO_{2peak}) exercises [86]. SEPs were obtained by stimulating the median nerve at the wrist and recorded from the contralateral S1. We observed a significant reduction in N20 latency after moderate-intensity exercise, but not after low-intensity exercise. As the latency of the initial SEP component reflects the nerve conduction velocity from the peripherally stimulated site to BA 3b, aerobic exercise facilitates the nerve conduction velocity of this pathway. An increase in body temperature by exercise is a plausible candidate for exercise-induced shortening of SEP latency.

4.2.2 Effects of Exercise on PPI

Because acute aerobic exercise decreases GABAergic inhibition in the M1 (i.e., SICI), it may also decrease the inhibitory function in the S1 (i.e., reduction of PPI). This is the first study to examine the effects of acute aerobic exercise on the inhibitory function of S1. We investigated PPI changes by low-intensity (30 min, 30%VO_{2peak}) and moderate-intensity (20 min, 50%VO_{2peak}) pedaling exercises for 60 min after the intervention (Fig. 2) [86]. Paired-pulse stimulation was applied to the median nerve at the wrist using three ISI (5, 30, and 100 ms). Our results revealed that moderate-intensity exercise decreased PPI at 30 ms ISI, but not at 5 and 100 ms ISI, 20 min after exercise. Low-intensity exercise did not modulate ISI for PPI. These results suggest that PPI reduction induced by acute aerobic exercise is intensity- and ISI-dependent. The reduction in PPI after moderate-intensity exercise is assumed to be caused by the modulation of the cortical inhibitory mechanism produced by the input of the first stimulus into the S1. As described above, S1 pyramidal cells receive various inhibitory inputs from inhibitory interneurons. However, the pathway involved in evoking PPI modulated by aerobic exercise remains unclear.

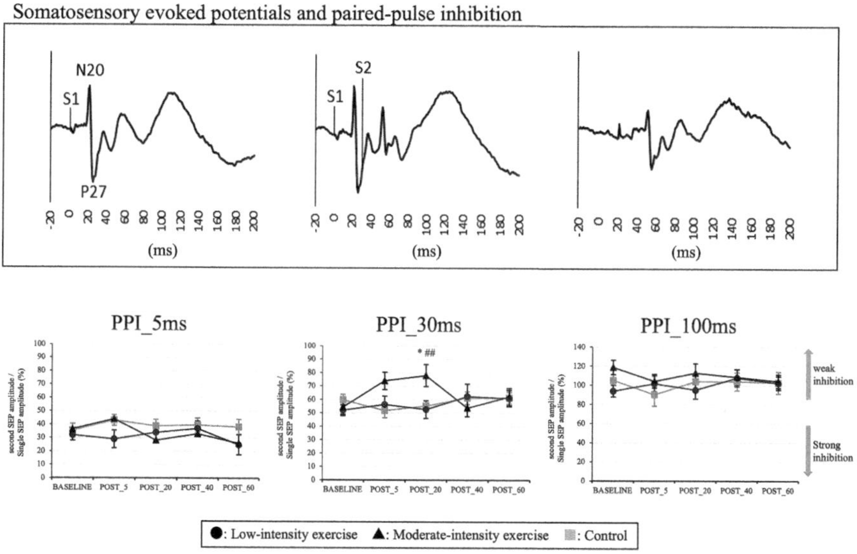

Fig. 2 Exercise-induced changes in S1 inhibitory function. Upper panel: Typical SEP waveform extracted from single participant. Single-pulse (left), paired-pulse (middle), and subtracted SEP (right). The lower panel indicates changes in PPI ratio (SEP amplitude of second pulse from subtracted waveform/single SEP amplitude) before and after each condition. The black triangle indicates moderate-intensity exercise condition (50%VO_{2peak}, 20 min). The dark gray circle indicates low-intensity exercise condition (30%VO_{2peak}, 30 min). The light gray rectangle indicates control condition. *$p < 0.05$ moderate-intensity exercise condition vs. low-intensity exercise condition, #$p < 0.05$ moderate-intensity exercise condition vs. control condition. SEP somatosensory-evoked potential, PPI paired-pulse inhibition

4.3 Summary

Aerobic exercise above moderate intensity may change cortical excitability in S1, although only one study has been conducted on this topic. The reduction in PPI_30 ms observed after moderate-intensity exercise is often demonstrated after excitatory NIBS, such as 5 Hz rTMS, iTBS, and anodal transcranial direct current stimulation of the S1 [87–89]. These results suggest that the E/I balance of the S1 shifts in the excitatory direction after moderate-intensity exercise. However, unlike the influence on M1, low-intensity exercise may not be sufficient to elicit a shift in the E/I balance toward excitation in S1.

5 Potential Candidates for Exercise-Induced Modulation of the Sensorimotor Cortex Excitability

In this section, we introduced gating theory as a neurophysiological mechanism by which acute aerobic exercise enhances the neural plasticity of the sensorimotor cortex. We also summarized the changes in intracortical circuits related to gating induced by acute aerobic exercise. This section discusses the potential mechanisms of the effects of exercise that alter intracortical excitability, possibly through gating.

5.1 Neurotransmitters

The excitability of intracortical circuits is controlled by glutamatergic and GABAergic E/I neurotransmissions. The exercise-induced modulation of intracortical excitability may be caused by changes in the concentration and/or receptor activity of these neurotransmitters. Magnetic resonance spectroscopy (MRS) helps elucidate the changes in the concentration of these neurotransmitters during acute aerobic exercise. Several studies have reported that acute aerobic exercise, particularly high-intensity exercise, increases the concentration of GABA in the anterior cingulate and primary visual and sensorimotor cortices but not in the dorsolateral prefrontal cortex [90–92]. Intuitively, these changes contradict the results of neurophysiological studies that indicate that exercise decreases intracortical inhibition. The exercise-induced decrease in intracortical inhibition may be attributed to an alteration in the activity of postsynaptic receptors rather than a change in the concentration of neurotransmitters. Since previous MRS studies have mainly employed high-intensity exercise, the dependency of exercise intensity on the concentration of neurotransmitters remains unclear.

5.2 Neuromodulators

Neuromodulators modulate glutamatergic and GABAergic activity with complexities, and alter the excitability of corticospinal and intracortical circuits. Monoamines (catecholamine and serotonin) and Ach, the primary neurotransmitters in the brain, are involved in neuroplasticity [93–95], and exercise stimulates their secretion [96]. Previous animal and human pharmacological studies have demonstrated that these neurotransmitters modulate intracortical excitability in the sensorimotor cortex and are associated with neuroplasticity. The pharmacological modulation of NIBS-induced plastic changes has been studied in humans. Although these studies suggest the involvement of neuromodulators in plastic changes in M1, the interpretation of this effect is not straightforward because the influence of neuromodulators on neuroplasticity depends on the receptor subtypes, direction of plasticity (LTP or LTD), and methods of induction of plastic changes (focal or non-focal) [97–106]. Similarly, the influence of neuromodulators on intracortical excitability is also complex [66, 107–112]. However, many studies have consistently indicated that noradrenaline shifts the E/I balance in the human cortex toward excitation (e.g., increases MEPs and decreases PPI) [113–116]. As higher-intensity exercise, particularly above the lactate threshold, increases noradrenaline secretion [117–119], the noradrenergic mechanism may contribute to a reduction in PPI and enhancement of neuroplasticity following higher-intensity exercise. Additionally, a recent critical study has reported that sulpiride, a D2 receptor blocker, abolished exercise-induced modulation of the intracortical E/I balance [120], suggesting the involvement of dopaminergic mechanism in exercise-induced modulation of intracortical excitability. Ach is secreted during very low-intensity exercises such as walking or locomotion, and acts on the cerebral cortex and hippocampus [121, 122]. Furthermore, Ach released from the basal forebrain inputs vasoactive intestinal peptide (VIP) neurons in layer I and disinhibits pyramidal neurons in layer V by activating the VIP-somatostatin GABAergic pathways during exercise [123]. Considering these two findings, Ach may play a role in exercise-induced cortical excitability and neuroplasticity in the sensorimotor cortex in humans.

Although Curtin et al. [120] have provided significant insights into the contribution of neurotransmitters to exercise-induced changes in intracortical excitability, knowledge of the contribution of neurotransmitters to the effects of exercise is still lacking. Exercise intensity is essential when considering the influence of aerobic-exercise-induced neuromodulators on cortical excitability. In general, the exercise-induced secretion of noradrenaline and dopamine is intensity-dependent and is increased above the ventilatory and lactate thresholds, which initiate stress responses [117–119] (note that noradrenaline and dopamine are also secreted even during non-stress low-intensity exercise [124]). Exercise-induced changes in intracortical excitability were observed from low to high intensities. However, whether changes in intracortical excitability due to different exercise intensities are influenced by the same neuromodulator or by different types and degrees of contribution depending on the exercise intensity remains unclear. Moreover, whether neuromodulators secreted during exercise affect intracortical excitability after exercise, whether the secretion of neuromodulators remains even after exercise, whether neuromodulator turnover from the synaptic cleft decreases, or whether the modulated states of synaptic activity persist even when the secretion terminates is unknown.

5.3 Neurotrophic Factors

Another potential factor related to exercise-induced alterations in intracortical excitability is the secretion of brain-derived neurotrophic factor (BDNF). BDNF contributes to neuronal survival, differentiation, and growth and is well known to be the critical mechanism of structural and functional brain improvement by long-term exercise. Moreover, BDNF works rapidly for neural activity. For example, it reduces inhibitory postsynaptic currents in the hippocampal CA1 region via Trk B activation [125]. In addition, BDNF reduces the surface expression of GABAA receptors, leading to a decrease in IPSC in hippocampal neuronal cultures [126]. Although this evidence came from hippocampal neurons, BDNF secreted during aerobic exercise may reduce IPSP and decrease intracortical inhibition, which is controlled by the GABAA receptor. However, the exercise-induced secretion of peripheral BDNF and the difference in BDNF gene polymorphisms do not affect intracortical excitability by acute aerobic exercise [42, 62].

5.4 Neuromodulation from Other Cortical and Non-Cortical Regions

Spatial maps of neural activity evoked by somatosensory stimuli for the median and tibial nerves overlap in the parietal operculum and insula, regions related to sensory integration, and bodily awareness [127]. Additionally, although the cerebellum and supplementary motor area (SMA) exhibit somatotopic organization, the somatotopic representations of the hand and leg areas within these regions overlap [128, 129]. Using dual-coil TMS methods, previous studies have demonstrated that the dorsal premotor cortex and SMA-M1 network dynamics for hand regions are altered during foot movement, suggesting the involvement of these networks in hand–foot coordination [130]. The dispersion of neural activities associated with motor and somatosensory functions in leg areas into hand areas from these overlapping regions may be one of the candidates to alter intracortical excitability in exercised and non-exercised limbs. Several studies have demonstrated that aerobic exercise changes the excitability of the cerebellum and SMA [22, 131], which are candidate modulators of the E/I balance of M1.

5.5 Multiple Sensory Inputs from Peripheral Areas

Somatosensory inputs from peripheral receptors are potential modulators of corticospinal tract and sensorimotor cortex excitability. Continuous sensory inputs from the exercised limbs, including tactile, proprioceptive, and inputs from the metabolic receptors, are present during aerobic exercise. Whether the sensory inputs that occur during aerobic exercise are associated with changes in sensorimotor cortex excitability observed after exercise is unknown.

Numerous studies employing PES have demonstrated the impact of somatosensory inputs from peripheral sites on the excitability of the corticospinal tract and sensorimotor cortex [132–134]. However, the excitability changes induced by PES are complex because they depend on various stimulus parameters, such as intensity, duration, frequency, and duty cycle. The presence or absence of muscle contraction can modulate CSE in the opposite direction [135]. Whether somatosensory inputs, such as those generated during aerobic exercise, affect sensorimotor cortical excitability remains uncertain. Yamaguchi et al. [41] found that passive pedaling using the lower limb did not alter MEP amplitude or SICI in healthy participants [41], suggesting that somatosensory inputs elicited by passive pedaling alone may not be sufficient to modify corticospinal and intracortical excitability in exercised regions of M1. In addition to somatosensory inputs, inputs from metabolic receptors in exercised muscles reach the brain through group III/IV afferent fibers. This input interacts with fatigue and modulates motor cortex excitability [136, 137]. Similar to metabolic information, information on body temperature from the muscles and skin is also conveyed through group III/IV fibers. Changes in body temperature can modulate brain activity, including thalamus, insula, or somatosensory cortex [138, 139]. Although increased body temperature resulting from aerobic exercise has been shown to shorten SEP latency [85, 86], further effects of elevated body temperature have not been thoroughly examined.

Taken together, although each sensory input has the potential to modulate the excitability of the sensorimotor cortex, how it contributes to the effects of aerobic exercise remains unclear. Further studies are required to comprehensively investigate this point.

6 Effects of Alteration of Intracortical Excitability by Acute Aerobic Exercise on Behavior

The impact of acute aerobic exercise on motor learning has increased since Roig et al. [140]. They demonstrated that a single bout of HIIT facilitates consolidation of learned performance in a visuomotor tracking task [140]. The effect was greater when exercise was performed after learning than before, and consolidation enhancement was observed even a month after the learning period. Subsequent investigations supported and replicated these findings, showing that performing HIIT before or after motor learning enhances skill consolidation [21, 46, 140–146] (but also see [147]). An enhancement of memory consolidation is related to changes in event-related desynchronization of β band in the sensorimotor cortex [144], an increase in lactate, BDNF, and anandamide in blood [141, 145], and an increase in the hippocampus and striatum activities [145]. Regarding cortical excitability addressed in this chapter, Stavrinos and Coxon reported a relationship between a decrease in SICI due to HIIT and motor memory consolidation enhancement [46]. Additionally, Ostadan et al. demonstrated a relationship between increased CSE and motor memory consolidation [148]. Previous studies have suggested that acute aerobic exercise, particularly HIIT, has a beneficial effect on motor memory consolidation, and that this effect may have individual differences that underlie exercise-induced changes in cortical excitability, brain activity, and secretion of neurotrophic factors and metabolites. Recent studies have also suggested the possible contribution of dopamine polymorphisms to motor memory consolidation [149, 150]. The optimal timing for exercise may be immediately after learning because the effect disappears when the interval between exercise and learning exceeds an hour [142, 151]. Some studies have also reported that a single bout of exercise after learning can protect learned skills from subsequent interference tasks [152, 153]. Several learning tasks have been used in previous research, and many have used hands and fingers, which are non-exercised limbs. As in NIBS studies, the effects of exercise on motor learning and consolidation are believed to extend to non-exercised limbs. Unlike previous studies, in which exercise before NIBS enhanced plastic changes, many studies have indicated that aerobic exercise after learning promotes memory consolidation. No studies have reported that acute aerobic exercise facilitates sensorimotor plasticity, as demonstrated by the NIBS study, directly affects learning-induced plastic changes, and enhances the learning rate.

Although many studies have identified beneficial effects of acute aerobic exercise on motor memory consolidation, its effects on motor learning are inconsistent, with some studies indicating improved learning [21, 60, 154] and others reporting no effect [140, 141, 155, 156]. Considering that aerobic exercise increases intracortical excitability, exercise-induced modulation of intracortical excitability may have less of an effect on motor learning or other side effects of exercise (e.g., fatigue or hyperarousal), abolishing positive effects. Although the details are unclear, in any case, the findings on the effects of acute aerobic exercise on motor learning are not robust compared to those on motor memory consolidation. Most of these studies have investigated the effects of exercise by learning tasks using the hands or fingers of non-exercised limbs. Few studies have used leg motor learning (e.g., walking on a split-belt treadmill). One study examined the effect of exercise on motor learning in the exercised leg but identified no facilitative effects [157].

Whether modulation of the E/I balance induced by acute aerobic exercise affects subsequent motor learning remains unclear. In addition, the influence of the inhibitory function of S1 on motor learning is unknown. However, several studies have shown that the contribution shifts from S1 to M1 as the learning stage progresses and suggest that S1 may be particularly involved in the early stages of motor learning [8, 9] (a contribution to consolidation has also been reported [10]). This idea is also supported by the NIBS study, which showed that prior suppression of S1 excitability attenuated the acquisition of subsequent motor learning [12]. In addition, PPI are altered after repetitive complex visuomotor tasks, but not after simple tasks [158]. Nevertheless, whether the inhibitory function in the S1 prior to learning affects subsequent skill acquisition or consolidation is unknown.

Whether exercise-induced PPI reduction alters somatosensory function or neuroplasticity in S1 remains unclear. Regarding the former, we did not observe an improvement in somatosensory temporal discrimination with low- or moderate-intensity exercise [86]. However, a recent meta-analysis has indicated that PPI modulation with 30 ms ISI is a good indicator for predicting the effect of NIBS to S1 on neuroplasticity [159]. This suggests that moderate-intensity exercise, which reduces PPI_30 ms ISI (Sect. 4.2.2), can potentially enhance neuroplasticity in the S1. We hope that further evidence on the effects of acute aerobic exercise on S1 neuroplasticity or somatosensory function will accumulate, beginning with our findings.

7 Conclusion

Acute aerobic exercise enhances neuroplasticity in the sensorimotor cortex, potentially through a gating mechanism driven by modulation of the E/I balance of intracortical excitability, particularly a reduction in intracortical inhibition. Several factors may contribute to the reduction in intracortical inhibition caused by acute aerobic exercise (Fig. 3). Future studies should explore the mechanisms underlying the reduction in intracortical inhibition. Additionally, the direct relationship between exercise-induced modulation of the E/I balance and enhancement of neuroplasticity needs to be elucidated. To enable this, a comprehensive research that combines neuroimaging and biochemical techniques as well as investigations that extend to brain regions beyond the sensorimotor cortex, in addition to conventional neurophysiological methods, should be conducted. As the effects of exercise are ascertained, increasing evidence that exercise acts as "brain stimulation" and "medicine" with beneficial effects that transcend cognition and memory is expected.

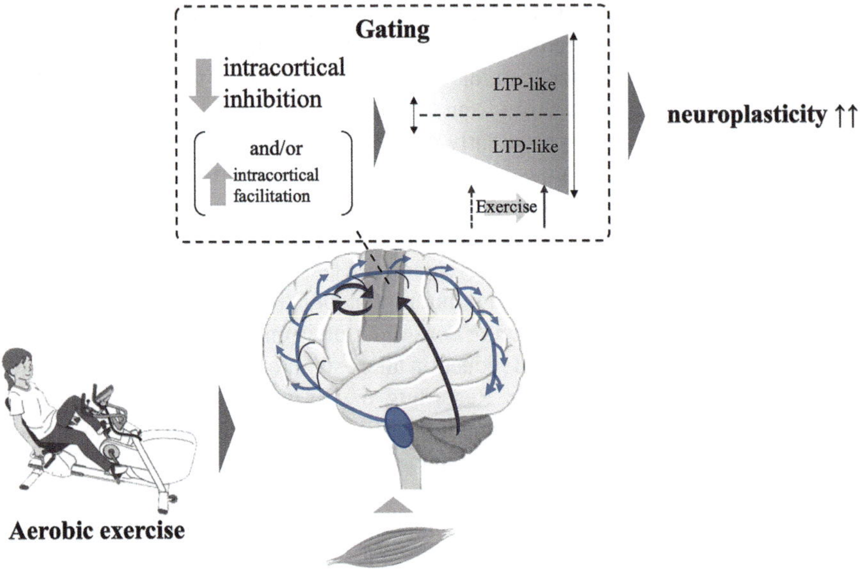

Fig. 3 Potential pathway of exercise-induced enhancement of neuroplasticity in the sensorimotor cortex. Acute aerobic exercise decreases intracortical inhibition in the sensorimotor cortex via variable factors, such as secretion of neurotransmitters, afferent inputs from peripheral muscles and organs, and inputs from other brain regions, such as the cerebellum, premotor cortex, supplementary motor area, and basal ganglia. A decrease in intracortical inhibition (and/or an increase in intracortical facilitation) enhances the neuroplasticity induced by NIBS and motor learning via a gating mechanism, which is one of the concepts of metaplasticity. NIBS, non-invasive brain stimulation

References

1. Hess G, Donoghue JP (1994) Long-term potentiation of horizontal connections provides a mechanism to reorganize cortical motor maps. J Neurophysiol 71:2543–2547
2. Hess G, Aizenman CD, Donoghue JP (1996) Conditions for the induction of long-term potentiation in layer II/III horizontal connections of the rat motor cortex. J Neurophysiol 75:1765–1778
3. Nudo RJ, Milliken GW, Jenkins WM, Merzenich MM (1996) Use-dependent alterations of movement representations in primary motor cortex of adult squirrel monkeys. J Neurosci 16:785–807
4. Rioult-Pedotti MS, Friedman D, Hess G, Donoghue JP (1998) Strengthening of horizontal cortical connections following skill learning. Nat Neurosci 1:230–234
5. Rioult-Pedotti MS, Friedman D, Donoghue JP (2000) Learning-induced LTP in neocortex. Science 290:533–536
6. Pleger B, Schwenkreis P, Dinse HR, Ragert P, Höffken O, Malin JP, Tegenthoff M (2003) Pharmacological suppression of plastic changes in human primary somatosensory cortex after motor learning. Exp Brain Res 148:525–532
7. Andrew D, Haavik H, Dancey E, Yielder P, Murphy B (2015) Somatosensory evoked potentials show plastic changes following a novel motor training task with the thumb. Clin Neurophysiol 126:575–580

8. Bernardi NF, Darainy M, Ostry DJ (2015) Somatosensory contribution to the initial stages of human motor learning. J Neurosci 35:14316–14326
9. Ohashi H, Gribble PL, Ostry DJ (2019) Somatosensory cortical excitability changes precede those in motor cortex during human motor learning. J Neurophysiol 122:1397–1405
10. Kumar N, Manning TF, Ostry DJ (2019) Somatosensory cortex participates in the consolidation of human motor memory. PLoS Biol 17:e3000469
11. Pavlides C, Miyashita E, Asanuma H (1993) Projection from the sensory to the motor cortex is important in learning motor skills in the monkey. J Neurophysiol 70:733–741
12. Vidoni ED, Acerra NE, Dao E, Meehan SK, Boyd LA (2010) Role of the primary somatosensory cortex in motor learning: an rTMS study. Neurobiol Learn Mem 93:532–539
13. Mathis MW, Mathis A, Uchida N (2017) Somatosensory cortex plays an essential role in forelimb motor adaptation in mice. Neuron 93:1493-1503
14. Stefan K, Kunesch E, Cohen LG, Benecke R, Classen J (2000) Induction of plasticity in the human motor cortex by paired associative stimulation. Brain 123(Pt 3):572–584
15. Stefan K, Kunesch E, Benecke R, Cohen LG, Classen J (2002) Mechanisms of enhancement of human motor cortex excitability induced by interventional paired associative stimulation. J Physiol 543:699–708
16. Cirillo J, Lavender AP, Ridding MC, Semmler JG (2009) Motor cortex plasticity induced by paired associative stimulation is enhanced in physically active individuals. J Physiol 587:5831–5842
17. Huang Y-Z, Rothwell JC (2004) The effect of short-duration bursts of high-frequency, low-intensity transcranial magnetic stimulation on the human motor cortex. Clin Neurophysiol 115:1069–1075
18. Huang YZ, Edwards MJ, Rounis E, Bhatia KP, Rothwell JC (2005) Theta burst stimulation of the human motor cortex. Neuron 45:201–206
19. McDonnell MN, Buckley JD, Opie GM, Ridding MC, Semmler JG (2013) A single bout of aerobic exercise promotes motor cortical neuroplasticity. J Appl Physiol 114:1174–1182
20. Singh AM, Neva JL, Staines WR (2014) Acute exercise enhances the response to paired associative stimulation-induced plasticity in the primary motor cortex. Exp Brain Res 232:3675–3685
21. Mang CS, Snow NJ, Campbell KL, Ross CJD, Boyd LA (2014) A single bout of high-intensity aerobic exercise facilitates response to paired associative stimulation and promotes sequence-specific implicit motor learning. J Appl Physiol 117:1325–1336
22. Mang CS, Brown KE, Neva JL, Snow NJ, Campbell KL, Boyd LA (2016) Promoting motor cortical plasticity with acute aerobic exercise: a role for cerebellar circuits. Neural Plast. 2016:6797928
23. Andrews SC, Curtin D, Hawi Z, Wongtrakun J, Stout JC, Coxon JP (2020) Intensity matters: high-intensity interval exercise enhances motor cortex plasticity more than moderate exercise. Cereb Cortex 30:101-112
24. Hamada M, Murase N, Hasan A, Balaratnam M, Rothwell JC (2013) The role of interneuron networks in driving human motor cortical plasticity. Cereb Cortex 23:1593–1605
25. López-Alonso V, Cheeran B, Río-Rodríguez D, Fernández-Del-Olmo M (2014) Inter-individual variability in response to non-invasive brain stimulation paradigms. Brain Stimul 7:372–380
26. Wiethoff S, Hamada M, Rothwell JC (2014) Variability in response to transcranial direct current stimulation of the motor cortex. Brain Stimul 7:468–475
27. Singh AM, Duncan RE, Staines WR (2016) Aerobic exercise abolishes cTBS-induced suppression of motor cortical excitability. Neurosci Lett 633:215–219
28. Abraham WC, Bear MF (1996) Metaplasticity: the plasticity of synaptic plasticity. Trends Neurosci 19:126–130
29. Abraham WC (2008) Metaplasticity: tuning synapses and networks for plasticity. Nat Rev Neurosci 9:387
30. Ziemann U, Siebner HR (2008) Modifying motor learning through gating and homeostatic metaplasticity. Brain Stimul 1:60–66

31. Müller-Dahlhaus F, Ziemann U (2015) Metaplasticity in human cortex. Neuroscientist 21:185–202
32. Bienenstock EL, Cooper LN, Munro PW (1982) Theory for the development of neuron selectivity: orientation specificity and binocular interaction in visual cortex. J Neurosci 2:32–48
33. Siebner HR (2010) A primer on priming the human motor cortex. Clin Neurophysiol 121:461–463
34. Ziemann U, Muellbacher W, Hallett M, Cohen LG (2001) Modulation of practice-dependent plasticity in human motor cortex. Brain 124:1171–1181
35. Ragert P, Dinse HR, Pleger B, Wilimzig C, Frombach E, Schwenkreis P, Tegenthoff M (2003) Combination of 5 Hz repetitive transcranial magnetic stimulation (rTMS) and tactile coactivation boosts tactile discrimination in humans. Neurosci Lett 348:105–108
36. Kim Y-H, Park J-W, Ko M-H, Jang SH, Lee PKW (2004) Facilitative effect of high frequency subthreshold repetitive transcranial magnetic stimulation on complex sequential motor learning in humans. Neurosci Lett 367:181–185
37. Delvendahl I, Jung NH, Mainberger F, Kuhnke NG, Cronjaeger M, Mall V (2010) Occlusion of bidirectional plasticity by preceding low-frequency stimulation in the human motor cortex. Clin Neurophysiol 121:594–602
38. Di Lazzaro V, Ziemann U, Lemon RN (2008) State of the art: physiology of transcranial motor cortex stimulation. Brain Stimul 1:345–362
39. Ziemann U, Reis J, Schwenkreis P, Rosanova M, Strafella A, Badawy R, Müller-Dahlhaus F (2015) TMS and drugs revisited 2014. Clin Neurophysiol 126:1847–1868
40. Stagg CJ, Bestmann S, Constantinescu AO, Moreno Moreno L, Allman C, Mekle R, Woolrich M, Near J, Johansen-Berg H, Rothwell JC (2011) Relationship between physiological measures of excitability and levels of glutamate and GABA in the human motor cortex. J Physiol 589:5845–5855
41. Yamaguchi T, Fujiwara T, Liu W, Liu M (2012) Effects of pedaling exercise on the intracortical inhibition of cortical leg area. Exp Brain Res 218:401–406
42. Singh AM, Duncan RE, Neva JL, Staines WR (2014) Aerobic exercise modulates intracortical inhibition and facilitation in a nonexercised upper limb muscle. BMC Sports Sci Med Rehabil 6:1–10
43. Smith AE, Goldsworthy MR, Garside T, Wood FM, Ridding MC (2014) The influence of a single bout of aerobic exercise on short-interval intracortical excitability. Exp Brain Res 232:1875–1882
44. Mooney RA, Coxon JP, Cirillo J, Glenny H, Gant N, Byblow WD (2016) Acute aerobic exercise modulates primary motor cortex inhibition. Exp Brain Res 234:3669–3676
45. Neva JL, Brown KE, Mang CS, Francisco BA, Boyd LA (2017) An acute bout of exercise modulates both intracortical and interhemispheric excitability. Eur J Neurosci 45:1343–1355
46. Stavrinos EL, Coxon JP (2017) High-intensity interval exercise promotes motor cortex disinhibition and early motor skill consolidation. J Cogn Neurosci 29:593–604
47. Yamazaki Y, Sato D, Yamashiro K, Nakano S, Onishi H, Maruyama A (2019) Acute low-intensity aerobic exercise modulates intracortical inhibitory and excitatory circuits in an exercised and a non-exercised muscle in the primary motor cortex. Front Physiol 10:1361
48. Neva JL, Greeley B, Chau B, Ferris JK, Jones CB, Denyer R, Hayward KS, Campbell KL, Boyd LA (2022) Acute high-intensity interval exercise modulates corticospinal excitability in older adults. Med Sci Sports Exerc 54:673–682
49. Lulic T, El-Sayes J, Fassett HJ, Nelson AJ (2017) Physical activity levels determine exerciseinduced changes in brain excitability. PLoS One 12:1–18
50. MacDonald MA, Khan H, Kraeutner SN, Usai F, Rogers EA, Kimmerly DS, Dechman G, Boe SG (2019) Intensity of acute aerobic exercise but not aerobic fitness impacts on corticospinal excitability. Appl Physiol Nutr Metab 44:869–878
51. Kujirai T, Caramia MD, Rothwell JC, Day BL, Thompson PD, Ferbert A, Wroe S, Asselman P, Marsden CD (1993) Corticocortical inhibition in human motor cortex. J Physiol 471:501–519
52. Ziemann U, Rothwell JC, Ridding MC (1996) Interaction between intracortical inhibition and facilitation in human motor cortex. J Physiol 496(Pt 3):873–881

53. Fisher RJ, Nakamura Y, Bestmann S, Rothwell JC, Bostock H (2002) Two phases of intracortical inhibition revealed by transcranial magnetic threshold tracking. Exp Brain Res 143:240–248
54. Di Lazzaro V, Pilato F, Dileone M, Profice P, Ranieri F, Ricci V, Bria P, Tonali PA, Ziemann U (2007) Segregating two inhibitory circuits in human motor cortex at the level of GABAA receptor subtypes: a TMS study. Clin Neurophysiol 118:2207–2214
55. Valls-Solé J, Pascual-Leone A, Wassermann EM, Hallett M (1992) Human motor evoked responses to paired transcranial magnetic stimuli. Electroencephalogr Clin Neurophysiol 85:355–364
56. McDonnell MN, Orekhov Y, Ziemann U (2006) The role of GABAB receptors in intracortical inhibition in the human motor cortex. Exp Brain Res 173:86–93
57. Mohammadi B, Krampfl K, Petri S, Bogdanova D, Kossev A, Bufler J, Dengler R (2006) Selective and nonselective benzodiazepine agonists have different effects on motor cortex excitability. Muscle Nerve 33:778–784
58. Teo JTH, Terranova C, Swayne O, Greenwood RJ, Rothwell JC (2009) Differing effects of intracortical circuits on plasticity. Exp Brain Res 193:555–563
59. Werhahn KJ, Kunesch E, Noachtar S, Benecke R, Classen J (1999) Differential effects on motorcortical inhibition induced by blockade of GABA uptake in humans. J Physiol 517(Pt 2):591–597
60. Opie GM, Semmler JG (2019) Acute exercise at different intensities influences corticomotor excitability and performance of a ballistic thumb training task. Neuroscience 412:29–39
61. Hendy AM, Andrushko JW, Della Gatta PA, Teo W-P (2022) Acute effects of high-intensity aerobic exercise on motor cortical excitability and inhibition in sedentary adults. Front Psychol 13:814633
62. Nicolini C, Michalski B, Toepp SL, Turco CV, D'Hoine T, Harasym D, Gibala MJ, Fahnestock M, Nelson AJ (2020) A single bout of high-intensity interval exercise increases corticospinal excitability, brain-derived neurotrophic factor, and uncarboxylated osteolcalcin in sedentary, healthy males. Neuroscience 437:242–255
63. Tokimura H, Di Lazzaro V, Tokimura Y, Oliviero A, Profice P, Insola A, Mazzone P, Tonali P, Rothwell JC (2000) Short latency inhibition of human hand motor cortex by somatosensory input from the hand. J Physiol 523(Pt 2):503–513
64. Di Lazzaro V, Oliviero A, Profice P, Pennisi MA, Di Giovanni S, Zito G, Tonali P, Rothwell JC (2000) Muscarinic receptor blockade has differential effects on the excitability of intracortical circuits in the human motor cortex. Exp Brain Res 135:455–461
65. Di Lazzaro V, Oliviero A, Pilato F, Saturno E, Dileone M, Marra C, Ghirlanda S, Ranieri F, Gainotti G, Tonali P (2005) Neurophysiological predictors of long term response to AChE inhibitors in AD patients. J Neurol Neurosurg Psychiatry 76:1064–1069
66. Grundey J, Freznosa S, Klinker F, Lang N, Paulus W, Nitsche MA (2013) Cortical excitability in smoking and not smoking individuals with and without nicotine. Psychopharmacology 229:653–664
67. Oliviero A, León AM, Holler I, Vila JF, Siebner HR, Della Marca G, Di Lazzaro V, Alvarez JT (2005) Reduced sensorimotor inhibition in the ipsilesional motor cortex in a patient with chronic stroke of the paramedian thalamus. Clin Neurophysiol 116:2592–2598
68. Dubbioso R, Pellegrino G, Antenora A, De Michele G, Filla A, Santoro L, Manganelli F (2015) The effect of cerebellar degeneration on human sensori-motor plasticity. Brain Stimul 8:1144–1150
69. Yildiz FG, Saka E, Elibol B, Temucin CM (2018) Modulation of cerebellar-cortical connections in multiple system atrophy type c by cerebellar repetitive transcranial magnetic stimulation. Neuromodulation 21:402–408
70. Devanne H, Degardin A, Tyvaert L, Bocquillon P, Houdayer E, Manceaux A, Derambure P, Cassim F (2009) Afferent-induced facilitation of primary motor cortex excitability in the region controlling hand muscles in humans. Eur J Neurosci 30:439–448
71. Chen R, Corwell B, Hallett M (1999) Modulation of motor cortex excitability by median nerve and digit stimulation. Exp Brain Res 129:77–86

72. Sailer A, Molnar GF, Cunic DI, Chen R (2002) Effects of peripheral sensory input on cortical inhibition in humans. J Physiol 544:617–629
73. Sailer A, Molnar GF, Paradiso G, Gunraj CA, Lang AE, Chen R (2003) Short and long latency afferent inhibition in Parkinson's disease. Brain 126:1883–1894
74. Brown KE, Neva JL, Mang CS, Chau B, Chiu LK, Francisco BA, Staines WR, Boyd LA (2020) The influence of an acute bout of moderate-intensity cycling exercise on sensorimotor integration. Eur J Neurosci 52:4779–4790
75. Turco CV, El-Sayes J, Locke MB, Chen R, Baker S, Nelson AJ (2018) Effects of lorazepam and baclofen on short- and long-latency afferent inhibition. J Physiol 596:5267–5280
76. Hashimoto I, Mashiko T, Imada T (1996) Somatic evoked high-frequency magnetic oscillations reflect activity of inhibitory interneurons in the human somatosensory cortex. Electroencephalogr Clin Neurophysiol – Evoked Potentials 100:189–203
77. Allison T, McCarthy G, Wood CC, Darcey TM, Spencer DD, Williamson PD (1989) Human cortical potentials evoked by stimulation of the median nerve. I. Cytoarchitectonic areas generating short-latency activity. J Neurophysiol 62:694–710
78. Inui K, Wang X, Tamura Y, Kaneoke Y, Kakigi R (2004) Serial processing in the human somatosensory system. Cereb Cortex 14:851–857
79. Feldmeyer D, Qi G, Emmenegger V, Staiger JF (2018) Inhibitory interneurons and their circuit motifs in the many layers of the barrel cortex. Neuroscience 368:132–151
80. Chowdhury SA, Rasmusson DD (2003) Corticocortical inhibition of peripheral inputs within primary somatosensory cortex: the role of GABA(A) and GABA(B) receptors. J Neurophysiol 90:851–856
81. Stude P, Lenz M, Höffken O, Tegenthoff M, Dinse H (2016) A single dose of lorazepam reduces paired-pulse suppression of median nerve evoked somatosensory evoked potentials. Eur J Neurosci 43:1156–1160
82. Wehr M, Zador AM (2005) Synaptic mechanisms of forward suppression in rat auditory cortex. Neuron 47:437–445
83. Höffken O, Lenz M, Tegenthoff M, Schwenkreis P (2010) Multichannel SEP-recording after paired median nerve stimulation suggests origin of paired-pulse inhibition rostral of the brainstem. Neurosci Lett 468:308–311
84. Perciavalle V, Alagona G, De Maria G, Rapisarda G, Costanzo E, Perciavalle V, Coco M (2015) Somatosensory evoked potentials and blood lactate levels. Neurol Sci 36:1597–1601
85. Nakata H, Oshiro M, Namba M, Shibasaki M (2016) Effects of aerobic exercise under different thermal conditions on human somatosensory processing. Am J Physiol Regul Integr Comp Physiol 311:R629–R636
86. Yamazaki Y, Yamashiro K, Onishi H, Otsuru N, Kojima S, Saito K, Sato D (2020) Modulation of inhibitory function in the primary somatosensory cortex and temporal discrimination threshold induced by acute aerobic exercise. Behav Brain Res 377:112253
87. Ragert P, Becker M, Tegenthoff M, Pleger B, Dinse HR (2004) Sustained increase of somatosensory cortex excitability by 5 Hz repetitive transcranial magnetic stimulation studied by paired median nerve stimulation in humans. Neurosci Lett 356:91–94
88. Ragert P, Franzkowiak S, Schwenkreis P, Tegenthoff M, Dinse HR (2008) Improvement of tactile perception and enhancement of cortical excitability through intermittent theta burst rTMS over human primary somatosensory cortex. Exp Brain Res 184:1–11
89. Rehmann R, Sczesny-Kaiser M, Lenz M, Gucia T, Schliesing A, Schwenkreis P, Tegenthoff M, Höffken O (2016) Polarity-specific cortical effects of transcranial direct current stimulation in primary somatosensory cortex of healthy humans. Front Hum Neurosci 10:1–7
90. Maddock RJ, Casazza GA, Buonocore MH, Tanase C (2011) Vigorous exercise increases brain lactate and Glx (glutamate+glutamine): a dynamic 1H-MRS study. NeuroImage 57:1324–1330
91. Maddock RJ, Casazza GA, Fernandez DH, Maddock MI (2016) Acute modulation of cortical glutamate and GABA content by physical activity. J Neurosci 36:2449–2457

92. Coxon JP, Cash RFH, Hendrikse JJ, Rogasch NC, Stavrinos E, Suo C, Yücel M (2018) GABA concentration in sensorimotor cortex following high-intensity exercise and relationship to lactate levels. J Physiol 596:691–702
93. Rasmusson DD (2000) The role of acetylcholine in cortical synaptic plasticity. Behav Brain Res 115:205–218
94. Gu Q (2002) Neuromodulatory transmitter systems in the cortex and their role in cortical plasticity. Neuroscience 111:815–835
95. Otani S, Daniel H, Roisin M-P, Crepel F (2003) Dopaminergic modulation of long-term synaptic plasticity in rat prefrontal neurons. Cereb Cortex 13:1251–1256
96. Meeusen R, Piacentini MF, De Meirleir K (2001) Brain microdialysis in exercise research. Sports Med 31:965–983
97. Nitsche MA, Lampe C, Antal A, Liebetanz D, Lang N, Tergau F, Paulus W (2006) Dopaminergic modulation of long-lasting direct current-induced cortical excitability changes in the human motor cortex. Eur J Neurosci 23:1651–1657
98. Kuo M-F, Paulus W, Nitsche MA (2008) Boosting focally-induced brain plasticity by dopamine. Cereb Cortex 18:648–651
99. Lang N, Speck S, Harms J, Rothkegel H, Paulus W, Sommer M (2008) Dopaminergic potentiation of rTMS-induced motor cortex inhibition. Biol Psychiatry 63:231–233
100. Nitsche MA, Kuo M-F, Karrasch R, Wächter B, Liebetanz D, Paulus W (2009) Serotonin affects transcranial direct current-induced neuroplasticity in humans. Biol Psychiatry 66:503–508
101. Monte-Silva K, Liebetanz D, Grundey J, Paulus W, Nitsche MA (2010) Dosage-dependent non-linear effect of L-dopa on human motor cortex plasticity. J Physiol 588:3415–3424
102. Korchounov A, Ziemann U (2011) Neuromodulatory neurotransmitters influence LTP-like plasticity in human cortex: a pharmaco-TMS study. Neuropsychopharmacology 36:1894–1902
103. Monte-Silva K, Ruge D, Teo JT, Paulus W, Rothwell JC, Nitsche MA (2011) D2 receptor block abolishes θ burst stimulation-induced neuroplasticity in the human motor cortex. Neuropsychopharmacology 36:2097–2102
104. Batsikadze G, Paulus W, Kuo MF, Nitsche MA (2013) Effect of serotonin on paired associative stimulation-induced plasticity in the human motor cortex. Neuropsychopharmacology 38:2260–2267
105. Kuo HI, Paulus W, Batsikadze G, Jamil A, Kuo MF, Nitsche MA (2017) Acute and chronic effects of noradrenergic enhancement on transcranial direct current stimulation-induced neuroplasticity in humans. J Physiol 595:1305–1314
106. Grundey J, Barlay J, Batsikadze G, Kuo M-F, Paulus W, Nitsche M (2018) Nicotine modulates human brain plasticity via calcium-dependent mechanisms. J Physiol 596:5429–5441
107. Werhahn KJ, Förderreuther S, Straube A (1998) Effects of the serotonin1B/1D receptor agonist zolmitriptan on motor cortical excitability in humans. Neurology 51:896–898
108. Ilic TV, Korchounov A, Ziemann U (2002) Complex modulation of human motor cortex excitability by the specific serotonin re-uptake inhibitor sertraline. Neurosci Lett 319:116–120
109. Eichhammer P, Langguth B, Wiegand R, Kharraz A, Frick U, Hajak G (2003) Allelic variation in the serotonin transporter promoter affects neuromodulatory effects of a selective serotonin transporter reuptake inhibitor (SSRI). Psychopharmacology 166:294–297
110. Robol E, Fiaschi A, Manganotti P (2004) Effects of citalopram on the excitability of the human motor cortex: a paired magnetic stimulation study. J Neurol Sci 221:41–46
111. Korchounov A, Ilic TV, Schwinge T, Ziemann U (2005) Modification of motor cortical excitability by an acetylcholinesterase inhibitor. Exp Brain Res 164:399–405
112. Korchounov A, Ilić TV, Ziemann U (2007) TMS-assisted neurophysiological profiling of the dopamine receptor agonist cabergoline in human motor cortex. J Neural Transm 114:223–229
113. Herwig U, Bräuer K, Connemann B, Spitzer M, Schönfeldt-Lecuona C (2002) Intracortical excitability is modulated by a norepinephrine-reuptake inhibitor as measured with paired-pulse transcranial magnetic stimulation. Psychopharmacology 164:228–232

114. Ilic TV, Alexei K, Ulf ZCA (2003) Methylphenidate facilitates and disinhibits the motor cortex in intact humans. NeuroReport 14:773–776
115. Höffken O, Lenz M, Höckelmann N, Dinse HR, Tegenthoff M (2012) Noradrenergic modulation of human visual cortex excitability assessed by paired-pulse visual-evoked potentials. Neuroreport 23:707–711
116. Kuo H-I, Paulus W, Batsikadze G, Jamil A, Kuo M-F, Nitsche MA (2017) Acute and chronic noradrenergic effects on cortical excitability in healthy humans. Int J Neuropsychopharmacol 20:634–643
117. Coplan NL, Gleim GW, Nicholas JA (1989) Exercise-related changes in serum catecholamines and potassium: effect of sustained exercise above and below lactate threshold. Am Heart J 117:1070–1075
118. Kraemer WJ, Dziados JE, Gordon SE, Marchitelli LJ, Fry AC, Reynolds KL (1990) The effects of graded exercise on plasma proenkephalin peptide F and catecholamine responses at sea level. Eur J Appl Physiol Occup Physiol 61:214–217
119. McMorris T (2016) Developing the catecholamines hypothesis for the acute exercise-cognition interaction in humans: lessons from animal studies. Physiol Behav 165:291–299
120. Curtin D, Taylor EM, Bellgrove MA, Chong TT-J, Coxon JP (2023) D2 receptor blockade eliminates exercise-induced changes in cortical inhibition and excitation. Brain Stimul 16:727–733
121. Kurosawa M, Okada K, Sato A, Uchida S (1993) Extracellular release of acetylcholine, noradrenaline and serotonin increases in the cerebral cortex during walking in conscious rats. Neurosci Lett 161:73–76
122. Nakajima K, Uchida S, Suzuki A, Hotta H, Aikawa Y (2003) The effect of walking on regional blood flow and acetylcholine in the hippocampus in conscious rats. Auton Neurosci 103:83–92
123. Fu Y, Tucciarone JM, Espinosa JS, Sheng N, Darcy DP, Nicoll RA, Huang ZJ, Stryker MP (2014) A cortical circuit for gain control by behavioral state. Cell 156:1139–1152
124. Hiraga T, Hata T, Soya S, Shimoda R, Takahashi K, Soya M, Inoue K, Johansen JP, Okamoto M, Soya H (2024) Light-exercise-induced dopaminergic and noradrenergic stimulation in the dorsal hippocampus: Using a rat physiological exercise model. FASEB J 38:e70215
125. Tanaka T, Saito H, Matsuki N (1997) Inhibition of GABAA synaptic responses by brain-derived neurotrophic factor (BDNF) in rat hippocampus. J Neurosci 17:2959–2966
126. Brünig I, Penschuck S, Berninger B, Benson J, Fritschy JM (2001) BDNF reduces miniature inhibitory postsynaptic currents by rapid downregulation of GABA(A) receptor surface expression. Eur J Neurosci 13:1320–1328
127. Avanzini P, Pelliccia V, Lo Russo G, Orban GA, Rizzolatti G (2018) Multiple time courses of somatosensory responses in human cortex. NeuroImage 169:212–226
128. Fink GR, Frackowiak RSJ, Pietrzyk U, Passingham RE (1997) Multiple nonprimary motor areas in the human cortex. J Neurophysiol 77:2164–2174
129. Kuper M, Thurling M, Stefanescu R, Maderwald S, Roths J, Elles HG, Ladd ME, Diedrichsen J, Timmann D (2012) Evidence for a motor somatotopy in the cerebellar dentate nucleus—an FMRI study in humans. Hum Brain Mapp 33:2741–2749
130. Byblow WD, Coxon JP, Stinear CM, Fleming MK, Williams G, Müller JFM, Ziemann U (2007) Functional connectivity between secondary and primary motor areas underlying hand–foot coordination. J Neurophysiol 98:414–422
131. Coco M, Perciavalle V, Cavallari P, Perciavalle V (2016) Effects of an exhaustive exercise on motor skill learning and on the excitability of primary motor cortex and supplementary motor area. Medicine 95:1–9
132. Hamdy S, Rothwell JC, Aziz Q, Singh KD, Thompson DG (1998) Long-term reorganization of human motor cortex driven by short-term sensory stimulation. Nat Neurosci 1:64–68
133. Ridding MC, Brouwer B, Miles TS, Pitcher JB, Thompson PD (2000) Changes in muscle responses to stimulation of the motor cortex induced by peripheral nerve stimulation in human subjects. Exp Brain Res 131:135–143

134. Schabrun SM, Ridding MC, Galea MP, Hodges PW, Chipchase LS (2012) Primary sensory and motor cortex excitability are co-modulated in response to peripheral electrical nerve stimulation. PLoS One 7:1–7
135. Sasaki R, Kotan S, Nakagawa M, Miyaguchi S, Kojima S, Saito K, Inukai Y, Onishi H (2017) Presence and absence of muscle contraction elicited by peripheral nerve electrical stimulation differentially modulate primary motor cortex excitability. Front Hum Neurosci 11:146
136. Sidhu SK, Weavil JC, Venturelli M, Garten RS, Rossman MJ, Richardson RS, Gmelch BS, Morgan DE, Amann M (2014) Spinal μ-opioid receptor-sensitive lower limb muscle afferents determine corticospinal responsiveness and promote central fatigue in upper limb muscle. J Physiol 592:5011–5024
137. Sidhu SK, Weavil JC, Mangum TS, Jessop JE, Richardson RS, Morgan DE, Amann M (2017) Group III/IV locomotor muscle afferents alter motor cortical and corticospinal excitability and promote central fatigue during cycling exercise. Clin Neurophysiol 128:44–55
138. Egan GF, Johnson J, Farrell M, McAllen R, Zamarripa F, McKinley MJ, Lancaster J, Denton D, Fox PT (2005) Cortical, thalamic, and hypothalamic responses to cooling and warming the skin in awake humans: a positron-emission tomography study. Proc Natl Acad Sci USA 102:5262–5267
139. Muzik O, Diwadkar VA (2016) In vivo correlates of thermoregulatory defense in humans: temporal course of sub-cortical and cortical responses assessed with fMRI. Hum Brain Mapp 37:3188–3202
140. Roig M, Skriver K, Lundbye-Jensen J, Kiens B, Nielsen JB (2012) A single bout of exercise improves motor memory. PLoS One 7:28–32
141. Skriver K, Roig M, Lundbye-Jensen J, Pingel J, Helge JW, Kiens B, Nielsen JB (2014) Acute exercise improves motor memory: exploring potential biomarkers. Neurobiol Learn Mem 116:46–58
142. Thomas R, Beck MM, Lind RR, Korsgaard Johnsen L, Geertsen SS, Christiansen L, Ritz C, Roig M, Lundbye-Jensen J (2016) Acute exercise and motor memory consolidation: the role of exercise timing. Neural Plast 2016:6205452
143. Thomas R, Flindtgaard M, Skriver K et al (2017) Acute exercise and motor memory consolidation: does exercise type play a role? Scand J Med Sci Sports 27:1523-1532
144. Dal Maso F, Desormeau B, Boudrias MH, Roig M (2018) Acute cardiovascular exercise promotes functional changes in cortico-motor networks during the early stages of motor memory consolidation. NeuroImage 174:380–392
145. Marin Bosch B, Bringard A, Logrieco MG, Lauer E, Imobersteg N, Thomas A, Ferretti G, Schwartz S, Igloi K (2020) Effect of acute physical exercise on motor sequence memory. Sci Rep 10:15322
146. Lundbye-Jensen J, Skriver K, Nielsen JB, Roig M (2017) Acute exercise improves motor memory consolidation in preadolescent children. Front Hum Neurosci 11:1–10
147. Hung A, Roig M, Gillen JB, Sabiston CM, Swardfager W, Chen JL (2021) Aerobic exercise and aerobic fitness level do not modify motor learning. Sci Rep 11:5366
148. Ostadan F, Centeno C, Daloze JF, Frenn M, Lundbye-Jensen J, Roig M (2016) Changes in corticospinal excitability during consolidation predict acute exercise-induced off-line gains in procedural memory. Neurobiol Learn Mem 136:196–203
149. Mang CS, McEwen LM, MacIsaac JL, Snow NJ, Campbell KL, Kobor MS, Ross CJD, Boyd LA (2017) Exploring genetic influences underlying acute aerobic exercise effects on motor learning. Sci Rep 7:12123
150. Christiansen L, Thomas R, Beck MM, Pingel J, Andersen JD, Mang CS, Madsen MAJ, Roig M, Lundbye-Jensen J (2019) The beneficial effect of acute exercise on motor memory consolidation is modulated by dopaminergic gene profile. J Clin Med Res 8:578
151. Statton MA, Encarnacion M, Celnik P, Bastian AJ (2015) A single bout of moderate aerobic exercise improves motor skill acquisition. PLoS One 10:1–13
152. Rhee J, Chen J, Riechman SM, Handa A, Bhatia S, Wright DL (2016) An acute bout of aerobic exercise can protect immediate offline motor sequence gains. Psychol Res 80:518–531
153. Lauber B, Franke S, Taube W, Gollhofer A (2017) The effects of a single bout of exercise on motor memory interference in the trained and untrained hemisphere. Neuroscience 347:57–64

154. Neva JL, Ma JA, Orsholits D, Boisgontier MP, Boyd LA (2019) The effects of acute exercise on visuomotor adaptation, learning, and inter-limb transfer. Exp Brain Res 237:1109–1127
155. Snow NJ, Mang CS, Roig M, McDonnell MN, Campbell KL, Boyd LA (2016) The effect of an acute bout of moderate-intensity aerobic exercise on motor learning of a continuous tracking task. PLoS One 11:1–16
156. Baird JF, Gaughan ME, Saffer HM, Sarzynski MA, Herter TM, Fritz SL, den Ouden DB, Stewart JC (2018) The effect of energy-matched exercise intensity on brain-derived neurotrophic factor and motor learning. Neurobiol Learn Mem 156:33–44
157. Charalambous CC, French MA, Morton SM, Reisman DS (2019) A single high-intensity exercise bout during early consolidation does not influence retention or relearning of sensorimotor locomotor long-term memories. Exp Brain Res 237:2799-2810
158. Van Pham M, Saito K, Miyaguchi S, Watanabe H, Ikarashi H, Nagasaka K, Yokota H, Kojima S, Inukai Y, Otsuru N, Onishi H (2022) Changes in excitability and GABAergic neuronal activity of the primary somatosensory cortex after motor learning. Front Neurosci 16:794173
159. Sasaki R, Watanabe H, Onishi H (2022) Therapeutic benefits of noninvasive somatosensory cortex stimulation on cortical plasticity and somatosensory function: a systematic review. Eur J Neurosci 56:4669-4698

Index

H. Soya (ed.), *Exercise Brain Stimulation for Cognitive Function and Mental Health*, Advances in Neurobiology 44,
https://doi.org/10.1007/978-981-95-0066-6